Handbook of Chemical and Biological Sensors

Handbook of Chemical and Biological Sensors

Edited by

Richard F Taylor

Arthur D Little Inc.

Jerome S Schultz

University of Pittsburgh

Institute of Physics Publishing
Bristol and Philadelphia

British Library Cataloguing-in-Publication Data

A catalogue record for this book is available from the British Library.

ISBN 0 7503 0323 9

Library of Congress Cataloging-in-Publication Data are available

Published by Institute of Physics Publishing, wholly owned by The Institute of Physics, London
Institute of Physics Publishing, Techno House, Redcliffe Way, Bristol BS1 6NX, UK
US Editorial Office: Institute of Physics Publishing, The Public Ledger Building, Suite 1035, 150 South Independence Mall West, Philadelphia, PA 19106, USA

Typeset in TeX using the IOP Bookmaker Macros
Printed in the UK by J W Arrowsmith Ltd, Bristol

Contents

Preface

Measurement represents one of the oldest methods used by man to better understand and control his life and his world. Since antiquity, new methods have evolved and replaced old to allow better, faster and more accurate measurements of materials, both chemical and biological. The driving force in the evolution of measurement methods is to gain and apply information in real time. Characteristics such as specificity, sensitivity, speed and cost all contribute to the success or failure of a new measurement technology.

Chemical and biological sensors are the evolved products of many measurement systems and many different technologies. Based on physical transduction methods and drawing on diverse disciplines such as polymer chemistry, physics, electronics and molecular biology, chemical and biological sensors represent multidisciplinary hybrid products of the physical, chemical and biological sciences. As a result, chemical and biological sensors are able to recognize a specific molecular species or event, converting this recognition event into an electrical signal or some other useful output.

The 1980s witnessed the evolution of early chemical and biological sensors into more sophisticated and complex measurement devices. The first electrode-based sensors were improved and became commercially viable products while new chemical sensors and biosensors based on direct binding interactions were reduced to practical prototypes and first-generation products. Now, in the 1990s, both types of sensors are being improved to new performance levels. By the turn of the century, chemical and biological sensors will be used routinely for medical, food, chemical and environmental applications and will, themselves, be the evolutionary precursors to more advanced microsensors and biocomputing devices.

The diversity of chemical and biological sensors in both type and application appeals to a broad group of scientists and engineers with interests ranging from basic sensor research and development to the application of sensors in the field and on processing lines. To date, no text has attempted to address the needs of this broad audience.

The *Handbook of Chemical and Biological Sensors* is aimed at all scientists and engineers who are interested in or are developing these sensors. The scope of the book includes both chemical and biological sensors, as well as the basic technologies associated with physical sensors which form the basis for them. The text is divided into two major parts, the first dealing with basic sensor technologies and the second with sensor applications. This approach allows the reader to first review the scientific basis for sensor transducers, surfaces, and signal output; these basic technologies are then extended to actual, functional sensors, many of them commercial products. The *Handbook*, then, is intended to be both a teaching and a reference tool for those interested in developing and using chemical and biological sensors.

It is our hope that the *Handbook* will be useful both to those who are new to the sensor area and to experienced sensor scientists and engineers who wish to broaden their knowledge of the wide-ranging sensor field. It is our purpose to present the many disciplines required for sensor development to this audience and to illustrate the current sensor state of the art. Finally, this text addresses the hard realities of sensor commercialization since the practical use and application of chemical and biological sensors is key to driving their further evolution. It is further hoped that this text, and projected updated editions, will become a standard reference text for those working with chemical and biological sensors.

Richard F Taylor
Jerome S Schultz

1

Introduction to chemical and biological sensors

Jerome S Schultz and Richard F Taylor

1.1 INTRODUCTION

The coming 21st century is being heralded as the era of information: expanding capabilities for computer-assisted management of information, increased capabilities in decision making and process control, and automated health care will all add to the pace and quality of life in this new era.

Our new abilities to simultaneously handle multiple information sources as well as vastly more efficient methods for classifying, sorting, and retrieving information will put increasing demands on the technologies and instruments used for obtaining information in a timely and continuous manner. Today, this capability is exemplified by physical sensing and measurement systems, i.e., systems able to detect and measure parameters such as temperature, pressure, electric charge, viscosity, and light intensity (see chapter 2 of this text).

During the 1980s and now the 1990s, it has become apparent that more sophisticated measurement devices are necessary to collect the information which can be processed in new management systems. Chemical and biological sensors have emerged as the means to this end. The basic technologies begun in the 1980s and being developed in the 1990s will result in chemical and biological sensors with near-infinite capabilities for analyte detection. This new generation of sensors will, by the end of this century, become an integral part of collection and control systems in nearly every industry and marketplace.

1.1.1 Definition of chemical and biological sensors

Chemical and *biological sensors* (the latter are also called *biosensors*) are more complex extensions of physical sensors. In many cases, the transducer technologies developed and commercialized for physical sensors are the basis for chemical sensors and biosensors. As used throughout this text, chemical sensors and biosensors are defined as *measurement devices which utilize chemical or biological reactions to detect and quantify a specific analyte or event.* Such sensors differ, therefore, from physical sensors which measure physical parameters.

The distinction between chemical sensors and biosensors is more complex. Many authors attempt to define a sensor based on the nature of the analyte detected. This approach can be misleading since nearly all analytes measured by a chemical or biosensor are chemicals or *bio*chemicals, the exception being sensors which detect whole cells. Other authors attempt to define a chemical of biosensor by the nature of the reaction which leads to the detection event. Again, this is confusing since all reactions at chemical and biosensor surfaces are chemical (or *bio*chemical) reactions.

In this text, we distinguish between chemical sensors and biosensors according to the nature of their *reactive surface*. By this definition, chemical sensors utilize specific polymeric membranes, either *per se* or containing doping agents, or are coated with non-biological (usually low-molecular-weight) materials. These polymeric layers or specific chemicals, attached to the layers or directly to the transducer, interact with and measure the analyte of interest. The nature of the analyte or the reaction which takes place is not limited with such chemical sensors.

We define biosensors in this text as sensors which contain a biomolecule (such as an enzyme, antibody, or receptor) or a cell as the active detection component. Again, the nature of the analyte and the reaction which leads to detection are not limited in this definition.

Given these definitions, we can further define the basic components of a chemical sensor or biosensor. These include the *active surface*, the *transducer*, and the *electronics/software* as shown in figure 1.1.

The active surface of a chemical or biosensor contains the detection component as described above, e.g., a polymeric layer or an immobilized biomolecule. Examples of these layers are given throughout this text (e.g., see chapters 7–10). It is the interaction between the active layer and the analyte(s) being measured that is detected by the transducer. Examples of transducers are described in table 1.1 and in chapters 3–6 of this text. The change in the transducer due to the active surface event is expressed as a specific signal which may include changes in impedance, voltage, light intensity, reflectance, weight, color, or temperature. This signal is then detected, amplified, and processed by the electronics/software module (see chapter 12).

An example of a sensor utilizing these three components is illustrated in figure 1.2, which shows a schematic of a typical enzyme electrode for the detection of glucose (also see chapter 17 of this text). The bioactive surface consists of immobilized glucose oxidase (GOD) sandwiched between a polycarbonate and cellulose acetate membrane. The transducer is a platinum electrode and the electronics are those typically found in any polarograph, i.e. an electronic system to measure low currents (on the order of microamperes) at a fixed voltage bias on the platinum electrode. The action of glucose

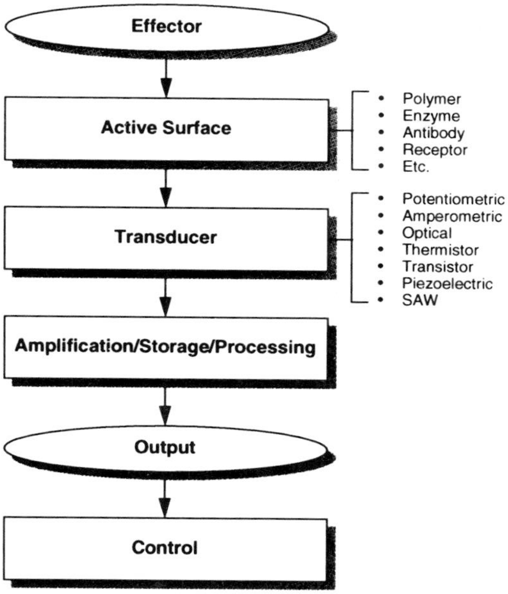

Figure 1.1 Basic components of a sensor.

oxidase on glucose results in oxygen depletion, resulting in a depression of the oxygen concentration in the immediate vicinity of the polarographic electrode transducer. The resulting reduction in steady state current is detected and translated to a millivolt output. This output (i.e. reduced availability of oxygen) can then be related to increases in glucose concentration.

While enzyme electrodes represent a successful commercial application of sensor technology, their dependence on enzymatic activity sets them apart from most other sensors. The majority of chemical sensors and biosensors function by binding the analyte(s) to the active surface. Such binding results in a change in the transducer output voltage, impedance, light scattering, etc. These types of *affinity* or *binding* sensors are discussed in chapters 14, 15, 18, and 19 of this text.

1.1.2 Historical perspective of chemical and biological sensors

Chemical sensors and biosensors are relatively new measurement devices. Up until approximately 30 years ago, the glass pH electrode could be considered the only portable chemical sensor sufficiently reliable for measuring a chemical parameter. Even this sensor, which has been under continuous development since it invention in 1922 (table 1.2), needs to be recalibrated on a daily basis and is limited to measurements in solutions or on wet surfaces.

Other sensing technologies based on oxidation–reduction reactions at electrodes were extensively pursued in the 1940s and 1950s providing analytical

Table 1.1 Major sensor transducer technologies.

Technology/example	Output change	Examples[a]
Electronic		
Amperometric	Applied current	Polymer enzyme, antibody, and whole-cell electrodes
Potentiometric	Voltage	Polymeric and enzyme electrodes, FETs, ENFETs
Capacitance/impedance	Impedance	Conductimeters, interdigitated electrode capacitors
Photometric		
Light absorption or scattering; refractive index	Light intensity, color, or emission	Ellipsometry, internal reflectometry, laser light scattering
Fluorescence of luminescence activation, quenching or polarization	Fluorescence or chemiluminescence	Surface plasmon resonance, fiber optic wave guides, fluorescence polarization
Acoustical/mechanical		
Acoustical	Amplitude, phase or frequency (acoustic wave)	SAW devices
Mass, density	Weight	Piezoelectric devices
Calorimetric		
Thermistor	Temperature	Enzyme and immunoenzyme reactors

[a] Abbreviations: FET, field effect transistor, ENFET, enzyme FET; SAW, surface acoustic wave.

methods for the detection of metallic ions and some organic compounds. The first application of these electrochemical techniques to make a sensor was for the measurement of oxygen content in tissues and physiological fluids [4]. Ion selective electrodes have provided new measurement capabilities but face the same limitations as well as less selectivity than the pH electrode.

In spite of this limitation, ion selective electrodes have provided some of the first transducers for chemical sensors and biosensors (see chapters 13 and 17 of this text). For example, many chemical sensors use a selective membrane containing a specific capture molecule (or dopant) to provide a selective permeability for the ion the electrode detects. Thus, a K^+ electrode can use valinomycin as its capture molecule as illustrated in figure 1.3 [21]. These types of chemical sensor provide a basis for the development of a wide range of sensors based on specific capture molecules.

The key concept for the adaptation of these electroanalytical techniques was Clark's idea of encapsulating the electrodes and supporting chemical components

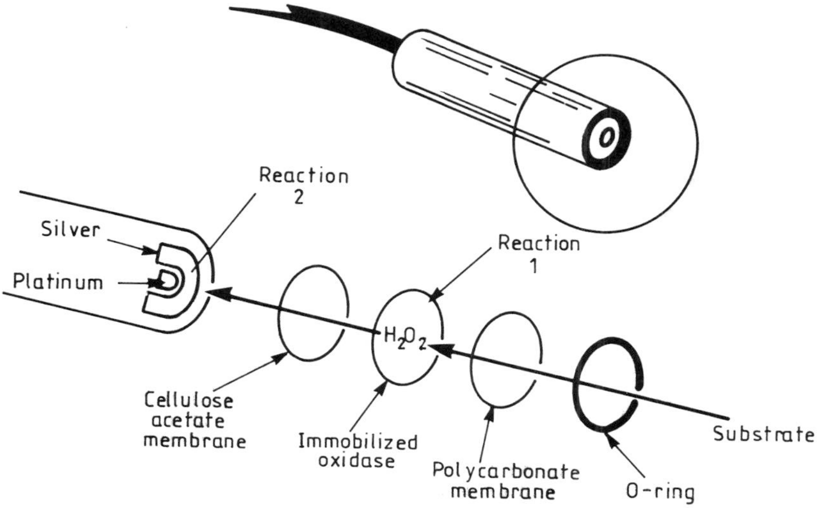

Figure 1.2 The basic components of a commercial enzyme electrode (reprinted with permission of YSI, Inc.).

by a semipermeable membrane. This allows the analyte to diffuse freely within the sensor without the loss of critical components, e.g. the enzyme [4]. The concept of interposing membrane layers between the solution and the electrode also provided the basis for the first biosensor, a glucose biosensor invented by Clark and Lyons [6] and its commercial product illustrated in figure 1.2. The first glucose sensor was the precursor to the development of other glucose sensors based on a wide range of transduction technologies and utilizing glucose oxidase as the active surface detection component. It is notable from figure 1.4 [22] that a sensor can be developed based on the measurement of any of the reactants or products of the glucose/GOD reaction.

The invention of enzyme electrodes, new transduction technologies, and new means to immobilize polymers and biomolecules onto transducers has led to the rapid evolution of chemical sensor and biosensor technology during the 1970s and 1980s (table 1.2). Examples of these technologies are found throughout this text and the reduction of these technologies to practical sensors is found in chapters 13–22. Sensor technology is being further advanced by new discoveries in conductive polymers, enzyme modification, and development of organic and biological components for organic computing devices (chapters 11 and 23). These advances will result in the commercialization of a variety of chemical and biosensors by the first years of the next century.

Table 1.2 Historical landmarks in the development of chemical sensors and biosensors.

Date	Event	Reference
1916	First report on the immobilization of proteins: adsorption of invertase to activated charcoal	[1]
1922	First glass pH electrode	[2]
1925	First blood pH electrode	[3]
1954	Invention of the oxygen electrode	[4]
	Invention of the pCO_2 electrode	[5]
1962	First amperometric biosensor: glucose oxidase-based enzyme electrode for glucose	[6]
	Method for generating lipid bilayer membranes	[7]
1964	Coated piezoelectric quartz crystals as sensors for water, hydrocarbons, polar molecules, and hydrogen sulfide	[8]
1969	First potentiometric biosensor: acrylamide-immobilized urease on an ammonia electrode to detect urea	[9]
1972–74	First commercial enzyme electrode (for glucose) and glucose analyzer using the electrode (Yellow Springs Instruments)	
1975	First microbe-based biosensor: immobilized *Acetobacterxylinium* in cellulose on an oxygen electrode for ethanol	[10]
	First binding-protein biosensor: immobilized concanavalin A in a polyvinyl chloride membrane on a platinum wire electrode to measure yeast mannan	[11]
	Invention of the pCO_2/pO_2 optrode	[12]
1975–76	First immunosensors: ovalbumin on a platinum wire electrode for ovalbumin antibody; antibody to human immunoglobulin G (hIgG) in an acetylcellulose membrane on a platinum electrode for hIgG measurements	[11, 13]
1979	Surface acoustic wave sensors for gases	[14]
1980	Fiber optic pH sensor for *in vivo* blood gases	[17]
1982	Fiber-optic-based biosensor for glucose	[15]
1983	Molecular level fabrication techniques and theory for molecular level electronic devices	[16]
1986	First tissue-based biosensor: antennules from blue crabs mounted in a chamber with a platinum electrode to detect amino acids	[18]
1987	First receptor-based biosensor: acetylcholine receptor on a capacitance transducer for cholinergics	[19]
	Electrically conductive redox enzymes	[20]

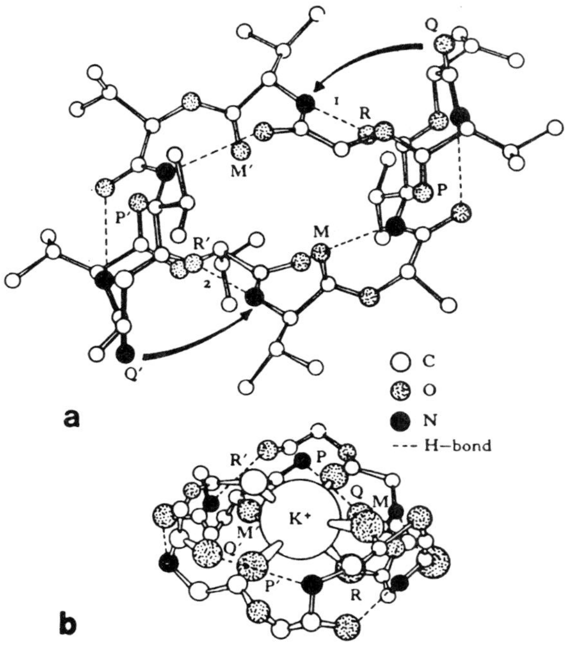

Figure 1.3 Structure of (a) free valinomycin and (b) its complex with K^+. Note the cooperation of the inner oxygen atoms P, M, and R in the complexation process. (Reprinted from [11] with permission of Cambridge University Press.)

1.1.3 The purpose of this handbook

The multiplicity of methods which have been applied to development of glucose sensors shown in figure 1.4 illustrates a major problem with chemical sensor and biosensor development. During the past two decades, as the need for these sensors has emerged, there has been more effort spent on the repeated demonstration of chemical sensor and biosensor designs than on focused efforts to bring sensors through mass production to commercialization.

This handbook is designed to address the needs of both development and commercialization in one text. The first 12 chapters focus on basic technology and methods for developing sensors. These include preparation of the active surface, the different types of transducer available for sensors and signal output and processing. These aim of these chapters is to provide a knowledge of the basic technologies and methods used in sensor development.

Chapters 13–22 deal with specific examples of sensors and their practical reduction to practice. The sensors addressed in these chapters are still mainly in the advanced prototype stage, still requiring final transfer to mass manufacturing. These sensors represent, however, the initial technologies and products which

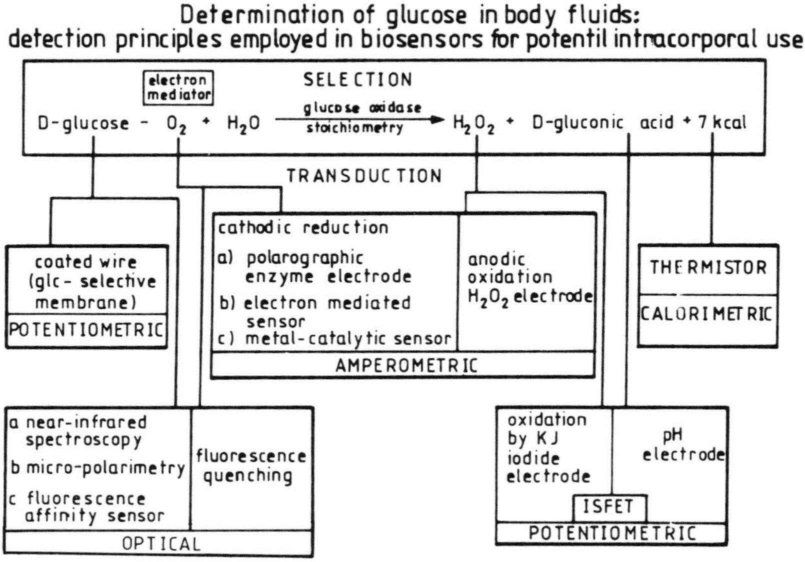

Figure 1.4 Determination of glucose in body fluids: detection principles employed in biosensors for potential intracorporal use.

will be launched into the 21st century.

The last chapter in this text deals with sensor commercialization and markets. Chemical and biosensors are following a commercialization pathway similar to other detection and measurement devices such as analytical chemistry instrumentation in the 1950s and 1960s, and immunoassay in the 1970s and 1980s. Common to these products, sensors will achieve a critical mass which will push them into large-scale commercialization by the first part of the 21st century.

REFERENCES

[1] Nelson J M and Griffin E G 1916 Adsorption of invertase *J. Am. Chem. Soc.* **38** 1109–15

[2] Hughes W S 1922 The potential difference between glass and electrolytes in contact with water *J. Am. Chem. Soc.* **44** 2860–6

[3] Kerridge P T 1925 The use of the glass electrode in biochemistry *Biochem. J.* **19** 611–7

[4] Clark L C Jr 1956 Monitor and control of blood tissue O_2 tensions *Trans. Am. Soc. Artif. Intern. Organs* **2** 41–8

[5] Stow R W and Randall B F 1954 Electrical measurement of the pCO_2 of blood *Abstract Am. J. Physiol.* **179** 678

[6] Clark L C Jr and Lyons C 1962 Electrode system for continuous monitoring in cardiovascular surgery *Ann. NY Acad. Sci.* **148** 133–53

[7] Mueller P, Rudin D O, Tien H T and Wescott W 1962 Reconstruction of excitable cell membrane structure in vitro *Circulation* **26** 1167–71

[8] King W H Jr 1964 Piezoelectric sorption detector *Anal. Chem.* **36** 1735–9

[9] Guilbault G and Montalvo J 1969 A urea specific enzyme electrode *J. Am. Chem. Soc.* **91** 2164–5

[10] Davies C 1975 Ethanol oxidation by an *Acetobacter xylinium* microbial electrode *Ann. Microbiol.* A **126** 175–86

[11] Janata J 1975 An immunoelectrode *J. Am. Chem. Soc.* **97** 2914–6

[12] Lübbers D W and Opitz N 1975 Die pCO_2/pO_2-Optrode: Eine neue pCO_2-bzw. pO_2-Messonde zur Messung des pCO_2 oder pO_2 von Gasen und Flüssigkeiten *Z. Naturf.* c **30** 532–3

[13] Aizawa M, Morioka A, Matsuoka H, Suzuki S, Nagamura Y, Shinohara R and Ishiguro I 1976 An enzyme immunosensor for IgG *J. Solid-Phase Biochem.* **1** 319–28

[14] Wohltjen H and Dessey R 1979 Surface acoustic wave probe for chemical analysis I. Introduction and instrument description *Anal. Chem.* **51** 1458–64

[15] Schultz J S, Mansouri S and Goldstein I J 1982 Affinity glucose sensor *Diabetes Care* **5** 245–53

[16] Carter F L 1983 Molecular level fabrication techniques and molecular electronic devices *J. Vac. Sci. Technol.* B **1** 959–68

[17] Peterson J I, Goldstein S R, Fitzgerald R V and Buckhold D K 1980 Fiber optic pH probe for physiological use *Anal. Chem.* **52** 864–9

[18] Belli S L and Rechnitz G A 1986 Prototype potentiometric biosensor using intact chemoreceptor structures *Anal. Lett.* **19** 403–16

[19] Taylor R F, Marenchic I G and Cook E J 1987 Receptor-based biosensors *US Patent* 5 001 048

[20] Heller A 1990 Electrical wiring of redox enzymes *Accounts Chem. Res.* **23** 1280034

[21] Janata J 1989 *Principles of Chemical Sensors* (New York: Plenum) p 317

[22] Fischer U, Rebin K, v Woedtke T and Abel P 1994 Clinical usefulness of the glucose concentration in the subcutaneous tissue—properties and pitfalls of electrochemical sensors *Horm. Metab. Res.* **26** 515–22

2

Physical Sensors

Robert A Peura and Stevan Kun

Physical sensors may be defined as devices that are used for measurements of physical parameters, such as displacement, pressure and temperature. For a long time physicians have been using their senses to determine various physical parameters of the patient, such as skin color and texture, temperature, pulse strength and rate and position and size of body organs. Measurements of physical parameters strictly limited to acquisition by human senses were a long time ago shown to be inadequate in providing the physician with enough data on the condition of the patient. Therefore, we have been witnessing an ongoing and probably never ending struggle of physicians and scientists, to provide faster, more accurate and less invasive means for acquiring data significant for evaluation of the patient's condition.

Physical sensors are some of the devices that are fundamental in the process of measurement and acquisition of parameters of living systems. Their development, that includes increased application of technology to clinical and basic biomedical research, is considered to be a prerequisite for further enhancement of medical practice.

The first part of this chapter presents the basic sensing principles (of physical sensors) that are used in biomedical instruments. Sensors that operate on these principles convert physical parameters into electric signals. An output from a sensor in the form of an electric signal is preferable because of the advantages of subsequent processing of electrical signals. Basic physical sensors include piezoelectric, resistive, inductive and capacitive sensors.

The second part of this chapter contains examples of the applications of the described sensors in measurements of displacement, blood pressure, etc. Temperature measurements are covered in a separate section of this volume.

2.1 PIEZOELECTRIC SENSORS

Piezoelectric sensors are devices that transduce the measured physical parameter, that is in the form of mechanical strain, into variations of electric charge. Piezoelectric devices are used to measure physiological displacements and pressures and record heart sounds, as well as for the generation and reception of ultrasound (used to visualize body organs). When mechanically strained, piezoelectric materials generate an electric charge and a potential. Conversely, when an electric potential is applied to the piezoelectric material, it physically deforms. When an asymmetrical piezoelectric crystal lattice is distorted, a charge reorientation takes place, which causes a relative displacement of negative and positive charges. These displaced internal charges induce surface charges of opposite polarity on opposite crystal sides. Surface charge is determined by the potential difference between the surfaces.

The induced charge q is directly proportional to the applied force f:

$$q = kf \tag{2.1}$$

where k is the piezoelectric constant, in units of coulombs per newton. The voltage change is determined by modeling the system as a parallel plate capacitor in which the capacitor voltage v is calculated as charge q divided by capacitance C. By substitution of (2.1), we find

$$v = kf/C = kfx/\varepsilon_0\varepsilon_r A \tag{2.2}$$

where ε_r, A and x describe the equivalent plate capacitor.

Typical values for k are 2.3 pC N^{-1} for quartz and 140 pC N^{-1} for barium titanate. For a 1 cm^2 area and 1 mm thickness piezoelectric sensor with an applied force of 0.1 N, the output voltage v is 0.23 mV and 14 mV for the quartz and barium titanate crystals, respectively. Piezoelectric constants are given in the literature [1, 2].

There are various operation modes for piezoelectric sensors, depending on the crystallographic orientation of the plate and the material [1]. These modes include transversal compression, thickness or longitudinal compression, thickness shear action and face shear action. Also available are piezoelectric polymeric films, which are very thin, lightweight and pliant, such as polyvinylidene fluoride (PVDF) [3, 4]. These films can be cut easily and adapted to uneven surfaces. Resonance applications are not possible with PVDFs because of their low mechanical quality factor. However, they can be used in acoustical broad-band applications for microphones and loudspeakers.

Piezoelectric materials have very high but finite resistance. Thus, if a static deflection x is applied, charge leaks through the leakage resistor (of the order of 100 GΩ). In order to preserve the signal, it is important that the input impedance of the external voltage measuring device be an order of magnitude higher than that of the piezoelectric sensor. An equivalent circuit for the piezoelectric sensor is given in figure 2.1, which is useful to quantify its dynamic response characteristics.

In the equivalent circuit, a charge generator q, defined by

$$q = Kx \tag{2.3}$$

drives the circuit, where K is the proportionality constant ($\mathrm{C\,m^{-1}}$) and x is the deflection (m).

Converting the charge generator to a current generator, i_g, simplifies the circuit:

$$i_g = \mathrm{d}q/\mathrm{d}x = K\,\mathrm{d}x/\mathrm{d}t. \tag{2.4}$$

The modified circuit in figure 2.1(b) shows equivalent combined resistances and capacitances. If we assume that the amplifier does not draw any current,

$$i_g = i_c + i_R \tag{2.5}$$

$$v_0 = v_c = \frac{1}{C}\int i_c\,\mathrm{d}t \tag{2.6}$$

$$i_g - i_R = C\,\mathrm{d}v_0/\mathrm{d}t = K\,\mathrm{d}x/\mathrm{d}t - v_0/R \tag{2.7}$$

or

$$V_0(\mathrm{i}\omega)/X(\mathrm{i}\omega) = K_s \mathrm{i}\omega\tau/(\mathrm{i}\omega\tau + 1) \tag{2.8}$$

where $K_S = K/C$ is the sensitivity ($\mathrm{V\,m^{-1}}$) and $\tau = RC$ is the time constant (s).

It is important to be able to calculate the system parameters from the model of the sensor. Let us determine the cut-off frequency for a piezoelectric sensor which has $C = 400$ pF and leakage resistance of 10 GΩ. The amplifier input impedance is 10 MΩ. If we use the modified piezoelectric sensor equivalent circuit (figure 2.1(b)) for this calculation, we find that the cut-off frequency for this circuit is

$$f_c = 1/(2\pi RC) = 1/[2\pi(10 \times 10^6)(400 \times 10^{-12})] = 40\ \mathrm{Hz}.$$

Note that if we increased the amplifier input impedance by a factor of 100, we would lower the low-corner frequency to 0.40 Hz.

The voltage output response of a piezoelectric sensor to a step displacement x is shown in figure 2.2. Due to the finite internal resistance of the piezoelectric material and finite input resistance of the amplifier, the output decays exponentially. If at time T the force is released, a displacement restoration will occur that is equal and opposite to the original displacement. A sudden decrease in voltage of magnitude Kx/C occurs, with a resulting undershoot equal to the exponential decay prior to the release of the displacement. Increasing the time constant, $\tau = RC$ will minimize the decay and undershoot. The easiest

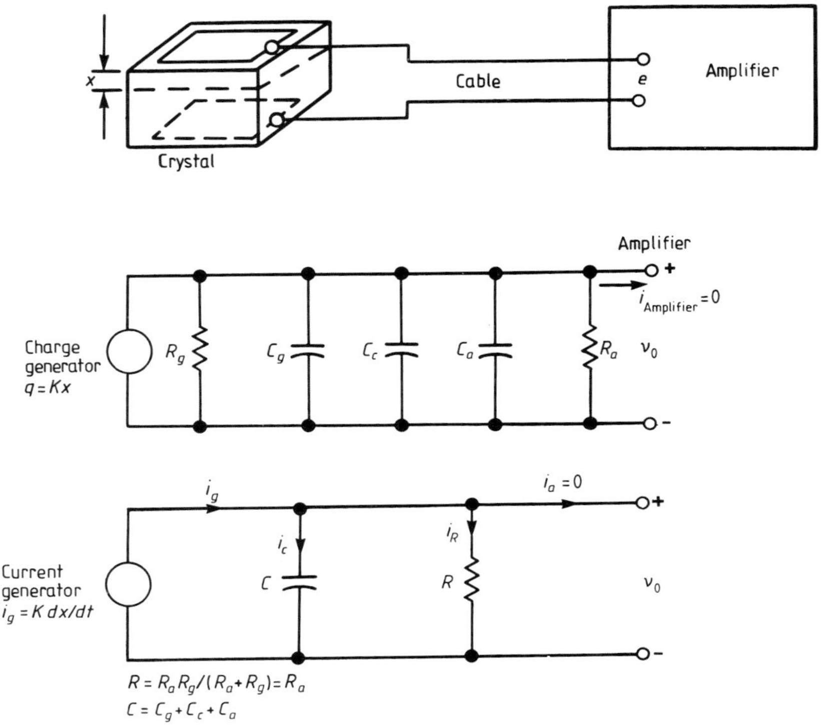

Figure 2.1 (a) Piezoelectric sensor equivalent circuit, where R_g is the crystal leakage resistance, C_g the crystal capacitance, C_g the cable capacitance, C_a the amplifier input capacitance, R_a the amplifier input resistance and q the charge generator. (b) An altered equivalent circuit with a current generator replacing the charge generator. (From Doebelin E O *Measurement Systems: Application and Design*, copyright © 1990 by McGraw-Hill, Inc.)

approach to increasing τ is to add a parallel capacitor. However, this causes a reduction in the sensitivity in the midband frequencies (see equation (2.8)). Another approach to improving the low-frequency response is to use a charge or electrometer amplifier (with a very high input impedance $> 10^{12}\ \Omega$).

The high-frequency equivalent circuit for a piezoelectric sensor is complex because of its mechanical resonance. This can be modeled by adding a series *RLC* circuit in parallel with the sensor capacitance and leakage resistance. The high-frequency equivalent circuit and its frequency response are shown in figure 2.3. In some applications, the mechanical resonance is desirable for accurate frequency control, as in the case of crystal filters.

Piezoelectric sensors are used in cardiovascular applications for external (body surface) and internal (intracardiac) phonocardiography. They are also used in the detection of Korotkoff sounds for indirect blood pressure

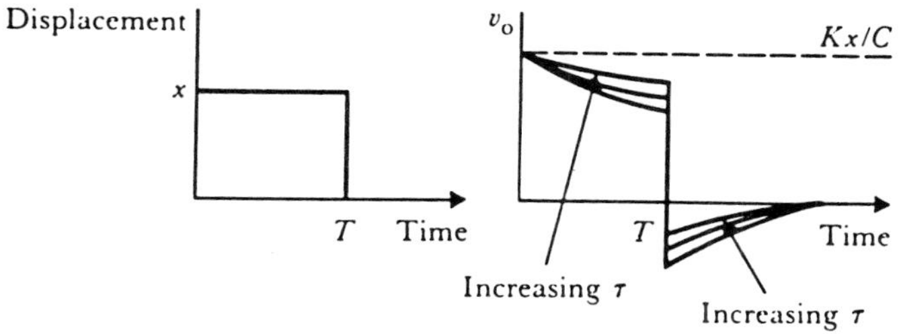

Figure 2.2 Piezoelectric sensor response to a step displacement. (Doebelin E O *Measurement Systems: Application and Design*, copyright © 1990 by McGraw-Hill, Inc.)

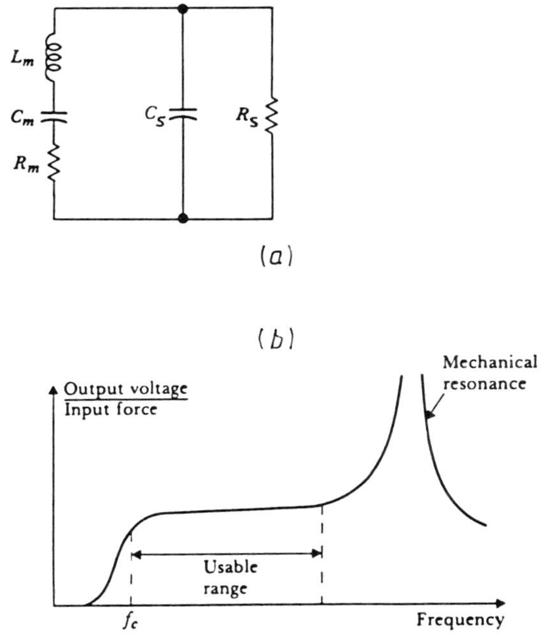

Figure 2.3 (a) The high-frequency circuit model for a piezoelectric sensor. R_s is the sensor leakage resistance and C_s the capacitance. L_m, C_m and R_m represent the mechanical system. (b) The piezoelectric sensor frequency response. (From R S C Cobbold *Transducers for Biomedical Measurements: Principles and Applications*, copyright © 1974, John Wiley and Sons, Inc.)

measurements. Additional applications of piezoelectric sensors involve their use in measurements of physiological accelerations such as a control parameter for a cardiac pacemaker. A piezoelectric sensor and circuit can be used to measure the acceleration due to human movements, providing an estimate of energy expenditure [5]. Another application, in which the piezoelectric element operates at mechanical resonance and emits and senses high-frequency sounds, is ultrasonic blood flow meters.

2.2 RESISTIVE SENSORS

Resistive sensors are devices that transduce the measured physical parameter into a change in resistance. This change is then measured by electrical means. Typically, resistivity change is induced by some kind of a displacement. There are two major groups of resistive sensors: potentiometers and strain gages.

2.2.1 Potentiometers

Potentiometers are passive three-port electric devices in which the linear or rotational mechanical movement of the central port produces variations in resistance measured between that central port and the other two ports. They are used for measuring displacement; there are three types of potentiometric device as shown in figure 2.4. The potentiometer in figure 2.4(a) measures translational displacements typically from 2 to 500 mm. Small rotational displacements from 10° to less than 250° can be detected by single-turn potentiometers, as shown in figure 2.4(b). Multiturn potentiometers are used for measurements of rotational displacements with a dynamic range of >250°—see figure 2.4(c). To produce an electrical output, the resistive elements, composed of wire-wound, carbon film, metal film, conducting plastic or ceramic materials, may be excited by DC or AC voltages.

Their major advantage is that a linear electrical output is produced as a function of displacement within 0.01% of full scale, without the use of additional hardware or signal conditioning. This makes potentiometers very easy to use, simple to design and inexpensive. Linearity results if the potentiometer is isolated from the load (which is easy to accomplish). The construction of these potentiometers determines their resolution, their temperature stability and noise levels. The major disadvantage of potentiometers is that they contain mechanical moving parts, that are subject to wear. Also, the frictional and inertial components of these potentiometers should be kept low in order to minimize dynamic system distortion caused by mechanically loading the source of the displacement movement.

2.2.2 Strain gages

Strain gages are passive two-port electric devices, in which a force-induced dimensional change of the strain gage material produces variations in resistance

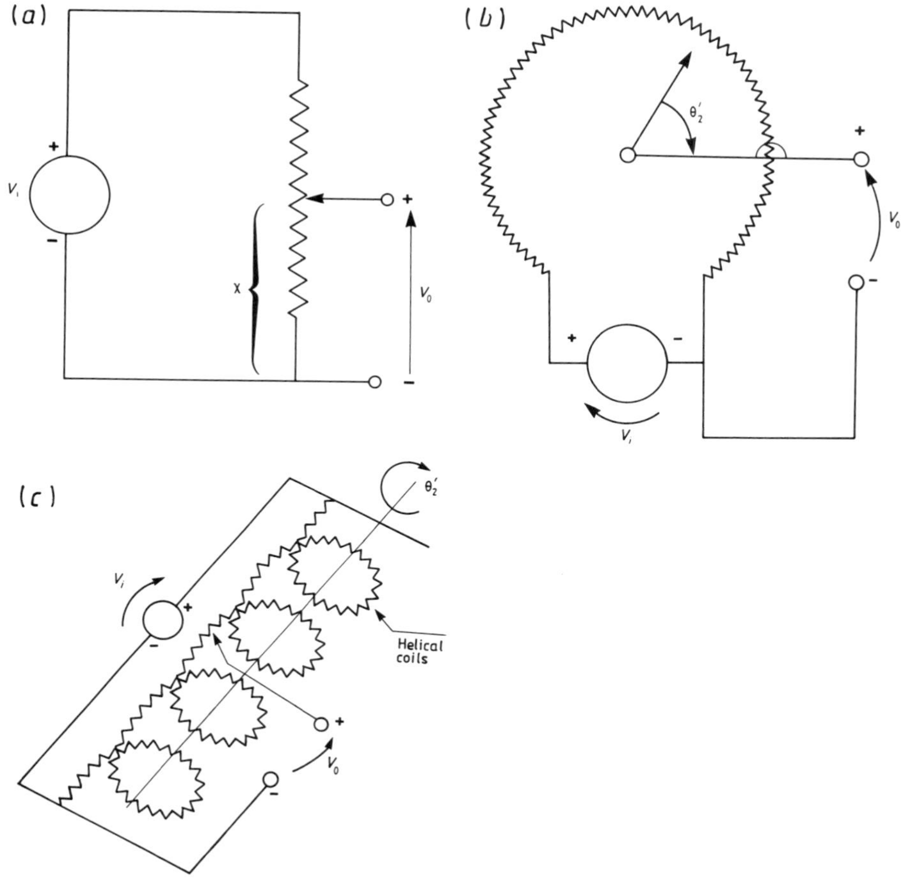

Figure 2.4 Three types of potentiometric device for measuring displacements: (a) translational; (b) single turn rotational; (c) multiturn rotational.

measured between the two ports. In addition to directly measuring extremely small displacements, of the order of nanometers, a strain gage can also be used to measure force or pressure. Biomedical examples of strain gage measurements are given below.

The principle of the strain gage operation is as follows: as a material is elongated or compressed, it exerts an equal and opposite force to the applied force. The strain (ε) is a dimensionless unit that can be calculated as the ratio of change of length to original length,

$$\varepsilon = \Delta L/L \tag{2.9}$$

where L is the object's original length along an axis and ΔL is the change in

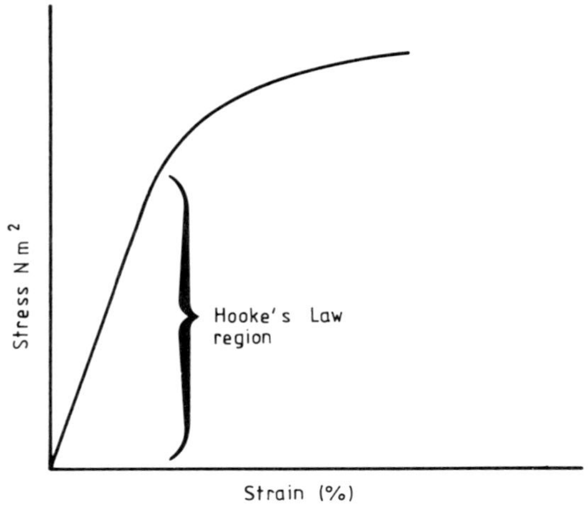

Figure 2.5 A plot of stress versus strain. The linear region of the curve is where Hooke's law applies.

length along that axis.

Strain measurements are normally stated as microstrain ($\mu\varepsilon$), where $1.0\,\mu\varepsilon = 1 \times 10^{-6}\,\varepsilon$. Compressive (or negative) strain shortens the object. Tensile (or positive) strain lengthens an object.

Stress (σ) is the resistance to strain. Figure 2.5 shows a typical stress–strain curve for a material. For a material whose stress–strain curve is similar to that of figure 2.5, the relationship between stress and strain is linear for small values of strain. Hooke's law applies to this linear region and is expressed as

$$\sigma = E\varepsilon \tag{2.10}$$

where E is the modulus of elasticity or Young's modulus.

Since the strain properties of an object may vary with position, strain gages are commonly used to measure strain on a small section of the object's surface area. The deformation of an object can be measured by using a series of strain gages attached to the object. In this way, the magnitude of load causing axial shearing, bending or twisting can be measured. A load cell is used, for example, to measure axially loaded forces.

As a fine wire (diameter$\approx$ 25 μm) is strained within the elastic limits, its resistance changes because of variations in the wire's diameter, length and resistivity. The derivation below shows how each of the parameters effects the resistance change. The resistance R of a wire with resistivity ρ (Ω m), length L (m) and cross-sectional area A (m^2) is given by

$$R = \rho L/A. \tag{2.11}$$

Differential change in R is found by taking the differential

$$\mathrm{d}R = \rho\,\mathrm{d}L/A - \rho A^{-2} L\,\mathrm{d}A + L\,\mathrm{d}\rho/A. \tag{2.12}$$

This expression should be rearranged in such a way that it represents finite changes in the parameters and is a function of standard mechanical coefficients. Dividing elements of (2.12) by corresponding elements of (2.11) and introducing incremental values, we obtain

$$\Delta R/R = \Delta L/L - \Delta A/A + \Delta\rho/\rho. \tag{2.13}$$

Assuming a constant wire volume, Poisson's ratio μ details the change in diameter D compared to the change in length, $\Delta D/D = -\mu\,\Delta L/L$. Substituting this into (2.13) gives

$$\Delta R/R = (1 + 2\mu)\Delta L/L + \Delta\rho/\rho. \tag{2.14}$$

It can be seen that the change in resistance is expressed as a function of changes in length ($\Delta L/L$) and area ($-2\mu\ \Delta L/L$) plus a strain-induced term due to changes in the lattice structure of the material, $\Delta\rho/\rho$. The gage factor G, found by dividing (2.14) by $\Delta L/L$, is useful for comparing various strain gage materials:

$$G = (\Delta R/R)/(\Delta L/L) = (1 + 2\mu) + (\Delta\rho/\rho)/(\Delta L/L). \tag{2.15}$$

Gage factors and temperature coefficients of resistivity of various strain gage materials are given in table 2.1. Semiconductor materials have a gage factor that is approximately 50–70 times that of metals. The gage factor for metals is primarily a function of dimensional effects. Poisson's ratio for most metals is 0.3 and thus G is at least 1.6. However, for semiconductors, the piezoresistive effect is dominant. Unfortunately, the desirable higher gage factors for semiconductor devices is offset by their higher resistivity–temperature coefficient, which makes them more sensitive to temperature variations. Consequently, designs for instruments that use semiconductor materials must incorporate temperature compensation.

There are two types of strain gage, unbonded or bonded. Figure 2.6(a) shows an unbonded strain gage unit. Four sets of strain-sensitive wires are connected to form a Wheatstone bridge (figure 2.6(b)). These unbonded strain gage wires are mounted under stress between the frame and the movable armature such that preload is greater than expected external compressive load. This unbonded strain gage type of sensor may be used to convert blood pressure to diaphragm movement, then to resistance change and finally to an electric signal.

A bonded strain gage element can be fabricated by using a metallic wire, etched foil, vacuum-deposited film or semiconductor bar, which is cemented to the strained surface (figure 2.7). The deviation from linearity is approximately

Table 2.1 Properties of strain gage materials.

Material	Composition (%)	Gage factor	Temperature coefficient of resistivity (per 10^5 °C)
Constantan (advance)	Ni_{45}, Cu_{55}	2.1	±2
Isoelastic	Ni_{36}, Cr_8, (Mn, Si, Mo)$_4$, Fe_{52}	3.52–3.6	+17
Karma	Ni_{74}, Cr_{20}, Fe_3, Cu_3	2.1	±2
Manganin	Cu_{84}, Mn_{12}, Ni_4	0.3–0.47	±2
Alloy 479	Pt_{92}, W_8	3.6–4.4	±24
Nickel	Pure	−12––20	670
Nichrome V	Ni_{80}, Cr_{20}	2.1–2.63	10
Silicon	(p type)	100–170	70–700
Silicon	(n type)	−100––140	70–700
Germanium	(p type)	102	
Germanium	(n type)	−150	

SOURCE: From R S C Cobbold, *Transducers for Biomedical Measurements* 1974, Wiley.

1% for the bonded strain gage. A useful method of temperature compensation to cancel the natural temperature sensitivity of bonded strain gages involves using a second strain gage as a dummy element that is also exposed to the temperature variation, but not to strain. The four-arm bridge (figure 2.6) measurement configuration should be used, not only because it provides temperature compensation but it also yields four times greater output when all four arms contain active gages, as explained below. Dental bite force can be measured using four bonded metal strain gages mounted on cantilever beams [6].

The semiconductor strain gage element has the advantage of having a high gage factor (table 2.1). Unfortunately, because the piezoresistive component varies with strain, it is more temperature sensitive and inherently more non-linear than metal strain gages. Semiconductor elements can be fabricated as bonded, unbonded or integrated strain gage units. Typical semiconductor strain gage units are illustrated in figure 2.8.

As mentioned, there are metallic and semiconductor, bonded and unbonded strain gages. All four combinations have very high sensitivity and have great potentials for miniaturization and inexpensive mass production. These characteristics make them ideal for use in medical instruments.

The major disadvantages of strain gages are their low linearity/small dynamic range, and temperature dependence. However, these disadvantages do not limit their use, but rather require strict design approaches when used in medical devices.

2.2.2.1 Elastic resistance strain gages. Elastic resistance strain gages are extensively used in biomedical plethysmography (volume measuring), especially in cardiovascular and respiratory monitoring. These systems employ a narrow

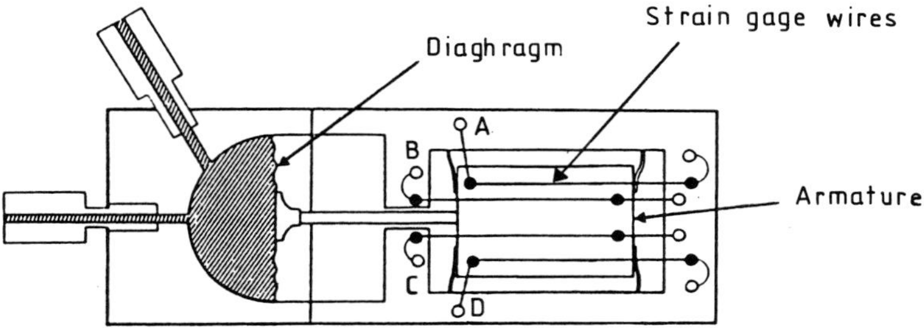

(a)

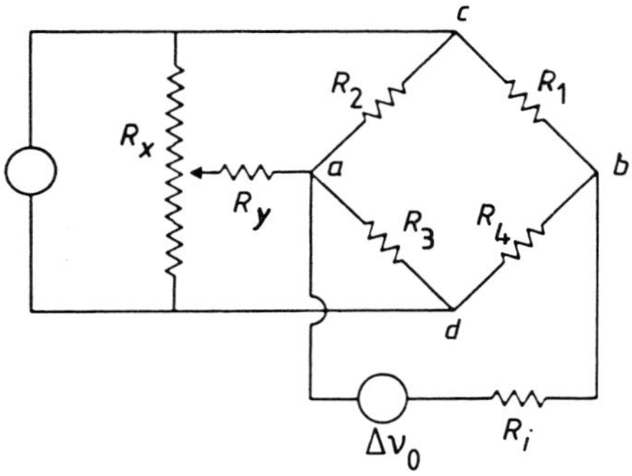

(b)

Figure 2.6 (a) An unbonded strain gage pressure sensor. The diaphragm is directly coupled by an armature to an unbonded strain-gage system. With increasing pressure, the strain on gage pair B and C is increased, while that on gage pair A and D is decreased. (b) A Wheatstone bridge with four active elements. $R_1 = \text{B}$, $R_2 = \text{A}$, $R_3 = \text{D}$ and $R_4 = \text{C}$ when the unbonded strain gage is connected for translational motion. Resistor R_y and potentiometer R_x are used to initially balance the bridge. v_i is the applied voltage and Δv_0 is the output voltage on a voltmeter or similar device with an internal resistance of R_i. (From Peura R A and Webster J G *Basic sensors and principles in Medical Instrumentation: Application and Design*, 2nd edition, edited by J G Webster, copyright © 1992 Houghton Mifflin Co., Boston.)

(0.5 mm ID, 2 mm OD), from 3 to 25 cm long silicone rubber tube which is filled with mercury, an electrolyte or conductive paste. Electrodes (amalgamated copper, silver or platinum) seal the ends of the tube. As the tube stretches, the diameter of the tube decreases and the length increases with a resultant increase in longitudinal resistance. Typical gages have a resistance per unit length of approximately 0.02–2 Ω cm^{-1}. These units are designed to measure higher displacements than the previously described types of strain gage.

Elastic strain gages have a linearity within 1% for 10% of the maximal extension. For extensions of up to 30% of the maximum, the non-linearity increases to 4%. Another problem is the dead band or initial non-linearity due

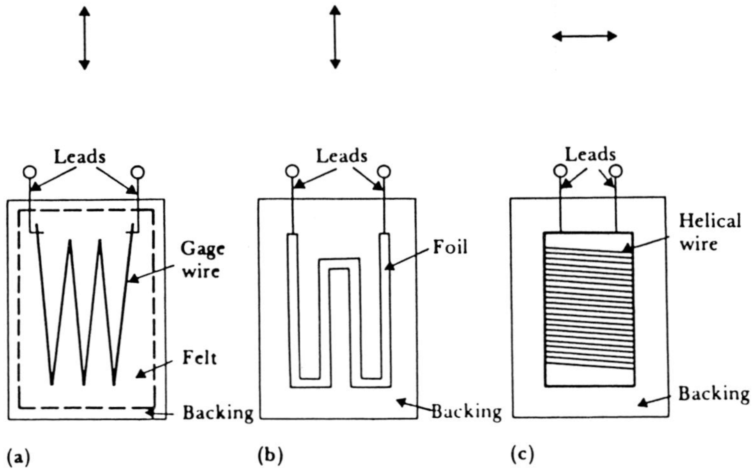

Figure 2.7 Typical bonded strain gage units: (a) resistance wire type; (b) foil type; (c) helical wire type. The direction of maximal sensitivity to strain is given by the arrows. (Parts (a) and (b) are modified from Lion K S *Instrumentation in Scientific Research*, copyright © 1959 by McGraw-Hill, Inc.)

to slackness of the elastic strain gage. Long-term creep is another disadvantage that is due to the properties of the rubber tubing. These disadvantages make elastic strain gages unsuitable for accurate absolute measurements. However, their advantages, such as ruggedness and ease of operation, make them quite useful for dynamic measurements.

Operational problems with the elastic strain gages include ensuring continuity of the mercury column, maintaining a good contact between the mercury column and electrodes and drift due to a large temperature coefficient. Due to the mass–elasticity and stress–strain relations of the tissue–strain gage interface, accurate calibration is difficult. When compared with other strain gage systems, the elastic strain gage requires more power since the resistance of the strain gage is lower.

The dynamic response of elastic strain gages was found to produce constant amplitude and phase up to 10 Hz [7]. They found that significant distortion occurred for frequencies higher than 30 Hz. A problem which is not fully appreciated is that the gage does not fully distend during pulsations when the diameter of the vessel is being measured [2]. The mass of the gage and its finite mechanical resistance can cause the gage to dig into the vessel wall as the vessel expands, such that the output is several times lower than that measured using other non-invasive approaches such as ultrasound or cineangiography.

A mercury in rubber strain gage can be used to measure the circumference of the human calf (figure 2.9) or chest. Bridge circuits are used to transfer the changes in strain resistance into measurable electrical signals. Dimension variations are determined by means of comparisons with calibration runs. An

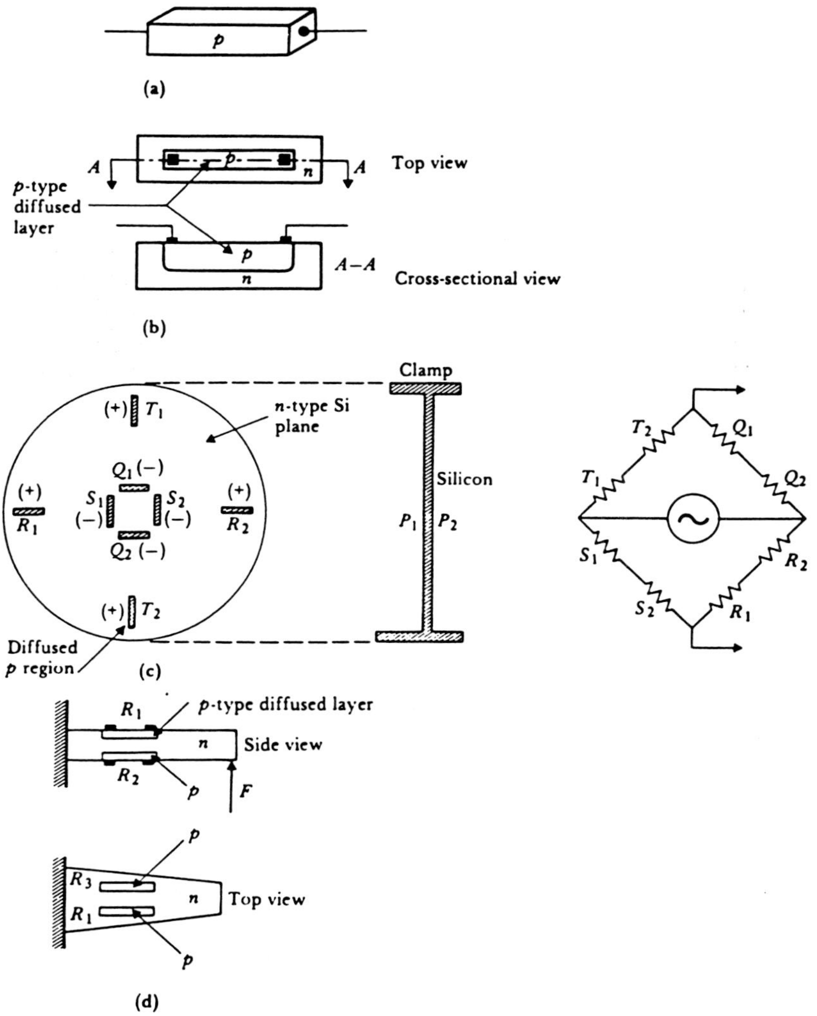

Figure 2.8 Typical semiconductor strain gage units: (a) unbonded, uniformly doped; (b) diffused p-type gage; (c) integrated pressure sensor; (d) integrated cantilever beam force sensor. (From R S C Cobbold *Transducers for Medical Measurements: Application and Design*, copyright © 1974, John Wiley and Sons, Inc.)

electrically calibrated mercury in rubber strain gage has been developed by Hokanson *et al* [8]. Hokanson's design eliminates lead wire errors, common with these devices, because the problems due to the low resistance of the strain gage, by effectively placing the strain gage at the corners of the measurement bridge. In addition, they used a constant-current source to drive the circuit which produces a linear output for large changes in gage resistance. The elastic strain

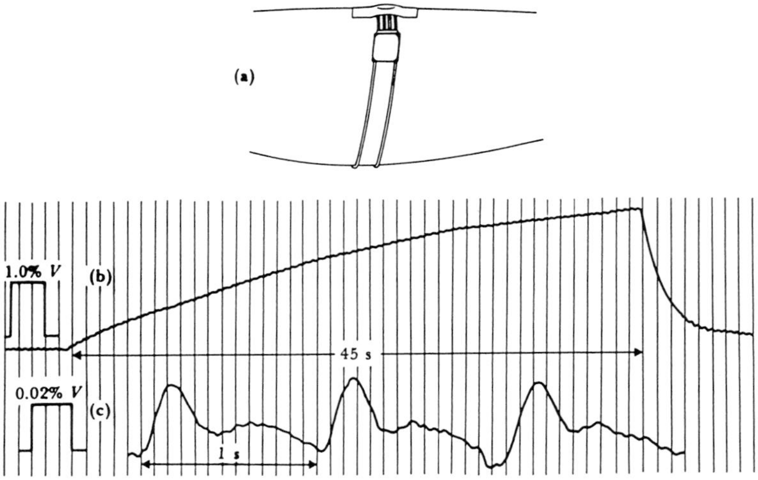

Figure 2.9 Mercury in rubber strain gage plethysmography: (a) four-lead mercury in rubber strain gage plethysmography applied to the human calf; (b) the bridge output for venous occlusion plethysmography with the venous circulation occluded for 45 s by means of a thigh cuff inflated above venous pressure, but below arterial pressure; (c) the bridge output showing arterial pulsations for arterial pulse plethysmography.

gage device and its output when applied to the human calf are illustrated in figure 2.9.

2.3 INDUCTIVE SENSORS

Inductive sensors are devices that transduce the change in the measured physical parameter, that is in the form of a mechanical displacement, into variations of a coil's inductance. The inductance, L, of a coil can be expressed as:

$$L = n^2 G \mu \tag{2.16}$$

where μ is the effective medium permeability, G the geometric form factor and n the number of coil turns.

Each of these parameters can be changed by mechanical means, therefore displacement can be measured by varying any one of these three parameters.

Figure 2.10 illustrates different types of inductive displacement sensor: (a) self-inductance, (b) mutual inductance and (c) differential transformer. In most cases a mutual inductance system can be converted into a self-inductance system by series or parallel connections of the coils. The mutual inductance device (figure 2.10(b)) becomes a self-inductance device (figure 2.10(a)) when terminals b and c are connected.

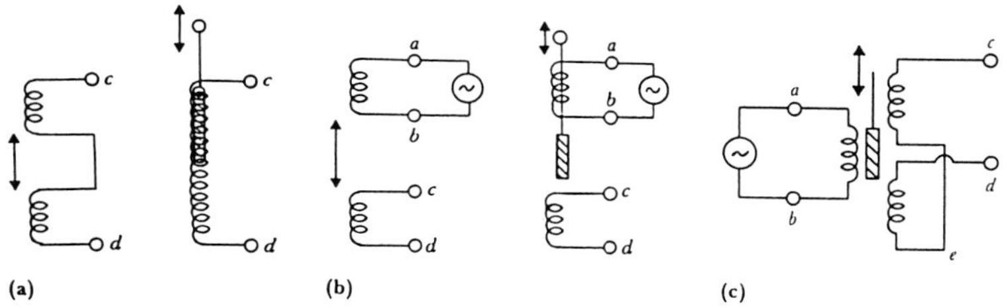

Figure 2.10 Inductive displacement sensors: (a) self-inductance (two examples); (b) mutual inductance (two examples); (c) differential transformer. (From Peura R A and Webster J G *Basic sensors and principles in Medical Instrumentation: Application and Design*, 2nd edition, edited by J G Webster, copyright © 1992 Houghton Mifflin Co., Boston.)

An advantage of an inductive sensor is that it is not affected by the dielectric properties of its environment. A disadvantage is that it can be affected by external magnetic fields due to magnetic materials in its proximity.

Figure 2.10(a) shows the variable-inductance sensor with a single displaceable core, which works on the principle that alterations in the self-inductance of a coil may be produced by changing the position of a magnetic core within the coil, or the geometric form factor. For this device, the change in inductance is not linearly related to displacement. These devices have low power requirements and produce large variations in inductance which make them useful for radiotelemetry applications. An intracardiac pressure sensor was developed by Allard [9] who used a single coil with a movable Mu-metal core. This device has a frequency response that extends beyond 1 kHz and thus can be used to measure both pressures and heart sounds.

Figure 2.10(b) illustrates a mutual-inductance sensor which employs two separate coils and uses the variation in their mutual magnetic coupling to measure displacement. These devices can be used to measure cardiac dimensions, monitor infant respiration and ascertain arterial diameters [2].

Applications of mutual inductance transformers in measuring changes in dimension of internal organs (kidney, major blood vessels and left ventricle) are presented by van Citters [10]. The secondary-coil-induced voltage is a function of the geometry of the coils (separation and axial alignment), the number of primary and secondary turns and the frequency and amplitude of the excitation voltage. The secondary voltage is a non-linear function of the separation of the coils. A frequency is selected to cause resonance in the secondary coil and thus a maximum output signal results. A demodulator and amplifier circuit is used to detect the output voltage.

A widely used sensor in physiological research and clinical medicine is the linear variable differential transformer (LVDT). It is used to measure pressure, displacement and force (Reddy and Kesavan 1988). The LVDT is composed of a primary coil (terminals a and b) and two secondary coils (c–e and d–e) connected in series (figure 2.10(c)). Coupling between these two coils is varied by the motion of a high-permeability alloy slug linking them. To achieve a wider region of linearity, the two secondary coils are connected in opposition.

A sinusoidal frequency between 60 Hz and 20 kHz is used as the primary-coil excitation. The alternating magnetic field induces nearly equal secondary voltages v_{ce} and v_{de}; the output voltage is $v_{cd} = v_{ce} - v_{ed}$. When the slug is symmetrically placed, the two secondary voltages are equal and the output signal is zero.

The output voltage v_{cd} of the LVDT is linear over a large range, exhibits a 180° change of phase when the core passes through the center position and saturates when the slug is at the two extreme positions. Commercial LVDT specifications include sensitivities of the order of 0.5–2 mV for a displacement of 0.01 mm/V of primary voltage, full-scale displacement of 0.1–250 mm and linearity of ±0.25%. Much higher sensitivities can be achieved with LVDTs than with strain gages.

An LVDT has the disadvantage that it requires complex signal processing instrumentation. This is illustrated in figure 2.11 in which essentially the same magnitude of output voltage results from two very different input displacements. Displacement direction may be determined by observing that there is a 180° phase shift when the core passes through the null position. Phase-sensitive demodulation can be used to determine the direction of displacement. A ring demodulator system can be used with the LVDT for this purpose. LVDTs are very popular devices; semiconductor manufacturers produce integrated circuits–LVDT drivers, that include both excitation and demodulation modules. They provide for easy, inexpensive and high-quality designs. An example of such a circuit is the AD598.

An interesting example of the use of inductive sensors for measurements of biomechanical parameters relates to obtaining knowledge of the biomechanical behavior of the human knee. Beynnon *et al* [12] indicate that the anterior cruciate ligament, ACL, is the most frequently totally disrupted knee ligament. Reconstruction of the disrupted ACL has the objective of restoring normal kinematic movement of the knee joint. Athletes have frequently used functional knee braces to augment recovery following surgical treatment. Strain measurements of the ACL help to determine the efficacy of functional braces during normal biomechanical behavior of the knee joint. Beynnon *et al* [12] described an arthroscopically implanted sensor capable of measuring displacement in the *in vivo* environment of the ligamentous unit. This sensor has sensitivity for measuring displacement within small segments of the ligament.

Beynnon uses a differential voltage reluctance transducer, DVRT, produced by Micro Strain, Burlington, VT, to measure ligament strain. The DVRT is a form of LVDT which was described above. The miniature DVRT displacement transducer allows access to delicate and hard-to-reach structures. It is much smaller than other transducers and measures only 1.5 mm in diameter. The

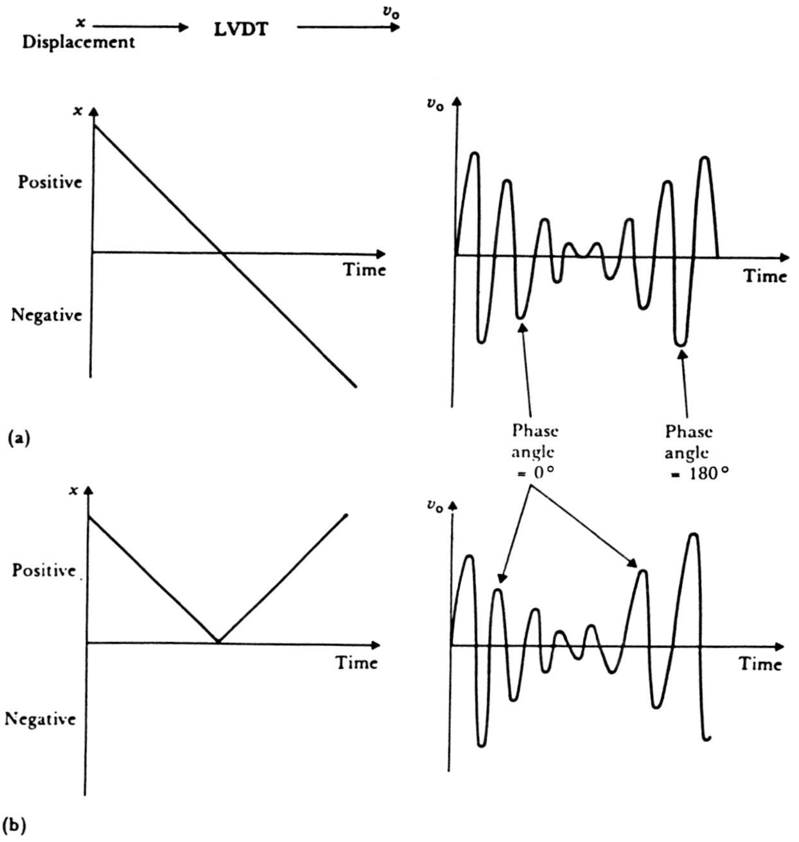

Figure 2.11 (a) As x moves through the null position, the phase changes by 180°, while the magnitude of v_0 is proportional to the magnitude of x. (b) An ordinary rectifier demodulator cannot distinguish between (a) and (b), so a phase-sensitive demodulator is required. (From Peura R A and Webster J G *Basic sensors and principles in Medical Instrumentation: Application and Design*, second edition, edited by J G Webster, copyright © 1992 Houghton Mifflin Co., Boston.)

stainless-steel-encased device is comprised of two coils and a free sliding, magnetically permeable, stainless steel core. The core position is detected by measuring the coils' differential reluctance by sinusoidal excitation and synchronous detection. Core movements cause the reluctance of one coil to increase, while the other is decreased. The reluctance difference is a sensitive measure of core position. Temperature changes are canceled because the reluctance of each coil changes similarly, thereby canceling out temperature effects. The sensor's electrical connections are potted in vacuum-pumped, biocompatible epoxy, within the cylindrical casing. Figure 2.12 shows the

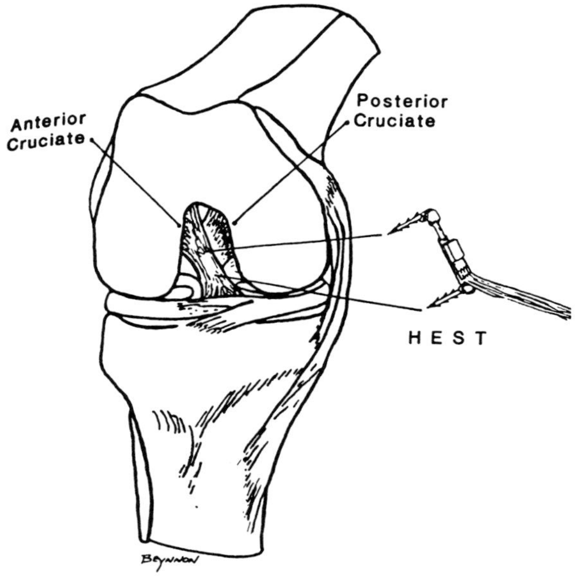

Figure 2.12 Orientation of the strain gage sensor on the anteromedial aspect of the anterior cruciate ligament. (From B D Beynnon, M H Pope, C M Werthheimer, R J Johnson, B F Fleming, C E Nichols and J G Howe, *The effect of functional knee-braces on strain on the anterior cruciate ligament in vivo*, 1992 *The Journal of Bone and Joint Surgery* **74** A, no 9, 1298–1312.)

attachment and orientation of the strain sensor.

Barbs on the end of the sensor system are the attachment points to the ligamentous tissue. Figure 2.13 illustrates the difference in the strain on the ACL for initial and repeated testing with and without a brace.

2.4 CAPACITIVE SENSORS

Capacitive sensors are devices that transduce the measured physical parameter, that is in the form of a mechanical displacement, into variations of a capacitor's capacitance. The capacitance between two parallel plates, separated by distance x, is

$$C = \varepsilon_0 \varepsilon_r A/x \tag{2.17}$$

where A is the area of the plates, x is the distance between the plates, ε_0 is the dielectric constant of free space and ε_r is the relative dielectric constant of the insulator (1.0 for air). Mechanical displacement can be used to vary any of the

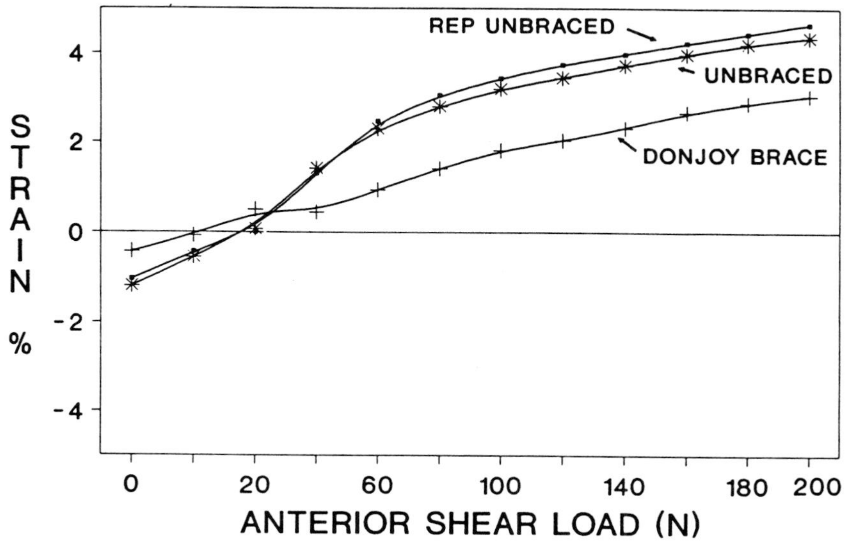

Figure 2.13 Results from anterior shear loading applied with the knee in 30° of flexion. The percentage of strain is plotted against anterior shear load (in newtons). The mean values for strain, averaged across four loading cycles, for testing without a brace, with the DonJoy brace and repeated testing without a brace, REP = repeated. (From B D Beynnon, M H Pope, C M Werthheimer, R J Johnson, B F Fleming, C E Nichols and J G Howe *The effect of functional knee-braces on strain on the anterior cruciate ligament in vivo*, *The Journal of Bone and Joint Surgery* **74** A, no 9, 1992, 1298–1312.)

three parameters: ε_r, A or x. The easiest method to implement, and the most commonly used, is to vary the separation between the plates.

Differentiating the capacitance (2.17) with respect to plate separation gives the sensitivity K of a capacitive sensor.

$$K = \Delta C/\Delta x = -\varepsilon_0\varepsilon_r A/x^2. \qquad (2.18)$$

Highest sensitivity occurs for small plate separations.

Substituting (2.17) into (2.18) gives an expression for the percentage change in C about any neutral point, which is equal to the change per unit in x for small displacements. Thus

$$\mathrm{d}C/\mathrm{d}x = -C/x \qquad (2.19)$$

or

$$\mathrm{d}C/C = -\mathrm{d}x/x. \qquad (2.20)$$

The capacitance microphone illustrated in figure 2.14 is a good example of a straightforward method for detecting variation in capacitance [2, 13]. A DC-excited circuit is used in this illustration, thus current does not flow when the

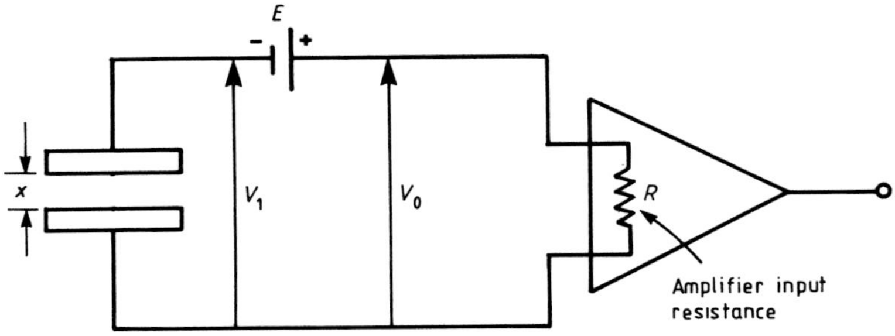

Figure 2.14 A capacitance sensor for measuring dynamic displacement changes.

capacitor is stationary (with separation x_0) resulting in $V_1 = E$. However, a change in plate separation distance $\Delta x = x_1 - x_0$ produces a voltage $V_0 = V_1 - E$. The output voltage V_0 is a function of x,

$$V_0(\mathrm{i}\omega)/X_1(\mathrm{i}\omega) = (E/x_0)\mathrm{i}\omega\tau/(\mathrm{i}\omega\tau + 1) \tag{2.21}$$

where the time constant $\tau = RC = R\varepsilon_0\varepsilon_r A/x_0$.

The frequency response of the system is a high-pass filter, since, for $\omega\tau \gg 1$, $V_0(\mathrm{i}\omega)/X_1(\mathrm{i}\omega) = E/x_0$, which is a constant. However, the response drops off for low frequencies, and it is zero when $\omega = 0$. This frequency response is sufficient for a microphone that does not measure sound pressures at frequencies below 20 Hz. The input impedance of the read-out device must be high (10 MΩ or higher) in order to achieve a required low-frequency bandwidth. Capacitance sensors are not suitable for measuring most physiological signals because the frequency spectra of these signals have dominant low-frequency components.

A capacitive patient movement detector may be constructed by employing compliant plastics of different dielectric constants which are placed between foil layers to form a capacitive mat on a bed. Movement of the patient changes the capacitance, which is detected to display respiratory movements from the lungs and ballistographic movements from the heart [14].

The capacitance sensor frequency response may be extended to DC by means of a guarded parallel plate capacitance sensor, shown in figure 2.15(a) [13]. This technique, which uses a high-gain feedback amplifier, provides a linear relationship between displacement and capacitance. The changes in capacitance are sensed by placing the displacement varying capacitor in the feedback loop of an operational amplifier (op amp) circuit (figure 2.15(b)).

The gain of the inverting op amp circuit (figure 2.15) is a negative of the ratio of the feedback to the input impedance. Thus

$$V_0(\mathrm{i}\omega)/V_i(\mathrm{i}\omega) = -Z_f(\mathrm{i}\omega)/Z_i(\mathrm{i}\omega) \tag{2.22}$$

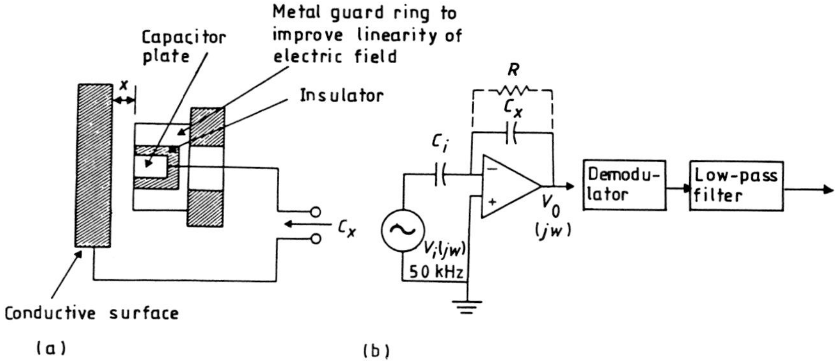

Figure 2.15 (a) A guarded parallel plate displacement sensor. (b) An instrumentation system with output proportional to capacitance displacement. (Part (a) is from *Measurement Systems: Application and Design* by E O Doebelin, copyright © 1990 by McGraw Hill, Inc.)

$$= -(1/\mathrm{i}\omega C_x)/(1/\mathrm{i}\omega C_i) \tag{2.23}$$

$$= -C_i/C_x. \tag{2.24}$$

Substituting equation (2.17) in (2.24) yields

$$V_0(\mathrm{i}\omega) = C_i x V_i(\mathrm{i}\omega)/\varepsilon_0\varepsilon_r A = Kx. \tag{2.25}$$

Output voltage is linearly related to the plate separation x (equation (2.25)). A high-frequency source (e.g. 50 kHz) $V_i(\mathrm{i}\omega)$ is used. The circuit produces an amplitude-modulated output voltage $V_0(\mathrm{i}\omega)$. The mean value of $V_0(\mathrm{i}\omega)$, proportional to x, is determined by demodulation and low-pass filtering (10 kHz corner frequency). In order to provide bias current for the amplifier (which should be an FET op amp), a discharge resistor R must be connected in parallel with C_x. The value of the feedback resistance should be high with respect to the reactance of C_x.

Such devices have been used for recordings of chest wall motions, apex motion, heart sounds and brachial and radial pulses [15]. This sensor has the advantages that it is non-contacting (skin is used as one side of the capacitor) and linear and has a wide frequency response. Sensor problems relate to the proper isolation of the patient from the high-amplitude AC excitation voltages, mechanical positioning of the probe and respiratory motion artifacts.

A capacitance sensor can be built using layers of mica insulators sandwiched between corrugated metal layers. Pressure flattens the corrugations, moving the metallic plates closer to each other, thus decreasing the plate separation which increases the capacitance. This sensor is not damaged by large overloads, since

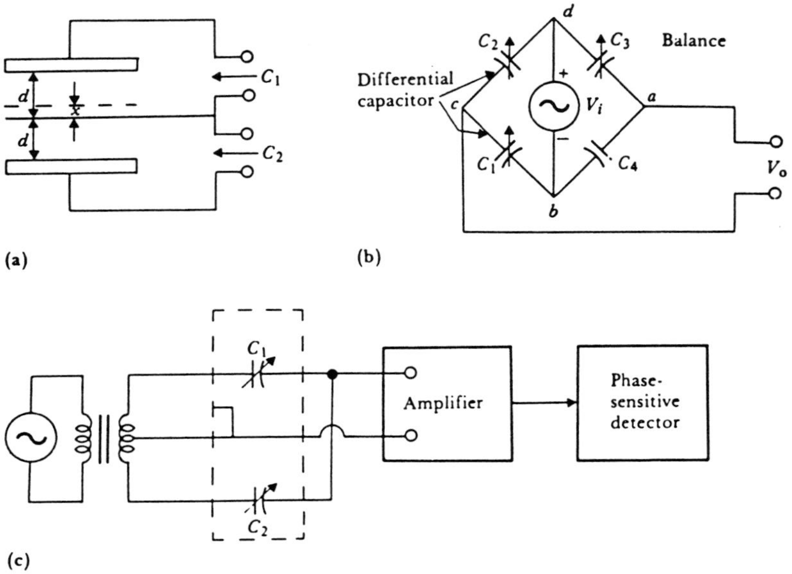

Figure 2.16 (a) A differential three-terminal capacitor. (b) A capacitance bridge circuit with output proportional to fractional difference in capacitance. (c) A transformer ratio arm bridge. (From R S C Cobbold R S C *Transducers for Biomedical Measurements: Principles and Applications*, copyright © 1974, John Wiley and Sons, Inc.)

flattening of the corrugations does not cause the metal to yield. It can be used to measure the pressure between the foot and the shoe [16].

Differential capacitor systems provide accurate measurements of displacement [2]. A differential three-terminal capacitor linearly relates displacement to $(C_1 - C_2)/(C_1 + C_2)$ (figure 2.16(a)). This advantage is seen by letting d be equal to the equilibrium displacement and x the displacement (positive direction up). Then

$$C_1 = \varepsilon_0\varepsilon_r A/(d - x) \quad \text{and} \quad C_2 = \varepsilon_0\varepsilon_r A/(d + x) \tag{2.26}$$

or

$$x/d = (C_1 - C_2)/(C_1 + C_2). \tag{2.27}$$

The bridge circuit (figure 2.16(b)) provides an output voltage proportional to the fractional difference in capacitance indicated by equation (2.27). Because at the equilibrium position the differential capacitor values C_1 and C_2 are equal and C_3 is balanced to equal C_4, the output voltage is

$$V_0 = (V_i/2)\,(C_1 - C_2)/(C_1 + C_2) \tag{2.28}$$

or

$$V_0 = (V_i/2d)x. \tag{2.29}$$

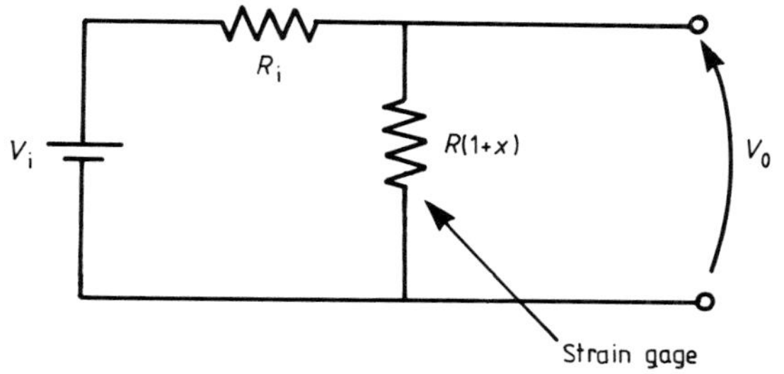

Figure 2.17 A voltage divider circuit used to measure the strain gage resistance change.

A transformer ratio arm bridge (figure 2.16(c)) can also be used to solve equation (2.27). The degree of bridge unbalance ($C_1 - C_2$) is a function of amplifier current which may be found from equation (2.26).

$$C_1 - C_2 = 2A\varepsilon_0\varepsilon_r x/(d^2 - x^2). \tag{2.30}$$

This expression is linear with x for the case $d \gg x$, that is a normal operating condition. This bridge circuit has high accuracy and sensitivity. Bridge balance is independent of the third terminal shield, which allows measurement of capacitance at various distances from the bridge.

2.5 BRIDGE CIRCUITS

One of the best and most frequently used techniques to measure small changes in resistance is by using a Wheatstone bridge circuit. The Wheatstone bridge was developed by S H Christie in 1933. The following example illustrates why it is important to employ a bridge circuit when performing physiological measurements using a strain gage. The strain gage will exhibit a small change in resistance when the specimen is strained. Figure 2.17 shows a simple voltage divider (not a bridge) circuit which could be used to measure the strain gage resistance change. To energize the strain gage, a constant voltage is applied to the gage. Since the voltage across a strain gage resistor is proportional to the value of its resistance and current flowing, a change in a voltage across the strain gage occurs which is proportional to the change in strain gage resistance. However, this change of strain gage voltage would be superimposed upon a much larger standing voltage, namely the voltage across the unstrained resistance of the gage. Such a small change (perhaps one part in 10 000) in a large standing value is difficult to detect directly and is subject to error due to other small fractional changes or artifacts (i.e. source voltage change and noise).

Assume that for a non-strained gage, initially $x = 0$. If $V_i = 10$ V, and $R = R_i$, then $V_0 = 5.0$ V. Now if the strain gage is subjected to a load, and

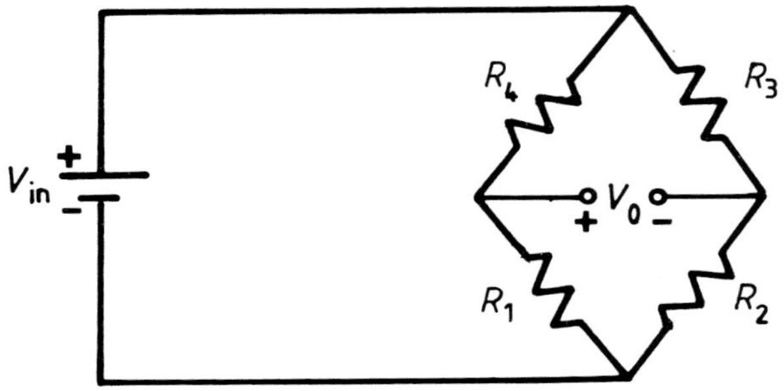

Figure 2.18 A Wheatstone bridge with an applied DC voltage of V_{in} and an output voltage V_0. Using the voltage-divider approach, V_0 is zero if the bridge is balanced—i.e. $R_1/R_4 = R_2/R_3$.

$x = 0.0001$, then $V_0 = 5.0005$ V. If we look at it another way, we have 0.5 mV riding on the 5.0 V signal. This is very difficult to differentiate from noise that may be present.

Bridge circuits overcome this described problem in such a way that it is possible to adjust their output voltage to zero when the strain gage is balanced or unstressed. The source of excitation can be either a voltage or current source (V or I) which is connected to the diagonals of the bridge, in which four two-terminal circuit elements form a quadrilateral. A Wheatstone bridge is shown (figure 2.18) with an applied DC voltage of V_{in} and an output voltage V_0. Using the voltage divider approach, V_0 is zero if the bridge is balanced—i.e. $R_1/R_4 = R_2/R_3$.

Resistance variation can be detected by either null-balance or deflection-balance bridge circuits. In a null-balance bridge the sensor resistance change is balanced (zero output) by a variable resistance in a bridge adjacent arm. The calibrated null adjustment is an indication of the change in sensor resistance. The deflection-balance method, on the other hand, makes use of the amount of bridge unbalance in order to determine the change in sensor resistance.

$$V_0 = R_1/(R_1 + R_4)V_{in} - R_2/(R_2 + R_3)V_{in} \tag{2.31}$$

$$V_0 = (R_1/R_4 - R_2/R_3)V_{in}/[(1 + R_1/R_4)(1 + R_2/R_3)] \tag{2.32}$$

at balance

$$V_0 = 0 \quad \text{if} \quad R_1/R_4 = R_2/R_3. \tag{2.33}$$

The bridge is at a null irrespective of excitation (current or voltage, AC or DC) and the input impedance of the detector (voltmeter).

The bridge circuit works in the following way.

(i) If the ratio R_2/R_3 is fixed at K, a null is achieved when $R_1 = KR_4$.

(ii) If R_1 is unknown (strain gage) and R_4 is an accurately adjustable variable resistor, the magnitude of R_1 is found by adjusting R_4 until null is achieved.

In the majority of sensor applications employing bridges, the deviation of one or more resistor in the bridge from an initial value must be measured as an indication of the magnitude or change of measurement. In figure 2.18, if we let $R_1 = R(1+x)$, and $R_2 = R_3 = R_4 = R$, where x is the fractional deviation around zero, as a function of (say) strain, then,

$$V_0 = [R(1+x)/(R+R(1+x))]V_{in} - 0.5V_{in} \quad (2.34)$$

$$V_0 = [(2+2x-2-x)/(2(2+x))]V_{in} \quad (2.35)$$

or

$$V_0 = (V_{in}/4)(x/(1+x/2)) \quad (2.36)$$

$$V_0 = (V_{in}/4)x \quad \text{for } x \ll 1. \quad (2.37)$$

As the equation (2.36) indicates, the relationship between bridge output and x is *not linear*, but for small ranges of x it is sufficiently linear for many purposes (equation (2.37)). For example, if $V_{in} = 10\,\text{V}$ and the maximum value of x is ± 0.002, the bridge output will be linear within 0.1% for the range of outputs from 0 to $\pm 5\,\text{mV}$, and to 1% for the output range 0 to $\pm 50\,\text{mV}$ (± 0.02 range for x).

It should be noted that the bridge sensitivity can be doubled if two identical variable elements can be used, at positions R_3 and R_1, i.e. two identically oriented strain gage resistances aligned in a single pattern. Note that bridge output is doubled, but the same degree of non-linearity exists. In figure 2.18, if we let $R_1 = R_3 = R(1+x)$, and $R_2 = R_4 = R$, then

$$V_0 = [R(1+x)/(R+R(1+x)) - R/(R+R(1+x))]V_{in} \quad (2.38)$$

$$V_0 = [x/(2+x)]V_{in} \quad (2.39)$$

or

$$V_0 = (V_{in}/2)x/(1+x/2) \quad (2.40)$$

$$V_0 = (V_{in}/2)x \quad \text{for } x \ll 1. \quad (2.41)$$

Another doubling of output can be achieved by using two resistors which increase and two resistors which decrease in the same ratio. An example is two identical two-element strain gages, attached to opposite faces of a thin carrier to measure its bending strains. The output is four times the output from a single-element bridge. It should be noted that a linear output results because of the complementary nature of the resistance change. In figure 2.18, if we let $R_1 = R_3 = R(1+x)$, and $R_2 = R_4 = R(1-x)$, then

$$V_0 = [R(1+x)/2R - R(1-x)/2R]V_{in} \quad (2.42)$$

$$V_0 = [(1+x-1+x)/2]V_{in} \quad (2.43)$$

or

$$V_0 = x\,V_{in} \quad (2.44)$$

which is linear with x.

2.6 DISPLACEMENT MEASUREMENTS

Biomedical scientists and physicians are interested in measuring the size, shape and position of the organs and tissues of the body. Changes in these parameters are important in differentiating normal from abnormal function. All four previously described sensor groups are used in both direct and indirect displacement measurements. An example of direct measurements is the determination of the change in dimensions of the cardiac chambers. Indirect displacement measurements can be used to quantify movements of cardiac blood. An example is the movement of a microphone diaphragm placed on the chest wall which indirectly measures the movement of the heart and the resulting heart murmurs due to cardiac blood flow and physiological effects.

An example of the use of piezoelectric sensors for measuring dimensional changes in the heart was described by Glower *et al* [17]. Their goal was to measure the intrinsic myocardial performance of the intact heart. Myocardial function was assessed by calculating segmental cardiac stroke work and end-diastolic segmental length after acute ischemic injury to the heart. They measured the end-diastolic segment length (figure 2.19) by implanting piezoelectric dimension transducers in the left ventricular subendocardium to assess regional myocardial segment length. One pair was placed in the anterior descending coronary distribution distal to a coronary occluder. The other pair of piezoelectric transducers was positioned in the distribution of the circumflex coronary artery. Ventricular pressure was determined by means of a solid state strain gage pressure sensor mounted on the tip of a catheter placed inside the ventricular cavity (see below for a description of the solid state pressure sensor).

A typical recording of myocardial segment length, segment elastance (ventricular pressure/segment length), transmural ventricular pressure (difference between the ventricular and pleural pressure) and first derivative of the transmural ventricular pressure during the cardiac cycle are shown in figure 2.20.

Piezoelectric sensors are useful for assessment of myocardial function after acute ischemic injury (as a function of occluding regional coronary blood flow). Figure 2.21 shows preischemic base line conditions and postischemic myocardial dysfunction. Ventricular pressure–segmental length loops and end-systolic pressure–length relationships under control conditions (preischemia) and 15 min after coronary occlusion (ischemia) are illustrated.

2.7 BLOOD PRESSURE MEASUREMENTS

This final section on physical sensors describes their use for measurements of blood pressure. Different kinds of sensor element may be used for this purpose; they include strain gages, LVDTs, variable inductances, variable capacitances, optoelectronics and piezoelectric and semiconductor devices. The principles of

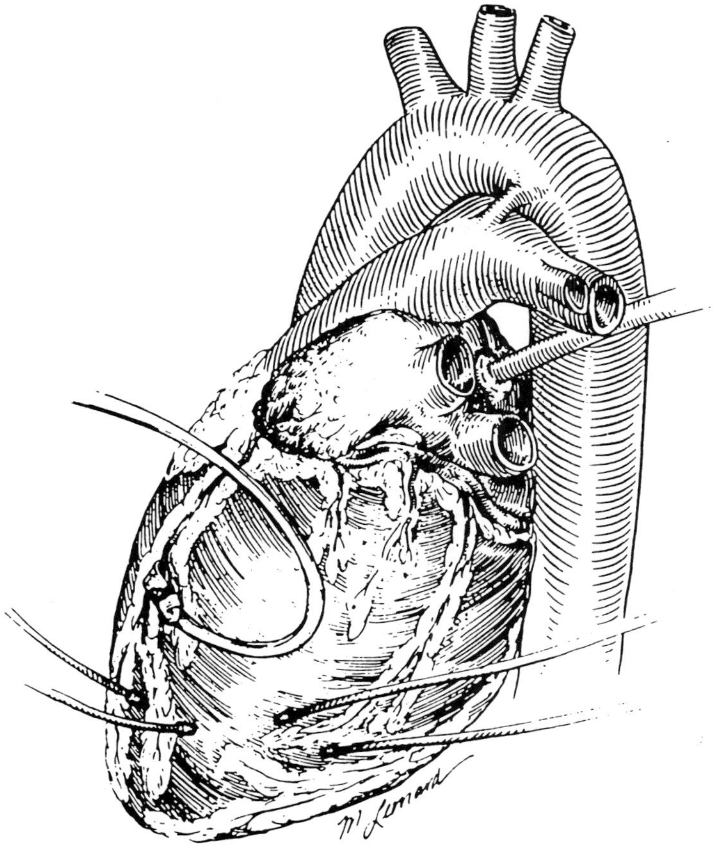

Figure 2.19 An experimental preparation with left anterior descending coronary pneumatic occluder, subendocardial piezoelectric dimensional sensors in ischemic and non-ischemic zones and left ventricular and pleural pressure sensors. (From D D Glower, J A Spratt, J S Kabas, J W Davis and J S Rankin *Quantification of regional myocardial dysfunction after acute ischemic injury*, 1988 *Am. J. Physiol.* **255** (*Heart Circ. Physiol.* **24**) H85-H93 1988.)

operation of extravascular and intravascular blood pressure measuring systems are described in the following sections.

2.7.1 Extravascular sensors

The most common clinical method for directly measuring pressure is to couple the vascular pressure to an external sensor element via a heparinized saline-filled catheter. The catheter has at one end a sensing port, while the other end is connected to a three-way stopcock attached to the dome of the pressure sensor

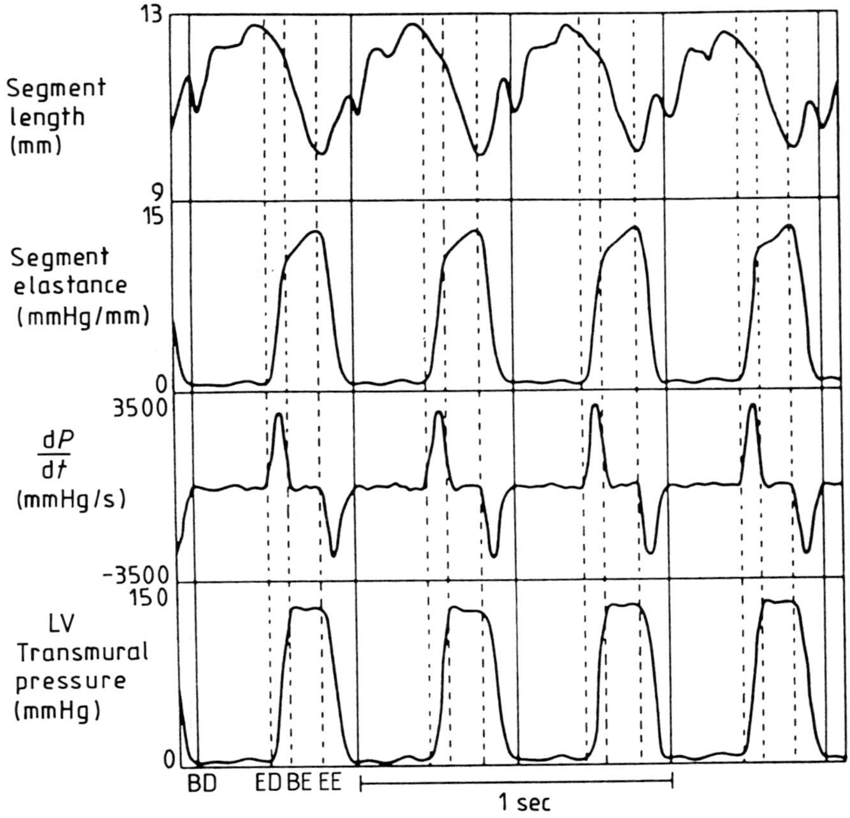

Figure 2.20 Recordings during the cardiac cycle showing myocardial segment length, segment elastance (ventricular pressure/segment length), transmural ventricular pressure (difference between the ventricular and pleural pressure) and first derivative of the transmural ventricular pressure. BD, beginning diastole; ED, end diastole; BE, beginning ejection; and EE, end ejection. (From D D Glower, J A Spratt, J S Kabas, J W Davis and J S Rankin, *Quantification of regional myocardial dysfunction after acute ischemic injury*, 1988 *Am. J. Physiol.* **255** (*Heart Circ. Physiol.* **24**) H85-H93 1988.)

(figure 2.22). The major advantage of the extravascular pressure measuring systems is that they have high stability and sensitivity.

The catheter is inserted either by means of a surgical cutdown using a special needle or guide wire technique. Heparinized saline is flushed through the catheter every few minutes to prevent blood from clotting at the tip. The patient's blood pressure is transmitted through the catheter liquid and sensor dome to the diaphragm. The diaphragm is deflected (figure 2.6(a)) and the displacement of the diaphragm is transmitted to a system composed of a moving armature and

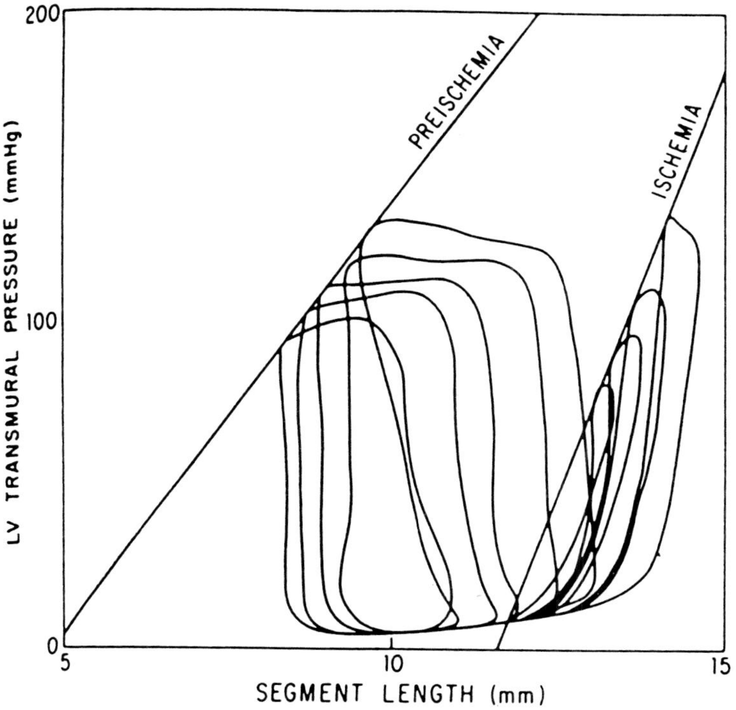

Figure 2.21 Preischemic base line and postischemic myocardial dysfunction. Typical ventricular pressure–segmental length loops and end-systolic pressure–length relationships under base line conditions (preischemia) and 15 min after coronary occlusion (ischemia). (From D D Glower, J A Spratt, J S Kabas, J W Davis and J S Rankin, *Quantification of regional myocardial dysfunction after acute ischemic injury*, 1988 *Am. J. Physiol.* **255** (*Heart Circ. Physiol.* **24**) H85-H93 1988.)

an unbonded strain gage in a bridge circuit, which produces an output voltage.

As presented in figure 2.6(a), there are four strain gages connected to the diaphragm. As the pressure is applied, the strain on the first pair of strain gages (B and C) increases, whereas that on the other pair (A and D) decreases. This strain gage system connected in a Wheatstone bridge circuit (figure 2.6(b)) is inherently temperature stable, and it provides linear output.

2.7.2 Intravascular sensors

With intravascular blood pressure measuring systems, the liquid coupling between the blood and the pressure transducer is eliminated by incorporating the sensor into the tip of a catheter that is placed in the vascular system [18].

Catheter tip sensors have the advantage that there is no hydraulic catheter connection between the pressure source and the sensor [18]. Thus the frequency

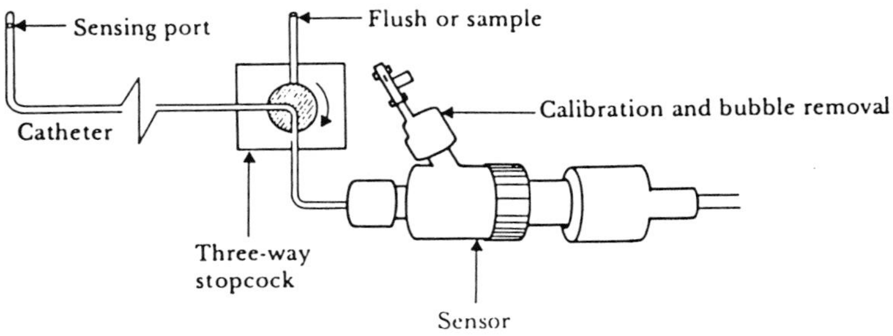

Figure 2.22 An extravascular pressure sensor system. A catheter couples heparinized saline through a three-way stopcock to the extravascular sensor element. The three-way stopcock is used to flush the catheter and to take blood samples. (From Peura R A and Webster J G *Basic sensors and principles in Medical Instrumentation: Application and Design*, 2nd edition, edited by J G Webster, copyright © 1992 Houghton Mifflin Co., Boston.)

response of the catheter sensor system is not limited by the hydraulic properties of the system. Detection of pressures at the tip of the catheter without the use of a liquid coupling system allows the physician to obtain a high-frequency response and eliminate time delays when the pressure pulse is transmitted through a hydraulic catheter sensor system.

Several basic types of sensor are used commercially for the detection of pressure in the catheter tip. These include various types of bonded strain gage system attached to a flexible diaphragm. The smallest gages of this type are available in F5 catheter (1.67 mm OD) size. Challenges exist in developing small catheters due to problems of temperature and electric drift, fragility and non-destructive sterilization. A disadvantage of the catheter tip pressure sensor is that it is more expensive than extravascular pressure sensors and may break after only a few uses, due to its fragile nature.

With an integrated-type sensor, the pressure sensor is fabricated by using a silicon substrate for the structural material of the diaphragm. The semiconductor gages are diffused directly onto the diaphragm. A radial stress component occurs at the edge when pressure is applied to the diaphragm. Placement of the eight diffused strain gage units (figure 2.8(c)) gives high sensitivity and good temperature compensation [2].

Fiber optic intravascular pressure sensors can be made in sizes comparable to piezoresistive ones, but at a lower cost. The fiber optic device determines the diaphragm displacement optically by measuring the varying reflection of light from the back of the deflecting diaphragm. These optical devices are inherently safer electrically, but lack a convenient way to measure relative pressure without an additional lumen either atmospherically vented or connected to a second pressure sensor.

A fiber optic microtip sensor for *in vivo* measurements is shown in

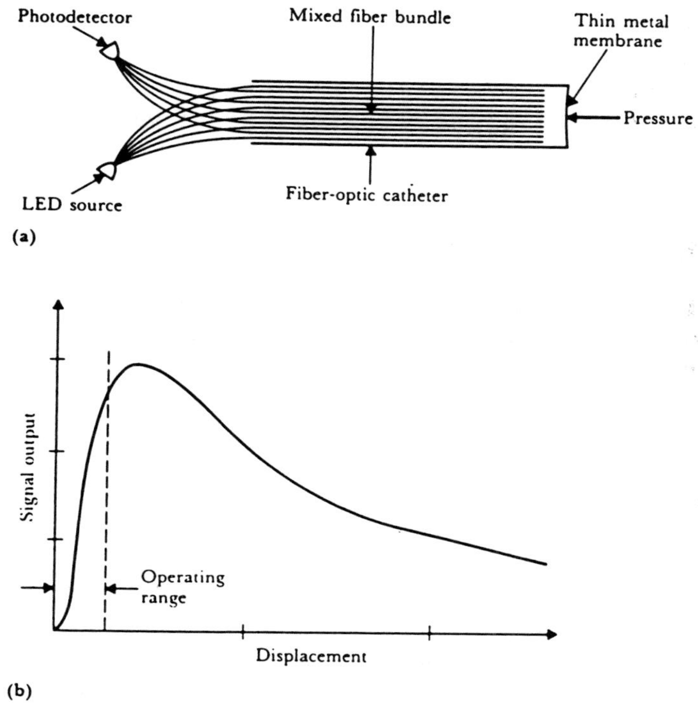

Figure 2.23 (a) A schematic diagram of an intravascular fiber optic pressure sensor. Pressure causes deflection in a thin metal membrane that modulates the coupling between the source and detector fibers. (b) A characteristic curve for the fiber optic pressure sensor. (From Peura R A and Webster J G *Basic sensors and principles in Medical Instrumentation: Application and Design*, 2nd edition, edited by J G Webster, copyright © 1992 Houghton Mifflin Co., Boston.)

figure 2.23(a). One leg of a bifurcated optical fiber bundle is connected to a light emitting diode (LED) source and the other to a photodetector. A thin metal membrane mounted at the common end of the mixed fiber bundle is the pressure sensor tip. The pressure of blood or other fluids deflects the membrane, causing a variation in the coupling between the LED source and the photodetector.

The output signal versus membrane deflection is illustrated in figure 2.23(b). The coupling between LED source and detector is related to the overlap of their acceptance angles on the pressure sensor membrane. The left-hand slope region is the operation portion of the curve because the characteristic is steepest and most linear.

2.7.3 Disposable pressure sensors

Traditionally, physiological pressure sensors have been reusable devices;

however, currently most hospitals have adopted inexpensive, disposable intravascular blood pressure sensors, in order to lower the risk of patient cross-contamination and reduce the amount of handling and damage to pressure sensors by hospital personnel. In addition, reusable pressure sensors, which are subject to the abuses of reprocessing and repeated user handling, tend to be less reliable than disposable sensors. The disposable pressure sensor system is composed of a silicon chip incorporated into a disposable pressure monitoring tubing system. The system also contains a thick-film resistor network that is laser trimmed to remove offset voltages, provide temperature compensation and set the same sensitivity for similar disposable pressure sensors. The resistance of the bridge elements is usually high in order to prevent self-heating, which causes erroneous results. A high output impedance for the device results, requiring the use of a pressure monitor with a high input impedance.

REFERENCES

[1] Lion K S 1959 *Instrumentation in Scientific Research* (New York: McGraw-Hill)

[2] Cobbold R S C 1974 *Transducers for Biomedical Measurements* (New York: Wiley)

[3] Hennig E M 1988 Piezoelectric sensors *Encyclopedia of Medical Devices and Instrumentation* ed J G Webster (New York: Wiley) pp 2310–9

[4] Webster J G (ed) 1988 *Tactile Sensors for Robotics and Medicine* (New York: Wiley)

[5] Servais S B and Webster J G 1984 Estimating human energy expenditure using an accelerometer device *J. Clin. Eng.* **9** 159–71

[6] Dechow P C 1988 Strain gages *Encyclopedia of Medical Devices and Instrumentation* ed J G Webster (New York: Wiley) pp 2715–21

[7] Lawton R W and Collins C C 1959 Calibration of an aortic circumference gauge *J. Appl. Physiol.* **14** 465–7

[8] Hokanson D E, Sumner D S and Strandness D E Jr 1975 An electrically calibrated plethysmograph for direct measurement of limb blood flow *IEEE Trans. Biomed. Eng.* **BME-22** 25–9

[9] Allard E M 1962 Sound and pressure signals obtained from a single intracardiac transducer *IRE Trans. Bio-Med. Electron.* **BME-9** 74–7

[10] van Citters R L 1966 Mutual inductance transducers *Methods in Medical Research* vol XI, ed R F Rushmer (Chicago: Year Book), pp 26–30

[11] Reddy N P and Kesavan S K 1988 Linear variable differential transformers *Encyclopedia of Medical Devices and Instrumentation* ed J G Webster (New York: Wiley) pp 1800–6

[12] Beynnon B D, Pope M H, Werthheimer C M, Johnson R J, Fleming B F, Nichols C E and Howe J G 1992 The effect of functional knee-braces on strain on the anterior cruciate ligament in vivo *J. Bone Joint Surg.* A **74** 1298–312

[13] Doebelin E O 1990 *Measurement Systems: Application and Design* 4th edn (New York: McGraw-Hill)

[14] Alihanka J, Vaahtoranta K and Bjorkqvist S E 1982 Apparatus in medicine for the monitoring and/or recording of the body movements of a person on a bed, for instance of a patient *United States Patent* 4 320 766

[15] Podolak E, Kinn J B and Westura E E 1969 Biomedical applications of a commercial capacitance transducer *IEEE Trans. Biomed. Eng.* **BME-16** 40–4

[16] Patel A, Kothari M, Webster J G, Tompkins W J and Wertsch J J 1989 A capacitance pressure sensor using a phase-locked loop *J. Rehabil. Res. Dev.* **26** 5562

[17] Glower D D, Spratt J A, Kabas J S, Davis J W and Rankin J S 1988 Quantification of regional myocardial dysfunction after acute ischemic injury *Am. J. Physiol.* **255** (*Heart Circ. Physiol.* **24**) H85–93

[18] Peura R A 1992 Blood pressure and sound *Medical Instrumentation: Application and Design* 2nd edn, ed J G Webster (Boston, MA: Houghton Mifflin)

3

Integrated circuit manufacturing techniques applied to microfabrication

Marc Madou and Hyunok Lynn Kim

3.1 INTRODUCTION

Microfabrication, micromachining, or micromanufacturing is comprised of the use of a set of manufacturing tools, mostly based on electronic integrated circuit (IC) and thick-film (hybrid) manufacturing technologies, which are often enhanced or modified for building devices such as sensors, actuators, and other microcomponents and microsystems.

In figure 3.1, typical processes involved in making a three-dimensional microstructure are schematically represented. In what follows, each of these steps is described in detail. It is the purpose of this chapter to make the reader familiar with the micromachining terms and to elucidate the different tools that are available to build biosensors. Special emphasis is given to those processes that make biosensor construction different from IC manufacture. Both micromachining and microelectronic fabrication begin with lithography: the technique used to transfer copies of a master pattern onto the surface of a solid material such as a silicon wafer. The most widely used form of lithography is photolithography.

3.2 PHOTOLITHOGRAPHY

3.2.1 Masks

A photomask is a nearly optically flat glass (transparent to near UV) or quartz plate (transparent to deep UV) with a metal absorber pattern printed on one side. The metal pattern is typically a 0.1 μm thick chromium layer. The metal absorber pattern on a photomask is opaque to ultraviolet radiation, whereas glass

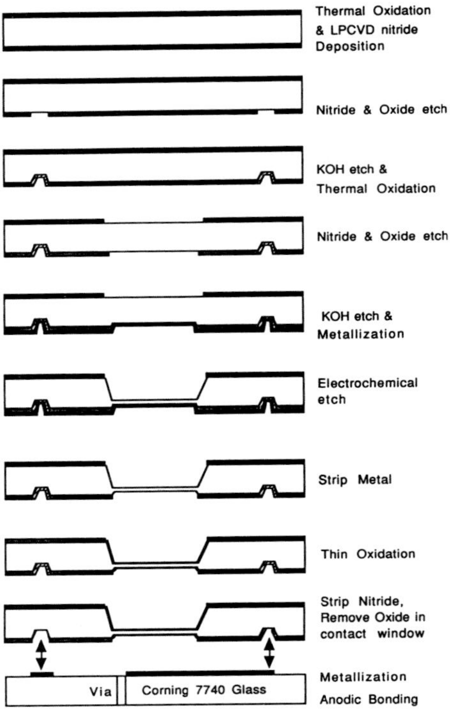

Figure 3.1 A typical process sequence used in microfabricating a 3D microstructure. The fabrication tool used is bulk micromachining.

or quartz is transparent. Such a photomask is placed in direct contact with or near a photoresist-coated surface, and the wafer is exposed to ultraviolet radiation. This procedure results in a 1:1 image transfer of the entire mask onto the silicon wafer. The masks that make direct physical contact (also referred to as 'hard contact') to the substrate are called 'contact masks' and they provide a sharper image transfer than the non-contact, proximity masks (also referred to as 'soft-contact' masks) which are raised 10–20 μm above the wafer. A disadvantage of the contact masks is that they degrade faster through wear, but they offer higher resolution. Contact mask and proximity mask printing are collectively known as shadow printing. In projection printing, the photomask is imaged one to one or reduced five to 10 times by a high-resolution lens system and then projected by stepping or scanning that image over the resist. In the latter case, mask lifetime is only limited by operator handling rather than by substrate contact.

3.2.2 Mask alignment

A major cause of product failure is poor alignment between the image being projected and the pre-existing patterns on the wafer. The pattern registration capability is the degree to which the pattern being printed 'fits' relative to mask alignment marks and to the previously printed patterns. Alignment or registration marks such as optical Vernier patterns are created on the different levels to be aligned. Generally, the collective misalignment should not exceed a quarter of the minimum feature size. Alignment errors are only one part of the total device ground rule tolerances. Typical mask errors range from 0.1 to 0.3 μm, etch errors from 0.2 to 0.4 μm, and alignment errors from 0.2 to 0.5 μm. Thus the tolerances of current lithography equipment lie in the range of 0.5–0.8 μm [1].

For the next generation of submicrometer ground rules, a fourfold reduction in alignment and other errors will be necessary. Almost every type of mask aligner—contact/proximity, projection scanner, and step and repeat—is used for some of the lithography steps in the fabrication of state of the art integrated circuitry today. For a detailed description of mask alignment equipment we refer to the literature [2, 3].

In micromachines, alignment is often even more complex than in IC manufacture. Indeed one not only deals with high-aspect-ratio 3D features, but often needs to align the 3D features on both sides of the wafer, as in the case of wafer to wafer (fusion bonding) [4] or wafer to glass bonding (anodic bonding) [5], where front to backside alignment is required. Because visible light cannot be used for such alignments, double-sided mask alignment with marks on both sides of the wafer is needed. Two major techniques to accomplish such alignment are the use of infrared radiation or the use of two light sources and mirrors. The alignment accuracy with that type of equipment is about 2 μm.

3.2.3 Spinning resist

Typically, the first step in photolithography, with Si as the substrate, is to grow a thin layer of oxide on the wafer surface by heating it to between 900 and 1150 °C in a humidified oxygen stream. Dry oxygen also works but steam is faster. The oxide layer will then be patterned and serve as a mask for subsequent etching or implantation of the base silicon wafer. As the first step in the lithography process itself, a thin layer of an organic polymer sensitive to ultraviolet radiation, a photoresist, is deposited on the oxide surface. The photoresist is dispensed from a viscous solution of the polymer onto the wafer lying on a wafer platen in a resist spinner. The wafer is then spun at high speeds between 1500 rpm and 8000 rpm depending on the viscosity and the required film thickness, to make a uniform film. The resist film's uniformity across a single substrate and from substrate to substrate must be within $\pm$15 nm (for a 1.5 μm film that is $\pm$1%) in order to ensure reproducible linewidths and development times in subsequent steps. The coating thickness of these thin glassy films depends on the chemical resistance required for image transfer and the fineness of the lines and spaces to be resolved. For silicon ICs the resist thickness can be between 0.5 and 2 μm after baking. For microfabricated structures, thicknesses up to 1 mm are used in the deep-x-ray lithography-based LIGA technique (see below). In the latter case, techniques such as casting or *in situ* polymerization of the resist are employed.

3.2.4 Exposure and development

3.2.4.1 Exposure. Before the coated wafers are exposed, they undergo a mild bake (called a soft bake) in order to remove the solvent of the resist and to suppress mechanical stress. Once the wafers are soft baked, they are transferred to some type of illumination or exposure system. In the simplest case, an exposure system is a UV lamp illuminating the resist-coated wafer through a mask without any lenses in between. The purpose of the illumination system is to deliver light to the wafer with the proper intensity, directionality, spectral characteristics, and uniformity across the wafer so as to transfer (also print) the mask image as perfectly as possible onto the resist in the form of a latent image.

In photolithography, wavelengths of the light source, used for flood exposure of the resist-coated wafer, range from deep ultraviolet (DUV) i.e. 150–300 nm to near UV i.e. 350–500 nm. In the near UV, one typically uses the g (436 nm) or i line (365 nm) of a mercury lamp. At shorter wavelengths light sources are less intense and optics are less transparent. Consequently a higher-sensitivity resist or an unconventional DUV source producing a very high flux of DUV radiation must be used to obtain a high throughput of printed wafers. An example is a KrF excimer laser with a short wavelength of 249 nm and a power of 10–20 W at that wavelength.

3.2.4.2 Development. Development transforms the latent image into an image in the resist which will serve as a mask for further additive and subtractive steps. During the development process, selective dissolving of the resist takes place. Two main technologies are available for development: wet development, widely used in industry, and dry development, mostly in the exploratory stage. Wet development by solvents can be based on three different types of exposure-induced change: (i) variation in molecular weight of the polymers (by cross-linking or by chain scission), (ii) reactivity change, or (iii) polarity change of the polymer. There are two major types of wet-developing system, i.e. immersion and spray developers. In immersion developing, cassette-loaded wafers are batch immersed for a timed period in a bath of developer and agitated at a specific temperature. In spray development, fresh developing solution is directed across wafer surfaces by fan-type sprayers. The use of solvents may lead to a swelling of the resist and a lack of adhesion of the resist to the substrate. These problems may be solved by dry developing which is either based on a vapor phase process or a plasma process (usually an oxygen plasma).

3.2.5 Resist tone

There are two main categories of photoresist: positive tone and negative tone. Upon exposure, the photochemical reaction weakens the positive resist, and the exposed region becomes more soluble in developing solutions by approximately an order of magnitude. The weakening occurs due to a rupture or scission of the main and side polymer chains. The photochemical reaction has the opposite effect on negative resists. It strengthens the polymer by random cross-linking of main chains or pendent side chains, and the resist becomes less soluble. Exposure and development sequences for negative and positive resists are shown in figure 3.2(a). With both the positive and negative resists, the principal components of photoresists are a polymer (base resin), a sensitizer, and a casting solvent. If no sensitizers are added, the resist is referred to as a single-component or one-component system; when sensitizers are added, the resist is referred to as a two-component system. Solvents and other additives are not always counted as components since they are not directly related to the resist's photoactivity.

In a UV exposure, not all of the photons strike the resist film in an orthogonal fashion. Scattering e.g. at the substrate/resist interface causes a broadening of the exposed region. The scattered radiation profiles for positive and negative resists are shown in figure 3.2(b). In figure 3.3, the prevalent photoresist profiles are compared. R in figure 3.3 is defined as the development rate of the exposed region, and R_0 is the development rate of the unexposed region. The scattered radiation profile for a positive resist (especially pronounced with overexposure) can, depending on the development mode, lead to a lip or overcut in the case of a fast developer ($R/R_0 > 10$) or to a straight resist profile with a quenched developer ($R/R_0 = 5$–10). In a developer-dominated process (also 'force' developing with $R/R_0 < 5$) the resist profile (also undercut) recedes and thinning of the entire resist layer occurs. With a negative resist, however, the exposed region becomes highly insoluble and this thinning does not occur. This characteristic renders the negative resists less susceptible to overdevelopment, but some uncontrollable swelling of the exposed resist can occur. The swelling in negative resists is one of the reasons they are limited to the manufacture of devices with minimum feature size greater than 3 μm. Positive resists do not exhibit swelling due to a different dissolution mechanism. Development of a positive resist is time dependent and enables the operator to tailor resist profiles to his needs. For example, a lip or overcut profile is desirable for lift-off which is an important process for patterning membranes and catalytic metals such as Pt. These are frequently used sensor materials, and they are difficult to etch directly, but by using the lift-off process they can be patterned with relative ease.

A further comparison of negative and positive photoresist features is presented in table 3.1. This table is not exhaustive and is meant only as a practical guide for selection of resist tone. The choice might depend on a variety of considerations such as cost, speed, resolution etc. The choice of resist tone might even depend on the specific intended pattern geometry.

Table 3.1 A comparison of negative and positive photoresist.

Characteristic	Positive resist	Negative resist
Adhesion to Si	Fair	Excellent
Lift-off	Yes	No
Swelling in developer	No	Yes
Photospeed	Slower	Faster
Minimum feature	$\leqslant 1 \mu m$	$\pm 2\ \mu m$
Wet chemical resistance	Fair	Excellent
Plasma etch resistance	Very good	Not very good
Thermal stability	Good	Fair
Available compositions	Many	Vast
Contrast g	Higher e.g. 2.2	Lower e.g. 1.5
Image width to resist thickness	1:1	3:1
Pinholes in mask	Prints mask pinholes	Not so sensitive to mask pinholes
Sensitizer quantum yield F	0.2–0.3	0.5–1
Opaque dirt on mask	Not very sensitive to it	Causes to print pinholes
Cost	More expensive	Less expensive
Developer	Aqueous based (more ecologically sound)	Organic solvent
Influence of oxygen	No	Yes
Proximity effect	Prints isolated holes or trenches better	Prints isolated lines better
Developer process window	Small	Very wide, insensitive to overdeveloping

Today's resist chemistry is quite a bit more complex than the simple picture above might convey; for further improvements in resist performance, plasticizers, adhesion promoters, speed enhancers, and non-ionic surfactants are routinely added to the resists. In addition to resist modifications, the Si wafers are usually vapor primed with reactive silicone primers before spin coating to further improve resist adhesion. An important step even before wafer priming is wafer cleaning which has become a scientific discipline in its own right with journals (e.g. *Microcontamination, the Magazine for Ultraclean Manufacturing Technology*), books, and conferences (*The Microcontamination Conference*) dedicated to contamination issues.

3.2.6 Critical dimension (CD) and resolution (R)

A lithographic system's practical resolution, R, can be determined by line width measurement made with devices such as a scanning electron microscope. Correct feature size must be maintained within a wafer and from wafer to wafer because device performance depends on the absolute size of the patterned structures. The term critical dimension (CD) refers to a specific feature size and is a measure

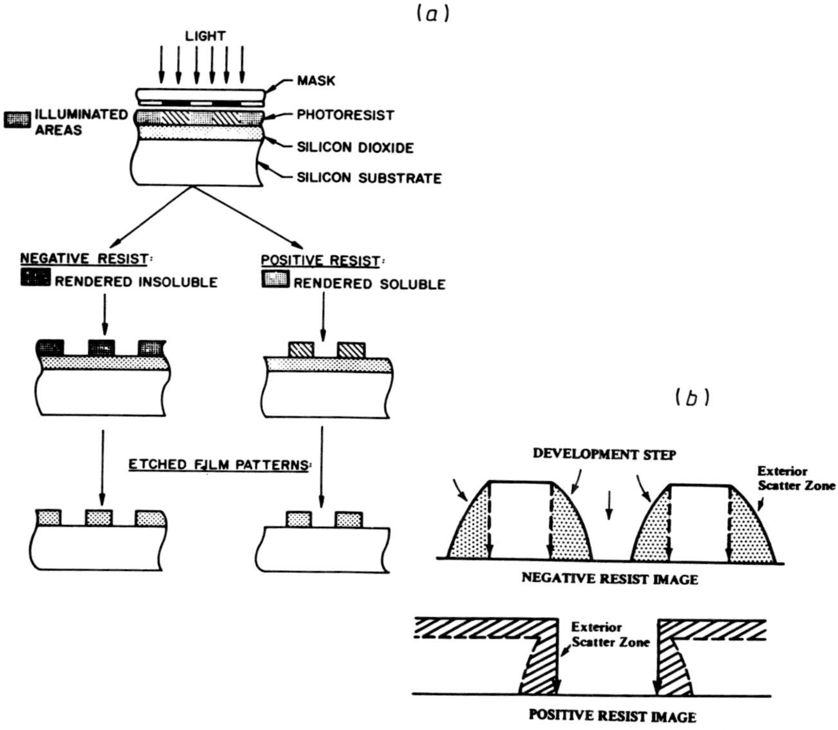

Figure 3.2 A positive and negative resist comparison. (a) Positive and negative resists: exposure and development. Positive resists develop in the exposed region and usually remain soluble for lift-off. Negative resists remain in the exposed region but are insoluble and not suitable for lift-off. (b) Edge-scattered radiation for negative and positive resists.

of the practical resolution of a lithographic process.

In the shadow printing mode (contact and proximity), optical lithography has a resolution with limits set by such factors as diffraction of light at the edge of opaque features in the mask, alignment of wafer to mask, non-uniformities in wafer flatness, and debris between mask and wafer. Diffraction causes the image of a perfectly delineated edge to become blurred or diffused. The theoretical resolution (also called R), i.e. minimum resolved feature size (b_{min}) of a grating mask (with $2b$ as the grating period) imaged on a conventional resist, is given by

$$R = b_{min} = \tfrac{3}{2}\sqrt{\lambda(s + \tfrac{1}{2}d)}. \tag{3.1}$$

In this equation, b_{min} is half the grating period, s is the gap between the mask and the photoresist surface, λ is the wavelength of the exposing radiation, and d is the photoresist thickness. For contact printing $s = 0$ and from equation (3.1), with $\lambda = 400$ nm and a 1 μm thick resist, the maximum resolution is slightly

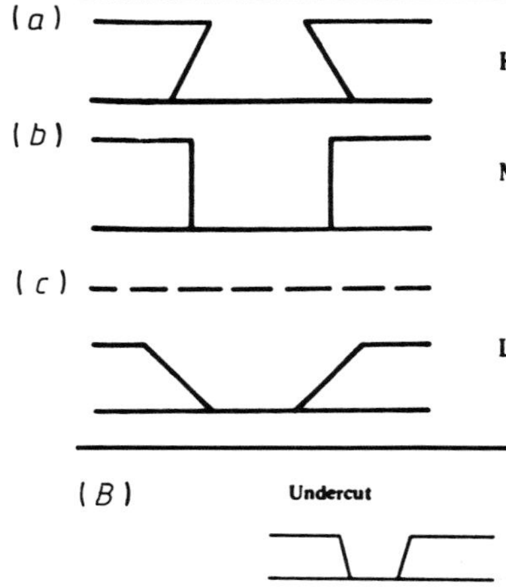

(A)

Profile	Dose	Developer influence	R/R_0	Uses
(a)	High	Low	>10	Lift-off Ion implant
(b)	Medium	Moderate	5–10	Dry etch Lift-off Wet etch
(c)	Low	Dominant	<5	Wet etch

(B)

Undercut	Low Dose Negative Resist	Wet Etch Plasma Etch

Figure 3.3 Photoresist profiles. (A) Positive resist: (a) desired resist profile for lift-off i.e. exposure-controlled profile, also called overcut; (b) perfect image transfer by applying a normal exposure dose and relying moderately on the developer; (c) receding photoresist structure with thinning of the resist layer i.e. developer control, also called undercut. (B) Negative resist: the profile is mainly determined by the exposure. Development swells the resist a little but has otherwise no influence on the wall profile.

less than 1 μm. Equation (3.1) clarifies why shorter wavelengths must be used in order to achieve higher resolution.

In proximity printing spacing the mask away from the substrate minimizes defects that result from contact, but increases the diffraction of the transmitted light which reduces the resolution. The degree of reduction in resolution and image distortion depends on the mask to substrate distance which may vary across the wafer. For proximity printing, equation (3.1) can be rewritten as

$$R = b_{min} = \pm\tfrac{3}{2}\sqrt{\lambda s}. \tag{3.2}$$

On the basis of equation (3.2), for a gap of 10 μm, using 400 nm exposing radiation, the resolution limit is about 3 μm which is poorer than that of contact printing.

The best resolution can be expected when using an *in situ* deposited mask i.e. a self-aligned or conformable mask. Projection printing can also be used to obtain very good resolution as the reduction step makes errors in the mask less significant and a higher-resolution lens with a smaller exposure field, i.e. higher numerical aperture (NA), can be used.

For more detailed information on resists such as multilayer resists (MLRs—often used to avoid thin-film interference effects), new types of resist such as polyimides, chemically amplified resists, resist tone reversal, lithography with hydrogels and ion selective membranes, resist monitoring, dry resists, surface imaging resists (SIRs), and other resist aspects not touched upon here or only mentioned briefly, we refer to [1] and [6–9]. Also *Solid State Technology*, *Semiconductor International*, and *Microlithography World* carry excellent tutorials on these topics.

3.2.7 Resist stripping

3.2.7.1 Wet stripping. After exposure, the wafers are rinsed in a developing solution or sprayed with a spray developer to remove either the exposed areas (positive tone) or the unexposed areas of the photoresist (negative tone), leaving a pattern of bare and photoresist-coated oxide on the wafer surface. Typically, the next process step after development is to bake the resist at a higher temperature than that used for the soft bake. This second bake, also called the hard bake, improves the hardness and interfacial adhesion of the resist weakened by developer penetration or by swelling. Improved hardness increases the resistance of the photoresist to the subsequent etching process. The wafers are then placed in a solution of HF, which attacks the oxide but not the photoresist or the underlying silicon. The photoresist protects the oxide areas it covers thereby transferring the mask pattern on the oxide layer. Once the exposed oxide has been etched away, the remaining photoresist can be stripped off in a number of ways. A strong acid such as H_2SO_4 or an acid–oxidant combination such as Cr_2O_3–H_2SO_4 can be used. Other commonly used liquid strippers are organic solvent strippers and alkaline strippers with or without oxidants.

3.2.7.2 Dry stripping. There are a few variations of dry stripping in terms of the composition of the plasma, but the basic mechanism is for the plasma to react with the organic photoresist after which the gaseous products formed are pumped away. Oxygen in its molecular form is commonly used for this process. Plasma stripping (also called ashing) has become more and more popular as it poses fewer disposal problems with toxic, flammable, or otherwise dangerous chemicals and does not lose potency over time. Wet-stripping baths lose potency with use, causing stripping rates to change with time due to accumulation of reaction products and contamination. Also liquid phase surface tension and mass transport tend to make photoresist removal difficult and uneven in wet stripping baths. Dry stripping is more controllable than liquid stripping, less corrosive with respect to metal features on the wafer, and, most importantly, under the right conditions, it leaves a cleaner surface. When the stripping of a resist layer on an oxide-covered wafer is completed, it results in a pattern of oxide on the wafer surface that duplicates the photoresist pattern and is either a positive or a negative copy of the pattern on the photomask. The oxide itself serves as a mask in subsequent processing steps.

3.2.8 Expanding the limits of lithography

Sometime after the year 2000, the smallest feature on an IC that advanced lithographies will be able to write will be about 0.1 μm. Today's DRAMs have smallest feature sizes of 0.8 μm. The current approach to improving the resolution of industrial lithography techniques has been to reduce the wavelength of the light source and increase the numerical aperture (NA) of the imaging lens. By combining shorter wavelengths with improved mask technology, better optics, and more sophisticated resist chemistries, photolithography will be usable to 0.25 μm and features slightly below 0.2 μm will become possible.

3.2.8.1 Phase shifting masks. One method for improving the photolithography resolution at a given wavelength and numerical aperture is by carefully controlling light diffraction. This can be achieved using constructive and destructive interference to help create a circuit pattern with phase shifting masks. In a phase shifting mask, one controls both the amplitude and phase of the light and, in particular, one arranges the mask so that light with opposite phase emerges from adjoining mask features. In such a case, destructive interference can be used to cancel some of the image spreading effects of diffraction [10].

3.2.8.2 Langmuir–Blodgett films. Another way to further improve resolution of photolithography is to use thinner resist layers. In IC technology, thinner imaging layers are employed so that a reduced depth of focus (DOF) of higher-resolution lenses is less critical and submicrometer features can be processed anyway. It is in this context that even ultrathin Langmuir–Blodgett (LB) resist films are being considered for the IC industry. LB films are prepared by transferring organic monolayers floating on a water surface onto solid substrates. These films are mainly being investigated as potential e-beam resists.

3.2.8.3 Inorganic resists. The contrast, g, of organic-based resists limits the photolithography resolution. By using an inorganic resist with g of 6.8 rather than say two for organic resists, higher resolutions can be obtained. An example of such a resist system is Ge_xSe_{1-x}, with a thin layer of Ag covering it [11]. Other advantages of these inorganic resists are: no swelling in development, possessing a broad-band spectral response to all regions of the UV spectrum, and being resistant to oxygen plasma allowing use in dry-resist processing.

3.2.8.4 X-ray, e-beam and ion beam lithography. In the IC industry, continuous improvements in optical lithography have postponed the industrial adoption of alternative lithographies. Because of the huge financial investment in photolithography equipment, this situation will remain unchanged for at least another five to 10 years.

With micromachines, the need to incorporate non-traditional materials, the need for large aspect ratios and perfectly vertical walls, and the need for a technology with large dimensional bandwidth in the x, y, and z dimensions are more important than feature size. Also batch fabrication is not always a prerequisite, especially since microinstruments may cost more than ICs. Consequently, in micromachining, there are other compelling reasons besides feature size and cost to switch to other lithography techniques. It is under these circumstances that alternative lithography techniques such as x-ray lithography, e-beam and ion-beam lithography, or even laser-based processes may become important manufacturing tools earlier in micromachining than in the IC industry.

X-ray technology, for example, has a big technological edge over other photolithography-based micromachining techniques because of the virtually infinite depth of focus, insensitivity to organic dust, and capability to draw parallel lines even into very thick resist, etc. These advantages are utilized in 'LIGA', an x-ray lithography, electrodeposition, and molding-based technique in which high-aspect-ratio photoresist structures are used as electroforming molds resulting in high-aspect-ratio metal shapes. The electroformed metal parts in turn are used as molds for high-aspect-ratio secondary plastic shapes which in turn can be used for secondary metal shapes [12]. Several plastic molding processes have been tested including reaction injection molding, thermoplastic injection molding, and hot embossing [13].

The development of better e-beam and ion beam sources continues to widen the scope of possibilities for nanoscale engineering through lithography, etching, depositing, and modification of a wide range of materials. Two major advantages of e-beam lithography are its ability to register accurately over small areas of a wafer and lower defect densities, the latter advantage resulting from the lack of the need of intermediate masks. Some of the disadvantages of e-beam lithography are that electrons scatter quickly in a solid, limiting practical resolution to dimensions greater than 10 nm. Electrons, being charged particles, also need to be held in a vacuum, making the apparatus more complex than for photolithography. The other major difficulties with scanning e-beam lithography are the relatively slow exposure speed (an e-beam must be scanned across the entire wafer) and high system cost. This has limited the use of e-beam lithography to mask making and direct writing on wafers for specialized applications e.g. small batches of custom ICs. An e-beam can also be used as an alternative way to build microstructures directly without the use of a mask, i.e. in a direct-write-type microfabrication method. For example, e-beam-induced metal deposition from a metal organic gas (e.g. W deposition from $W(CO)_6$) has been used for the formation of microstructures of various geometries [14]. These devices are made one by one rather than in a large batch. Usually this type of slow, expensive fabrication technique is commercially unacceptable, but some microstructures, especially intricate microsystems, might warrant a big price tag.

Just as the deposition of a material can be e-beam induced, the removal of a material from a special site in a device can be accomplished with a focused ion beam. The focused ion beam (FIB) can be seen as yet another tool in the microfabrication arsenal with applications well beyond lithography. As a machining tool, however, FIB is very slow and will probably not become a 'micromachining' tool outside research.

3.2.8.5 Scanning tunneling microscopy, atomic-force-based lithography. Micromachined structures may provide a solution to make direct chip e-beam writing a cost-effective manufacturing proposition. At Cornell's National Nanofabrication Facility (NNF), for example, a research group has been proposing arrays of microfabricated, miniaturized SEMs based on scanning tunneling microscopes (STMs). In this STM aligned field emission (SAFE) system, the physical dimensions of the column (length and diameter) are of the order of millimeters. A field emission tip is mounted onto an STM, and the STM feedback principle is used for precision x, y, and z piezoelectric alignment of the tip to a miniaturized electron lens forming a focused probe of electrons. Since many electron–optic aberrations scale with size, microfabrication techniques enable lenses with negligible aberrations, resulting in exceptionally high brightness and resolution. The STM controls also allow the stability of the emission to be controlled by automatically adjusting the z position through the piezoelement. An array of these microcolumns each with a field emission tip as the source with individual STM sensors and controls can be used to generate patterns, one or more columns per chip, in parallel. The low voltage of operation of these tips might obviate the need for proximity effect corrections [15].

The same equipment can be used to locally modify surfaces [16], and thus can be used in yet another mode to perform lithography. The electric field strength in the vicinity of a probe tip is very strong and non-homogeneous (say a field of 2 V $Å^{-1}$, and concentrated around the probe tip). This field can be used to manipulate atoms, including sliding of atoms over surfaces and transferring atoms by pick and place. Since one can manipulate individual atoms with these techniques, the theoretical resolution of a lithography technique based on these atomic probes is a single atom. In practice, using STM, lines of 100 Å in width have been written. For reference, a single memory bit can be stored in an area that measures 100 Å on a side. This enables bit storage of 10^{12} bits cm^{-2} as compared to 10^9 bits cm^{-2} with conventional technology [17].

One major negative factor to bear in mind with atom placing or removing techniques is the time involved in generating even the simplest of features. Say for example you want to write a line 10 μm long, 1 μm in width, and 0.5 μm high. This may contain 10^{16} atoms; even at a deposition rate of 10^9 atoms s^{-1} it will take more than 100 days. So writing or removing clusters of atoms and parallel processing will be essential for these approaches to ever become viable. Nature, working with similarly small building blocks (amino acids and proteins), to circumvent the time problem, has indeed a lot of redundancy and parallel processing built in. Microfabrication like the IC industry is extremely dynamic;

Table 3.2 A partial list of subtractive methods.

	Application
Wet chemical etching (isotropic and anisotropic)	Suspended membranes
Electrochemical etching	Etch stop
Photoetching	When a bias is undesirable
Photoassisted electrochemical etching	P–n junction etch stop
Dry chemical etching	Isotropic profile is desired
Physical–chemical etching	Perfect pattern transfer
Physical dry etching or sputter etching and ion milling	Anisotropic profile is desired
Focused ion beam milling (FIB)	Very local removal of material
Laser and ultrasonic drilling, laser ablation	Vias
Mechanical grinding, polishing and sawing	Si/glass sandwich

new tools are continuously emerging. Some of these new tools are STM, ion beams, synchrotron radiation, laser deposition, and cluster deposition. Several of these tools will play important roles in lithography or other micromachining aspects such as deposition and etching.

3.3 SUBTRACTIVE TECHNIQUES

3.3.1 Overview

Lithography steps are followed by a number of subtractive and additive processes, transferring the lithography patterns into ICs or 3D micromachines. Table 3.2 is a partial list of subtractive steps used in building microstructures. In subtractive processes material is removed from the device under construction, usually very selectively, through the use of a resist or other mask pattern (e.g. an oxide or a nitride).

3.3.2 Dry etching

Dry-etching techniques, in general, are methods by which a solid state surface is etched *physically* by ion bombardment or *chemically* by a chemical reaction with a reactive species at the surface or combined *physical and chemical* mechanisms. Under chemical methods, one distinguishes between wet etching (solvent, vapor, electrochemical) and dry etching in the gas phase. Depending on the mechanism, isotropic or anisotropic (directional) etch profiles are obtained.

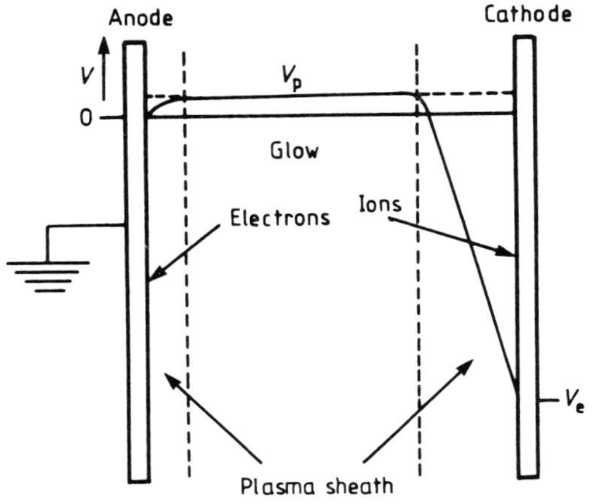

Figure 3.4 The structure of the glow discharge in a DC diode system.

3.3.2.1 Plasmas or discharges. Tables 3.3 and 3.4 represent an overview of some of the different popular dry-etching techniques. Some of the plasma etching acronyms and jargon are explained in tables 3.3 and table 3.4 respectively. The described techniques find their common base in discharges or plasmas, which are areas of high-energy electric and magnetic fields that will rapidly dissociate any gases present to form energetic ions, photons, electrons, and highly reactive radicals and molecules. Plasmas are formed by filling a reaction chamber with an inert gas such as argon at a reduced pressure (say 1 Torr) and applying a voltage between the anode and the cathode. An example of a reaction chamber is the DC diode glow discharge set-up shown in figure 3.4. Electrical breakdown of the argon gas in this reactor will occur when electrons, accelerated in the existing field, transfer an amount of kinetic energy, greater than the argon ionization potential (i.e. 15.7 eV), to the argon neutrals. Such energetic collisions generate a second free electron and a positive ion for each successful strike. Both free electrons can become energized again creating an avalanche of ions and electrons resulting in a gas breakdown. Once dynamic equilibrium is reached, the glow region of the plasma, being a good electrical conductor, hardly sustains any field and its potential is almost constant. All of the potential drop is at electrode surfaces where electrical double layers are formed in so-called sheath fields, counteracting the loss of electrons from the plasma. A striking characteristic of a plasma is that it is always positively charged with respect to the electrodes as a result of the random motion of the electrons and ions.

3.3.2.2 Physical etching: sputtering or ion etching. Simple ion bombardment of a surface with inert ions such as argon ions is referred to as ion etching or sputter etching. In the simplest case of ion sputtering, the substrates to be etched are laid on the cathode (target) of a discharge reactor. When ions of sufficient energy impinge vertically on a surface, momentum transfer (sputtering) causes

Table 3.3 Acronyms.

Acronym	Technique	Category
RIE	Reactive ion etching	Chemical–physical
MERIE	Magnetically enhanced reactive ion etching	Chemical–physical
ICP	Inductively coupled plasma	Chemical–physical
RIBE	Reactive ion beam etching	Chemical–physical
ECR	Electron cyclotron resonance	Chemical–physical
CAIBE	Chemically assisted ion beam etching	Chemical–physical

Table 3.4 Dry etching jargon.

Mechanical	Chemical	Chemical–physical
Sputter etch	Dry chemical etching	Reactive ion etching (RIE)
Sputtering	Ashing	Reactive ion beam etching (RIBE)
Ion beam milling	Plasma etching	Chemically assisted ion beam etching (CAIBE)
Ion etching		Reactive sputter etching (RSE)
Plasma sputtering		
Magnetron sputtering		
Ion sputtering		

bond breakage and ballistic material ejection, throwing the bombarded material across the reactor to an opposing collecting surface. A low pressure and a long mean free path are required for material to leave the vicinity of the sputtered surface without being backscattered and redeposited. It is the kinetic energy of the incoming particles that largely dictates which events are most likely to take place i.e. physisorption, surface damage, substrate heating, reflection, sputtering, or ion implantation. At energies below 5 eV, incoming particles are either reflected or physisorbed. At energies between 5 and 10 eV surface migration and surface damage results. At energies greater than 10 eV substrate heating, surface damage, and material ejection, i.e sputtering or ion etching, takes place. At yet higher energies greater than 10 keV ion implantation, i.e. doping, takes place (see below).

In short, with physical etching, ion etching, or sputtering, ions such as argon ions are accelerated in an electrical field towards the substrate where etching is purely impact controlled. Both DC and rf plasma can be used depending on the system. Sputtering is inherently unselective because the ion energy required to eject material is large compared to differences in surface bond energies and chemical reactivity. The method is also slow compared to other etching means with etch rates limited to several hundreds of ångströms per minute compared to thousands of ångströms per minute and higher with chemical and ion-assisted etching.

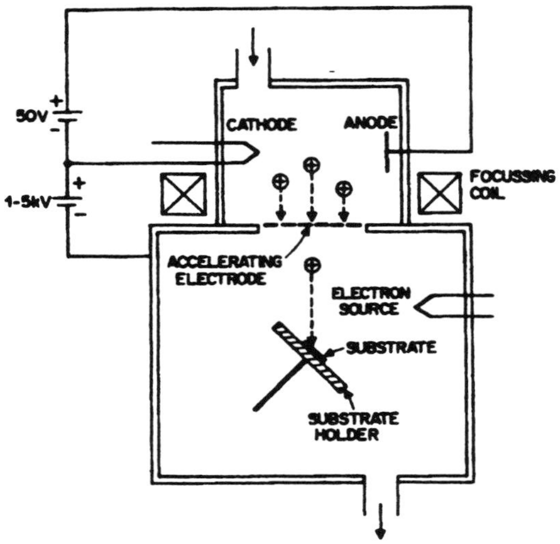

Figure 3.5 Ion beam etching apparatus: a triode. In an ion beam apparatus the beam diameter is approximately 8 cm. The substrates are mounted on a moveable holder allowing etching of the larger substrate areas. The coils focus the ion beam and densify the ion flux.

As an alternate set-up for sputtering, the plasma source for ion etching can be decoupled from the substrates and placed on a third electrode. This is referred to as ion beam milling, and the equipment, called a triode set-up, is shown in figure 3.5. Electrons are emitted by a hot filament (typically Ta or W) and accelerated by a potential difference between the anode and the cathode. Ions are extracted from the upper chamber by the sievelike electrode, formed into a beam, accelerated, and fired into the lower chamber where they strike the substrate. The advantages of this set-up are that the energy and flux of the ions to the substrates are quasi-independently controllable, and the argon pressure in the upper portion of the chamber can be quite low, say 10^{-4} Torr, so that the mean free path of the ions is large. Inert ion beam etching is, in principle, capable of very high resolution (<100 Å).

3.3.2.3 Etching profiles for physical etching. The ideal result in dry or wet etching is the exact transfer of the mask pattern to the substrate, with no distortion of the critical dimensions (CDs). Isotropic etching (dry or wet) always enlarges features and thus distorts the CDs. Chemical anisotropic etching is crystallographic and, as a consequence, critical dimensions can be maintained only if features are strategically aligned along certain lattice planes (see below). With sputtering, the anisotropy is controllable by the plasma conditions. Sputtering is far from ideal for high-aspect-ratio etching or for the creation of vertical walls. Ion sputtering indeed exhibits a tendency to develop a facet on

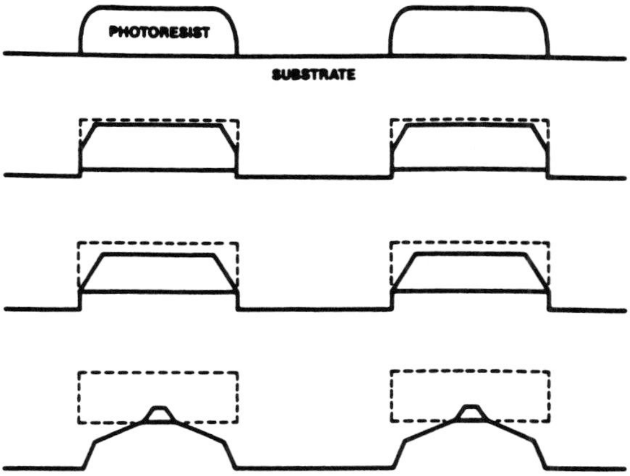

Figure 3.6 Facetting. Sputtering creates angled features. An angled facet (~60°) in the resist propagates and the mask is eroded away.

the mask edge at the angle of maximum etch rate. This corner facetting is shown in figure 3.6. Facetting at the mask corner occurs because the sputter yield for materials is usually a strong function of the angle at which ions are directed at the surface. The sputter etch rate of resist, for example, reaches a maximum at an incidence angle of about 60°, more than twice the rate at normal incidence [18]. When resist is sputtered, the rapid removal at this critical 'facet' exposes more substrate surface, resulting in a sloping step in the substrate material. The facetting of the substrate itself will proceed along its own preferred sputtering direction angle, and it is more pronounced with an applied bias due to the increased electric field at corners. This facetting can be minimized or eliminated by a more ideal resist profile, with a flat top surface oriented at 0° and with perpendicular sidewalls at 90°. Another sputtering limitation is the redeposition of the involatile products on the step edges. By tilting and rotating the substrate during etching, etch profiles can be improved [19]. In micromachining, where we create 3D features with very high aspect ratios, tilting and rotating of substrates is a recurring topic. It is one of the desirable modifications of standard equipment one should look for when dealing with micromachining applications.

3.3.2.4 Dry chemical etching. In reactive plasma etching, reactive neutral chemical species such as chlorine or fluorine atoms and molecular species generated in the plasma diffuse to the substrate where they form volatile products with the layer to be removed. The plasma is present only as a means to supply the gaseous, reactive etchant species. Consequently, if the feed gas were itself reactive enough no plasma would be needed. The surface is etched almost exclusively by chemically active, neutral species formed in the plasma. At pressures of greater than 10^{-3} Torr, the neutrals strike the surface at random angles, leading to isotropic, rounded features. This dry-chemical-etching regime can be established by operating at voltages low enough to prevent high-energy ions which can impinge on the sample and sputter the surface. The reaction products, volatile gases, are removed by a vacuum system. This volatility aspect is a major difference with the sputtering case where the fragments are ejected billiard-ball-wise and often redeposited close by.

The different popular reactor configurations for plasma etching are barrel etcher, downstream stripper, and parallel reactor. In the case of resist plasma etching, one also refers to ashing. Depending on the configuration, high-energy ion bombardment of the substrate can more or less be prevented and plasma-induced device damage avoided. For example, in a barrel reactor the substrates are shielded by a perforated metal shield to reduce substrate exposure to charged highly energetic species in the plasma. In a triode set-up used for ion milling, the reactor allows reactant generation and stripping to take place in two physically separated zones. In parallel plate strippers, the substrates are placed directly inside the plasma source and consequently the ion damage is higher compared to the two previous methods. Often the diode-like, parallel plate stripper is called a reactive ion etcher (RIE), although ions rarely reactively 'eat' away the substrate.

In glow discharge, a gas such as CF_4 dissociates to some degree by impact with energetic particles such as plasma electrons that can have an average energy distribution between 1 and 10 eV. With CF_4 the impact forms species such as CF_3^+, $CF_3^\bullet$, and F. Some dissociation reactions occur in the gas phase (homogeneous reactions), and others occur at the surface (heterogeneous reactions). These dissociated species then react with the surface layer to be removed and form volatiles. In the absence of crystallographic effects (typically seen with III–V compounds but not with Si) chemical dry etching leads to isotropic profiles. This isotropic etching of the reactive species unfortunately leads to mask undercutting. However, because of the chemical nature of the etching process, a high degree of control over the relative etch rates of different materials, i.e. selectivity, can be obtained by the choice of suitable reactive gases.

3.3.2.5 Physical–chemical etching. The most useful plasma etching is neither entirely chemical nor physical. By adding a physical component to a purely chemical etching mechanism, the shortcomings of both sputter-based and purely chemical dry-etching processes can be surmounted. One example of chemical–physical etching is reactive ion beam etching (RIBE), a rather exceptional case where ions are reactive and etch the surface directly. At very low pressures in RIBE systems, the reactive ions can replace the Ar ions and directly sustain a modest etch rate. In a more general picture, ion bombardment of a substrate, in the presence of a reactive etchant species, often leads to a synergism in which fast directional material removal rates greatly exceed the sum of separate chemical attack and sputtering rates. Once this type of ion-assisted etching was thought to be caused by chemical reactions between the ions and the surface material (as in RIBE). However, for etch rates in most practical situations, i.e. from 1000 to 10 000 Å min^{-1}, this is not even a theoretical possibility as the ion flux is much lower than the actual surface removal rates [18]. The ability of an ion to stimulate surface reactions appears to depend much more on the ion's energy and mass than on its chemical identity.

An important negative characteristic of all types of dry etching is the dependence of etch rate and uniformity on wafer loading. This is the result of the gas phase etchant being depleted by reaction at the surface of the substrate material.

3.3.3 Wet etching

Wet etching is the selective removal of layers using chemical solutions. The nature of the resulting features is determined by such parameters as temperature, composition, agitation, orientation of the single crystal, size of etching feature, and level of accumulations in the bath itself. Wet etching can be done either isotropically or anisotropically depending on the composition of the bath and the material to be etched.

3.3.3.1 Anisotropic and isotropic etching. Anisotropic etching is possible with crystalline silicon due to its non-equivalent orientations in its diamond cubic structure. The most common orientations used in the IC industry are the ⟨100⟩ or ⟨111⟩ orientation; in micromachining ⟨110⟩ wafers are used quite often as well. Certain chemicals, typically alkaline types, will etch away crystalline silicon at different rates depending on the orientation of the exposed crystal plane. A wide variety of etchants such as aqueous solutions of KOH, NaOH, LiOH, CsOH, NH_4OH, and quaternary ammonium hydroxides have been used for anisotropic etching. Etching occurs without the application of an external voltage and is dopant insensitive over several orders of magnitude. Aqueous solutions containing ethylenediamine, choline, or hydrazine with additives such as pyrocathechol and pyrazine are used for anisotropic etching as well. It has been well established that addition of certain compounds enhances or otherwise controls the characteristics of the etchant so additives are routinely used both in industry and in research [20].

The rate of anisotropic etching is diffusion controlled because the hydroxide ion must diffuse through the layer of complexed silicon reaction products. During etching the silicon wafer is placed in a holder and the solution is vigorously agitated in order to minimize the diffusion layer thickness. The etch rate for all planes increases with temperature and the surface roughness decreases with increasing temperature. In practice an etch temperature of 80–85 °C is used to avoid solvent evaporation and temperature gradients in the solution. The principal characteristics of three different anisotropic etchants are listed in table 3.5. One of the parameters used to characterize an anisotropic etchant is the 'anisotropy ratio' (AR). It is defined as AR = (etch rate plane 1)/(etch rate

Table 3.5 Principal characteristics of three different anisotropic etchants [9]. (EDP = ethylenediamine pyrocathechol.)

	KOH	EDP	N_2H_4
Composition	7 M in H_2O	300 ml ED 80 g P 100 ml H_2O	160 ml N_2H_4 40 ml H_2O
Temperature (°C)	80	95 (115)	100
Etch rate (μm min^{-1})	1.2	0.8 (1.5)	1.5
AR value	400:1	40:1	
Etch mask	Si_2N_4	SiO_2	SiO_2
Surface finish	very smooth	rough	smooth
Boron etch stop	$>10^{20}$ cm^{-3}	$>7 \times 10^{19}$ cm^{-3}	no dependence
Explosive	no	yes	extremely
Toxic	no	yes	yes

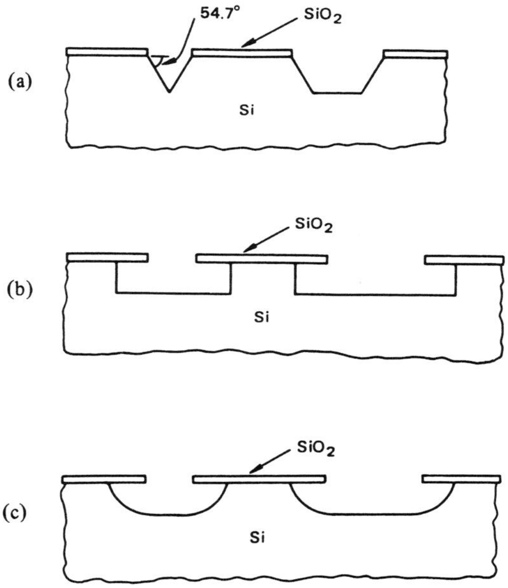

Figure 3.7 (a), (b) Anisotropic and (c) isotropic etched features.

plane 2) so that this ratio is equal to unity for isotropic etchants and can be as high as 400 for anisotropic etchants. There are several models used to understand the chemistry of wet etching, and the reader is referred to [21–25].

When using more aggressive acidic etchants, preferential etching does not occur, and rounded isotropic patterns are created. An example is the isotropic etching of silicon under a patterned mask. The etchant moves down and out from an opening in the oxide mask, undercutting the mask and enlarging the etched pit while deepening it (figure 3.7). With isotropic etchants for silicon one uses as a mask a non-etching metal e.g. Au, Si_3N_4, and sometimes even photoresist or SiO_2 when only a shallow etch is needed. Isotropic etchants are so strong that the activation barriers associated with etching the different Si planes are not differentiated as they are with the anisotropic etchants: all planes etch equally fast. The most commonly used isotropic etchant is a mixture of HF and nitric acids, sometimes with acetic acid added as a diluent (HNA etchants). Etch rates of more than 50 μm min^{-1} can be obtained with HNA; however, with such fast etch rates, good control of the etch process is difficult. In many microfabrication applications anisotropic etching is followed by a short isotropic etch to smooth surfaces or to round corners.

3.3.3.2 Etch stop techniques. In many cases it is desirable to stop etching in silicon when a certain cavity depth or a certain membrane thickness is reached. Moreover one wants thicknesses to be uniform and reproducible. Non-uniformity in resulting devices due to non-uniformity of the silicon wafer thickness can be quite high. Taper of double-polished wafers can be as high as 40 μm! Even with the highest wafer quality, the wafer taper is still around 2 μm. The taper together with the variation in etch depth due to processing conditions can lead to intolerable thickness variations for many applications. With good temperature, etchant concentration, and stirring control, the variation in etch depth is typically 1%. Although for less demanding applications one can time the etching process to control the thickness, it is much more convenient to work with etch stop techniques which not only control the thickness but also smooth out the taper. High-resolution silicon micromachining is dependent on the availability of effective etch stop layers. It is actually the existence of impurity-based etch stops in silicon that has allowed micromachining to become a high-yield production process.

The most widely used etch stop technique is based on the fact that anisotropic etchants, especially ethylene diamine pyrocathechol (EDP), do not attack heavily boron- (p^+-) doped layers. This allows the fabrication of structures such as beams and diaphragms with a level of control not possible before. A simple boron diffusion or implantation can be introduced from the front of the wafer, and, after patterning, the wafer can be etched from the backside to free the structure. Figure 3.8 illustrates this process for the fabrication of a micronozzle [26]. One disadvantage with this etch stop technique is that the extremely high boron concentrations are not compatible with standard CMOS or bipolar techniques, so they can only be used for microstructures without integrated electronics. However, with improved techniques for controlling epitaxial layers, the highly doped boron layer can be buried under an epitaxial layer of more lightly doped Si, and the problem of incompatibility with active circuitry can be avoided.

There are a number of other etch stop techniques that are commonly used. The electrochemical diode technique utilizes a lightly doped p–n junction as the etch stop by applying a bias between the wafer and the etch solution. An

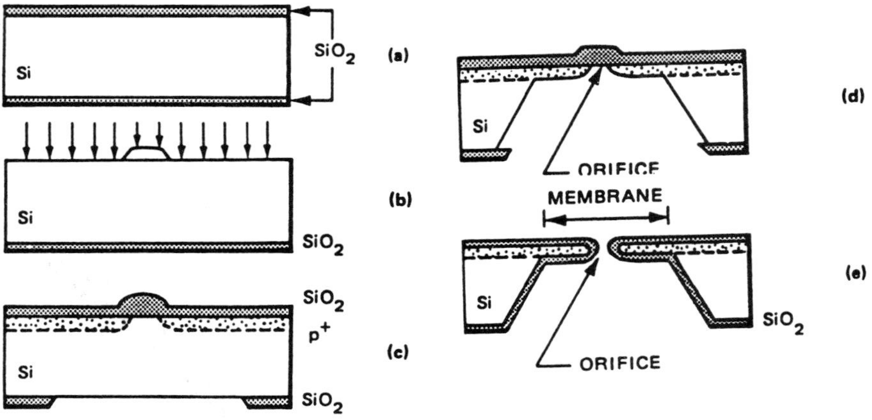

Figure 3.8 An illustration of the boron etch stop in the fabrication of a nozzle [26]: (a) oxidation; (b) lithography and development; (c) boron diffusion; (d) anisotropic etching; (e) stripping and reoxidation.

external electrode is needed for this process. Another etch stop technique, the photoassisted electrochemical etch stop, is a variation of the diode technique and is specific to n-type Si. Both types of electrochemical etching require the application of a metal electrode to the silicon in order to apply the bias. The application of such a metal electrode often induces contamination and obviously constitutes at least one extra process step as well as requiring additional equipment. An etch stop technique specific to p-type Si, photoinduced preferential anodization (PIPA), does not require metal deposition on to the silicon. The enabling mechanism here is that illumination of a p–n junction biases the p-type Si anodically and p-Si converts into porous Si while the n-type Si acts as a cathode for the reaction. The final etch stop technique discussed here is perhaps the most important of all. This technique uses silicon on insulator materials (SOI). A layer of SiO_2 is buried between two layers of crystalline silicon and forms an excellent etch stop because of the good selectivity of many etchants for Si over SiO_2. As with the PIPA no metal contacts are needed with an SOI etch stop. Until recently, SOI wafers were extremely difficult to produce. This made the cost of these wafers prohibitively high for micromachining despite the excellent etch stop characteristics. With the current improvements in manufacturing methods for the SOI wafers, it is now reasonable to use them as the basis for micromachined devices.

3.3.4 Comparison of dry- and wet-etch techniques

The popularity of dry etching stems from a variety of advantages over wet-etching methods: fewer disposal problems, better process control, less corrosion of metal features in the structure, less undercutting and broadening of photoresist features i.e. better CD control, and, under the right circumstances, a cleaner resulting surface. Also, with the current trend towards submicrometer geometries, surface tension might preclude a wet etchant from reaching down between submicrometer photoresist features whereas dry etching excludes any problem of that nature.

However, many problems with and concerns about dry etching need solving, such as the low etch rate, relatively low selectivity, and high sensitivity to operating parameters. A more recent concern is the search for alternative chemistries since production of chlorofluorocarbons (CFCs) will be banned after 1995 (according to the Montreal Protocol all ozone depleting CFCs must be eliminated by the year 2000; this includes etchants such as $CFCl_3$, CCl_2F_2, CF_3Cl, C_2F_5Cl, and CF_3Br) [27]. Environmental issues helped cause a switch from wet to dry etching for most IC applications. Another more decisive concern is the need for better CD control. More stringent rules are now causing even a reconsideration of the gases used in dry etching.

In bulk micromachining, requiring more extreme topologies (more z axis), wet etching of crystalline Si still dominates the state of the art. In the latter case anisotropic etching of Si results in atomically smooth planes and atomically sharp edges, properties impossible to obtain with dry etching. Much more R&D is required to come up with new dry-etching schemes that would come close to these performances. In surface micromachining, with processes more similar to the ones used in the IC industry, both isotropic wet etchants and anisotropic dry etchants are used (see table 3.11). Table 3.6 shows a comparison of dry- and wet-etching techniques.

3.4 ADDITIVE TECHNIQUES

3.4.1 Overview

Adding to an exposed and developed area on a substrate involves either deposition or growth. In additive processes, a solid or a precursor material is deposited through a resist or other mask onto a substrate. The solid is deposited from a liquid, a gas, or the solid state. Most processes, especially in the thin-film arena, are identical for IC applications and micromachines. In table 3.7 a partial list of additive processes is provided with typical applications.

3.4.2 Growth

Growth of an oxide involves both consumption and addition of material but is listed in table 3.7 as an additive process since its result is a new material. Oxidation involves heating the wafer in a wet or dry oxygen stream at elevated temperatures (between 600 and 1250 °C). Wet oxidation in steam is much faster than dry oxidation but the quality of the oxide is somewhat lower,

Table 3.6 A comparison of dry- against wet-etching techniques.

Parameter	Dry etching	Wet etching
Directionality	Can be highly directional with any material	Only directional with single-crystal materials
Production line automation	Better	Poor
Environmental impact	Less	More
Cost of chemicals	Lower	Higher
Selectivity	Poor	Can be very good
Materials that can be etched	Only certain materials can be etched (not e.g. Fe, Ni, Co)	All
Radiation damage	Can be severe	None
Process scale-up	Difficult	Easy
Cleanliness	Good under the right operational conditions	Good to very good
CD control	Very good	Poor
Equipment	Expensive	Inexpensive
Submicrometer features	Applicable	Not applicable
Typical etch rate	Slow	Fast
Theory	Very complex, not well understood	Better understood
Operating parameters	Many	Few
Control of etch rate	Better due to slow etching	Difficult

and water causes a loosening effect on the SiO_2, making it more prone to impurity diffusion. Both types of oxidation are carried out in a quartz tube. Oxide thicknesses of a few tenths of a micrometer are frequently used, with 1–2 μm being the upper limit for conventional thermal oxides. Silicon dioxide formed by thermal oxidation of silicon is used as a common insulating layer, as a mask, and as a sacrificial material. A good further introduction to silicon oxidation by Fair can be found in [11].

3.4.3 Physical vapor deposition

Physical vapor deposition (PVD) is a direct line of sight impingement deposition technique. At the low pressures employed in a PVD reactor, the vaporized material encounters few intermolecular collisions while traveling to the substrate, and modeling of deposition rates is a relatively straightforward exercise in geometry.

3.4.3.1 Evaporation. Evaporation is based on the boiling off (or sublimating) of a heated metal onto a substrate in a vacuum. In laboratory settings, the metal is usually evaporated by passing a high current through a highly refractory metal containment structure (e.g. a tungsten boat or filament) holding the metal to be

Table 3.7 A partial list of additive technologies and examples.

	Application example
Material transformation (oxidation, nitridation, etc)	Growth of SiO_2 on Si
Thermal evaporation (physical vapor deposition)	Aluminum on glass
Sputter deposition (physical vapor deposition)	Gold on silicon
Chemical vapor deposition	Tungsten on metal
Epitaxy (CVD)	GaAs
Molecular beam epitaxy (PVD)	GaAs
Spray pyrolysis (CVD)	CdS on metal
Spin-on	Thin resist (0.1–2 μm)
Casting	Thick resist (10–1000 μm)
Screen printing	Resistors, hydrogels
Droplet delivery systems	Epoxy
Electroless deposition	Contact vias
Electrochemical deposition	Copper on steel
Electrophoresis	Coating of insulation on heater wires
Electrostatic toning	Xerography
Thermal spray deposition from plasmas or flames	Aircraft engine parts
Thermomigration	Aluminum contacts through silicon
Ion implantation and diffusion of dopants	Boron into silicon
Bonding techniques	7740 glass to silicon
Ion cluster deposition	Magnesium deposition
Ion plating	Very hard TiN coatings
Laser deposition	Superconductor compounds

evaporated. This method is called 'resistive heating'. In industrial applications, resistive heating has been surpassed by e-beam and rf induction evaporation. In the e-beam mode of operation a high-intensity e-beam gun (up to 15 keV) is focused on the target material, and placed in a recess in a water-cooled copper hearth to evaporate the material onto the substrate.

In table 3.8 the most important characteristics of evaporation and sputtering are compared.

3.4.3.2 Sputtering. Sputtering is preferred over evaporation in many applications due to a wider choice of materials to work with, better step coverage, and better adhesion to the substrate. In a sputtering process, the target (a disk of the material to be deposited) is bombarded at a high negative potential with positive argon ions (other inert gases such as xenon can be used as well) from the plasma (also glow discharge). The target material is sputtered away by momentum transfer and ejected surface atoms are deposited onto the substrate which is placed on the anode. For conductors a DC sputtering set-up can be used;

Table 3.8 Comparison of evaporation and sputtering technology.

	Evaporation	Sputtering
Rate	1000 atomic layers s^{-1} (e.g. 0.5 μm min^{-1} for Al)	1 atomic layer s^{-1}
Choice of materials	Limited	Almost unlimited
Purity	Better (no gas inclusions, very high vacuum)	Possibility of incorporating impurities (low–medium-vacuum range)
Substrate heating	Very low	Unless magnetron is used substrate heating can be substantial
Surface damage	Very low; with e-beam x-ray damage is possible	Ionic bombardment damage
Insitu cleaning	Not an option	Easily done with a sputter etch
Alloy compositions, stoichiometry	Little or no control	Alloy composition can be more tightly controlled
X-ray damage	Only with e-beam evaporation	All types of radiation and particle damage are possible
Changes in source material	Easy	Expensive
Decomposition of material	High	Low
Scaling up	Difficult	Good
Uniformity	Difficult	Easy over large areas
Capital equipment	Low cost	More expensive
Number of depositions	Only one deposition per charge	Many depositions can be carried out per target
Thickness control	Not easy to control	Several controls possible
Adhesion	Often poor	Excellent
Shadowing effect	Large	Small
Film properties (e.g. grain size and step coverage)	Difficult to control	Control by bias, pressure, substrate heat

for insulating materials an rf power supply is required. Higher ion densities permit higher sputter rates. In a magnetron sputtering apparatus, the crossed electric and magnetic fields contain the electrons and force them into long, helical paths thus increasing the probability of an ionizing collision with an argon atom.

3.4.3.3 Molecular beam epitaxy. Epitaxial techniques are techniques of arranging atoms in single-crystal fashion on crystalline substrates so that the lattice of the newly grown film duplicates that of the substrate. If the film is of the same material as the substrate, the process is called homoepitaxy, epitaxy, or simply epi. The most important applications here are Si epi on Si substrates and GaAs epi on GaAs substrates. If the deposit is made on a substrate that is chemically different, the process is termed heteroepitaxy. An important application is the deposition of silicon on an insulator (SOI) e.g. with sapphire (Al_2O_3) as the insulator in the silicon on sapphire (SOS) process.

In molecular beam epitaxy (MBE), the heated single-crystal sample is placed in an ultrahigh vacuum (10^{-11} Torr) in the path of streams of atoms from heated cells that contain the materials of interest. These atomic streams impinge on the surface, creating layers whose structure is controlled by the crystal structure of the surface, the thermodynamics of the constituents, and the sample temperature. The deposition rate of MBE is very low i.e. about 1 μm h^{-1} or 1 monolayer s^{-1}, and the deposition is controllable with fast-acting shutters. The low growth temperatures reduce diffusion and autodoping effects. Precise control of layer thickness and doping profile on an atomic layer level is possible, and novel structures such as silicon/insulator/metal sandwiches and superlattices can be made.

3.4.3.4 Laser ablation deposition. Laser ablation deposition uses intense laser radiation to erode a target and deposit the eroded material onto a substrate. A high-energy focused laser beam avoids the x-ray damage to the substrate that is encountered with e-beam evaporation. This technique is particularly useful when dealing with complex compounds as in the case of the deposition of high-temperature superconductor films (HTSCs) e.g. $YBa_2Cu_3O_{7-x}$. Normally the laser-deposited films are amorphous. The energy necessary to crystallize the film is provided by heating the substrate (700–900 °C) and by the energy transferred from the intense laser beam to the substrate via atomic clusters.

3.4.4 Chemical vapor deposition

Chemical vapor deposition (CVD) is based on diffusive–convective mass transfer. At the relatively high pressures of a CVD reactor, the mass transport to the substrate is more complex than for PVD, involving intermolecular momentum transfer.

3.4.4.1 AP CVD and LP CVD. In CVD, the constituents of a vapor phase, often diluted with an inert carrier gas, react at a hot surface to deposit a solid film. The sample surface chemistry, its temperature, and its thermodynamics determine the compounds deposited. The reactions forming the solid materials do not always occur on or close to the heated substrate (heterogeneous reactions). These reactions can also occur in the gas phase (homogeneous reactions). Heterogeneous reactions are preferred, as homogeneous reactions lead to gas phase cluster deposition which results in poor adhesion, low density, and high-defect films. Depending on the gas pressure either the gas phase or the surface processes can be rate determining. In a low-pressure CVD reactor (LP CVD) ($\sim$1 Torr) the diffusivity of the gas species is increased by a factor of 1000 over that of atmospheric-pressure CVD (AP CVD), resulting in a one order of magnitude increase in the transport of reactants to the substrate, and the rate limiting step becomes the surface reaction.

The CVD method is very versatile and can work at low or atmospheric pressure and at relatively low temperatures. Amorphous, polycrystalline, epitaxial, and uniaxially oriented polycrystalline layers can be deposited with a high degree of purity, control, and economy. CVD is used extensively in the semiconductor industry and has played an important role in past transistor miniaturization by making it possible to deposit very thin films of silicon. CVD also constitutes the principal building technique in surface micromachining (see below).

3.4.4.2 PE CVD. In high-temperature CVD reactors the thermal energy is the sole driving force, but for lower-temperature deposition an additional energy source is needed. Radiofrequency, photoradiation or laser radiation can be used to enhance the process and it is known as plasma-enhanced CVD (PE CVD), photon-assisted CVD, or laser-assisted CVD respectively. We will only discuss PE CVD here. Plasma activation provides the radicals that result in the deposited films, and ion bombardment of the substrate is often needed to further provide the energy required to arrive at the stable desired end products. The PE CVD method is surface reaction limited and adequate substrate temperature control is needed to ensure film thickness uniformity. Microstructure, stress, density and other film properties also show marked response to ion bombardment during deposition. In general the more ion bombardment the better the film quality. PE CVD enables dielectric films such as oxides, nitrides, and oxynitrides to be deposited on wafers with small feature sizes and linewidths at low temperatures on devices unable to withstand high temperatures. The temperatures are lower with PE CVD because a part of the activation energy needed for deposition comes from the plasma. The most important application is probably the deposition of SiO_2 or Si_3N_4 over metal lines.

Unlike bulk micromachining, where the substrate silicon is the machining element, surface micromachining utilizes deposited films, such as polysilicon, silicon nitride, and nickel. In surface micromachining plasma settings are an

Table 3.9 Critical temperatures of often-used CVD-deposited surface micromachining materials.

	Temperature (°C)	Material
LP CVD deposition	450	Low-temperature oxide (LTO)
	610	Low-stress poly-Si
	650	Doped poly-Si
	800	Nitride
Annealing	1050	Poly-Si stress annealing

important controlling parameter of intrinsic stress in CVD films. Some important surface micromachining materials and deposition temperatures are listed in table 3.9.

3.4.4.3 Spray pyrolysis. The CVD technologies discussed so far all evolved around the IC industry. Spray pyrolysis has never been a contender in that arena, but it is a viable technology for large-area devices such as solar cells and antireflective window coatings. We believe it is also a good candidate for the manufacture of hybrid gas sensors and ion-selective electrodes. In spray pyrolysis, the simplest form of CVD, a reagent dissolved in a carrier liquid is sprayed on a hot surface in the form of tiny droplets. The reagent decomposes or reacts with oxygen on the hot surface to deposit a stable residue. A simple spray pyrolysis set-up is shown in figure 3.9. In all spraying processes, the significant variables are the substrate temperature, ambient temperature, chemical composition of the carrier gas and/or environment, carrier gas flow rate, nozzle to substrate distance, droplet radius, solution concentration, solution flow rate, and—for continuous processes—substrate motion [28]. The process produces relatively thick films, is difficult to control, and is not compatible with IC processing.

Because individual droplets evaporate and react quickly, grain sizes are very small, usually less than 0.1 μm. The small grains are a disadvantage for most semiconductor applications but not necessarily for sensor applications e.g. in gas sensors, where surface area is important.

3.4.5 Electrodeposition and electroless deposition

In general, for the deposition of thin films used in microelectronics, low-pressure processes, such as sputtering and low-pressure chemical vapor deposition (LP CVD), are preferred over the older, conventional chemical methods such as electroplating because deposition from an aqueous solution produces poorer quality films. On the other hand electrochemical and electroless metal deposition from solution are gaining renewed interest with micromachinists because of their emerging importance to LIGA.

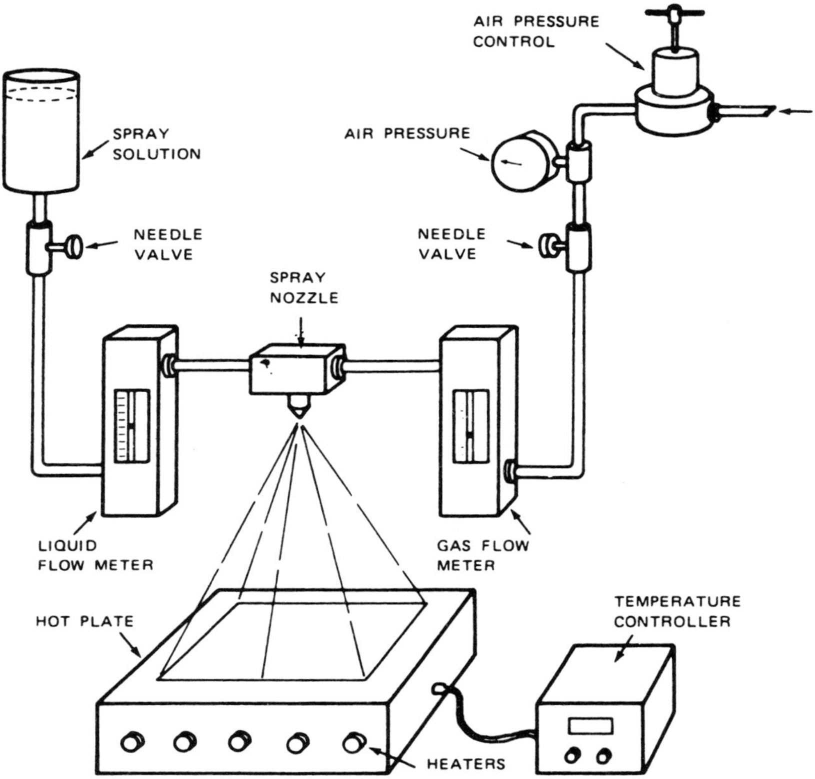

Figure 3.9 A spray pyrolysis set-up [29].

3.4.5.1 Electroless deposition. Electroless metal deposition occurs on surfaces catalytic to metal deposition without the need of a biased conductive substrate. The deposition is from a solution containing a metal salt and a reducing agent as well as various other additives such as stabilizers, surfactants, etc. Electroless plating is an inexpensive technique enabling plating of conductors and non-conductors alike (plastics such as ABS, polypropylene, Teflon, and polycarbonate are all plated in huge quantities today). A catalyzing procedure is necessary for electroless deposition on non-active surfaces such as plastics and ceramics, however. The most common method for sensitizing those surfaces is immersion in $SnCl_2/HCl$ or $PdCl_2/HCl$. This chemical treatment produces sites which provide a chemical path for the initiation of the plating process.

In the IC industry electroless metal deposition can be used for contact filling, via filling, and conductor patterns. In micromachining electroless deposition can be used for all of the same purposes but also to make structural microelements from a wide variety of metals, metal alloys, and even composite materials.

A key advantage of electroless deposition is that a metal can be deposited without the need to make electrical contact to a voltage source. This feature makes it extremely suitable for depositing metals on a wafer with CNMOS circuitry, which otherwise might be damaged in a conventional plating process.

3.4.5.2 Electrodeposition. In electroplating, a conductive substrate is used as the cathode for metal deposition. This method, although more complicated, allows for more operator control.

Plating of high-aspect-ratio features is of key interest to micromachinists in general and in particular for those interested in LIGA (see below). Both electroless and electrodeposition can be used for this application.

3.4.6 Chemical sensor fabrication technology

In chemical sensors one often encounters materials rarely dealt with in the IC industry. A particular class are organic membranes for the manufacture of ion selective electrodes, gas sensors, enzyme sensors, immunosensors, etc. Most chemical membranes are based on classical polymers, such as PVC, PVA, PHEMA, and silicone rubber, and on biological materials, such as enzymes, antigens, and antibodies. Thick-film hybrid technology is often the manufacturing option of choice for such sensors (e.g. for glucose sensors and ion selective electrodes). Techniques such as spin coating, dip coating, casting, LB, ink jet printing (i.e. drop delivery systems), and silk screening have been employed.

3.4.6.1 Silk screening or screen printing. Screen printing is more cost effective than IC technology for the fabrication of relatively low production volumes. For chemical and biosensors screen printing is a viable alternative to Si thin-film technologies. The technology was originally developed for the production of miniature, robust, and, above all, cheap electronic circuits. The up-front investment in a thick-film facility is low compared to the needs in IC manufacturing.

In silk screening, a paste or ink is pressed onto a substrate through openings in the emulsion on a stainless steel screen (see figure 3.10). The paste is a mixture of the material of interest, an organic binder, and a solvent. The lithographic pattern in the screen emulsion is transferred onto the substrate by forcing the paste through the mask openings with a squeegee. The resolution of the process is dependent on the openings in the screen and the nature of the pastes. With a 325-mesh screen (i.e. 40 μm holes) and a typical paste, a resolution of 100 μm can be obtained. After printing, the wet films are allowed to settle for 15 min to flatten the surface and are dried. This removes the solvents from the paste. Subsequent firing burns off the organic binder and metallic particles are reduced or oxidized and glass particles are sintered. Typical temperatures range from

Table 3.10 A comparison of thin- against thick-film technology.

Property	Si/thin film	Hybrid/thick film
Resolution	0.25 μm and better	12 μm
Minimum feature size	0.75 μm and better	90 μm
Temperature range	<125 °C	$\gg 125$ °C
Sensor size	Small	Quite small
Cost of mass production	Low	Moderate
Geometric accuracy	Very high	Poor
Cost of small production	High	Low
Reliability of non-encapsulated device	Low	High
Electronic compatibility	Good	Moderate
Versatility	Low	Very good
Roughness, purity, and porosity of deposited materials	Superior	Moderate
Energy consumption	Low	Moderate
Handling	Difficult	Easy
Capital equipment	Very large	Small

500 °C to 1000 °C. After firing, the thickness of the film ranges from 10 μm to 50 μm.

Almost all materials which are compatible with the high firing temperature and the other ink constituents can be used to screen print. Different pastes are available commercially: conductive (e.g. Au, Pt, Ag/Pd), resistive (e.g. RuO_2, IrO_2), overglaze, and dielectric pastes (e.g. Al_2O_3, ZrO_2).

More recently inks specifically developed for sensor applications have become available, for example SnO_2 pastes incorporating Pt, Pd, and Sb dopants for the construction of semiconductor gas sensors. For biosensor applications, thick-film technology based on polymer films is extremely important, and special grades of polymer pastes (carbon, Ag, and Ag/AgCl) are becoming available [30]. However, in the development of chemical and biosensors, new pastes must usually be developed. Polymer thick films can be screen printed on cheap polymer substrates with a thickness anywhere between 5 and 50 μm. Importantly, no high-temperature steps are involved in the deposition process. The first commercial planar electrochemical glucose sensor (the ExacTech by MediSense) was a screen-printed sensor. A comparison of thick-film deposition against thin-film deposition is shown in table 3.10.

Resolution and minimum feature size are particularly superior for thin films. Also the porosity, roughness, and purity of deposited metals is less reproducible with thick films compared to thin films. Finally, the geometric accuracy is poorer with thick films. On the other hand the thick-film method is very versatile and when small size is not too important, but cost in relatively small production volumes is, silk screening forms an excellent alternative. The size limitations clear from table 3.10 make thick-film sensors more appropriate for *in vitro* applications than *in vivo* applications. For *in vivo* applications where size is

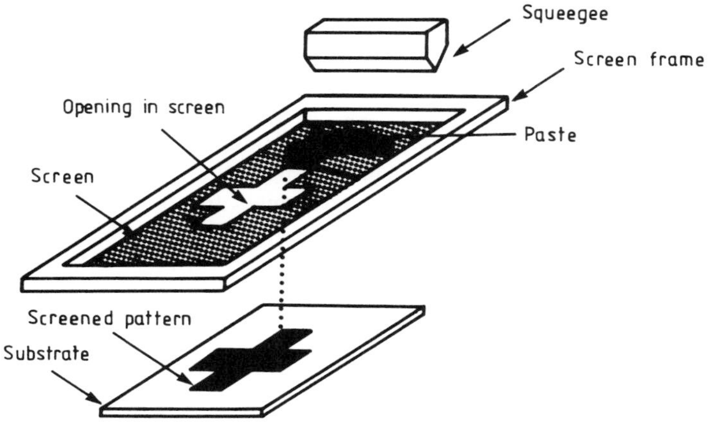

Figure 3.10 A schematic representation of the screen printing process.

more crucial, IC-based technologies might be more appropriate.

3.4.6.2 Spin coating. This technology has been optimized for deposition of thin layers of photoresist, say about 1 μm thick, on Si wafers which are round and almost ideally flat surfaces. Resists are applied by dropping the resist solution, a polymer, a sensitizer (for two-component resists), and a solvent on the wafer and rotating the wafer on a spinning wheel at high speed so that centrifugal forces force the excess solution over the edge of the wafer and the residue on the wafer is held there by surface tension. In this way films down to 0.1 μm can be made. Often biosensor substrates are neither round nor flat. Also the thickness of many chemical membranes is considerably thicker than 1 μm. This technology therefore is not necessarily the best candidate for thick chemical membranes on a variety of substrates.

3.4.6.3 Dip coating. Dip coating is the simplest method to apply a membrane to a substrate. Most biosensors today are made by dip coating. It is especially suited for wire-type electrodes in which case the membrane forrns a droplet at the end of the wire. For example, a chloridized silver wire is dipped into a solution containing the polymer and a solvent. After evaporation of the solvent, a thin membrane is formed on the surface of the sensor. To obtain pinhole-free membranes the dipping is repeated several times with a drying period in between dippings. Even though the eventual goal is a planar sensor structure an Ag wire is often used in the research phase to quickly evaluate a new membrane composition.

3.4.6.4 Casting. Casting is based on the application of a given amount of dissolved material on the surface of a mounted sensor and letting the solvent evaporate. Often a rim structure is fashioned around the substrate providing a flat beaker for the solution. This method gives a more uniform and a more reproducible membrane than dip coating.

3.4.6.5 Langmuir–Blodgett film approach. The need for making monolayer resist in the IC industry is to obtain yet higher resolution, and in sensor manufacture it is often to provide anchor points for subsequent membranes or immunosensitive layers. One method receiving considerable attention is LB deposition. In the LB process, a monolayer of film forming molecules (stearic acid is often used as a model molecule) on an aqueous surface is compressed into a compact floating film, and transferred to a solid substrate by passing the substrate through the water surface.

3.4.6.6 Ink jet printing and other drop dispensing systems. Ink jet drop delivery is based on the same principle as that used for commercial ink jet printing. On a somewhat larger scale, epoxy delivery systems used in the IC industry similarly deliver drops in a serial fashion on specific spots on a substrate. A typical commercial drop dispensing system (e.g. the Ivek Digispense 2000) delivers 0.20–0.50 μl in a drop and has a cycle time per dispense of one second. A vision system can be used to verify substrate position and accurate dispense location to within ± 25 μm of a specified location. In the case of an ink jet printing head, the ink jet nozzle, connected to a reservoir filled with the chemical membrane solution, is placed above a computer-controlled XY stage. Depending on the ink expulsion method, even temperature-sensitive enzyme formulations can be delivered this way. Different nozzles can be used to print different membranes in parallel. In the authors' experience, some of the problems are excessive splashing, clogging of the nozzle, and poor uniformity of the deposit. The clogging problem could be alleviated by using a solvent-saturated deposition environment. Although these drop delivery systems are serial, they can be very fast as evidenced by epoxy delivery stations in an IC manufacturing line.

3.4.6.7 Membranes deposited with a photolithography process. Spinning, UV exposure, and development of photosensitive materials are well known, low-cost, mass production procedures. One has to take into account, however, that photosensitized membrane materials needed in biosensors are not available commercially. To cope with this problem, one has to prepare the photosensitive material from high-purity materials.

Lift-off techniques for lithographically patterning membrane materials have been developed. A problem with lift-off for thick organic membranes is the thickness needed for the photoresist layer. A typical resist layer is about 1 μm thick. The thickness needed for chemical membranes can be as high as 50 μm, increasing the resist layer thickness that is necessary to pattern these thick layers. Also, lift-off can only be used with materials that are resistant to the solvent used to remove the resist. Although attractive for batch fabrication this technology can only be applied in few cases.

A comparison of the above-reviewed membrane deposition and patterning techniques might at first suggest that photolithography seems very promising for biosensor development, but the chemistry is complex and very few results have been published to date. Since a lot more development work is needed, most developers have opted for screen printing, i.e. the safest and least capital expensive approach.

In comparing different sensor technologies it should be mentioned that 60–80% of sensor fabrication cost consists of packaging costs, an aspect not addressed frequently enough. In comparison with CMOS-compatible sensors, the packaging of thick-film sensors is often easy. Since packaging expenses overshadow all other costs, this is a very important decision criterion.

3.4.7 Three main doping techniques

Fabrication of circuit elements and micromachines requires a method for selective n- or p-type doping of the silicon substrate. The three means of doping Si are diffusion, ion implantation, and neutron transformation or transmutation. In the past, dopants were diffused thermally into the substrate in a furnace at temperatures between 950 and 1280 °C; now they are implanted as high-energy ions or neutron transformation. Implantation offers the advantage of being able to place any ion at any depth in the sample, independent of the thermodynamics of diffusion and problems with solid solubility and precipitation. Ion beams do produce crystal damage that can reduce electrical conductivity; however, most of this damage can be eliminated by annealing at 700–1000 °C.

3.5 COMPARISON OF MICROMACHINING TOOLS

Micromachining is emerging as a set of new manufacturing tools at our disposal to solve specific industrial manufacturing problems which cannot be addressed with classical machining approaches. The number of micromachining tools made available to fabricate small devices has grown dramatically. The most important are compared in table 3.11. These techniques are based on combinations of the lithography, subtractive, and additive processes discussed above.

Wet bulk micromachining is restricted to very specific crystallographic orientations, obviously not offering freedom to create any desired shape in the wafer plane, i.e. the x–y plane. There are strong incentives to move away from wet etchants, as environmental concerns about their disposal have grown and demands for improved process control on finer features have increased. In dry bulk micromachining, one can freely create any desired shape in the x–y plane with large dimensions perpendicular to the wafer, i.e. in the z direction in mechanically superior crystalline materials.

Table 3.11 Machining tools.

Chemical sensors went from integrated to hybrid.

	Surface μ		Wet Bulk μ	Dry Bulk μ	LIGA**	Laser	Hybrid
	Poly-Si	SOI	Bulk μ didn't make instruments so LIGA and laser machining are attempted now.				
Z-Height	< 10 μm	up to 100 μm	Single wafer : 600 μm	up to 100 μm	Up to 1 mm was reported	Very wide range	Less than 20 μm is difficult
X, y-Shapes	Free	Free	Limited	Free	Free	Free	Free
IC Comp.	Good	Good	Poor	Good	Not demon-strated	Good	Not applicable
Materials	Poly-Si	SCS***	SCS	Free	Some choice	Free	A lot of choices
Maturity	Fair	New	Mature	New	New	New	Mature
Cost	Low	Fair	High	Fair	?(low with replication)	High	Low
Access	Good	Not yet	Good	Good	Poor	Fair	Good
Parallel (P)/ Serial (S)	P	P	P	P	P	S	P

In mechanical sensors there is a move towards surface μ-machining and more integration, especially SOI will become important and more dry etching will be used.

* Single Crystal Released Etched and Metallized
** From the German acronym for lithography, electro deposition and molding.
*** Single Crystal Silicon

Surface micromachining, although free to create any x–y shape, is usually quite limited in feature height. Specifically, with surface-micromachined LP CVD poly-Si, only a few micrometers in the z direction can be built up. On the other hand, 25–30 μm and beyond can readily be obtained when surface micromachining on insulator wafers (SOI). In the latter case, crystalline Si is deposited on top of the SiO_2 insulator by epitaxial techniques. Since the latter surface micromachining technique results in structural elements made from single-crystalline Si, a material which is mechanically superior to poly-Si, better sensors can be expected. The major power of poly-Si surface micromachining is in its CMOS compatibility. SOI micromachining features the same CMOS compatibility as poly-Si surface micromachining while at the same time relying on single-crystal Si for the performance excellence expected from bulk micromachining.

LIGA is the German acronym for lithography (*x-ray lithographie*), electroforming (*Galvanoformung*), and molding (*Abformtechnik*). Besides freedom in the x–y plane and the possibility of very high structures with unprecedented ARs this technique also enables the use of a large number of manufacturing materials ranging from metal to ceramics and plastics. In contrast to mechanical sensors where the excellent mechanical properties of single-crystalline Si often tend to favor Si technology, the choice of the optimum manufacturing technology for chemical sensors and microinstrumentation is far less evident. We can point out the following trends. Whereas mechanical sensors (pressure, acceleration, temperature, etc) are moving toward more integration embodied in more surface-micromachined products, chemical and biosensors are moving away from integration and toward hybrid technology. Where instrumentation is concerned there is a move toward using laser machining and some early exploration of LIGA use.

ACKNOWLEDGMENT

The authors would like to thank Mr John Hines of NASA Ames Sensors 2000! for his support in the writing of this chapter.

REFERENCES

[1] Moreau W M 1988 *Semiconductor Lithography* (New York: Plenum)
[2] 1993 *Solid State Technol.* **April** 58–61
[3] 1992 *Microlithogr. World* **March/April** 23–6
[4] Bower R W and Ismail M S 1993 *Appl. Phys. Lett.* **62** 28
[5] Pomerantz D I 1968 *US Patent* 3 397 278
[6] 1988 *Solid State Technol.* **June** 37–43
[7] SRI International 1990 *US Patent* 4 923 421
[8] 1992 *Microlithogr. World* **November/December** 7–14
[9] Lambrechts M and Sansen W 1992 *Biosensors* (Bristol: Institute of Physics)
[10] 1992 *Microlithogr. World* **March/April** 6–12
[11] Hess D W and Jensen K F (ed) 1989 *Microelectronics Processing, Chemical Engineering Aspects* (Washington, DC: ACS)
[12] Becker E W, Betz H, Ehrfeld W, Glashauser W, Heuberger A, Michel H J, Munchmeyer D, Pongratz S and von Siemens R 1982 *Naturwissenschaften 69* pp 520–3
[13] Hagmann P, Ehrfeld W and Vollmer H 1989 *Makromol. Chem. Macromol. Symp.* **24** 241–51
[14] Brunger *et al* 1993 *IEEE Micro Electro Mechanical Systems (Fort Lauderdale, FL, 1993)*
[15] *Solid State Technol.* **August** 25–6
[16] Stroscio J A and Eigler D M 1991 *Science* **253** 1319–26
[17] Quate F 1989 *Scanning Tunneling Microscopy and Related Methods (NATO ASI Series E: Applied Sciences 184)* ed R J Behm, N Garcia and H Rorher
[18] Flamm *et al* 1989 *Plasma Etching an Introduction* ed D M Manos and D L Flamm (New York: Academic)
[19] Harper J 1989 *Plasma Etching an Introduction* ed D M Manos and D L Flamm (New York: Academic)
[20] Linde H and Austin L 1992 *J. Electrochem. Soc.* **139** 1170–4
[21] Allongue P, Costa-Kieling V and Gerischer H 1993 *J. Electrochem. Soc.* **140** 1009–18, 1018–26
[22] Kern W 1978 *RCA Rev.* **29** 278
[23] Seidel H, Cspregi L, Heuberger A and Baumgartel H 1990 *J. Electrochem. Soc.* **137** 3612, 3626
[24] Raley N F, Sugiyama Y and Van Duzer T 1994 *J. Electrochem. Soc.* **131** 161–71

[25] Elwenspoek M 1993 *J. Electrochem. Soc.* **140** 2075–80; *IEEE Micro Electro Mechanical Systems (Oiso, 1994)*
[26] Brodie I and Muray J J 1982 *The Physics of Microfabrication* (New York: Plenum)
[27] 1992 *Semicond. Int.* **May** 68
[28] Tomar M S and Garcia F J 1981 *Prog. Cryst. Growth Charact.* 221–48
[29] Mooney J B and Radding S B 1982 *Annu. Rev. Mater. Sci.* **12** 81–101
[30] Acheson Colloids Company 1991 Product data sheets on screen-printable Ag, Ag/AgCl, and C pastes

4

Photometric transduction

Donald G Buerk

4.1 INTRODUCTION

The scientific foundations of modern optics began in the seventeenth century with the observations and experiments of de Fermat, Galilei, Grimaldi, Huygens, Snell, and Newton. Later work by Faraday, Fresnel, Lorentz, Maxwell, and Young laid the basis for the classical approach to understanding light and its interactions with matter. The modern era of relativity and quantum mechanics began with the theories of Bohr, Einstein, and Planck, which were later amplified and extended by de Broglie, Dirac, Heisenberg, and Schrödinger. The science of optics allowed mankind to invent telescopes that extended his vision into the heavens and microscopes that allowed him to turn inward to peer inside living cells. Photometric transduction has now extended our senses even further by allowing us to use light and electromagnetic energy to detect and measure other 'non-visible' quantities, even to peer into the chemical nanoenvironments of individual molecules.

Optical methods for biological measurements began with the early work of Otto Warburg in Germany before World War II. The principle of using a differential, dual-beam spectrophotometer had been developed by the English physicist J Tyndall in the late 1800s. This concept was applied to biological measurements by several laboratories, and a modified split beam spectrophotometer made commercially by Beckman Instruments became an important optical instrument for many scientists. Chance [1] has reviewed the development of modern optical methods for spectrophotometric measurements in tissue and provides a historical perspective on the evolution of optical techniques in biology.

This chapter will review some of the major photometric transduction methods. Representative examples will be described that use photometric techniques for biologically relevant measurements.

4.1.1 The wave nature of light

According to the classical electromagnetic viewpoint, light is energy in the form of transverse spherical waves. The most general definition of light would include the entire electromagnetic radiation spectrum. A ray of light propagating in the z direction has an electrical field (E) in the polarization plane, with a magnetic field (B) at a right angle to it. Both electrical and magnetic fields can be defined in space and time by Maxwell's equations. The two perpendicular fields in the x and y directions are related by

$$E_x = E_0 \cos[\omega(t - z/v)] = vB_y \tag{4.1}$$

where the field intensity E_0 is constant and ω is the angular frequency ($\omega = 2\pi f$). For a wave in free space with frequency f and velocity $v = c$, the wavelength is $\lambda = c/f$, where $c = 299\,792\,458$ m s^{-1} is the speed of light in a vacuum. In a lossless, isotropic medium with homogeneous properties, the velocity is slower, as given by

$$v = \frac{1}{\sqrt{\mu\epsilon}} < c \tag{4.2}$$

where μ is the permeability and ϵ is the dielectric constant (or capacitivity) for the medium. By treating light as a wave, many basic optical phenomena (e.g., interference, standing waves, reflected waves, polarized light) can be explained mathematically by simple superposition of complex (sinusoidal) waveforms.

The quantum theory approach views light as photons, which are conceptualized as massless particles that have an elemental packet of energy. Although massless, the photon has momentum and angular momentum. For a single mode (single wavelength, direction, and polarization) the quantum energy equals $h\nu$ where the Planck constant h is $\sim 6.626 \times 10^{-34}$ J s and ν is the frequency in Hz (s^{-1}). Light can only exchange energy in integral multiples of $h\nu$. There are some optical phenomena, such as spontaneous light emission, that can only be explained in terms of quantum theory.

4.1.2 Light sources

There have been significant developments in the light sources used for optical instruments. Numerous photometric transduction techniques developed that use conventional light sources where light is emitted from a hot body. These include incadescent lamps which radiate light from an electrically heated tungsten metal filament, and gas lamps (mercury, xenon) which radiate light after the ionized gas has been excited with an electrical current to a higher quantum state. Zinc lamps with predominant spectral lines at 214 and 308 nm have been used for ultraviolet measurements. Light from these conventional sources is incoherent, that is, there is a random phase distribution for the light waves. Also, these hot-body devices have limited lifetimes, and can change their spectral properties with age.

With advances in semiconductor technology, light emitting diodes (LEDs) have been developed. Non-coherent light is produced by electroluminescence at a forward-biased PN junction. The semiconductor can be doped with polymers or organic materials that fluoresce at certain wavelengths (typically red or green light), or doped with combinations of materials to produce a certain color (e.g., green and red for yellow). LED light sources are efficient and relatively inexpensive light sources and have the potential for miniaturization. For example, Granström *et al* [2] were able to fabricate LEDs with diameters as small as 100 nm. Nano-LEDs may be useful as light sources for near-field optical sensors, which are discussed later in this chapter.

The most significant advance in optical technology was the development of the laser, or light amplification by stimulated emission of radiation, in the late 1950s. One of the most useful characteristics of the laser is that the resulting light waves are planar waves that are in phase (coherent). The light spectrum that is produced by a laser will depend on the properties of the excited medium, which can be a solid, liquid, or gas. Early lasers used a mixture of helium and neon gases. The range of light wavelengths in the laser output spectrum can be tuned to some degree by changing the pressure of the gases. Higher output power can also be achieved with higher pressures. Typically, red light with a wavelength around 633 nm is produced from the helium:neon (He:Ne) laser. Other gases such as hydrogen cyanide, carbon monoxide, and carbon dioxide can be used for laser outputs in the infrared range. Visible and near-ultraviolet wavelengths can be generated with argon and krypton gas lasers. Solid state lasers include various types of glass (silicate, borate, fluorophosphate, fluoride, and others) doped with rare-earth elements which have trivalent ions. One of the earliest rare-earth elements used for solid state lasers was the neodymium (Nd^{3+}) ion. Similar doping of neodymium ions into rare-earth crystals such as yttrium aluminum garnet (YAG) or chromium (Cr^{3+}) in crystallized aluminum oxide (ruby) have also been used for lasers. Transition metals such as chromium have also been used as sensitizers which improve the efficiency of optical pumping by increasing the availability of higher-energy states when excited. Chromium ions increase the efficiency of the ruby laser. Other rare-earth elements used for lasers include holmium (Ho^{3+}) and erbium (Er^{3+}) ions in YAG crystals. The Ho:YAG laser emits infrared light with a wavelength around 2400 nm and the Er:YAG laser emits around 2940 nm.

Since light amplification can be precisely controlled, it is possible to generate very short light pulses with many types of laser. Another laser can be used as the exciting incident light. Chemically modified gases, such as hydrogen fluoride, can utilize very fast chemical reactions to produce short-duration light pulses. Excimer lasers exploit the high-energy states that can be reached with some rare gas monohalides, such as fluorides, chlorides, or bromides of krypton or xenon gases. Excimer lasers can produce pulses in the ultraviolet range. Properties for some of the more common types of continuous wave and pulsed lasers are summarized in table 4.1.

Semiconductor lasers, using materials such as gallium arsenide, have also been a relatively recent technological advance. Free electrons (anions) at PN junctions can combine with holes (cations) in the material and cause light emission. Spectral properties of the laser light can be modified by doping the semiconductor. Although the total power of semiconductor lasers is not particularly high, they can have relatively high power densities, and can have very high efficiencies, better than 50% for some types. It is possible to achieve higher power output by constructing an array of smaller lasers.

4.1.3 Optical waveguides

Photometric transducers require optical waveguides to either deliver light or to detect it with minimal distortion or loss of signal. The communications industry has led to the widespread availability of flexible, optically clear glass and plastic fibers that can transmit light with minimal losses over large distances. The internal waveguide of optically clear material (core) is usually cladded with a plastic material to provide mechanical strength. Often the core is made from silica glass with very low levels of impurities such as iron and water. Manufacturing methods have been developed to form long optic fibers with precise dimensional tolerances. Two common types of optic fiber are shown in figure 4.2. With a small-diameter core, there is minimal dispersion of the light wave and a single mode is guided. The output faithfully matches the input pulse, except for some minor attenuation of the signal, depending on the length of the fiber. With a larger-diameter core, multiple modes can be guided through the fiber since there can be many different internal scattering pathways. As a result, the output pulse from the multimode fiber will be broadened compared to the input pulse. This type of fiber is useful for guiding images over short distances. Other types of optic fiber have cores where the refractive index varies with diameter (graded refraction index). This design can reduced the broadening in multimode fibers. There are also optic fibers that maintain the polarization of an input signal. The use of single fibers or small bundles of fibers in optic cables has allowed miniaturization of quite a number of optical techniques in the past decade. With the development of lasers with emissions in the infrared range, optical fibers are required that do not absorb light in this range. Fluorophosphate and fluoride glasses have reasonably good transparency to infrared. Sapphire optical fibers have been particularly useful for Er:YAG lasers. Diamond has a very wide optical transmission range from the ultraviolet to far infrared, although there is some intrinsic absorption in the 2 to 6 μm range. The hardness of diamond is particularly useful for optical waveguides in harsh chemical environments. Also, the thermal conductivity of diamond is over 100 times greater than glass, so it is much less susceptible to heat damage at

Table 4.1

Laser	Wavelength (nm)	Laser medium	Pulse duration	Maximum energy (J per pulse)	Maximum repetition rate (Hz)	Maximum average power (W)	Maximum power
Pulsed laser properties							
Excimer	193	ArF gas	5–25 ns	0.5	1000	50	
	249	KrF gas	2–50 ns	1	500	100	
	308	XeCl gas	1–300 ns	1.5	500	150	
	351	XeF gas	1–30 ns	0.5	500	30	
Dye:Ar pumped	400–1000	Liquid dye	3–50 ns	0.1	100	10	
flash pumped	350–1000	Liquid dye	0.2–30 μs	50	50	50	
Metal vapor	628	Au vapor	15 ns	10^{-4}	10^4	1	
Semiconductor	750–1550	Single device	0.2–2 μs	10^{-4}	10^4	0.1	
(AlGaAs)		Array	2–200 μs		100	1	
Nd:YAG	1060	Solid state	10 ns	1	20	20	
Er:YAG	2940	Solid state	200 μs	2	20	30	
Continuous (CW) Laser Properties							
HeCd	325.0	Gas					50 mW
	442.0						150 mW
Ar ion	488.0	Gas					20 W
	514.5						
Kr ion	413.1	Gas					5 W
	530.9						
	647.1						
Dye (Ar pumped)	400–1000	Liquid					2 W
HeNe	632.8	Gas					50 mW
Semiconductor	750–900	Single device					100 mW
(AlGaAs)		Array					10 W
Nd:YAG	1060	Solid state					600 W
CO_2	10 600	Gas					100 W

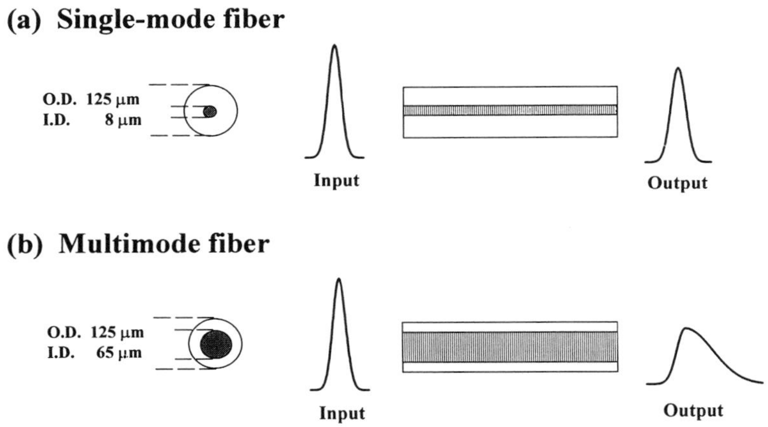

Figure 4.1 Two common types of fiber optic waveguides. (a) Light transmitted through the small-diameter core of a single-mode fiber has little broadening of the input. (b) The output from the larger core of a multimode fiber has more broadening.

high power. Vapor deposition techniques have recently been developed which may lead to more common use of diamonds as optical elements.

Planar waveguides are also used for photometric transducers. Recently, there have been applications of photolithographic methods to create miniaturized waveguides, or other microfabrication techniques to produce thin films of organic materials for optical waveguides. For example, Hanken and Corn [3] have investigated ultra-thin films of zirconium phosphonate produced by Longmuir–Blodgett (single monolayer) techniques. An ultra-thin waveguide for a specific wavelength can be designed by controlling the thickness and index of refraction using multi-layer fabrication methods. Healey *et al* [4] describe a photodeposition technique to produce patterned microstructures of photopolymerizable polymers on the distal end of a multimode optical fiber. Arrays of hemispherical polymer microspots could be fabricated with diameters around 2 μm and heights around 0.3 μm, spaced 4 μm apart. These microdots can be doped with other chemicals to alter optical properties, or loaded with chemiluminescent, fluorescent or phosphorescent indicators. The ultimate goal is to integrate the miniaturized waveguide with electronic circuitry for light detection and signal processing. Non-linear optical materials, where the refractive index or light polarization properties can be altered by applying electrical or magnetic fields, are also being investigated.

Technological advances continue to be made for conventional optical components. New types of filter for specific wavelengths of light are being developed. Computer-controlled grinding and polishing techniques have been developed for producing precise dimensions and curvatures. Multiple-dye interference coatings are being developed to improve the bandpass filtering of optical components. Efficiencies of the order of 75% are possible at the present time. Further advances can be expected and many more optical applications will be developed taking advantage of the wide dynamic ranges and broad spectral sensitivity of various light detection techniques.

4.1.4 Light detection

Vacuum tubes, such as phototubes and photomultiplier tubes, have been widely used for photometric transducers for a number of decades. The active detection area can be very large (>50 mm^2) with very high amplification (>10^7) and low background noise. The photocathode can be constructed from different materials to optimize responses from the near-ultraviolet to the near-infrared range. Photomultiplier tubes are capable of counting single photons, allowing the detection of very low light levels. Resolution can be increased by cooling the tube. In the past, the instrumentation for photomultiplier tubes has been heavy and bulky, since high-voltage circuits have been required. The glass tubes were also fragile and could be damaged. More recent vacuum tubes have been miniaturized and made more rugged.

Solid state alternatives to the photomultiplier tube have only recently been available. The large-area (~20 mm^2) avalanche photodiode was commercially available in 1991. The semiconductor can be reverse biased at a very high voltage (in kV range) and could achieve gains around 10^3. When light with wavelengths in the 300 to 1000 nm range strikes the active area, charge carriers (electrons and holes) are created. Each photon would produce one electron–hole pair. After photons strike the active area, the high bias voltage causes the electrons that are created to be accelerated toward the cathode, creating many more secondary carriers by impact ionization in the semiconductor. A modest increase in sensitivity could be achieved by cooling the device. The gain was substantially increased to 10^6 with the development of ceramic vacuum avalanche diodes. The devices are packaged under vacuum, and the cathode can be biased at 8 kV. The initial electrons produced by the incident photons can be multiplied by a factor of 2000 in this strong electrical field, before further amplification by the secondary carriers that are produced as the electrons flow through the semiconductor. Vacuum avalanche photodiodes can have lower noise and a wider and more linear dynamic range than photomultiplier tubes. Also, the semiconductor device is much less susceptible to interaction with magnetic fields, has a much greater quantum efficiency (~70% at 960 nm), can be operated at room temperature, and is smaller and more robust compared to vacuum tubes.

Digital imaging technology is another light detection method that has been particularly useful for biological measurements. Television cameras with cooled charge-coupled devices (CCDs) using silicon integrated circuits can be used for light detection at higher sensitivities and lower cost than some previous systems. Since its invention by AT&T Bell Telephone Laboratories around 1970, the CCD has been the primary technology for imaging. It has led to the development of highly sensitive digital imaging systems for microscopy. Optical fibers can also be coupled to CCDs for use as optical biosensors. The light intensity can be quantified for each picture element (pixel) of the entire digital image, or for a smaller region of image. The rate of light intensity change with time can be computed for each pixel. The intensity image, or an image based on the rate of intensity changes, can be displayed either on a gray level scale, or as pseudo-colored images. Color images are often easier to interpret by visual inspection, since the human eye cannot see subtle differences in gray level. High-resolution (1024 × 1024 pixels, or 1 million elements) CCDs are now routinely used with typical image rates of 30 frames per second. Higher-resolution CCDs (2048 × 2048 pixels, or 4 million elements) are also available. Ultrahigh-resolution CCDs (4096 × 4096, or 16 million elements) are now reaching the marketplace, and chips with 5120 × 5120 pixels will soon become available. To achieve the highest possible sensitivities, the electronics can be cooled with liquid nitrogen. It is now possible to have resolutions of 12 bits per pixel (4096 gray levels). Detailed spatial and temporal information can be derived from images captured from conventional microscopes, or from confocal imaging systems. Enormous numbers of data are generated, requiring a great deal of computer processing and data storage. Also, mathematical algorithms are being developed for correcting and improving images acquired by confocal microscopy.

Further advances in imaging technology are taking place with the development of high-definition television (HDTV). One of the new image detection methods that is being developed is based on an active-pixel sensor, which acts like a random memory access device. The sensor consists of several active transistors that are monolithically integrated into the pixel element. On-chip analog to digital conversion electronics and other types of processing electronics, such as real-time integration of the signal, will be part of the chip. As the HDTV becomes commercially viable, there is no doubt that the new imaging technology will be quickly adapted for biological and medical imaging uses.

4.2 PHOTOTRANSDUCTION BASED ON INTERACTIONS BETWEEN LIGHT AND MATTER

It is probably safe to say that there is really no satisfactory mathematical description that can fully account for all of the possible interactions between light and matter, particularly in complex biological media. Two common interactions are illustrated in figure 4.2. Although Newton incorrectly explained the principle of refraction, it is now understood that light changes speed and direction when passing from one medium to another. The index of refraction is defined as

$$n = \frac{v_1}{v_2} = \frac{\sin\theta_1}{\sin\theta_2} \tag{4.3}$$

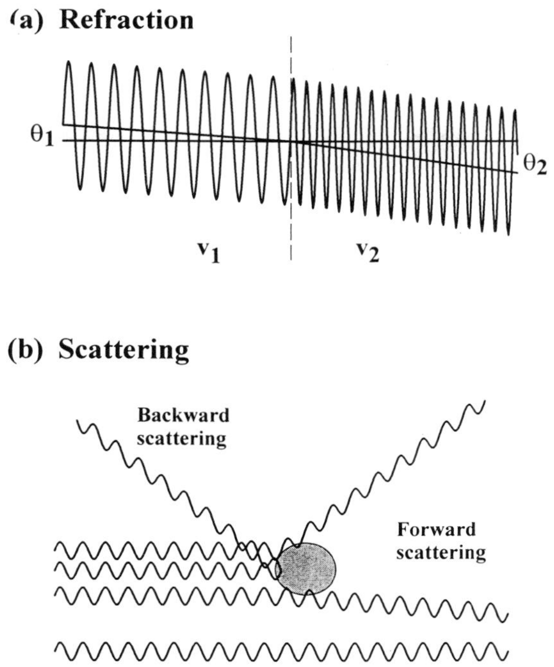

Figure 4.2 Two common interactions between light and matter. (a) At certain angles, light is refracted when passing from one medium to another (at dashed line), changing direction and velocity. (b) When light hits a particle, elastic collisions result in changes in direction and phase (Rayleigh scattering), whereas inelastic collisions result in frequency changes (Raman effect).

where θ_1 and θ_2 are the angles normal to the interface between the two media (dashed line) and v_1 and v_2 are the wave velocities. The velocities are related to the speed of light in vacuum by

$$c = n_1 v_1 = n_2 v_2. \tag{4.4}$$

There is a critical angle for refraction θ_c which is related to the index of refraction by

$$\sin \theta_c = \frac{1}{n}. \tag{4.5}$$

When $\theta_1 < \theta_c$, the light is refracted, and when $\theta_1 > \theta_c$, the light is reflected back.

Scattering, illustrated in figure 4.2(b), is more complex. Generally, it is characterized by the loss of energy when a light wave interacts with matter. Consequently, the light can change direction, velocity, and phase angle. It is assumed that the collisions between light and particles in the medium are elastic, and that there is no change in frequency for the scattered light. Rayleigh derived an equation, based on N_p particles in a volume V, for the intensity of scattered light $I(\theta)$ as a function of the scattering angle, given by

$$\frac{I(\theta)}{I_0} = \frac{\pi N_p V^2(1 + \cos^2 \theta)(n - 1)^2}{x^2 \lambda^4} \tag{4.6}$$

where x is the distance from the initial light at intensity I_0. It is assumed that the particles are smaller than the wavelength of light, and that there is no dependence between particles. These assumptions are not valid for larger particles, or for fluids where the particles and fluid movement are related. Theoretical models for scattering with larger particles have been developed by Tyndall, Mie, and others. Another type of scattering, known as the Raman effect, occurs with a distinct change in energy and frequency of the scattered light. This effect is generally much smaller than Rayleigh scattering, and was difficult to study until the advent of coherent laser light sources. The collision between light and molecules in the medium is inelastic, altering the vibrational and rotational energies of the molecules.

4.2.1 Absorption spectroscopy

In simplest terms, this method involves the measurement of light intensity or frequency after it has been transmitted through a medium. The medium can be either a gas, a liquid, or a solid. There are no perfectly transparent materials that would allow light or electromagnetic energy to pass through them without any change. Some energy must be absorbed, depending on the physical properties of the transmitting medium. For example, water is fairly transparent to visible and ultraviolet light, but strongly absorbs infrared radiation and begins to absorb in the far-ultraviolet range. The optical properties of a liquid medium can also be affected by the concentration (C, moles per liter) of other chemical solutes that are dissolved in it. The Lambert–Beer law is often used to characterize the intensity (I) of light transmitted through a uniform, homogenous medium as a function of the incident light (I_0). The intensity is given by

$$I = I_0 \, e^{-k' C \Delta x} \tag{4.7}$$

where k' is an extinction coefficient and Δx is the thickness of the medium. Light scattering in the sample is not considered. The optical density of the medium, is usually defined as

$$\text{O.D.} = \log_{10}(I_0/I) = \epsilon C \Delta x \tag{4.8}$$

where ϵ is the molar extinction coefficient ($\epsilon = k'/2.3030$). The optical density changes as the concentrations of different solutes are varied. A spectrum of light will be absorbed in some range of wavelengths, which would be distinct for each solute. Variations in light absorption are due to the vibrational and rotational movements for different chemical bonds in the molecules in the light path which absorb energy at different wavelengths.

Usually, a light intensity signal in the presence of a chemical species (I_C) is compared to a reference signal (I_R) which is measured when there is no chemical species present ($C = 0$) or at some other calibration level at a known concentration ($C = C_0$). Concentration can be monitored by comparing light spectrum changes with the reference spectrum. This comparison is usually non-linear. The measured concentration might be related to light intensities by a power law

$$C_{measured} = a + b(I_C/I_R)^n \tag{4.9}$$

or by a power series

$$C_{measured} = a + b(I_C/I_R) + c(I_C/I_R)^2 + \ldots \tag{4.10}$$

where a, b, and c are empirically determined constants, or by some other non-linear mathematical relationship.

There are many possible sources of error with the light absorption method, including light scattering and the turbidity of the medium, which places a physical limitation on the depth of light penetration that can be achieved. Care must be taken to minimize internal light reflections from surfaces within the sample chamber or in the optical components. When the measurement is intended for living tissues, the light power cannot be too great, to avoid thermal damage to the cells. Disturbances from external light sources can also influence the measurement. If the sampled region has non-uniform optical properties, or these properties change with time, it may be very difficult to interpret absorbance measurements. Also, water itself has high background absorption which may problematic for some molecules and in some light wavelength ranges, e.g., in the infrared around 975 nm.

4.2.2 Reflectance spectroscopy

There may be advantages to measuring the light reflected back from a surface or from deeper layers of a medium. Changes in the intensity of reflected light may accurately represent physical as well as chemical events that occur. The depth of light that penetrates into an ideal, non-scattering medium is given by

$$\Delta x = \frac{\lambda}{2\pi n_1 \sqrt{\sin^2\theta - (n_2/n_1)^2}} \tag{4.11}$$

when the relative refractive index of the transmission path (n_1) is greater than in the medium ($n_1 > n_2$). When illuminated at an oblique incidence angle θ, the incident light is totally reflected. It is possible to have an evanescent wave that only penetrates very short distances (<1000 Å) into the medium, allowing study

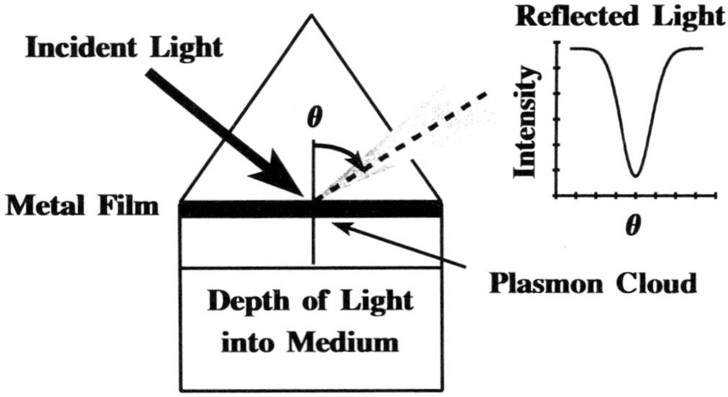

Figure 4.3 Penetration of light (evanescent wave) into medium. A plasmon cloud forms in the medium near the metal film on the waveguide, and the reflected light has a minimum at a certain angle (inset). Physical and chemical events within the evanescent wave can change the angle.

of localized surface phenomena. In other applications, it may be desirable to have the light penetrate as deeply as possible into the medium. In complex media, there may be multiple layers with different refractive indexes and coefficients for light absorption and scattering.

4.2.3 Evanescent waves and surface plasmon resonance

A plasmon has been defined as an oscillation of free electrons on the surface of a metal conductor (typically gold or silver). These electrons can be excited by applying an external electromagnetic field or by light at certain angles. A schematic drawing for a metal-film-coated glass prism waveguide is shown in figure 4.3. When a beam of light is applied at the appropriate oblique incident angle, total internal reflection will occur within the glass waveguide, and an evanescent wave will propagate out for a short distance, depending on the refractive indices for the two media as described by equation (4.11). Free electron clouds on the metal surface can resonate with the light and absorb some of its electromagnetic energy. This phenomenon is known as surface plasmon resonance. As a result, there is a decrease in the intensity of the reflected light, as shown in the inset of figure 4.3, that will occur at a specific resonance angle θ. If there are any changes in refractive index in the medium near the metal film surface, there will be a shift in the angle where this resonance occurs. If the reflected light intensity is measured by a photodetector array with sufficient spatial resolution, changes in the resonance angle can be followed from the shift in the location of the minimum light intensity. This information is useful for characterizing physical or chemical events in the medium near the metal surface.

4.2.4 Chemiluminescence

Luminescence is generally defined as the emission of light from atoms or molecules as a result of a transition from an electronically excited state to a lower-energy state. Some chemical reactions can create excited intermediates. The quantum yield of a luminescent reaction is defined as the ratio of the number of photons produced for a given number of molecules. There are a number of chemiluminescent reagents which can be used to detect specific biochemical reactions that occur, either from isolated test tube studies or directly from living plant and animal cells. In the process of completing these reactions, light is emitted. This phenomenon is often described as 'cold light' since photons are released without intermediate formation of heat. No external source of light is required to initiate this reaction, and the whole sample is involved.

A simplified model of chemiluminescence can be represented by two steps (activation and decay)

$$\mathrm{A} \overset{k_1}{\rightarrow} \mathrm{A}' \overset{k_2}{\rightarrow} h\nu \tag{4.12}$$

involving coupled first-order reactions, where k_1 is the rate of activation and k_2 is the rate of decay. The light intensity of the emitted signal is then given by

$$I = \text{constant} \times k_2 C_{A'} \tag{4.13}$$

where the scaling constant includes the amplifier gain and other factors associated with the measurement of the light signal. The time-dependent response for the light intensity is given by

$$I = \text{constant} \times C_{A_0} \frac{k_1 k_2}{k_2 - k_1} (\mathrm{e}^{-k_1 t} - \mathrm{e}^{-k_2 t}) \tag{4.14}$$

where C_{A_0} is the initial concentration of the chemiluminescent substrate. The maximum intensity (I_{max}) of emitted light is reached at the time

$$t_{max} = \frac{\ln(k_2/k_1)}{k_2 - k_1}. \tag{4.15}$$

I_{max} can be found by substituting t_{max} into equation (4.14).

Since chemiluminescent signals are normally very rapid, the photomultiplier and amplifying electronics must have a suitable bandwidth to faithfully record the event. The actual physical mixing of the substrate into the chemiluminescent reagent must also be very rapid. It is possible to manipulate the chemiluminescence decay rate either by changing the concentration of the enzyme, or by adding inhibitors of the enzyme. If the decay rate is slowed down, a quasi-steady-state analysis can be made, measuring the peak 'glow' (sustained light intensity) of the sample. This type of measurement is also less sensitive to the rate of mixing.

The firefly (*Photinus pyralis*) is a familiar example of chemiluminescence. The photochemical reaction requires the enzyme luciferase (EC 1.13.12.7). In the presence of Mg^{2+} and the substrate luciferin, ATP is utilized in the following reaction

$$\text{luciferin} + \text{ATP} + O_2 \rightarrow \text{oxyluciferin} + \text{AMP} + \text{PPi} + CO_2 + h\nu \quad (4.16)$$

catalyzed by luciferase. The light emission is yellow–green with a peak around 560 nm. The quantum yield is remarkably high for the firefly system, of the order of 0.9. There are many other natural luminescent proteins, including aequorin and obelin, from marine coelenterates. The enzyme luciferase is responsible for the luminescent properties of marine bacteria.

There are also synthetic compounds that emit light in the presence of H_2O_2, including luminol, lophine, lucigen, and derivatives of oxalates. Luminol reacts as follows

$$\text{luminol} + 2H_2O_2 + OH^- \rightarrow \text{3-aminophthalate} + N_2 + 3H_2O + h\nu \quad (4.17)$$

under alkaline conditions with an appropriate catalyst or co-oxidant. This reaction is much less efficient, with a quantum yield around 0.2.

4.2.5 Fluorescence

In contrast to chemiluminescence, an external source of light is required to initiate fluorescence. In general, a fluorophore initially in the ground state S_0 is excited by the light source to a new state S', as given by

$$S_0 + h\nu \rightarrow S'. \quad (4.18)$$

The absorption of a quantum of energy results in the unpairing of electrons in the ground state, producing an excited electronic state. Unpaired electrons will move from the ground state into either the first or second singlet state. There are a number of pathways for the excited electrons to move back to the ground state, as shown schematically by the Jablonski energy diagrams in figure 4.4. Three pathways are shown in figure 4.4(a) which do not result in any radiation. After excitation to the first singlet level (left), there is some energy loss (internal conversion) within the singlet state then a crossover with no loss in energy to a triplet state. After some energy loss in the triplet state, it is possible to cross into the ground state and the higher vibrational energy can quickly return to the ground state through collisions with other molecules. In the second pathway (middle), electrons move into the second singlet state and lose energy, then cross to the first singlet state and lose more energy, finally crossing to the ground state and returning to S_0. For the third pathway (right), the excited electrons follow the same path, then to the first triplet state before crossing to the ground state and returning to S_0.

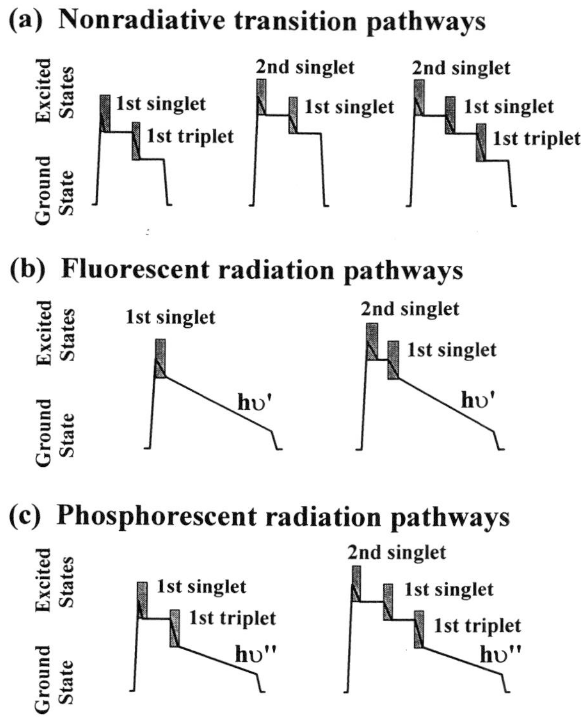

Figure 4.4 Jablonski-type energy diagrams for possible excited energy states when light interacts with matter. (a) Three possible transition pathways for return to ground state without radiation. (b) Two possible transition pathways with fluorescent light emission as final step on return to ground state. (c) Two possible transition pathways with phosphorescent light emission as final step on return to ground state.

As shown schematically by the Jablonski energy diagrams in figure 4.4(b) and (c), there are some molecules that allow the return of excited electrons to S_0 by radiating energy. The fluorescent pathway (figure 4.4(b)) returns to the ground state with the emission of light according to

$$S' \rightarrow S_0 + h\nu_1 \qquad (4.19)$$

where ν_1 is the frequency, which is generally lower (longer wavelength) than the excitation light source. The fluorescent light intensity I_f is given by

$$I_f = A[S'] \qquad (4.20)$$

where A is the Einstein coefficient for spontaneous emission. The fluorescent light is usually much weaker than the excitation signal. The path of the resulting fluorescent light through the medium is assumed to obey the Lambert–Beer law. Parker's law has also been used to represent the fluorescent light intensity I_f

$$I_f = \text{constant} \times C\epsilon\phi_f \Delta x \tag{4.21}$$

from a sample with thickness Δx, where ϵ is the extinction coefficient and ϕ_f is the quantum yield of the fluorescent dye. The energy from the excitation light source must not be destructive to the biological sample or cause photobleaching of the fluorescent dye in the sample. Fluorescent imaging systems generally employ pulsed light excitation using lasers or mechanical light chopping devices. The lifetime of the fluorescent signal, assuming a monoexponential decay, is defined as

$$\tau = (1 - 1/\text{e}) \tag{4.22}$$

and is typically <10 nanoseconds.

A number of fluorescent indicator dyes (BAPTA, QUIN1, QUIN2, bis-azo dyes) were developed in the late 1970s and early 1980s for following intracellular Ca^{2+} by fluorescence microscopy, as reviewed by Tsien [5]. More recently, other fluorescent dyes for Ca^{2+} including fura-2 and indo-1 have been used. Most Ca^{2+} imaging systems using CCD cameras can acquire fluorescent images every 33 milliseconds, although it may be necessary to average several frames to improve the signal to noise ratio. Using fluorescent probes with confocal laser microscopy, even more detailed three-dimensional images of intracellular concentrations and their variation with time are now possible.

There are technical difficulties in acquiring and interpreting fluorescent measurements. One problem is requirement for a relatively large excitation signal, which may have multiple peaks at other wavelengths that might be close to the wavelength of the emission. The emission amplitude is often very small, and can also have multiple resonant peaks at other wavelengths. Very small differences may exist between the major wavelength bands, which could be as close together as 30 nm. New techniques have been developed using multiple excitation wavelengths or, as discussed later in this section, by modulating the excitation light source.

In the past, fluorescence measurements were limited to the UV and visible portions of the spectrum. Mercury arc lamps, which have strong peaks in the UV range, have been effectively used for some fluorophores. Xenon arc lamps, which have more uniform spectra in the visible and UV ranges, may not be as effective. With the development of solid state lasers such as Nd:YAG lasers or new tuneable dye lasers, fluorescent measurements are now being extended out into the IR range where there is less possibility for interference with the emitted light. New fluorogenic dyes that emit light with wavelengths in the range from 0.7 to 1.5 μm are now being investigated. Since photomultiplier tubes are not as useful for wavelengths >1 μm, solid state detectors are more commonly used. The recording photodetector must have a wide dynamic response capable of

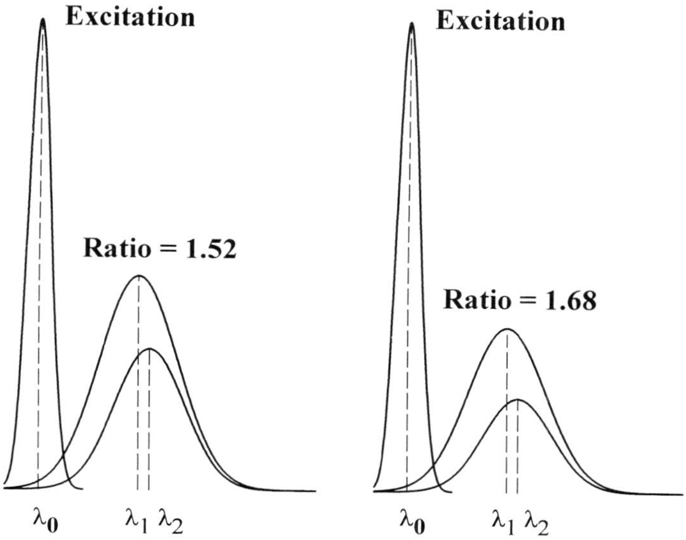

Figure 4.5 Principle of fluorescence ratio imaging. Two fluorophores emit light with wavelengths λ_1, λ_2 after excitation by a laser or other light source with wavelength λ_0. One of the fluorophores varies in light intensity with the concentration of the target analyte. However, the intensity of light emission can decrease with time due to photobleaching, as illustrated by the difference in emission signals from left to right. If the second fluorophore is not sensitive to the analyte, and rates of photobleaching are similar, the ratio of intensities can be used to derive information on the changes in analyte concentration.

quickly resolving huge differences in light intensity, on the order of a million times (six orders of magnitude) or more.

The maximum light intensity of the fluorescent signal and its subsequent decay with time will depend on the concentration of fluorophore present, and can also depend on the concentrations of reactants or coreactants. Heterogenous distributions can cause difficulties in quantifying the fluorescent signal. Other complications in interpreting fluorescent images include additional signals from natural autofluorescent biochemical processes, for example from NADH, that occur in plant and animal cells. Photobleaching can attenuate the signal, and the physical environment (temperature, ions) surrounding the fluorophore can alter its emission characteristics. Of course, the sensitivity and resolution of the detector and the optical properties of the microscope are also factors.

4.2.6 Ratio fluorescence microscopy

By using two different fluorescent indicators, some of the difficulties outlined above can be circumvented and quantitative information can be derived. Ratios

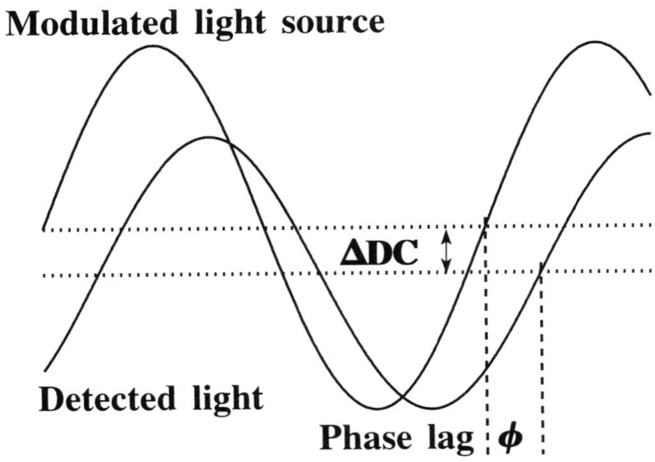

Figure 4.6 Principle of phase-resolved fluorescence. If the excitation source is modulated, the time delay in fluorescence causes a phase lag in the detected light signal, depending on the wavelength of the emitted light. With appropriate signal processing, it is possible to detect phase lags for multiple fluorescent indicators.

can be measured from the resulting intensities of fluorescent light at two different emission wavelengths, either by sequentially exciting the sample with a single wavelength or with two different wavelengths. A schematic drawing of two sequential fluorescence measurements is shown in figure 4.6 for two fluorophores with emissions at λ_1 and λ_2 for an excitation at λ_0. In this illustration, the intensities for both fluorophores have diminished in the second measurement (right) due to photobleaching. If the intensity data had been used for either fluorophore, a decrease in analyte would be calculated. However, the ratio of the two fluorophore intensities, as indicated in the figure, shows an increase. Measurements performed for a wide range of fluorophore concentrations and conditions should give the same ratio for the specific chemical species of interest, and provide accurate, quantitative results. Dunn *et al* [6] reviewed applications of this technique for cell biology. For example, to measure pH in living cells, the two fluorophores rhodamine (not sensitive to pH) and fluorescein (sensitive to pH) can be used. By conjugating transferrin to both fluorophores, they will be taken up within the endosomes of the cells under study. Since fluorescein is more susceptible to photobleaching than rhodamine, care must be taken to minimize the light intensity and duration of the excitation. The emission ratio of rhodamine to fluorescein increases as the environment becomes more acidic. In the opinion of Dunn *et al* [6], further advances in fluorescent microscopy require the development of new fluorophores. They feel that the optics, detection, and imaging technologies are already in place, although certainly improvements in these technologies will occur.

4.2.7 Phase-resolved fluorescence

Very fast fluorescence signals can be examined by a phase-resolved frequency domain technique. A high-frequency modulated light source is used to excite the sample according to

$$E_{ex}(t) = I_{ex}(1 + M_{ex} \sin \omega t) \tag{4.23}$$

where I_{ex} is the average light intensity, M_{ex} is the depth of modulation, and ω is the angular frequency. Frequencies are typically in the MHz to GHz range. The time-varying fluorescence signal is also sinusoidal, given by

$$F_f(t) = I_f\,[1 + M_{ex} M_d \sin(\omega t - \phi)] \tag{4.24}$$

where I_f is the average fluorescence intensity, M_d is a demodulation factor, and ϕ is the angle of the phase shift between the excitation source and the fluorescence signal. The relationship between the modulated light source and the detected fluorescence signal for a single fluorophore is shown in figure 4.6. There may also be a DC shift between the two waveforms. The phase shift ϕ can be detected with a lock-in amplifier, which produces a phase-resolved signal

$$S(\phi_d) = I_f M_{ex} M_d \cos(\phi_d - \phi) \tag{4.25}$$

where ϕ_d is the adjustable phase angle of the detector. The signal is zero when there is a 90° difference between the phase shift and detector angles.

Another feature of the phase-resolved method is that contributions for individual fluorophores can be examined if there is adequate separation in the different fluorescence excited lifetimes. When there are n fluorescent species present, the resulting signal from the lock-in amplifier

$$S(\phi_d) = \sum_{i=1}^{n} I_{f,i} M_{ex} M_{d,i} \cos(\phi_d - \phi_i) \tag{4.26}$$

contains information on all n species, where the average intensity ($I_{f,i}$) and demodulation factor ($M_{d,i}$) for each species is included in the amplitude term, and the phase shift (ϕ_i) is different for each fluorophore.

4.2.8 Phosphorescence

When a phosphorescent material is illuminated with light, light is absorbed, exciting it into a higher-energy state, as shown by the Jablonski diagram in figure 4.4(c). Since internal conversion can occur through the first singlet to the first triplet, the phosphorescent pathway is typically much slower than the fluorescence pathway, which does not require this additional step. After reaching the first triplet state, the energy must then be transferred by light emission, or by other non-radiative pathways. The time course of light emission can be accelerated when certain chemical species are present, through the process known as quenching. Practical applications are possible when the lifetime τ of the phosphorescent material in the presence of a given concentration C of the quenching chemical species is long enough to be measured.

The theory for the time course of phosphorescent or fluorescent light emission was described by Stern and Volmer [7]. The relationship is given by

$$\frac{\tau_0}{\tau} = 1 + k_q \tau_0 C \tag{4.27}$$

where τ is the lifetime at a given quencher concentration C, τ_0 is the lifetime in the absence of the quencher ($C = 0$), and the quenching parameter k_q is related to the combined rates of diffusional transport for the quenching agent and the phosphorescent or fluorescent material. Usually, the quenching agent is much more diffusible than the phosphorescent or fluorescent species, although the time response could be limited by the diffusion rate of the quenching species. The relative light intensity is

$$\frac{I_0}{I} = 1 + \left(K_{eq} + k_q \tau_0\right) C + K_{eq} k_q \tau_0 C^2 \tag{4.28}$$

where I_0 is the phosphorescent or fluorescent light intensity in the absence of quencher, and K_{eq} is the association constant for binding of the quencher to the phosphorescent or fluorescent species. When $k_q \tau_0 \gg K_{eq}$, or if quenching is purely dynamic ($K_{eq} = 0$), then equation (4.28) can be simplified to

$$\frac{I_0}{I} = \frac{\tau_0}{\tau} = 1 + k_q \tau_0 C \tag{4.29}$$

and both the lifetime and intensity are inversely related to the quencher concentration.

If a biosensor membrane of thickness Δx and diffusivity D is controlling the transport of quenching agent, then

$$k_q = \frac{2D}{\Delta x^2} \tag{4.30}$$

can be substituted into equations (4.27)–(4.29) as an approximation for the diffusion-limited process. To have the optimum response time, the membrane should be freely permeable to the quenching agent and should be as thin as possible.

A digital imaging method using phosphorescent lifetime measurements is illustrated in figure 4.7. Using a high-speed CCD, it is possible to capture a sequence of digital images after laser-induced excitation and examine the change in phosphorescent light intensity from each pixel. A schematic drawing

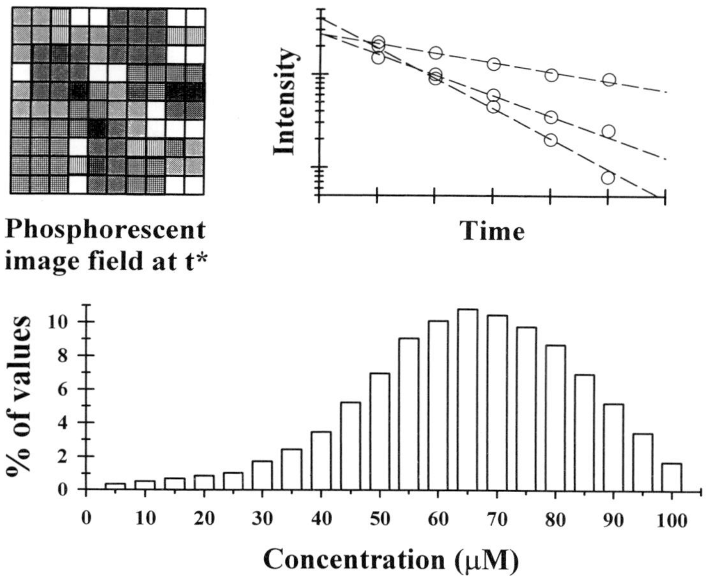

Figure 4.7 Phosphorescent lifetime measurements from digital imaging. By capturing a sequence of digital images after excitation, such as the field illustrated for time t^* in the top left corner, the change in phosphorescent light intensity can be used to calculate the concentration of a quencher. The semilog plots for the time course of intensity changes in three pixels from five sequential image frames are shown in the top right corner, using the Stern–Volmer equation (4.29) (dashed lines). In the bottom plot, a histogram of all computed concentration values in the image field is shown.

of one image is shown in the top left corner, with plots of the light intensity changes for three pixels from five sequential frames. From the slope of the semilog plot, which would be linear if equation (4.29) were valid, the lifetime τ for each pixel in the image sequence can be determined, and the quencher concentration calculated. This measurement is independent of the absolute value of the phosphorescent light intensity. With digital imaging, enormous numbers of data can be generated, with highly detailed spatial maps of the concentration. This information could be distilled as a histogram, as shown at the bottom of figure 4.7. Temporal changes can be quantified, with a separate histogram for each captured sequence.

Often, it has been observed that I_0/I is not linearly related to the inverse lifetime (equation (4.29)), and is often downward curved for many luminescent systems. This complicates the calibration of optical biosensors based on phosphorescence or fluorescence quenching. An empirical power law model

$$\frac{I_0}{I} = 1 + k_q \tau_0 C^n \tag{4.31}$$

has been used to correct for non-linearity.

Another mathematical model based on the relative contributions of static and dynamic quenching mechanisms in heterogeneous systems was derived by Carraway *et al* [8]. The light intensity in their model is described by a series of Stern–Volmer relationships

$$\frac{I_0}{I} = \left[\sum_{i=1}^{n} \frac{f_i}{1 + k_{qi}\tau_i C}\right]^{-1} \tag{4.32}$$

with individual fractions

$$f_{0i} = \frac{I_{0i}}{\sum_{i=1}^{n} I_{0i}} = \frac{\alpha_i \tau_i}{\sum_{i=1}^{n} \alpha_i \tau_i} \tag{4.33}$$

where α_i are weighting constants. A weighted mean lifetime

$$\tau_M = \frac{\sum_{i=1}^{n} \alpha_i \tau_i}{\sum_{i=1}^{n} \alpha_i} \tag{4.34}$$

allows comparison of lifetime and intensity measurements. A weighted mean lifetime in the absence of quencher τ_{M0} is found by substituting τ_{0i} for τ_i in equation (4.34). Computer simulations have confirmed the usefulness of the model for a wide range of heterogeneous systems.

4.3 APPLICATIONS FOR PHOTOMETRIC TRANSDUCERS

An exhaustive survey of all photometric transducers used for biological or medical applications would be a major undertaking, beyond the scope of this chapter. Instead, a selective survey is presented which illustrates the general optical techniques that have been outlined in the preceding sections.

4.3.1 Optodes

In the early 1970s, Lübbers and Opitz [9] coined the term ‘optodes’ as a description for photometric transducers, analogous to the use of the term ‘electrodes’ for electrochemical transducers. Often a single optical fiber is used. Bifurcated designs, with one fiber for excitation and the second for light detection, are also used. General designs for optodes using fiber optics are illustrated in figure 4.8. The most common design has a single indicator dye

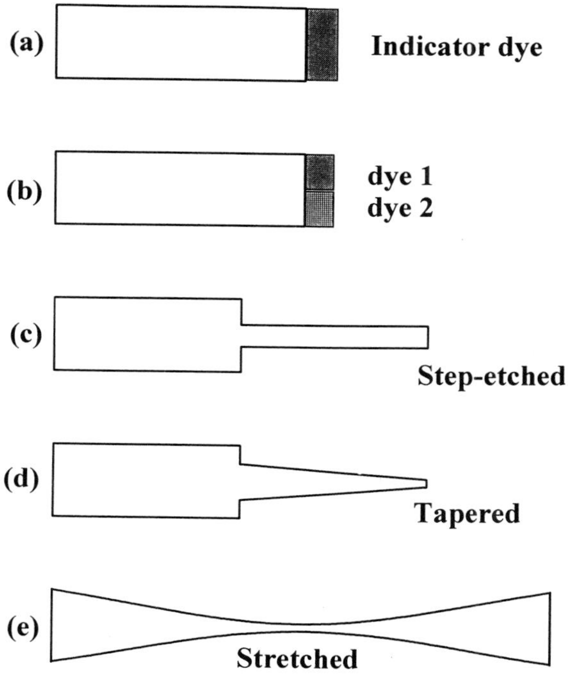

Figure 4.8 Common optode designs using fiber optics. (a) Chamber or membrane with a single bioluminescent, fluorescent, or phosphorescent indicator dye at the distal end of an optical fiber. (b) Differential or dual-analyte design with two separate indicator dyes. Evanescent waves are produced using (c) step-etched, (d) tapered, or (e) heat-stretched designs.

(a) which is either contained in a chamber near the tip, or within a membrane attached to the fiber. The membrane or chamber wall must be readily permeable to the analyte of interest. Differential or dual-analyte designs can use two different indicator dyes, or modify the second detection site with enzymes or other biological recognition methods. Either reflected or transmitted light can be used for detection. Recently, there has been interest in using evanescent waves produced with fiber optics. After removing the plastic cladding, the glass core can be altered by chemical etching with hydrofluoric acid. The geometry can be changed, either as an abrupt reduction in diameter (c) or with a more gentle taper (d). Another method is to heat stretch the glass fiber (e). When the dimensions of the optical probe are reduced, it is possible to have a better match between the waveguide properties and the refractive index of the measurement medium. Optimally designed waveguides with smaller dimensions are able to detect a greater signal from evanescent waves at the surface.

There has been considerable effort to develop fiber-optic-based O_2 optodes. Surgi [10] reviewed some of the indicator molecules which have been used to detect O_2. These include perylene dibutyrate, decacyclene, pyrene, 5-(4-bromo-1-naphthoyl)pentyl-trimethylammonium bromide, N-methylacridone, and Pt(II) or Ru(II) derivatives of α-diimine ligands. O_2 optodes based on N-methylacridone-coated silica on Teflon^{TM} membranes were described by Surgi [10]. The indicator was excited at 400 nm and fluorescence emission monitored at 460 nm. Fluorescence quenching was found to have a biexponential character, with a very short component (about 6.5 nanoseconds) and a much longer component (about 50 microseconds). The longer component is more suitable for practical measurements based on quenching lifetimes. The 90% response time of the optode was about 12 seconds when switching from N_2 to O_2, but was longer (about 42 seconds) when the opposite change was made. Peterson *et al* [11] also investigated fluorescent fiber optic probes for O_2. Seventy different fluorescent dyes were considered, but the best characteristics were determined to be for the non-toxic dye perylene dibutyrate (solvent green 5). The dye was encapsulated inside a rigid, gas-permeable tube of porous polypropylene with approximately 25 μm thick walls. Two 0.25 mm diameter optical fibers were placed in the dye chamber, and shielded from ambient light. One fiber was used to transmit the excitation light from a 60 W deuterium (blue) light source and the other fiber picked up the fluorescent emission (green light) from the dye. The pO_2 could be related to the ratios of the blue and green light intensities (I) by

$$pO_2 = \text{gain}\left(I_{blue}/I_{green} - 1\right)^m \tag{4.35}$$

where an exponent (m) was added to empirically correct the Stern–Volmer relationship (similar to equation (4.31)) for non-ideal curvature. The step response to 98% completion was about 2 min. The sensitivity of the dye declined by about 3% per day during dry storage, but was much less (0.1% per day) when stored in water. The probe was successfully tested in blood by inserting it into the carotid artery of a sheep.

Wolthuis *et al* [12] described an optode for O_2 based on a viologen compound in a 20% dipyridyl and 40% glycol film spin coated on glass. The viologen was excited by UV light, coupled though optical fibers. The absorption was measured with a photodiode. An advantage of this design is that the transducer can be sterilized. Gamma radiation (2.5 mRad) sterilization caused only a 20% loss in sensitivity. Nitrous oxide and halogenated anesthetic gases had no effect on the optode. This O_2 optode was envisioned as a disposable device and preliminary tests indicated that it had a lifetime capable of over 250 measurements.

Berndt and Lakowicz [13] used a violet light (454 nm) emitting solid state electroluminescent lamp (ELL) for an O_2 optode. This inexpensive light source can provide more power than blue LEDs, although aging effects were observed. They measured the quenching lifetime of tris(4, 7-diphenyl-1, 10-phenylanthroline)ruthenium(II) chloride embedded in a silicon matrix. A 10 kHz square wave was used to drive the ELL and modulate its light intensity. The emitted light was detected by a photodiode and the phase shift induced by the fluorescence was measured, as discussed earlier (equations (4.24)–(4.26)). The phase shift was independent of the light intensity and depended only on quenching lifetime. In the absence of O_2, a 40° phase shift was seen. In room air, an 8° shift was measured. They concluded that the ELL device could be very useful, especially if the lamp output could be stabilized to keep the light intensity constant as it ages. Under optimal conditions, they estimated that the ELL optode should be able to measure O_2 quenching decay times as short as 30 nanoseconds.

Phosphorescence quenching has also been used to measure O_2, and the technique has been used for O_2 optodes. Carraway *et al* [14] investigated the photochemistry of an O_2 sensor using a luminescent transition-metal complex Ru(4, 5-diphenyl-1, 10-phenanthroline)$_3^{2+}$ in silicon rubber. The Stern–Volmer plots for the data obtained with this sensor were found to be non-linear. Static quenching was assessed to be a minor factor. By fitting both the intensity and lifetime data, they found that equation (4.32) was more accurate than the empirical equation (4.31). Their data were consistent with a two-site model ($n = 2$ in equation (4.32)), with the second site much less reactive than the first. A two-site Stern–Volmer model was also found to describe O_2 quenching of pyrene in O_2 optodes designed by Xu *et al* [15] using polymer films that contained pyrene. They found that the polymers and cross-linking agents used to prepare the pyrene films had an effect on the O_2 quenching. They often found a deviation between intensity and lifetime data at higher O_2, above normal atmospheric levels (~21%). This suggests that more than two sites are necessary to interpret the data, due to the heterogeneity of quenching and the fact that the rapid components should be included in the mean lifetime used for quantifying the measurement.

Other types of optode have been designed, often using fluorescent dyes for detecting different analytes. A pH optode was described by Peterson *et al* [16, 17], who coupled a bifurcated fiber optic cable to a small cavity containing a pH-sensitive dye. Using the Lambert–Beer law, the pH can be related to the concentration of the dye (C_{dye}) and its buffering capacity (pK) by

$$\mathrm{pH} = \mathrm{p}K - \ln\left[\epsilon C_{dye} L / \ln\left(I_0/I_m\right) - 1\right] \qquad (4.36)$$

where I_0 is the incident light, I_m is the measured light, and is the extinction coefficient for a given light pathlength L. Besar *et al* [18] developed a LED-based fiber optic system using either phenol red or BHD universal indicator. The respective measuring and reference wavelengths were 565 and 635 nm for phenol red and 635 and 930 nm for the universal dye. Ultrabright red or green LEDs were used for the reference light sources with a photodiode for the detector. Accuracies between 0.015 and 0.03 pH units were achieved in the pH range from 5.5 to 8.5 with this optode. Hale and Payne [19] used a tapered optic fiber design (figure 4.8(e)) for a pH sensor. A single-mode optic fiber with an 80 μm diameter was heat stretched to a final diameter around 5 μm, with the internal core around 2 μm. The tapered region was coated with fluorescein and a photodiode was used to measure the fluorescent emission ($\lambda = 526$ nm) from the evanescent field excited by an argon laser ($\lambda = 488$ nm) as pH was varied. The pH sensitivity was improved with the tapered design. Based on theoretical considerations, they calculated that the tapered design should collect about 30 times more fluorescent signal than a polished multimode fiber with the same length.

Kawabata *et al* [20] described a pCO_2 sensor using fluorescein dye as a pH indicator, based on the pH change with formation of bicarbonate ions. A single-fiber design with the tip ground at an angle was used. A xenon arc lamp with wavelength adjusted to 490 nm was used as the excitation signal and a photomultipler tube as the detector, with a mechanical light chopper modulating the light signals at 80 Hz. The fluorescein was immobilized on the fiber optic tip in a mixture of high- and low-weight poly(ethylene glycol). The pH of the polymer was adjusted to 7, and the total thickness of the dried polymer was about 10 μm. A particular advantage of this design was that it did not need to use an aqueous buffer, which could evaporate and alter the optode sensitivity. Luo and Walt [21] described a penicillin optode based on fluorescence using immobilized penicillinase (EC 3.5.2.6, type I from *Bacillus cereus*) and with an ethyl butyrate optode using immobilized esterase (EC 3.1.1.1).

A dual-analyte fiber optic biosensor for O_2 and glucose was developed by Li and Walt [22] based on O_2 quenching of a phosphorescent ruthenium dye. Excitation was at $\lambda = 480$ nm, with fluorescent emission captured by a CCD camera. A relatively large (350 μm diameter) imaging fiber with 6000 elements was modified by attaching two separate drops of ruthenium dye encapsulated in poly(hydroxyethyl methacrylate) polymer (HEMA). The ruthenium dye allowed measurements of O_2 in both encapsulated drops, which were approximately 50 μm in diameter. A two-site Stern–Volmer quenching model (equation (4.32) with $n = 2$) was used to determine O_2 concentration from measurements of fluorescence intensity. One of the drops had the enzyme glucose oxidase (EC 1.1.3.4) in the HEMA membrane. Glucose oxidase catalyzes the oxidation of D-glucose according to the reaction

$$\text{D-glucose} + O_2 + H_2O \xrightarrow{\text{glucose oxidase}} H_2O_2 + \text{D-gluconic acid}. \tag{4.37}$$

In the drop without glucose oxidase, the O_2 level is in equilibrium with the medium. In the drop encapsulated with glucose oxidase, O_2 is lower, depending on the glucose concentration. The relationship between O_2 and glucose is non-linear since the reaction follows typical enzyme kinetics (equation (4.37)). The O_2 level in the glucose oxidase drop also depends on the diffusion properties of the membrane. The differential sensor design would allow simultaneous measurements of O_2 and glucose as both varied in a biological system.

4.3.2 Ultraminiature optical probes

Ultraminiaturized fiber optic sensors under 100 μm have only recently been fabricated. Tan *et al* [23, 24] developed a submicrometer optical fiber tip by pulling out silica fibers on a micropipette puller using a 25 W CO_2 infrared laser as a heat source. Tips as small as 0.1 μm could be reliably fabricated. After pulling, the tips were sputtered with aluminum in a vacuum chamber. This fabrication technique leaves a very small aperture at the tip, which can then be used as a near-field optical device (discussed in the next section).

The aluminum-coated tips were modified with a fluorescent dye (acryloylfluorescein) for use as an optical pH sensor. First the tips were silanized to activate the glass surface for attaching the fluorophore. Two methods of copolymerization were tested. Thermal copolymerization could not be easily controlled for reproducible coating thicknesses and was wasteful. Photopolymerization was found to be more practical and less wasteful. The tips were first sensitized by soaking in a benzophenone–cyclohexane solution for about 15 minutes. Then the tip was immersed at a controlled depth into the polymer solution and laser light from either a He:Cd laser (λ 442 nm) or an argon ion laser (λ 488 nm) was transmitted through the optical fiber. The photochemical reaction caused copolymerization only at the tip where the light was transmitted, and covalently immobilized the fluorophore on the glass tip. The thickness of the polymer coating depended on the light intensity, exposure time, temperature generated at the tip, and concentrations of the reactants in the polymer and fluorophore mixture.

Tests were conducted by inserting the ultraminiature optical pH probe into small pores of porous polycarbonate membranes which had been presoaked in nine different buffer solutions between pH 4 and 9. Tests were also conducted with different laser excitation intensities for laser powers between 0.3 and 30 mW. Higher fluorescent signals were elicited with the higher excitation powers as expected. However, there was also much greater photobleaching at higher powers, which was dependent on the local heating and the amount of dissolved oxygen in the sample. The local heating was apparently much greater when the probe was in air. When the tip was immersed in liquid, the loss in sensitivity at the highest laser power was about 10% over a 40 minute period. At lower laser powers, Tan *et al* [23, 24] estimated that tens of thousands of measurements could be made without significant loss of sensitivity due to photobleaching. Responses times ranged between 100 and 500 milliseconds, depending on diffusion and mixing in the test sample rather than through the very thin polymer and fluorophore coating. These response times are much faster than other pH optodes which have much larger dimensions and thicker membranes.

With such small amounts of fluorophore, photobleaching of individual molecules could have significant effects. To avoid these difficulties, Tan *et al* [23, 24] utilized a fluorescence intensity ratio technique by exciting the fluorophore at two different wavelengths using a 0.3 W argon ion laser ($\lambda = 488$ nm) and

30 mW He:Cd laser ($\lambda = 442$ nm) for the two excitation sources. The laser beams were optically coupled to the distal end of the optical fiber. The light flux measured at the tip of a 0.2 μm diameter ultraminiature optical pH probe was $\sim 10^{12}$ photons s^{-1}. When the tip was observed though a microscope, only a small, unresolved spot of light could be observed since the tip size was less than the diffraction limit of optical microscopy. Fluorescent spectra were obtained by imaging the tip through a commercial inverted frame fluorescence microscope at 600 times magnification. When excited at 442 nm, the fluorescence intensity decreased with higher pH. When excited at 488 nm, there was an opposite effect, with the fluorescence intensity increasing with higher pH. An isobestic point for the two wavelengths was found around 464 nm. These opposing changes in fluorescence intensity resulted in an enhancement factor of about 140% in the pH range from 7 to 8. Using the two laser excitation sources, the fluorescence ratios at 540/490 nm and at 540/610 nm were found to be large enough for measuring pH in physiologically relevant ranges. Tan *et al* [23, 24] tested the device in biological samples including blood cells, frog cells, and rat embryos at different stages of development. The ultraminiature optical sensor is capable of making intracellular pH measurements. Other types of fluorescent modification should be possible for detecting other analytes with excellent spatial resolution. A limitation of the design is that the tip must be visually imaged, so it would be restricted to thin biological samples or to penetrations into isolated cells in culture.

4.3.3 Near-field optical microscopy

Conventional microscopy has been limited to a maximum resolution of half of the wavelength of light (the Abbé diffraction limit $\lambda/2$), which is approximately 200 nm for visible light. This restriction has arisen with the traditional use of optical lenses, which must be placed at a distance from the subject. This distance is typically several orders of magnitude greater than the wavelength of visible light (a far-field measurement). Recently, the technique of scanning tunneling microscopy has been modified for optical microscopy, using extremely fine microprobes that are capable of making near-field measurements. Betzig and Trautman [25] describe the basic principles of the technique, which requires scanning a subject in very close proximity to the surface using a microprobe with an extremely small aperture (dimension $< \lambda$). A schematic drawing of a near-field optical probe is shown in figure 4.9, used for measurements either from a surface (a) or from an individual molecule (b). The probe is coated with a suitable material to prevent losses from the upper part of the probe and direct light to the aperture.

In order to use this technique, a mechanical device with exceptionally good stability must be used to maintain the probe position. A small-amplitude ($\sim$10 nm), high-frequency ($\sim$30 to 130 kHz) feedback resonant oscillation method has been successfully used to keep the tip positioned on a irregular

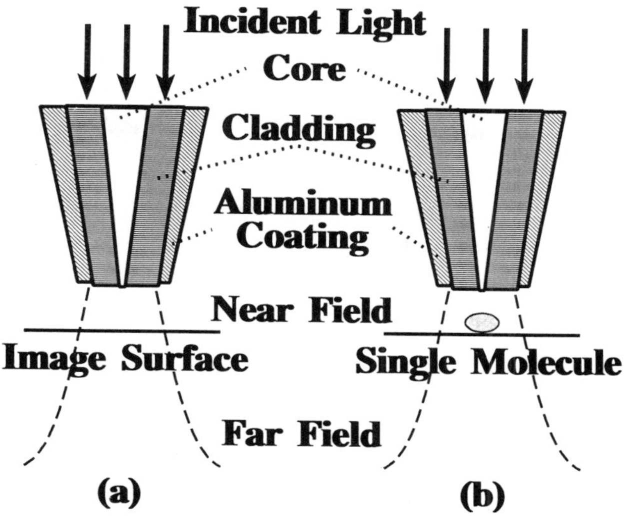

Figure 4.9 Schematic drawing of near-field optical probes. By pulling out an optical fiber and coating with aluminum, an optical probe with a very small aperture (<100 nm) is created. The probe can detect fluorescence or phosphorescence from objects in the near field close to the aperture, such as a thin film or flat surface (a) or from a labeled molecule (b). The probe is relatively insensitive to light in the medium in the far field further from the tip. The excitation light source can be transmitted through the probe or externally.

surface. Feedback positioning is based on the phase shift of the resonant oscillation as the tip comes into contact with the sample. Vertical spatial resolution can be ~1 nm. The optical probe can be used in the transmission mode, provided that the material around the tip is relatively transparent and does not strongly absorb the transmitted light. In the reflecting mode, the probe can detect very low fluorescent signals. The excitation light can be delivered through the probe, with appropriate time gating during detection to eliminate initial effects of the excitation pulse. The probe can be more efficient in collecting light than a far-field fluorescent measurement since the tip is so close to the fluorescent source. Another advantage is that the probe is relatively insensitive to background light in the far-field range. However, measurements from single molecules of the fluorescent dye sulforhodamine by Xie and Dunn [26] and by Ambrose *et al* [27] have shown that the position of the probe over the molecule is critical, since the metallic coating (aluminum) will quench the fluorescence. The time course of fluorescent decay of this dye was measured, with time scales in nanoseconds. Ambrose *et al* [27] found monoexponential decay rates with lifetimes between 3 and 4 nanoseconds, that were relatively independent of the excitation powers used (ranging from ~2 to 40 nW). Trautman *et al* [28] used a near-field optical probe for single-molecule spectroscopy of a fluorescent dye (1, 1′-dioctadecyl-3, 3, 3′, 3′-tetramethylindocarbocyanine) on a thin film of polymethylmethacrylate. In this study, they were able to obtain separate spectra from two molecules only ~150 nm apart, without interaction.

4.3.4 Optical immunosensors

Antibodies are Y-shaped protein molecules produced by the immune system (see chapter 7 for further discussion). They are composed of equal numbers of heavy and light polypeptide amino acid chains held together with disulfide bonds. These highly specialized proteins have a three-dimensional structure which is able to recognize and bind only certain types of antigen molecule. The majority of mammalian antibodies are in the immuno-gamma-globulin (IgG) class, with molecular weights around 160 000 Dalton. If an analyte can elicit an immune response, biosensors can be developed based on the antibodies generated by the immunogenic response. Monoclonal antibodies, which have a homogenous population, can be derived from animals exposed to the antigen. Polyclonal antibodies, a heterogenous mixture of immunoglobulins elicited from the immune response, can be even more effective by recognizing several different sites. Biotechnological advances in cloning and culturing hybrid cells have permitted the production of monoclonal and polyclonal antibodies which are specific to a chosen antigen or a group of closely related chemical species. Some antibodies can be manufactured in relatively large quantities without great expense by exposing cultured cells or microbes to an antigen. Genetically engineered antibodies are being developed, using site-directed mutagenesis to create new, synthetic antibodies with antigen specificities that are not normally possible. There may be an infinite number of possible immunosensors using antibodies that detect very specific antigens.

Immunosensors can be divided into two general categories: nonlabeled types which rely on some change in physical properties, and labeled immunoassays which rely on the direct detection of a specific label. Immunosensors can be developed by incorporating either the antigen or the antibody on the sensor surface, although the latter design is used most often. Since the immunocomplex reactions that occur with antigens and antibodies are highly specific, extremely precise sensitivities at the molecular level are possible. However, it is still possible to have non-specific interactions. These interactions, which can include adsorption of proteins, changes in fluid viscosity and density, or other physical changes near the surface, may be important for some types of transducer and less important for others. Another problem is that some immunoreactions are not reversible, so that only a single immunoassay is possible. Considerable research efforts have been directed towards the development of renewable antibody surfaces or for maintaining reagent concentrations so that repetitive immunoassays can be made without loss of sensitivity.

Many types of fiber optic or other optical waveguide have been used for immunosensors employing both single-fiber and bifurcated designs. The light source is delivered to the tip and the transmitted or reflected light is picked up by a light detector. Arnold [29] reviewed some of the principles for both enzyme- and antibody-based measurements with fiber optics, and the many chemical modifications for attaching the enzymes or antibodies on the surfaces on which the reactions take place. One design has the antibodies suspended in a microcavity behind a membrane that is permeable to the antigen. Optical changes that occur in the microcavity can be monitored. Other designs bind the antibody or antigen to the membrane surface where direct contact is made with the sample. Various optical techniques have been investigated for detecting changes in optical properties that result from the interaction of antibody and antigen on the surface. The measurements can be complicated by scattering and non-specific absorption processes.

Sutherland *et al* [30] measured changes in light intensity due to light scattering effects on a glass slide coated with antibodies. The slide was illuminated with light at a wavelength around 440 nm. In essence, the glass slide was acting as a light waveguide, with the immunoreaction on the surface causing the measured changes in light scattering. In another design, Sutherland *et al* [31] used a quartz optical waveguide with antibody covalently attached to the surface to study antibody binding of methotrexate. Light was transmitted through the waveguide at a wavelength of 310 nm, which was estimated to penetrate approximately 90 nm into the surface coating. With increased methotrexate binding, light was absorbed at a wavelength around 300 nm, primarily due to a benzoyl group in the molecule. The change in absorbance was related to methotrexate concentration, with a lower detection limit of 0.26 μM.

Bioluminescent conjugates of antibodies or antigens with enzymes such as luciferase have been developed which emit light during reaction. Jablonski [32] covalently crosslinked a marine bacterial luciferase from *V. harveyi* with either *Staphylococcus aureus* protein A, or with anti-human IgG. Both conjugates retained their light emitting ability. This bioluminescent immunoassay was compared with a radioimmunoassay method used commercially for detecting human rubella immunity, with favorable results. Other chemiluminescent labels, such as luminol, are described by Aizawa [33].

Fluorescent probes can be used to label antibodies or antigens. After a fluorescent label attaches to an immunosensor during immunoreaction, fluorescent emission can be measured after excitation by direct imaging though a microscope, through optical fibers or through optical waveguides. Since fluorescent signals are usually very weak, optimal coupling is required for detection. Improved sensitivities with evanescent wave biosensors have been obtained by scientists at the Naval Research Laboratory by reducing optic fiber dimensions with chemical etching. Characteristics of evanescent wave biosensors using rabbit antigoat IgG immobilized on the optic fiber in response to different concentrations of fluorescent labeled goat IgG. Results for a step-etched design (figure 4.8(c)) were reported by Anderson *et al* [34]. Results for a tapered design (figure 4.8(d)) were also reported by Golden *et al* [35]. Both designs were superior to distal type sensors (figure 4.8(a)). The tapered design was judged to be superior to the step-etched design. The power of the evanescent wave is increased since the geometry converts the excitation mode into higher order modes and eliminates mismatches between the wavelength of the emitted light and the physical properties of the waveguide.

Barnard and Walt [36] developed a competitive fluoroimmunoassay with a fiber optic sensor using the controlled release of labeled antibodies to maintain sensitivity. An antibody was labeled with fluorescein and an immunoglobulin G was labeled with Texas Red. When the antibody and IgG form an immunocomplex, the two fluorophor labels are very close to one another, allowing non-radiative transfer of energy. When the fluorescein molecule is excited by light, it donates energy to the Texas Red fluorophor, enhancing its fluorescent light emission. Fluorescein was excited at 480 nm and peak emission monitored at 520 nm. Texas Red was excited at 570 nm with peak emission monitored at 610 nm. A competitive reaction scheme was devised, where the analyte was unlabeled IgG. Two controlled release polymer reservoirs were fabricated using ethylene vinyl acetate. One reservoir continually released fluorescein-labeled antibody, while the second released Texas Red-labeled IgG into a reaction chamber. After an initial hydration period of approximately 2 days, the release rates from both reservoirs were constant and the immunosensors remained sensitive for 30 days. Longer lifetimes may be possible with greater loading or larger polymer reservoirs.

Choquette *et al* [37] used planar waveguide technology to measure antibody–antigen fluorescence, with liposomes to augment the antigen concentration entering immunoreaction. A small flow channel (20 mm $\times$ 1.2 mm $\times$ 0.4 mm) with total volume of 9.2 μl was constructed on the glass surface of a planar waveguide coated with a thin layer of dielectric film to reflect light. The waveguide was tested with a He:Ne laser (λ = 632.8 nm). An immunosensor for theophylline was tested, using an antibody prepared from mouse anti-theophylline IgG attached to the waveguide surface with 3-glycidoxypropyltrimethoxysilane. The concentration of antigen reaching the antibodies was increased by using liposomes, incorporating theophylline-conjugated phosphatidylethanolamine into the lipid membrane. Fluorescence experiments were conducted with an argon ion laser excitation source, with corrections for background scattering. The laser power was reduced to minimize photobleaching losses. Theophylline standards in the range from 2×10^{-10} to 2×10^{-5} M were tested. Non-specific adsorption was also tested with γ-globulin. Antigen-tagged liposomes were introduced by flow injection, then the flow was stopped to allow equilibrium. For the lowest concentrations, up to 1 hour was required to reach equilibrium. Antibody activity was regenerated by exposing the liposomes to the detergent 1-O-octyl-β-D-glucopyranoside to disrupt the membranes, followed by a thorough washout. There was a significant loss in sensitivity with repeated measurements at the lower concentrations, down by approximately 50% after 12 samples. However, the sensitivity at the upper concentration ranges was not affected as much. The immunosensor sensitivity to theophylline in the clinically important range from 10 to 20 mg l^{-1} was fairly stable.

A phase-resolved evanescent wave technique was investigated by Lundgren *et al* [38] for an immunosensor exposed to binary mixtures of fluorophores. After removing the cladding from a 600 μm diameter optic fiber, protein G was immobilized in a thin sol-gel layer on the distal end. The protein G was then reacted with antifluorescein antibodies. Ultraviolet light was modulated at frequencies between 1 to 275 MHz to provide a sinusoidally varying excitation. They were able to recover the two separate fluorescent spectra from binary mixtures, as predicted from theory (equation (4.26)), by nulling out the contribution of either component from the total fluorescent spectrum. Another advantage of the phase-resolved technique is that it can discriminate between fluorescence in the bulk medium and at the surface where the immunoreaction is occurring, as long as the emission lifetimes are different. In their study, free fluorescein had an emission lifetime around 4000 picoseconds in the bulk phase, but was much shorter, around 500 picoseconds, when bound to antibody on the surface. Static techniques based on fluorescent intensity would not be able to separate the contributions of surface and bulk fluorescence, unless there is a measureable difference in emission wavelengths for the free and bound fluorescent species.

Optical techniques based on surface plasmon resonance have been used with antibodies. Stenberg *et al* [39] demonstrated that the absolute concentration of protein could be determined by the surface plasmon resonance method. A gold-coated glass slide was coated with a layer of negatively charged dextran hydrogel and used in a flow-through system. The light from a high-output LED ($\lambda \sim$ 760 nm) was projected at an incident angle to produce an evanescent wave (as previously shown in figure 4.5) in the hydrogel layer and flowing fluid above the gold layer. The resulting shift in the resonant angle was measured as proteins, including monoclonal antibodies, were absorbed on the surface. The proteins were radiolabeled so that a quantitative measurement could be made of the total amount absorbed.

A commercial system based on surface plasmon resonance, called biospecific interaction analysis (BIA), has been developed (BIAcore™, Pharmacia Biosensor AB, Piscataway, NJ). Besides immunosensor applications, it has also been used to study the binding properties of biomolecules. The BIAcore™ system is modularly designed with four parallel flow channels on removable sample chambers. Each flow cell is 2.1 mm $\times$ 0.55 mm $\times$ 0.05 mm for a total volume of 60 nl. Sample volumes from 1 to 50 μl can be used. A fluid sample containing the bioactive material is delivered to the detecting area, which has been designed as an integrated unit on a removable module. The automated delivery of fluids and sensing is controlled by computer. A particular advantage of this system is that the surface of each immunosensor can be renewed for 50 to 100 cycles (depending on the stability of the linking ligand) by washing with hydrochloric acid. Typical analysis times are only 5 to 10 minutes, and reproducibilities are reported to be >95%. The time course of change can be followed, allowing estimation of kinetic parameters for binding or dissociation.

Another example of an optical biosensor using antibodies was described by Watts *et al* [40]. A resonant mirror device was used to detect a strain of *Staphylococcus aureus* (Cowan-1) cells in diluted milk samples. Since this strain of cells express protein-A at their surfaces, human immunoglobin G (IgG) antibody was used as the specific recognition element. As a control, another *Staphylococcus aureus* cell line (Wood-46) which does not express protein-A was also used to evaluate the optical device. Watts *et al* [40] estimated that the limit of detection for their system was about 6 to 8×10^6 cells ml^{-1} in the milk flowing over the resonant mirror device, based on the minimum resonant shift that could be measured with their instrumentation. Other techniques have been able to detect much lower cell concentrations. There are several reasons for the relatively poor resolution of the device. First, the *Staphylococcus aureus* cell size, around 0.8 μm in diameter, is several times larger than the major part of the evanescent field which only extends out 200 to 300 nm from the mirror surface. Second, the refractive index for the cell wall is low and not much different from the surrounding medium. Finally, the greatly reduced cell coverage suggests that the dextran coating may be interacting with the cell wall and hindering its binding to the human IgG antibody.

Watts *et al* [40] were able to improve the dynamic range of the resonant mirror device by using a gold conjugate sandwich assay technique. A 100:1 mixture of human IgG antibody to colloidal gold (30 nm diameter particles) was prepared, then further diluted in phosphate-buffered saline to a constant absorbance measured at 530 nm. The number of human IgG molecules absorbed by each gold particle was determined to be 52, based on a radiolabeling study. After binding Cowan-1 cells to the directly IgG-coated surface (without dextran), the colloidal gold and human IgG antibody solution was allowed to flow over the mirror. After washing excess gold away, the shift in resonance was then measured. The shift in resonance was proportional to a much lower Cowan-1 cell concentration, in the range from 8×10^3 to 1.6×10^6 cells ml^{-1} for the gold sandwich assay. However, the response began to plateau around 1×10^6 cells ml^{-1}. The Wood-46 cell strain binding was negligible for the sandwich assay. The sensitivity of the sandwich assay technique was further demonstrated with tests of whole milk contaminated by adding Cowan-1 cells to concentrations ranging from 4×10^3 to 4×10^6 cells ml^{-1}. The milk samples had to be diluted by half in order to reduce the refractive index of the sample.

4.3.5 DNA sequencing and analysis

With the recent major research funding of the Human Genome Project, there has been increasing interest in methods for improving DNA sequencing and techniques for analyzing nucleotides. Immunosensors can be developed for

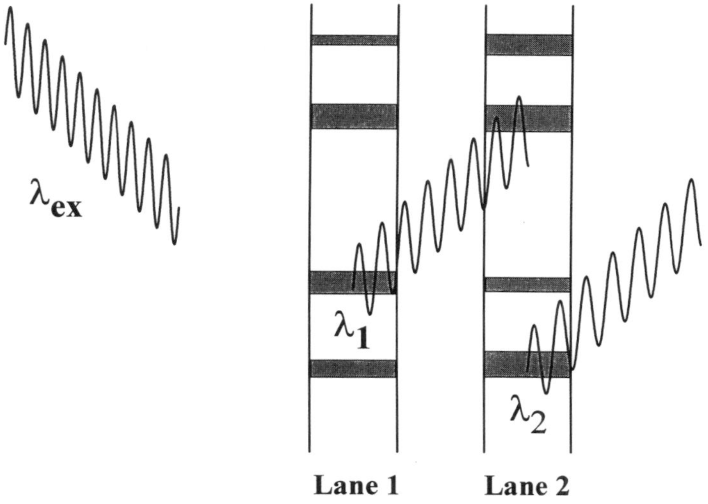

Figure 4.10 DNA sequencing by electrophoresis with laser-induced fluorescence. Using fluorescent labels for nucleotide bases which emit light at different wavelengths after excitation at wavelength λ_{ex}, the bases can be accurately identified after separating DNA or RNA fragments by gel or capillary electrophoresis.

specific DNA tests. For example, Vo-Dinh *et al* [41] developed a fiber optic system for measuring DNA from human placental samples. A monoclonal antibody from mouse spleen cells was used to detect benzo[a]pyrene tetrol after mild acid hydrolysis of human placental tissue samples. A 10 mW He:Cd laser ($\lambda = 325$ nm) was used with emission detected by a photomultiplier tube through a 40 nm bandpass filter centered at 400 nm. Recently, Chalfie *et al* [42] used a green fluorescent protein (GFP) which is produced by the jellyfish *Aequorea victoria* as a marker for gene expression and to localize target proteins. They were able to express GFP in *E. coli*, and in touch receptor neurons of the nematode *C. elegans.* A promoter of the *mec*-7 gene which encodes the protein β-tubulin was used to express GFP in the nematode neurons. The expressed GFP was very stable, and emitted green light ($\lambda = 508$ nm) when excited by blue or near-UV light ($\lambda = 395$ nm), with very little photobleaching. GFP was expressed intracellularly, did not require any exogenous substrates or cofactors to fluoresce, and did not appear to have any toxic effects. Also, GFP retains its fluorescent properties in tissues fixed with formaldehyde. Cubitt *et al* [43] list an extensive table of research studies which have used GFP. They also report on the progress on re-engineering the GFP by mutagenesis to improve its properties and to shift the excitation and/or emission wavelengths.

DNA sequencing techniques have relied heavily on electrophoresis to separate the DNA fragments. Autoradiography and optical density measurements have been used to help interpret the resulting bands in the electrophoresis gel.

Recently, commercial DNA sequencing systems have been developed using laser-induced fluorescence, as illustrated in figure 4.10. The fluorescent labels with emission wavelengths λ_1 and λ_2 are associated with two different bases in this example, which would aid in the identification of the nucleotides. Perkin–Elmer Applied Biosystems manufacture automated DNA sequencing instruments which use four fluorescent labels, one for each of the four bases in DNA. It may be possible to improve the sensitivities of this technique. Ju *et al* [44] investigated the use of fluorescence energy transfer to develop optimized labels with greater fluorescence and electrophoretic mobility. The four primers were excited by an argon laser ($\lambda_{ex} = 488$ nm) and had well separated emission peaks at $\lambda = 525, 555, 580$, and 605 nm which were twofold to 14-fold greater compared to the fluorescence without energy transfer.

Fluorescent dyes can also bind with nucleic acid ligands. McGown *et al* [35] describe the potential uses of nucleic acid ligands as tools for molecular recognition. They suggest that the relatively smaller nucleic acid ligands may be simpler to manufacture and utilize compared to larger, more complex antibodies and enzymes that have been utilized for previous biosensors. It may also be easier to control the conformation of the nucleic acid ligands, making it possible to regenerate the recognition element of the biosensor using relatively simple methods, such as changing the buffer-ion concentrations. Fluorescence polarization measurements may also be easier for certain analytes, since the relative increase in size after binding will be large compared to an antibody, which would have little change in size after combining with the analyte.

4.4 CONCLUDING REMARKS

Medical and biomedical research using optical technology has produced tremendous advances in understanding basic biological phenomena. These advances have also had a substantial impact on our ability to detect and successfully treat human diseases. However, much of the promise for improving human health using optical technology remains unfulfilled. Although highly sensitive and very specific biosensors have been demonstrated in the laboratory using present photometric transducers, relatively few optical biosensors have reached the marketplace. Perhaps one of the most successful commercial applications so far has been for determining blood (oxyhemoglobin) O_2 saturation in humans based on comparing light absorption or reflection at specific wavelengths. Optical and fiber-optic-based medical instruments do not have the safety problems that other technologies have, where electrical currents can be hazardous to patients and must be carefully controlled. We can anticipate that optical biosensors for measuring biologically relevant analytes in blood, tissue, and biological fluids will be developed. Optical biosensors for detecting infectious and viral diseases, for identifying genetic abnormalities, for detecting cancers, and for other medical tests will become available. The development of new pharmaceuticals will also benefit from the development of new optical devices. Future applications are possible in many areas besides healthcare, such as monitoring the environment and for food processing industries.

REFERENCES

[1] Chance B 1991 Optical method *Annu. Rev. Biophys. Biophys. Chem.* **20** 1–28

[2] Granström M, Berggren M and Ingan@s O 1995 Micrometer- and nanometer-sized polymeric light-emitting diodes *Science* **267** 1479–81

[3] Hanken D G and Corn R M 1995 Variable index of refraction ultrathin films formed from self-assembled zirconium phosphonate multilayers: Characterization by surface plasma resonance measurements and polarization/modulation FT-IR spectroscopy *Anal. Chem.* **67** 3767–74

[3] Healey B G, Foran S E and Walt D R 1995 Photodeposition of micrometer-scale polymer patterns on optical imaging fibers *Science* **269** 1078–80

[5] Tsien R Y 1983 Intracellular measurements of ion activities *Annu. Rev. Bioeng.* **12** 91–116

[6] Dunn K W, Mayor S, Myers J N and Maxfield F R 1994 Applications of ratio fluorescence microscopy in the study of cell physiology *FASEB J.* **8** 573–82

[7] Stern O and Volmer M 1919 Über die Abklingungszeit der Fluoreszenz *Phys. Z.* **20** 183

[8] Carraway E R, Demas J N and DeGraff B A 1991 Luminescence quenching mechanisms for microheterogeneous systems *Anal. Chem.* **63** 332–6

[9] Lübbers D W and Opitz N 1976 Quantitative fluorescence photometry with biological fluids and gases *Oxygen Transport to Tissue—II* ed J Grote, D Reneau and G Thews (New York: Plenum); *Adv. Exp. Med. Biol.* **75** 65–8

[10] Surgi M R 1989 Design and evaluation of a reversible fiber optic sensor for determination of oxygen *Applied Biosensors* ed D L Wise (Boston, MA: Butterworths) pp 249–90

[11] Peterson J I, Fitzgerald R V and Buckhold D K 1984 Fiber-optic probe for *in vivo* measurement of oxygen partial pressure *Anal. Chem.* **56** 62–7

[12] Wolthuis R S, McRae D, Hartl J C, Saaski E G, Mitchell L, Garcin K and Willard R 1992 Development of a medical fiber-optic oxygen sensor based on optical absorption change *IEEE Trans. Biomed. Eng.* **BME-39** 185–93

[13] Berndt K W and Lakowicz J R 1992 Electroluminescent lamp-based phase fluorometer and oxygen sensor *Anal. Biochem.* **201** 319–25

[14] Carraway E R, Demas J N, DeGraff B A and Bacon J R 1991 Photophysics and photochemistry of oxygen sensors based on luminescent transition-metal complexes *Anal. Chem.* **63** 337–42

[15] Xu W, Schmidt R, Whaley M, Demas J N, DeGraff B A, Karikari E K and Farmer B L 1995 Oxygen sensors based on luminescence quenching: interactions of pyrene with the polymer supports *Anal. Chem.* **67** 3172–80

[16] Peterson J I, Goldstein S R, Fitzgerald R V and Buckhold D K 1980 Fiber optic pH probe for physiological use *Anal. Chem.* **52** 864–9

[17] Peterson J I and Vurek G 1984 Fiber-optic sensors for biomedical applications *Science* **123** 123–7

[18] Besar S S A, Kelly S W and Greenhalgh P A 1989 Simple fibre optic spectrophotometric cell for pH determination *J. Biomed. Eng.* **11** 151–6

[19] Kawabata Y, Kamichika T, Imasaka T and Ishibashi N 1989 Fiber-optic sensor for carbon dioxide with a pH indicator dispersed in a poly(ethylene glycol) membrane *Anal. Chim. Acta* **219** 223–9

[20] Hale Z M and Payne F P 1994 Fluorescent sensors based on tapered single-mode optical fibres *Sensors Actuators* **17** 233–40

[21] Luo S and Walt D R 1989 Avidin–biotin coupling as a general method for preparing enzyme-based fiber-optic sensors *Anal. Chem.* **61** 1069–72

[22] Li L and Walt D R 1995 Dual-analyte fiber-optic sensor for the simultaneous and continuous measurement of glucose and oxygen *Anal. Chem.* **67** 3746–52

[23] Tan W, Shi Z-Y, Smith S, Birnbaum D and Kopelman R 1992 Submicrometer intracellular chemical optical fiber sensors *Science* **258** 778–81

[24] Tan W, Shi Z-Y and Kopelman R 1992 Development of submicron chemical fiber optic sensors *Anal. Chem.* **64** 2985–90

[25] Betzig E and Trautman J K 1992 Near field optics: microscopy, spectroscopy, and surface modification beyond the diffraction limit *Science* **257** 189–95

[26] Xie X S and Dunn R C 1994 Probing single molecule dynamics *Science* **265** 361–4

[27] Ambrose W P, Goodwin P M, Martin J C and Keller R A 1994 Alterations of single molecule fluorescence lifetimes in near-field optical microscopy *Science* **265** 364–7

[28] Trautman J K, Macklin J J, Brus L E and Betzig E 1994 Near-field spectroscopy of single molecules at room temperature *Nature* **369** 40–2

[29] Arnold M A 1991 Fluorophore- and chromophore-based fiber optic biosensors *Biosensor Principles and Applications* ed L J Blum and P R Coulet (New York: Dekker) pp 195–211

[30] Sutherland R M, Dahne C and Place J F 1984 Preliminary results obtained with a no-label, homogeneous immunoassay for human immunoglobulin G *Anal. Lett.* **17** 43–55

[31] Sutherland R M, Dahne C, Place J F and Ringrose A S 1984 Optical detection of antibody–antigen reactions at a glass–liquid interface *Clin. Chem.* **30** 1533–8

[32] Jablonski E 1985 The preparation of bacterial luciferase conjugates for immunoassay and application to rubella antibody detection *Anal. Biochem.* **148** 199–206

[33] Aizawa M 1991 Immunosensors *Biosensor Principles and Applications* ed L J Blum and P R Coulet (New York: Dekker) pp 249–66

[34] Barnard S M and Walt D R 1991 Chemical sensors based on controlled release polymer systems *Science* **251** 927–9

[35] Anderson G P, Golden J P and Ligler F S 1994 An evanescent wave biosensor—Part I: Fluorescent signal acquisition from step-etched fiber optic probes *IEEE Trans. Biomed. Eng.* **41** 578–84

[36] Golden J P, Anderson G P, Rabbany S Y and Ligler F S 1994 An evanescent wave biosensor—Part II: Fluorescent signal acquisition from tapered fiber optic probes *IEEE Trans. Biomed. Eng.* **41** 585–91

[37] Choquette S J, Locascio-Brown L and Durst R A 1992 Planar waveguide immunosensor with fluorescent liposome amplification *Anal. Chem.* **64** 55–60

[38] Stenberg E, Persson B, Roos H and Urbaniczky C 1991 Quantitative determination of surface concentration of protein with surface plasmon resonance using radiolabeled proteins *J. Colloid Interface Sci.* **143** 513–26

[39]] Lundgren J S, Bekos E J, Wang R and Bright F V 1994 Phase-resolved evanescent wave induced fluorescence. An *in situ* tool for studying heterogeneous interfaces *Anal. Chem.* **66** 2433–40

[40] Watts H J, Lowe C R and Pollard-Knight D V 1994 Optical biosensor for monitoring microbial cells *Anal. Chem.* **66** 2465–70

[41] Vo-Dinh T, Alarie J P, Johnson R W, Sepaniak M J and Santella R M 1991 Evaluation of the fiber-optic antibody-based fluoroimmunosensor for DNA adducts in human placenta samples *Clin. Chem.* **37** 532–5

[42] Chalfie M, Tu Y, Euskirchen G, Ward W W and Prasher D C 1994 Green fluorescent protein as a marker for gene expression *Science* **263** 802–5

[43] Cubitt A B, Heim R, Adams S R, Boyd A E, Gross L A and Tsien R Y 1995 Understanding, improving and using green fluorescent proteins *Trends Biol. Sci.* **20** 448–55

[44] Ju J, Kheterpal I, Scherer J R, Ruan C, Fuller C W, Glazer A N and Mathies R A 1995 Design and synthesis of fluorescence energy transfer dye-labeled primers and their application for DNA sequencing and analysis *Anal. Biochem.* **231** 131–40

[45] McGown L B, Joseph M J, Pitner J B, Vonk G P and Linn C P 1995 The nucleic acid ligand. A new tool for molecular recognition *Anal. Chem.* **66** 663A–8A

5

Electrochemical transduction

Joseph Wang

5.1 INTRODUCTION

Electrochemical sensors combine the specificity of biological or chemical recognition layers with the inherent advantages (sensitivity, speed, miniaturization, linearity) of electrochemical transduction. The electrochemical transducer converts the biological or chemical recognition process into a useful electrical signal (figure 5.1). The resulting electrical signal is related to the recognition process, and is proportional to the analyte concentration. Depending upon the nature of the electrical signal, electrochemical sensors fall into one of three categories: amperometric, potentiometric, or conductometric.

The purpose of this chapter is to summarize the principles of electrochemical sensors, with focus on the transduction mechanism. The most common form of electrochemical sensor, the enzyme electrode, will be used below for illustrating the principles of electrochemical transduction. Here, the biocatalytic recognition of the substrate, by an immobilized enzyme layer, is followed by electrochemical measurement of the product or depleted cofactor (figure 5.2). The choice of the appropriate electrochemical transducer is governed by the nature of the substrate, the desired analytical performance, and the shape and size of the device.

5.2 AMPEROMETRIC TRANSDUCTION

5.2.1 Electrode processes

Amperometric sensors are based on the detection of electroactive species involved in the recognition process. The transduction process is accomplished by controlling the potential of the working electrode at a fixed value (relative to a reference electrode) and monitoring the current as a function of time. The applied potential serves as the driving force for the electron transfer reaction of the electroactive species. It can be viewed as the 'electron pressure' which forces the species to gain or lose an electron. The resulting current is a direct measure of the rate of the electron transfer reaction. Accordingly, it reflects the rate of the recognition (enzymatic) reaction, and is thus proportional to the concentration of the target analyte (the substrate). Such constant-potential measurements have the advantage of being free of the charging background current and thus can result in extremely low detection limits.

The term working electrode is reserved for one of the electrodes at which the redox reaction of the electroactive species of interest occurs:

$$\mathrm{O} + ne \rightleftharpoons \mathrm{R} \qquad (5.1)$$

where O and R are the oxidized and reduced forms, respectively, of the electroactive species, and n is the number of electrons transferred. Such electron transfer will take place when the applied potential becomes sufficiently negative or positive for the electroactive species to receive or donate electrons from the surface. (At more negative or positive potentials the surface becomes a better electron source or sink, respectively.) The overall electrode reaction (5.1) is quite complicated and composed of a series of steps, including transport of O from the bulk solution to the surface, the actual electron transfer, chemical or surface reactions of O or R, and transport of R back to the bulk solution. Each of these steps has its own inherent rate. The slowest of these steps limits the net rate of the reaction, and hence controls the current signal. At sufficiently high potentials (above the characteristic standard potential E^0 for the redox reaction), the current response is limited by the rate of mass transport towards the transducer surface. Under these conditions, the operating potential drives the surface concentration of O to zero, and causes depletion of the electroactive species at the surface, as determined by the Nernst equation (at 25 °C):

$$E = E^\circ + (0.059/n)\log([C_\mathrm{O}]_\mathrm{s}/[C_\mathrm{R}]_\mathrm{s}) \qquad (5.2)$$

where $[C_\mathrm{O}]_\mathrm{s}$ and $[C_\mathrm{R}]_\mathrm{s}$ are the surface concentrations of O and R, respectively. Such depletion leads to a concentration gradient in the solution adjacent to the electrode (figure 5.3), in a region known as the diffusion layer. At the bulk of the solution (far from the surface), the composition remains uniform. Diffusional mass transport thus acts to 'remove' the concentration gradient by movement of species from the bulk to the lower-concentration surface region. The mass-transport-controlled current is determined by the flux (associated with the slope of the concentration gradient) and is given by

$$i = nFADC/\delta \qquad (5.3)$$

where F is the value of the faraday, A the electrode area, D the diffusion coefficient, and δ is the thickness of the diffusion layer. The smaller δ, the larger

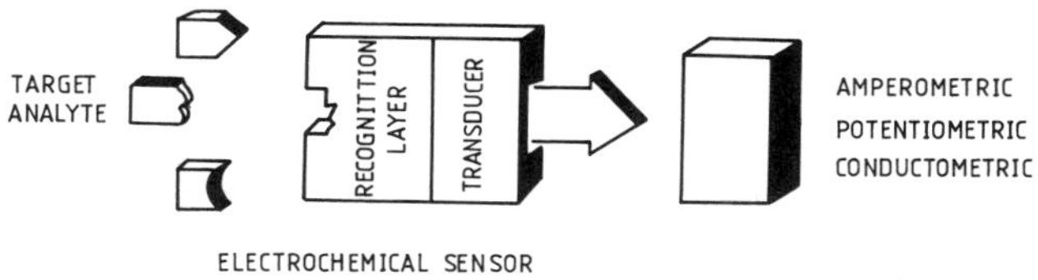

Figure 5.1 Processes involved in an electrochemical sensor system, including the analyte recognition, signal transduction, and readout.

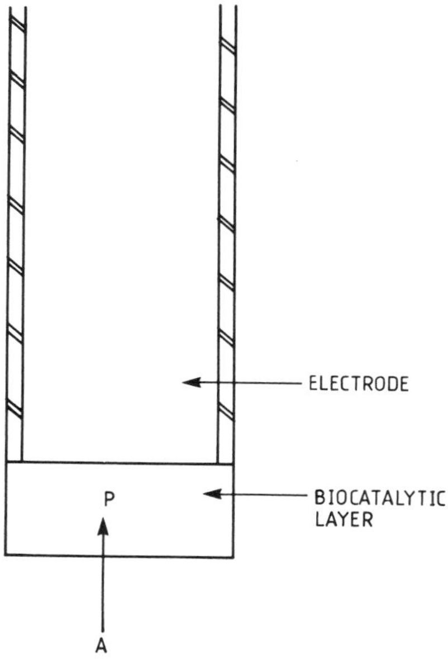

Figure 5.2 A schematic diagram of an enzyme electrode: A, analyte; P, product of the biocatalytic reaction.

the concentration gradient, and the higher the current. Higher sensitivity can thus be achieved through forced convection (e.g., solution stirring or electrode rotation) that minimizes the diffusion layer thickness. For reactions controlled by the rate of electron transfer (i.e., reactions with sufficiently fast mass transport), the current–potential relationship is given by the Butler–Volmer equation [1–3].

A complete understanding of electrode processes requires knowledge of kinetics, thermodynamics, hydrodynamics, solution and surface processes, and basic electrochemical principles. Several monographs [1–3] provide a comprehensive treatment of these topics, and are recommended for further reading.

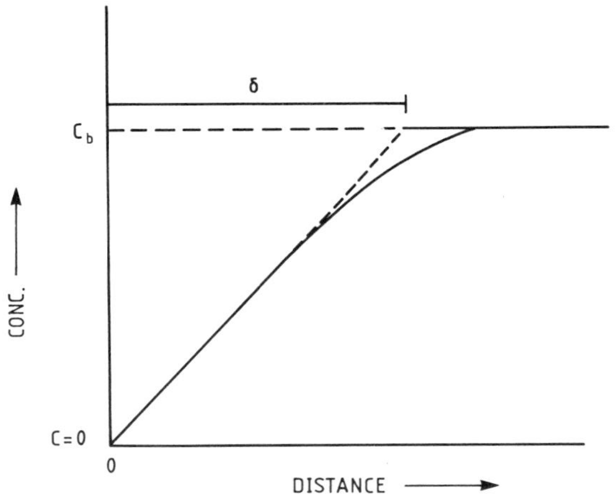

Figure 5.3 The concentration–distance profile in the solution adjacent to the electrode. C_b is the bulk concentration, while δ is the diffusion layer thickness.

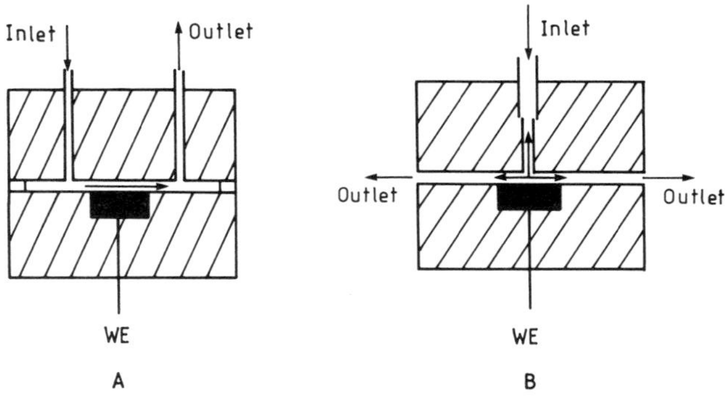

Figure 5.4 Common amperometric detectors: thin-layer (A) and wall jet (B) flow cells.

5.2.2 Practical considerations

The choice of the working electrode is critical to the success of amperometric sensing. The working electrode material chosen for a given application should provide high sensitivity, selectivity, and stability. The following factors are thus considered: the kinetics of the electron transfer reaction, the potential limits, background processes of coexisting electroactive species, and ease of immobilizing the biocomponent.

Solid electrodes, especially glassy carbon, platinum, gold, or carbon paste disks have been particularly useful in connection with most enzyme electrodes. Further improvements can be achieved by modifying or pretreating these

Table 5.1 Current suppliers of electrochemical analyzers.

Supplier	Address
Bioanalytical Systems	2701 Kent Avenue, W Lafayette, IN 47906, USA
Cypress	PO Box 3931, Lawrence, KS 66046, USA
ECO Chemie	PO Box 85163, 3508 AD Utrecht, The Netherlands
EG&G PAR	PO Box 2565, Princeton, NJ 08543, USA
ESA	45 Wiggins Ave, Bedford, MA 01730, USA
Metrohm	CH-9109 Herisau, Switzerland
Oxford Electrodes	Oxford, UK
Tacussel/Radiometer	27 rue d'Alscace, F-69627 Villeurbanne, France

electrode surfaces. This ability to manipulate the molecular structure of the transducer surface holds great promise for future designs of biosensors. Microelectrodes with one dimension smaller than 50 μm (e.g. carbon fibers) are useful for assays of microliter samples or for *in vivo* work, while flow-through configurations can be employed in high-speed automated systems and for on-line monitoring. The most widely used flow detectors are based on the thin-layer and wall jet configurations (figure 5.4). Easily oxidizable species can thus be detected down to the picogram level.

The tiny dimensions of ultramicroelectrodes also offer various fundamental advantages, such as enhanced mass transport (in motionless solutions) and reduced ohmic losses (in low-ionic-strength media). Arrays of microelectrodes can be used to couple these advantages of individual microelectrodes with large 'collective' currents. In addition to the working electrode, an amperometric sensor contains the reference electrode which provides a stable potential against which the potential of the working electrode is compared and often a third, auxiliary, electrode for carrying the current. Silver/silver chloride and platinum are the most commonly used reference and auxiliary electrodes, respectively.

The three-electrode system is connected to a potentiostat, which controls the potential of the working electrode while monitoring the resulting current. Besides the widely used fixed-potential measurements, such instruments can apply other potential–time waveforms such as potential step, linear scan, pulse excitations that may be useful in certain sensing applications. Such instruments are commercially available from various sources, listed in table 5.1. A three-

Table 5.2 Selected oxidase- or dehydrogenase-based electrodes.

Target analyte	Immobilized enzyme	Detected species
Amino acids	Amino acid oxidase	H_2O_2
Ethanol	Alcohol dehydrogenase	NADH
Galactose	Galactose oxidase	O_2
Glutamate	Glutamate oxidase	H_2O_2
Glucose	Glucose oxidase	H_2O_2
Lactate	Lactate dehydrogenase	NADH
Malate	Malate dehydrogenase	NADH
Oxalate	Oxalate oxidase	O_2
Sulfite	Sulfite oxidase	H_2O_2
Xanthine	Xanthine oxidase	H_2O_2

electrode system offers a more accurate control of the applied potential than a two-electrode one. Disposable strips, containing the various electrodes on a planar support, are often used for single-use applications [4]. Microfabrication techniques, such as screen printing thick-film technology or silicon-type thin-film lithography, are used for the large-scale mass production of these low-cost sensor strips. Such microfabricated sensors are currently being coupled to a new generation of hand-held, user-friendly, analyzers. Sample droplets (of 50–100 μl volume) can be applied in connection with single-use sensor strips or in flow injection biosensing schemes, while large volumes (5–10 ml) are employed in connection with conventional electrochemical cells. Samples of high ionic strength (>0.01 M) are desired to insure the conductive media which amperometry requires.

5.2.3 Amperometric enzyme electrodes

Because of the relatively large distance between the redox center of enzymes and the transducer surface, it is difficult to achieve electron transfer between redox enzymes and electrodes. Alternately, the biocatalytic generation or consumption of electroactive species make oxidoreductase enzymes most suitable for amperometric biosensing. In particular, a large number of hydrogen peroxide generating oxidases and NAD^+-dependent dehydrogenases have been particularly useful for the measurement of a wide range of substrates, based on the following general reactions:

$$\text{substrate} + O_2 \xrightarrow{\text{oxidase}} \text{product} + H_2O_2 \tag{5.4}$$

$$\text{substrate} + NAD^+ \xrightarrow{\text{dehydrogenase}} \text{product} + NADH. \tag{5.5}$$

For specific examples see table 5.2.

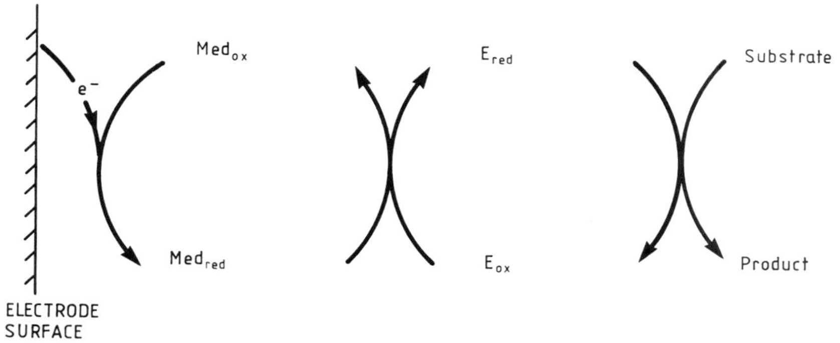

Figure 5.5 The sequence of events occurring in a mediated reaction. Med_{ox} and Med_{red} are the redox forms of the mediators, while E_{ox} and E_{red} are the redox forms of the enzyme. (Reproduced with permission from [6].)

The liberated peroxide or NADH species can be readily detected at relatively modest potentials (0.6–0.8 V against Ag/AgCl, depending upon the working electrode material). The overall reaction involves many steps and the actual response is determined by the rate limiting step in the overall reaction scheme. This involves diffusion of the substrate molecule towards the surface and its reaction with the immobilized enzyme, which in turn is regenerated into its native form by reaction with the (oxygen or NAD^+) cofactor.

This type of 'first-generation' bioelectrode may suffer from background contributions due to easily oxidizable sample constituents (such as ascorbic or uric acids). Such background interferences can be addressed by covering the transducer with a permselective/discriminative membrane barrier (e.g., cellulose acetate or Nafion) [5], or through the use of electron transfer mediators [6]. Discriminative films based on different transport properties (e.g., solute size, charge, polarity) have thus been used for controlling the access towards the transducer. Multilayers (based on overlaid films) can offer additional selectivity advantages. The mediator is a low-molecular-weight species which shuttles electrons between the redox center of the enzyme and the working electrode (figure 5.5). It should be easily reduced by the enzyme, and rapidly reoxidized at the working electrode. This occurs at low operating potentials where contributions of sample constituents are minimal. Useful sensors based on ferrocene or quinone derivatives, tetrathiafulvalene (TTF), ferricyanide, methyl viologen, or ruthenium amine have been reported. Such mediator-based sensors also address the oxygen deficiency (at high substrate concentration) of oxidase-based devices, hence leading to extended linearity and reduced dependence on the oxygen tension. Ferrocene mediators and ferricyanide are being routinely used for personal blood glucose monitoring (in connection with screen-printed glucose strips). Another promising avenue for facilitating the flow of electrons from the enzyme to the electrode is to use polycationic redox polymers that

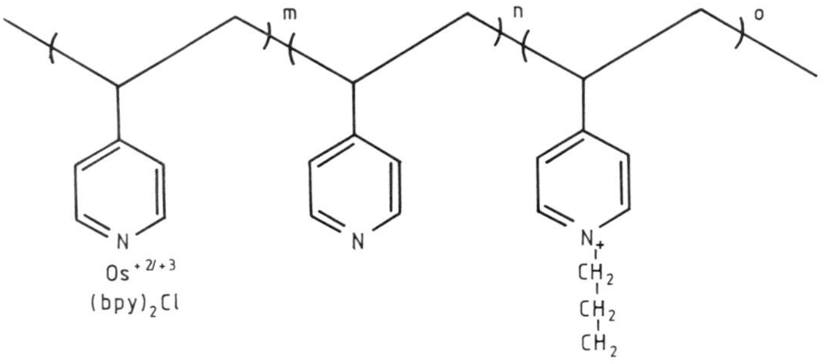

Figure 5.6 The composition of the electron relaying redox polymer. (Reproduced with permission from [7].)

electrically communicate between the two. In particular, Heller's group has developed an efficient electrical connection using Os-bpy complex electron relays on a polymer backbone (figure 5.6), which form a three-dimensional redox epoxy network with the immobilized enzyme on the transducer surface [8]. Coupled reactions of two or more enzymes can also be used to minimize interference, as well as to amplify the response and extend the scope of the enzyme electrode towards additional analytes. For example, peroxidases can be coupled with oxidases to allow low-potential detection of the liberated peroxide. Electrocatalytic surfaces, particularly those based on metallized carbon, represent a new and effective approach for minimizing electroactive interference [9]. Such strategy relies on the preferential electrocatalytic detection of the liberated peroxide or NADH species at rhodium or ruthenium dispersed carbon bioelectrodes.

5.3 POTENTIOMETRIC TRANSDUCTION

5.3.1 Introduction

Electrochemical transduction mechanisms other than electron transfer can expand the scope of biosensors towards target substrates other than those for redox enzymes. In particular, potentiometric measurements have led to many useful sensing devices. In such sensors the analytical information is obtained by converting the recognition process into a potential signal, which is proportional (in a logarithmic fashion) to the concentration (activity) of species generated or consumed in the recognition event. The biocatalytic liberation of ions or gases thus leads to the development of a potential across the sample/transducer interface. The potential is measured under conditions of essentially zero current and presented to the user as the analytical signal. Such a transduction mechanism is very attractive for the operation of biosensors because of its selectivity,

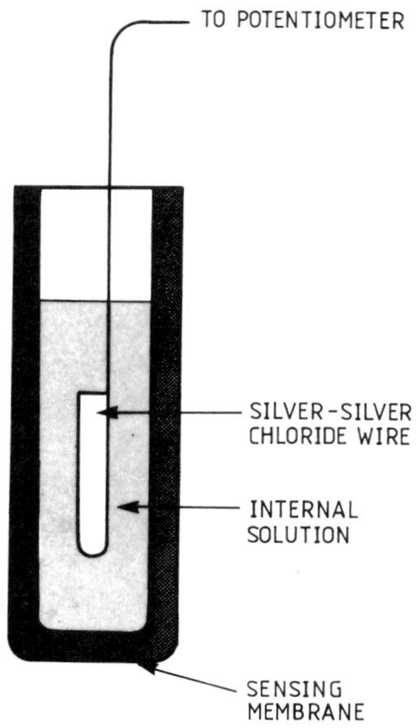

Figure 5.7 The general structure of an ISE.

simplicity, and low cost. It is, however, less sensitive and often slower than the amperometric counterpart.

5.3.2 Principles

Potentiometric biosensors rely on the use of ion selective electrodes (ISEs) for obtaining the analytical signal. Such a transducer is an indicator electrode that preferentially responds to one ionic species. ISEs are, in general, membrane-based devices, consisting of permselective ion conducting materials, which separate the sample from the inside of the electrode (figure 5.7). On the inside is an electrolyte solution containing the ion of interest at a constant activity. The composition of the membrane is designed to yield a potential that is primarily due to the ion of interest. The challenge is to find a membrane that will selectively bind the target ion. Most often the membrane contains a reagent (ionophore) for the ion of interest. The reagent–ion interaction at the membrane–sample interface results in a charge separation (figure 5.8), which is measured as a potential against the reference electrode. A membrane potential thus arises whenever the activity of the ion of interest in the sample differs from that in the inner electrolyte solution. Since the potential of the reference electrode is fixed, the measured cell potential can be related to the activity of the target ion. ISEs can be classified into different categories, according to the nature of the membrane material: (a) glass membrane electrodes; (b) liquid membrane electrodes; (c)

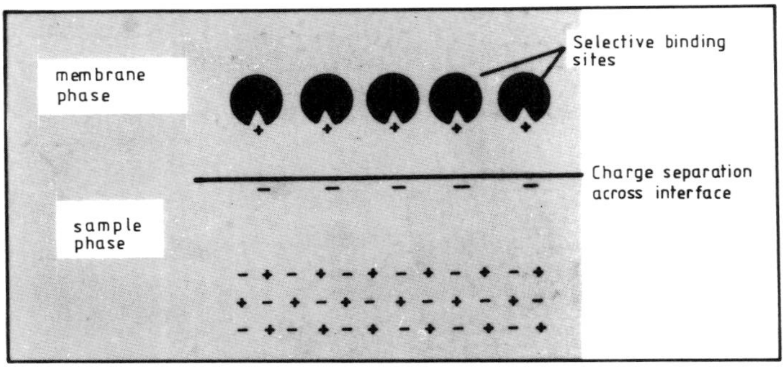

Figure 5.8 Potentiometric transduction: ISE mechanism. (Reproduced with permission from [10].)

solid state membrane electrodes. The pH sensor, based on a glass membrane, is the oldest and best known ISE.

Ion selective electrodes should ideally obey the equation

$$E = K + (2.303RT/ZF)\log a_i \tag{5.6}$$

where E is the potential, R the universal gas constant, T the absolute temperature, Z the ionic charge, and a_i the activity of ion i. K is a constant, containing contributions from various sources, e.g., several liquid junction potentials. Equation (5.6) predicts that the electrode potential is proportional to the logarithm of the activity of the ion monitored. For example, at room temperature a 59.1 mV change in the electrode potential should result from a 10-fold change in the activity of a monovalent ion ($Z = 1$). The activity of an ion i in solution is related to its concentration, c_i, by the equation

$$a_i = \gamma_i c_i \tag{5.7}$$

where γ_i is the activity coefficient. The activity coefficient depends on the types of ion present and on the total ionic strength of the solution.

According to equation (5.6), a semilog plot of the electrode potential against the activity of the ion of interest leads to a straight line, with a slope of approximately 59/Z mV (figure 5.9). Departure from such Nernstian behavior and linearity is commonly observed at low (micromolar) analyte concentrations due to the presence of coexisting ions. Such a contribution of interfering ions is given by the Nikolski–Eisenman equation:

$$E = K + (2.303RT/Z_iF)\log(a_i + k_{ij}a^{Z_i/Z_j}) \tag{5.8}$$

where k_{ij} is the selectivity coefficient.

5.3.3 Practical examples

The first potentiometric enzyme electrode, aimed at monitoring urea, was developed by Guilbault and Montalvo [11]. In this case, urease was entrapped

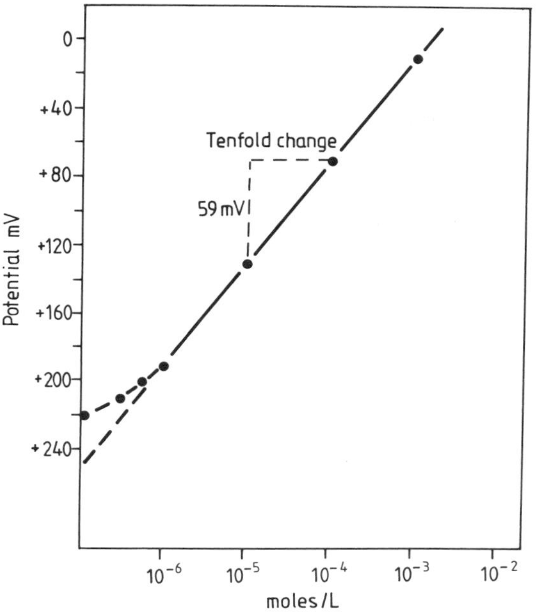

Figure 5.9 A typical calibration plot for an ISE for a monovalent ion.

in a gel matrix that is in contact with the surface on an ammonium ion selective electrode, which detects the liberated ammonium ions generated in the thin reaction layer:

$$\text{urea} \xrightarrow{\text{urease}} NH_4^+ + HCO_3^- . \tag{5.9}$$

The cation electrode was thus capable of measuring urea concentrations ranging from 5×10^{-5} M to 2×10^{-1} M. Miniaturization of this sensor has led to its deployment on a disposable biosensor chip used for near-patient monitoring of urea, as well as various electrolytes (figure 5.10).

Several potentiometric biosensors use a transducer scheme based on potentiometric monitoring of local pH changes. Any enzymatic pathway which leads to a change in the hydrogen ion activity can be applicable for this task. One useful example is the penicillin electrode, which relies on the immobilization of penicillin on a glass pH electrode [12]. The decrease in pH due to the production of penicilloic acid is sensed by the glass electrode:

$$\text{penicillin} \xrightarrow{\text{penicillinase}} \text{penicilloic acid} + H^+ . \tag{5.10}$$

The potential change is proportional to the logarithm of the penicillin concentration. Solid state fluoride or iodide selective electrodes have been employed for the biosensing of glucose or amino acids. These transducers measure the decreasing activity of iodide or fluoride (upon their reaction with the peroxide product).

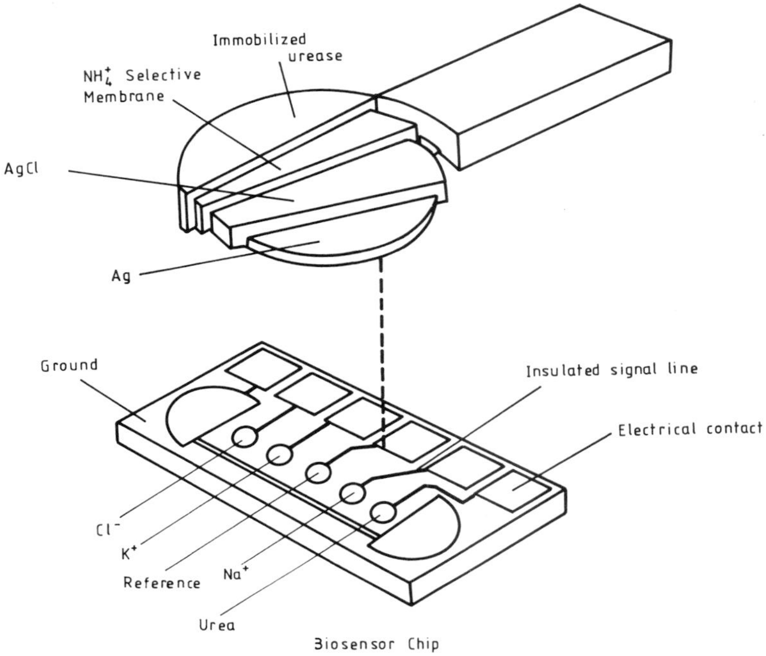

Figure 5.10 A silicon-based sensor array for monitoring urea and various electrolytes (courtesy of i-STAT Co.).

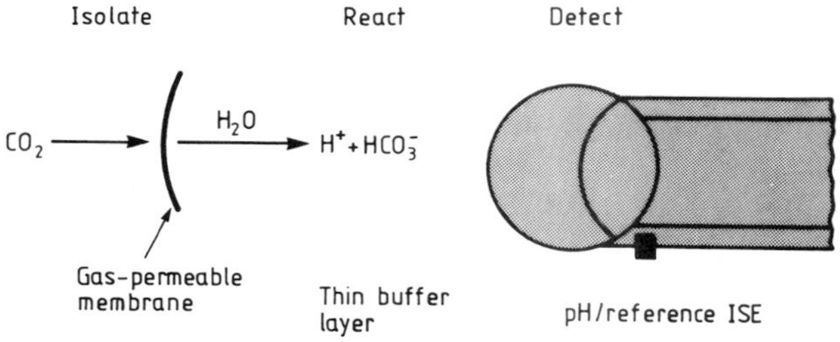

Figure 5.11 A potentiometric carbon dioxide electrode. (Reproduced with permission from [10].)

Gas sensing potentiometric electrodes can also be used to follow biocatalytic reactions that generate gaseous species. For example, several decarboxylases, such as lysine or tyrosine decarboxylases, generating carbon dioxide as a product, can be coupled to a carbon dioxide electrode, hence leading to the sensing of the corresponding lysine or tyrosine substrates. The CO_2 gas sensor incorporates a conventional pH glass electrode, surrounded by a thin film of a sodium bicarbonate solution and enclosed by a hydrophobic carbon dioxide

Table 5.3 Properties of electrochemical sensors based on different electrode transducers.

Transducer type	Transduction mechanism	Recognition event	Signal
Amperometric	Electron-transfer reaction	Generation or consumption of electroactive species	Current
Potentiometric	Change in ionic distribution across a membrane	Generation or consumption of ionic species	Potential
Conductimetric	Change in cell resistance	Modulation of conductivity	Conductance

membrane (figure 5.11). Such a membrane allows a relatively fast equilibration to be established between the sample and the internal solution. The enzymatically generated CO_2 thus diffuses through the membrane and undergoes a reaction with the internal solution that leads to local pH change. However, biosensors based on this principle suffer from relatively long response and recovery times.

5.4 CONDUCTIMETRIC TRANSDUCTION

Measurements of the cell conductance represent the third electrochemical transduction mode. The resulting sensing devices thus rely on the biological or chemical modulation of the surface conductivity. Despite the promise of conductimetric devices, their exploitation in modern biosensors is still in its infancy.

The conductance is equal to the reciprocal of the solution resistance:

$$L = 1/R \tag{5.11}$$

and hence varies with the cell resistance. Accordingly, various types of chemiresistor (utilizing a variety of electrode materials) have been described. Particularly attractive are those based on interdigitated electrode arrays. Such devices are covered with an appropriate recognition layer, that changes its admittance upon interaction with the sample solution. The actual conductance measurement can be carried out in the DC mode or with potential or current excitations.

The major advantage of the conductimetric transduction mode is that a large number of enzymatic reactions involves the consumption or production of charged species, and therefore leads to a change in the solution conductance. For example, a sensitive urea biosensing can be accomplished because the urease enzyme produces, from urea, four ions for each catalytic event [13]. Such a transduction mode is also very fast, inexpensive, does not require a reference electrode, and is suitable for miniaturization (in connection with modern lithographic procedures). A disadvantage of conductimetric measurements is the strong dependence of the response upon the buffer capacity. In addition, since conductance is an additive (non-selective) property, its measurement depends upon the number of ions present.

5.5 CONCLUSIONS

Table 5.3 summarizes the properties of electrochemical sensors based on different electrode transducers. Because of the inherent advantages of electrochemical transaction modes, the development of electrochemical sensors continues to be a rapidly growing area of research. Such devices meet many of the desired characteristics of an ideal sensing system, and hence were the first reaching the commercial stage. A great number of enzymes, and a number of other biological or chemical recognition elements, have already been combined with various electrochemical transducers. Such coupling holds great promise for clinical, environmental, and food analysis. We anticipate that significant developments, particularly the introduction of new tailored electrode surfaces, mass-producible nanoscopic devices, novel and efficient electron relays, or sensor arrays for multianalyte detection will further enhance the power and utility of electrochemical sensors.

REFERENCES

[1] Kissinger P T and Heineman W R (ed) 1984 *Laboratory Techniques in Electroanalytical Chemistry* (New York: Dekker)
[2] Wang J 1994 *Analytical Electrochemistry* (New York: VCH)
[3] Bard A J and Faulkner L 1980 *Electrochemical Methods* (New York: Wiley)
[4] Hart J and Wring S 1994 Screen-printed voltammetric and amperometric sensors for decentralized testing *Electroanalysis* **6** 617–24
[5] Wang J 1991 Modified electrodes for electrochemical sensors *Electroanalysis* **3** 255–9
[6] Frew J and Hill H A O 1987 Electrochemical biosensors *Anal. Chem.* **59** 933A–9A
[7] Vreeke M, Maiden R and Heller A 1992 Hydrogen peroxide and NADH sensing amperometric electrodes based on electrical connection of HRP to electrodes *Anal. Chem.* **64** 3084–90
[8] Heller A 1990 Electrical wiring of redox enzyme *Accounts Chem. Res.* **23** 128–34
[9] Wang J, Lu F, Angnes L, Liu J, Sakslund H, Chen Q, Pedrero M, Chen L and Hammerich O 1995 Remarkably selective metallized-carbon amperometric biosensors *Anal. Chim. Acta* **305** 3–7

[10] Czaban J 1985 Electrochemical sensors in clinical chemistry *Anal. Chem.* **57** 345A–50A

[11] Guilbault G and Montalvo J 1970 Potentiometric urea electrode *J. Am. Chem. Soc.* **29** 2533–5

[12] Papariello G, Mukherji A and Shearer C 1973 Penicillin enzyme electrode *Anal. Chem.* **45** 790–4

[13] Thompson J, Mazoh J, Hochberg A, Tseng S and Seago J 1991 Enzyme-amplified rate conductimetric immunoassay *Anal. Biochem.* **194** 295–301

6

Modification of sensor surfaces

P N Bartlett

6.1 INTRODUCTION

In this chapter we review the various approaches available for the chemical modification of surfaces and their applications to the field of chemical and biological sensors. Surface modification is important in many areas, but is particularly so in chemical and biological sensors because of the central role played by the interface between sensor and sample. Chemical modification of sensor surfaces is used in two major ways: first to attach selective groups (binding sites or catalysts) to the sensor surface in order to recognize target species in the sample, and second to increase the selectivity of the sensor by reduction of interferences arising from non-specific interactions. Although these two purposes are different the repertoire of chemical modification techniques used in the two applications is similar. In essence the surface chemistry is used to design the interface at the molecular level and thus to control the interactions between the surface and the different species present in the analyte sample.

Surface modification techniques can be divided up in a number of ways. For example we can distinguish between covalent and non-covalent surface modification strategies. In the covalent approaches a chemical bond is made between the surface and the attached species. This tends to be an irreversible process. The nature of this bond and the types of modification which can be carried out depend on the particular surface and the chemical functional groups present on the surface. In non-covalent approaches weaker, non-bonding, interactions between the surface and the adsorbed species are utilized. These arise from charge–charge, charge–dipole, dipole–dipole, dipole–induced dipole, and induced dipole-induced dipole interactions, also referred to as van der Waals interactions, between the surface and the adsorbed molecules [1]. In general these are less surface specific than the covalent approaches and non-covalent surface modification can often be more readily achieved than covalent modification.

Surface modification techniques can also be distinguished on the basis of the thickness of the modified film generated at the surface. Thus techniques can be used which produce either monomolecular layers or multilayers of material at the surface, going up to thicknesses of several micrometers or more. The choice between the two depends on the application.

There is an extensive literature on the use of surface modification and surface attachment relating to several areas including chromatography, solid phase peptide synthesis, protein modification [2, 3], and enzyme immobilization [4]. For example much of the original work within the area of chemically modified electrodes [5, 6] grew out of developments in surface modification as applied within chromatography, and expertise and experience in surface modification of chromatographic supports continues to find application in the design of gas sensor arrays [7–9].

The chemical and biochemical groups which can be attached to surfaces is almost unlimited, ranging from simple chemical species to large, and complex, macromolecules such as enzymes and antibodies. Obvious examples include the attachment of chelating groups, or binding sites, to provide molecular recognition at the sensor surface. Charged groups and charged polymeric films have been attached to surfaces to reduce interference from ionic species, for example in the form of layers to reduce interference and as films to reduce biological fouling of the surface. Redox active sites have been attached to surfaces to act as catalytic sites for the oxidation or reduction of target species; other catalytic sites have been attached to surfaces to enhance the response of the sensor towards particular species or classes of species.

In the following sections we discuss the different approaches towards the modification of surfaces and we describe representative applications of these approaches in chemical and biological sensors.

6.2 COVALENT MODIFICATION OF SURFACES

The covalent modification of a surface requires the formation of a bond to some functional group on the surface. The nature of the groups available for modification depends upon the material to be modified and pretreatments applied to this material. It the following sections we discuss the different surface modification reagents in turn.

6.2.1 Silanes

Reactive organosilanes have been widely used to modify surfaces by reaction between the organosilane and oxide or hydroxide groups on the surface [6]. Oxide surfaces form an important class of surfaces for modification and include not only materials such as Al_2O_3, SiO_2, TiO_2, RuO_2, and SnO_2, but also the oxidized surfaces of metals and semiconductors, although the latter are often not as well characterized. For example the surfaces of platinum, gold, germanium, silicon, and gallium arsenide can be modified by silane coupling provided they are first suitably treated in order to produce a thin oxide layer. Note that although carbon surfaces may be oxidized and hence possess oxygen containing functional groups (such as –OH and $-CO_2H$), silane modification does not generally work well at these surfaces and other methods are preferred for their modification (see below). Typically this thin oxide layer may be only one or two monolayers thick. For example, it is known that two monolayers of oxide can be formed on platinum but that only the outer layer reacts with the silane [10]. Furthermore electrochemical experiments show that the inner oxide layer can be reduced without loss of the bound silane.

The reaction between the organosilane (R–Si–X) and the oxidized surface (M–O–H) is generally carried out by treating the surface under anhydrous conditions with a solution of the silane in a suitable solvent, although treatment with gas phase silanes has also been used. Depending upon the reactivity of the silane and the conditions used one or more M–O–Si linkages per silane molecule can be formed at the surface. This leads to surface coverages of the silane of the order of $1–5 \times 10^{-10}$ mol cm^{-2}, thus the surfaces are only loosely packed with the attached groups in a semifluid state. The basic chemistry of the process is shown in figure 6.1.

The coupling of organosilanes to the surface alters the interfacial properties of the surface because the polar oxide and hydroxide groups originally present at the interface are replaced by organic groups. In the case of the metal–, or oxide–, solution interface this is manifest in a change in the interfacial capacitance. Silanization of the surface also alters the wettability (the hydrophobic/hydrophilic nature) of the surface. This effect has been utilized to guide the growth of thin films of conducting polymers across the surface of glass. For example Nishizawa *et al* [11] have used alkylsilane treatment of glass to promote the lateral growth of poly(pyrrole) films and have used this in the fabrication of pH-sensitive microarrays [12]. Subsequently the same group have demonstrated that by patterning the areas of substrate treated with silane they can guide the direction of poly(pyrrole) film growth from an electrode across the insulating glass substrate [13]. Systematic studies using silanes with different-length alkyl chains attached showed that the ability to guide polymer growth correlated with published data for the hydrophobicity of the silanized SiO_2 surfaces as determined by contact angle measurements [14]. In addition, when the growth solution was changed from water to acetonitrile, silanization was no longer effective in guiding growth.

Silane coupling is also widely used as a method of introducing other functional groups onto the surface to which covalent bonds can be made, and other molecules thus attached. For example alkylamine silanes have been widely used in this manner and a variety of molecules have been attached to surfaces using this strategy, including redox active groups, fluorescent groups, ion binding sites, and other types of molecular functionality [6]. Figure 6.2 shows examples

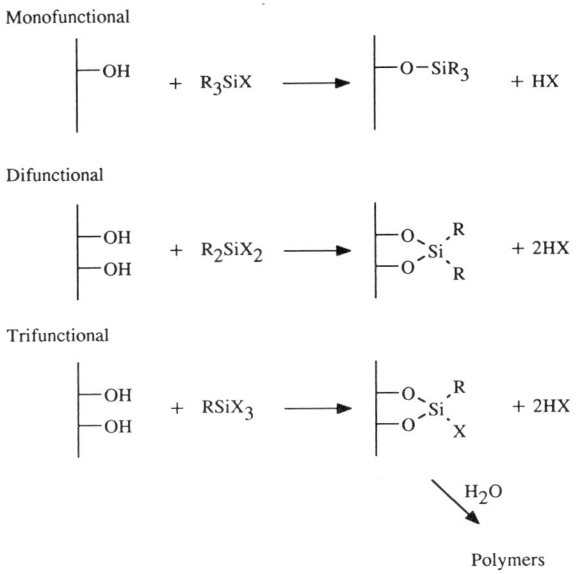

Monofunctional

$$\text{—OH} + R_3SiX \longrightarrow \text{—O—SiR}_3 + HX$$

Difunctional

$$\begin{matrix}\text{—OH}\\\text{—OH}\end{matrix} + R_2SiX_2 \longrightarrow \begin{matrix}\text{—O}\\\text{—O}\end{matrix}\!\!Si\!\!\begin{matrix}R\\R\end{matrix} + 2HX$$

Trifunctional

$$\begin{matrix}\text{—OH}\\\text{—OH}\end{matrix} + RSiX_3 \longrightarrow \begin{matrix}\text{—O}\\\text{—O}\end{matrix}\!\!Si\!\!\begin{matrix}R\\X\end{matrix} + 2HX$$

$$\downarrow H_2O$$

Polymers

Figure 6.1 The reaction between organosilanes and surface hydroxyl groups (silanization). In these structures X is a reactive group such as Cl or $O(CH_3)$. Depending on the number of reactive groups on each silane one or two links can be made to the surface. Trifunctional silanes can go on to give polymeric multilayers in the presence of traces of water.

of the chemistry which can be developed in this manner. In addition to the widespread use of alkylamine silanes other organosilane derivatives have also been used in similar ways [6], examples are shown in figure 6.3. It is also known that under some conditions cross-linking between the attached silanes can occur leading to the formation of polymeric silane films covalently bound to the surface (see below).

Silanization can be used to modify the surface of quartz crystal oscillators and surface acoustic wave (SAW) devices. For example Thompson and colleagues have used aminopropyltriethoxysilane to treat the surfaces of SAW devices to obtain sensitivity towards nitrobenzene derivatives [15]. In this case the selectivity arises through complementary hydrogen bonding interactions between the amine groups attached to the SAW surface and the nitro groups of the analyte. This same surface treatment has been used as a precoat to attach biomolecules to quartz crystal oscillators through modification of the surface-bound amine groups [16].

6.2.2 Surface esters

In addition to modification by silanization, oxide surfaces can also be modified by the formation of surface esters. This has been used with SnO_2 and TiO_2 to

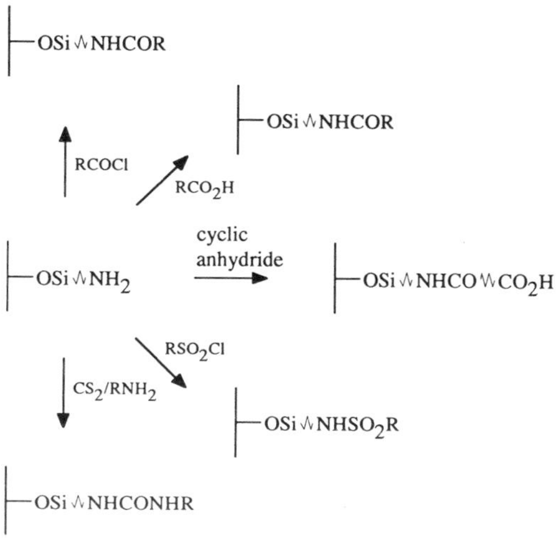

Figure 6.2 Development of surface chemistry starting from an alkylamine silane-modified surface. ╱╲ represents a generalized linking chemical group such as an alkyl chain.

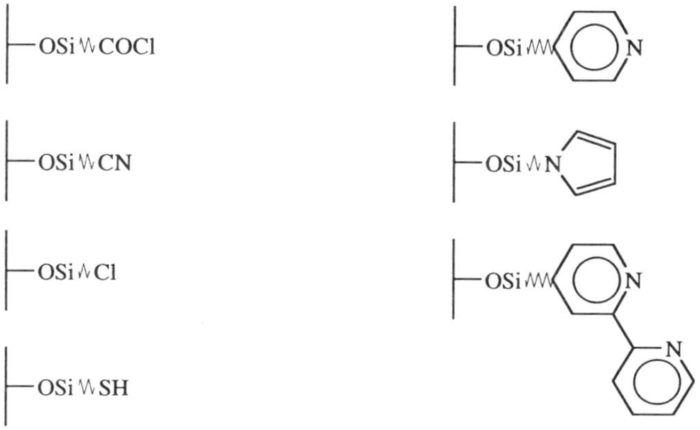

Figure 6.3 Some examples of functional groups which have been attached to surfaces by use of the corresponding reactive organosilanes. ╱╲ represents a generalized linking chemical group such as an alkyl chain.

attach species to the surface. Surface coverages appear to be around one-tenth of a monolayer and the modified surfaces appear less stable than those formed by silanization [6].

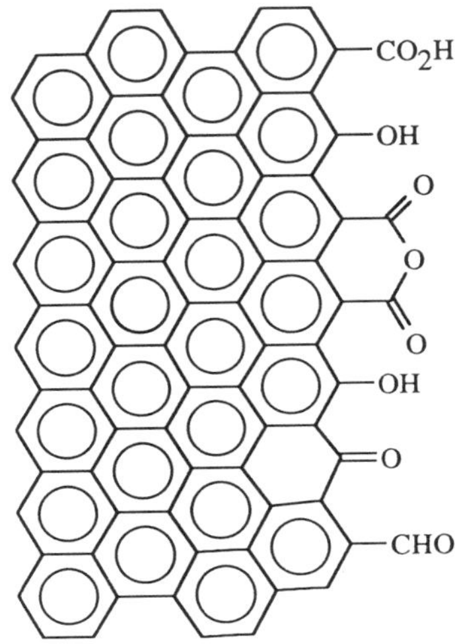

Figure 6.4 A schematic representation of a graphite surface showing the basal plane region and the types of oxygen functionality found at the edge plane.

6.2.3 Cyanuric chloride

Cyanuric chloride is a trifunctional reagent which can be used to react with surface hydroxylic groups. Although it has been used with metal oxide surfaces, it is generally less successful than organosilane modification for this purpose. However it is a useful modification procedure for use with carbon surfaces. The graphite structure comprises large sheets of fused aromatic rings in coplanar stacks. The basal plane surface of the material is of low polarity, π electron rich, and hydrophobic. However where these sheets terminate at the edge plane, or are disrupted, for example by polishing or mechanical damage, the surface is covered in oxygen containing functionalities (see figure 6.4). It is these oxygen containing surface functional groups which can be modified by various reagents. Carbon is available in many forms, including highly orientated pyrolytic graphite (HOPG), pyrolytic graphite, glassy carbon, and various compacted polycrystalline forms and powders, and these all have different surface oxygen contents. Furthermore the coverage of surface oxygen containing groups can be enhanced by oxidative chemical or electrochemical treatments. These produce a variety of surface functional groups such as carboxylic acids, quinones, lactones, hydroxyls, aldehydes, and ketones [17].

The use of cyanuric chloride to modify carbon surfaces occurs in two steps. First the surface is reacted with the cyanuric chloride to produce an activated surface. Then this activated surface is further reacted to attach the group of interest (see figure 6.5). The immobilized cyanuric group is reactive towards alcohols and amines [18] and has been used to attach enzymes [19], as well as

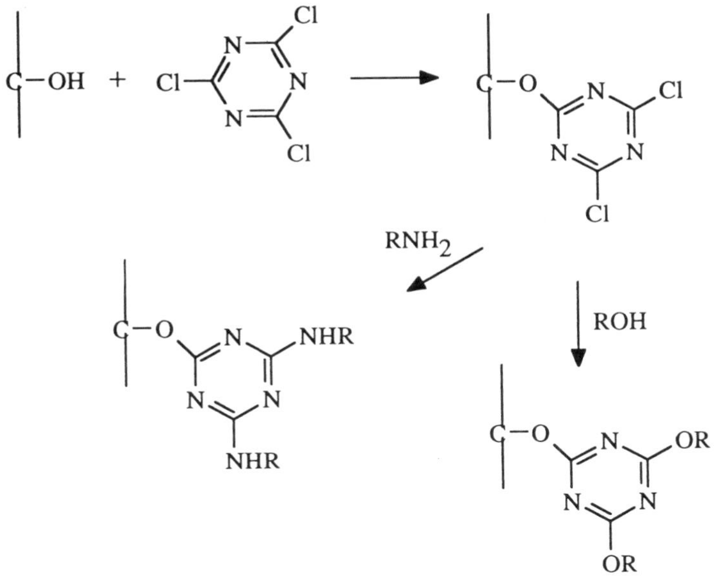

Figure 6.5 The use of cyanuric chloride to couple to hydroxyl groups on surfaces.

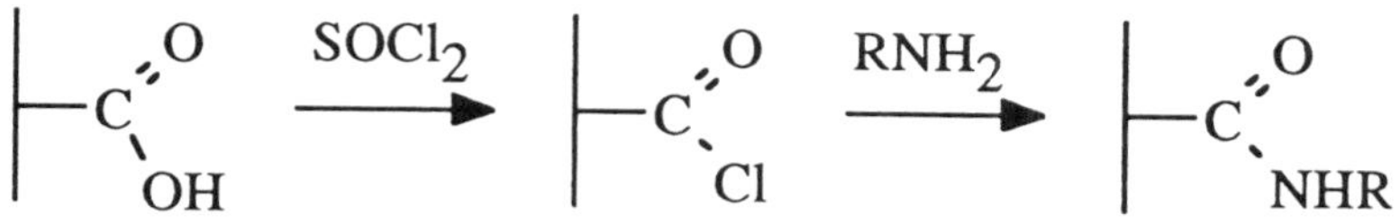

Figure 6.6 The use of thionyl chloride ($SOCl_2$) to activate polyacrylic acid surfaces.

a variety of other functional molecules [20, 21], to carbon surfaces. Cyanuric chloride can also be used to attach molecules to polymer surfaces provided they have suitable functionality. For example it has been used to attach proteins to poly(ethylene glycol) [22].

6.2.4 Other reagents

Thionyl chloride can be used to activate surface carboxylic groups for subsequent modification with amines and esters. This approach has been used with oxidized carbon surfaces [23, 24] and for the activation of poly(acrylic) acid surfaces prior to diazo coupling [25] (see figure 6.6).

Surface carboxylic acids can also be activated by using diimides, such as 1, 3-dicyclohexylcarbodiimide (DCC) or 1-ethyl-3-(dimethylaminopropyl)carbodiimide (EDC). Activation with diimides is a mild procedure which takes advantage of reagents developed for peptide synthesis (see figure 6.7). This approach can be used to couple simple molecules to surfaces or for the immobilization of larger entities such as enzymes. This is illustrated by a

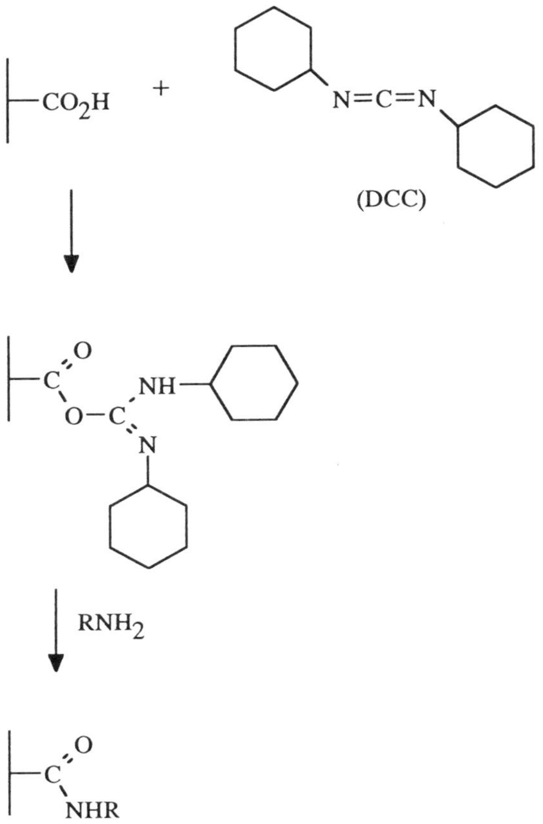

Figure 6.7 Activation of an oxidized carbon surface using a carbodiimide (in this case DCC).

recent paper by Wu *et al* [26] which describes the immobilization of glucose oxidase (molecular weight 180 000) at carbon surfaces using 1-cyclohexyl-3-(2-morpholinoethyl)carbodiimide metho-*p*-toluene sulfonate (CMCI) to form amide linkages between carboxylic acid groups on the carbon surface and amine residues on the protein. This approach was shown to work for oxidized surfaces of both HOPG and carbon fibers. As an alternative to the direct immobilization of the enzyme, Pantano and Kuhr [27, 28] have described the use of EDC to couple biotin, through a diamine chain, to carboxyl groups on carbon surfaces. These biotinylated surfaces were then modified with biotinylated glutamate dehydrogenase by using avidin to complex to both the biotin residues on the surface and on the modified enzyme (see figure 6.8). This approach makes use of the very strong interaction ($K_d = 10^{-15}$ mol l^{-1}) between each avidin molecule and up to four biotin molecules. Similar approaches have been demonstrated for immobilization of horseradish peroxidase [29].

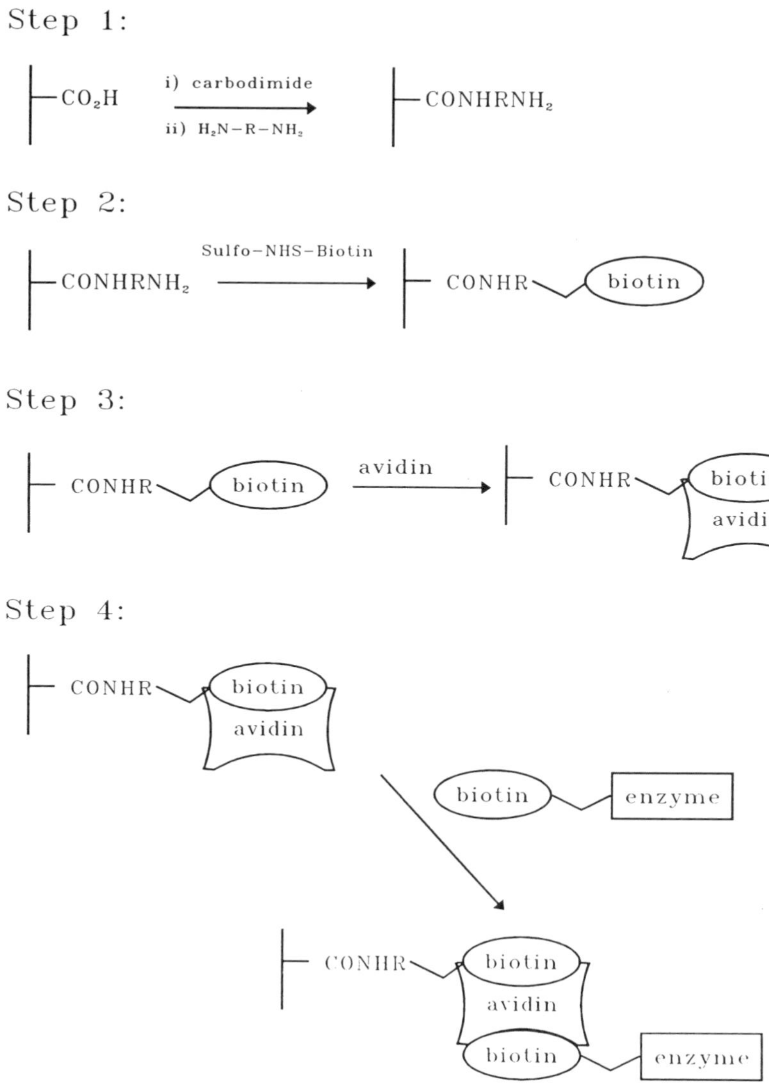

Figure 6.8 The scheme used by Pantano and Kuhr to attach glutamate dehydrogenase to the surface of a carbon fiber. Step 1, after electrochemical oxidation of the fiber to produce surface oxygen functionalities, a diamine is attached to the surface using carbodiimide coupling. Step 2, biotin is attached to the other end of the diamine. Step 3, the bound biotin is reacted with avidin to form a biotin–avidin complex attached to the carbon surface. Step 4, biotinylated enzyme is complexed to the avidin to complete construction of the enzyme modified carbon fiber surface. (Adapted from [27].)

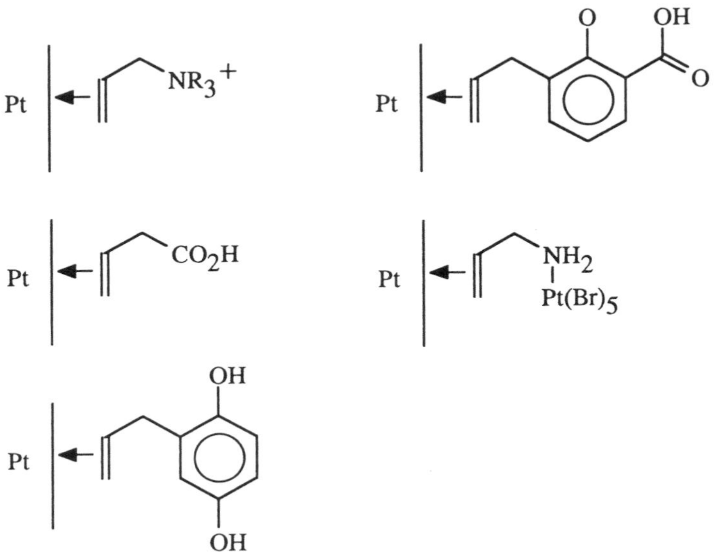

Figure 6.9 Examples of alkenes which undergo irreversible adsorption onto platinum surfaces through π electron interactions. (Adapted from [6].)

6.3 SELF-ASSEMBLED MONOLAYERS AND ADSORPTION

Covalent modification of surfaces is, in general, an irreversible process. The formation of self-assembled monolayers and adsorbed layers relies on weaker, non-bonding interactions between the surface and the adsorbing molecules. An advantage of this approach is that the adsorbed molecules often form close-packed arrays at the surface. A disadvantage is that the modified surfaces can be less stable because the interactions with the surface are weaker.

6.3.1 Chemisorption

Chemisorption onto metal and carbon surfaces has been widely investigated and the adsorbed layers characterized in some detail. For example the surface of platinum can be modified by chemisorption of substituted alkenes. This is an irreversible process in which the π electrons of the alkene interact strongly with the metal (figure 6.9). Chemisorption can also be used to modify carbon surfaces, although in this case it is generally advantageous to use molecules with aromatic π electron systems to interact with the π electrons on the graphitic carbon. This approach has been used to modify carbon electrodes with a variety of redox mediators for use as catalysts for NADH oxidation in biosensors [30–34] (figure 6.10). An advantage of the chemisorption approach is the ease with which the modification is achieved. The surface is first polished on wet, fine emery paper and then washed thoroughly with deionized water before exposure

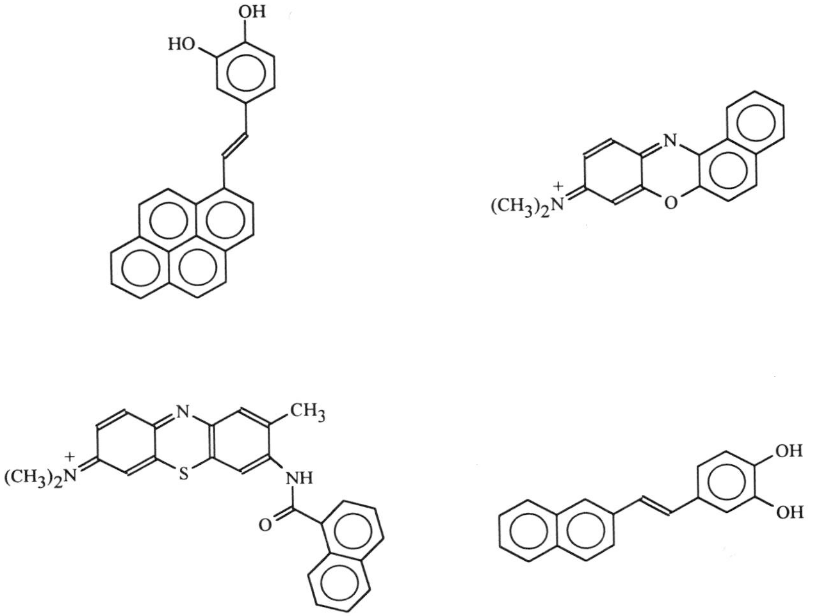

Figure 6.10 Examples of four redox dyes (4-[2-(1-pyrenyl)vinyl]catechol, meldola blue, 3-β-naphthoyl-toluidine blue O, and 4-[2-(2-naphthoyl)vinyl]catechol) used to modify carbon surfaces for the mediation of NADH oxidation. Each dye molecule has an aromatic substituent attached in order to promote adsorption onto the basal plane regions of graphite through π–π interaction. The examples are taken from [30, 31, 33, 34].

to a solution containing the adsorbate dissolved in ethanol. Adsorption of the surface modifying species then occurs spontaneously.

Chemisorption has been widely applied in the modification of electrode surfaces for protein electrochemistry following the pioneering work by Hill and his colleagues in this area [35]. They showed that by adsorption of suitable molecules at the electrode surface they were able to control and promote the direct electrochemistry of large redox proteins such as cytochrome *c*. The general features of the promoters used in these studies are shown in figure 6.11. They comprise one functional group to provide a chemisorptive link to the metal surface and a second functional group to interact with, and orient, the protein at the interface [36]. In this application it is important that the linkage between these two functional groups be rigid and of the correct geometry to orient the binding group towards the solution in order for it to interact with the protein. Adsorbed amino acids [37] and metal ions [38] have also been used to modify electrode surfaces for direct electrochemistry of redox proteins. Taking this further the promoter can also be used to immobilize the redox protein at the electrode surface as part of a sensor [39, 40].

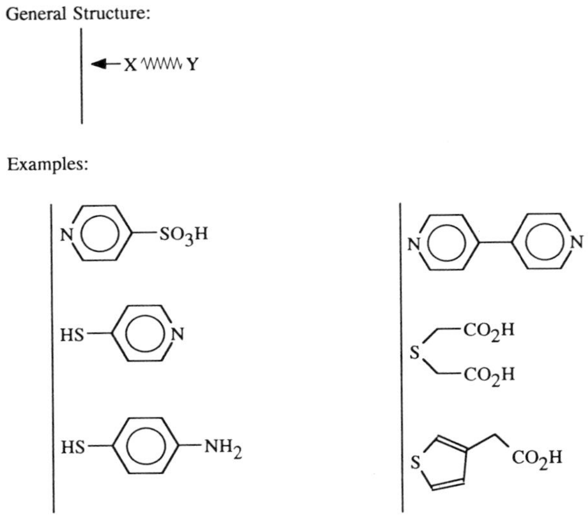

Figure 6.11 The general features of the bifunctional molecules shown by Hill *et al* to promote cytochrome *c* electrochemistry at gold electrodes. X is a functional group able to interact with the electrode surface and Y is a functional group which interacts with the protein. Also shown are some examples of successful promoters. (Adapted from [36].)

6.3.2 Self-assembly

Hydrophobic interactions between adsorbed molecules can lead to the formation of close-packed monolayers at surfaces. This has been exploited in studies of alkane thiol films formed at gold [41–43] and platinum [44] surfaces in order to produce organized 'self-assembled' films. The strong interaction between gold surfaces and sulfur (about 150 kJ mol^{-1}) [45] leads to irreversible adsorption of monolayers in which the hydrocarbon tails are oriented nearly perpendicular to the surface in an extended all-*trans* conformation driven by the hydrophobic interactions between the alkane tails. These types of monolayer have been exploited to introduce a number of different functional groups onto surfaces with the close-packed hydrocarbon chains blocking diffusion of species to the underlying gold surface (figure 6.12). One concern with these types of film is the degree of perfection of the film on the molecular scale. It is commonly observed that these films contain pinholes on the molecular scale and larger. These are observed through their effect on the interfacial capacitance and electrochemistry of the gold films [46, 47]. Using longer poly(methylene) chains generally gives better quality films, presumably because the hydrophilic interchain interactions are more significant for longer tails. Various strategies for filling in these pinhole defects in the films have been suggested including the use of electropolymerized films [48].

In principle these molecular scale pinholes can also be exploited to create

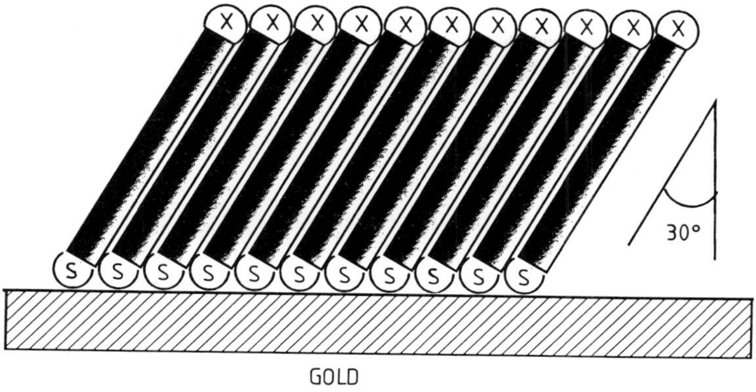

Figure 6.12 The structure of self-assembled alkylthiol films at a gold surface. In general the molecules form close-packed layers with the polymethylene chains tilted at about 30° from the vertical. The interfacial properties of the films are determined by the end groups, X, of the chains. (Adapted from [50].)

molecular recognition sites, but as yet this is in a very early stage [49] and problems of stability and reproducible control of the film fabrication need to be addressed further. These schemes for making functional monolayers rely on the ability to make mixed monolayers by using mixtures of alkanethiols to carry out coassembly. The successful coassembly of mixed layers has been demonstrated from mixtures of long-chain alkanethiols with different chain lengths; however, the compositions of the mixed layers are not identical to the compositions of the reagents in solution [50, 51]. Mixed layers have also been made using mixtures of functionalized alkanethiols with unfunctionalized alkanethiols used to 'dilute' the functionality on the modified surface [52].

The ability to form self-assembled monomolecular films and to control their composition has been applied in a number of ways as a method for controlling surface properties. For example Song *et al* used self-assembled films of alkanethiolates to form modified surfaces on gold for direct oxidation and reduction of cytochrome *c* [53, 54]. This type of approach has also been used to make ion-sensitive monomolecular layers, for example by using the pH dependence of blocking by diundecylphosphate dithiol films to make ion gating layers [55], or by immobilizing ion binding sites within alkanethiol films [56, 57]. Recent work has shown that it is possible to pattern self-assembled alkanethiol films using photochemical techniques to modify parts of the surface of the film [58]. Using patterned alkanethiol films it is possible to control cell adhesion by control of the surface chemistry [59], thus there is considerable scope for further developments in this area.

Other types of self-assembled film include those from long-chain siloxane monolayers on glass, quartz, Al, Ge, and ZnSe [60–64] and pyridine derivatives on platinum [65]. In addition it is possible to build up self-assembled films by following a sequence of self-assembly, chemical activation, and then further self-assembly [56, 57, 63, 66–72]. However the addition of each layer is accompanied by increasing disorder. Self-assembled films can also be prepared using *n*-alkanoic acids on alumina [73, 74].

6.3.3 Langmuir–Blodgett films

The more traditional and widely studied method for preparing ordered monolayer and multilayer films is the Langmuir–Blodgett technique [75] (see chapter 9). In this approach pinhole-free monomolecular layers are assembled at an air–water interface and then transferred onto a surface (figure 6.13). Strong interactions between the substrate and the film are not essential but such interactions do enhance film stability. Multilayer films can be built up by repeated dipping of the same surface through the air–water interface. Successful formation of organized monolayers by this technique requires the correct balance of interactions between the molecules and between the molecules and the water at the air–water interface. Typical Langmuir–Blodgett film forming molecules possess a hydrophilic head group and a hydrophobic tail. When these molecules are applied to the air–water interface they spread out with their hydrophilic head groups in the water phase and the hydrophobic tails sticking out from the water. Compression of the molecules spread on the water surface, using a Teflon or similar barrier, produces a two-dimensional close-packed molecular array which can then be transferred to a solid support. If a hydrophilic solid is used the head groups of the film will be in contact with the solid, if a hydrophobic support is used the film will be the other way round. An attraction of the Langmuir–Blodgett technique is the potential for control over film thickness, orientation, and composition at the molecular level. However this is not achieved without some problems arising from artifacts during preparation [76] and it is important to take precautions over cleanliness for successful film preparation.

The ability to control molecular film formation has led to studies of applications of these films in many areas, including their use as gating layers for ion detection [77–79] and as films containing active gate molecules [80, 81]. Enzymes can also be formed into active Langmuir–Blodgett films which can them be transferred to solid surfaces. Studies of Langmuir–Blodgett films of glucose oxidase have shown that the activity of the films can be increased by glutaraldehyde treatment of the enzyme before Langmuir–Blodgett film formation. This apparently prevents denaturing of the enzyme at the air–water and air–solid interfaces by intermolecular and intramolecular cross-linking [82]. The Langmuir–Blodgett technique has also been used to fabricate gas sensors by depositing layers of gas-sensitive materials, such as metallophthalocyanines and substituted phthalocyanines, at solid surfaces [83–85].

6.3.4 Phospholipid films

Surfaces can also be modified by adsorbed films of phospholipids and this is an

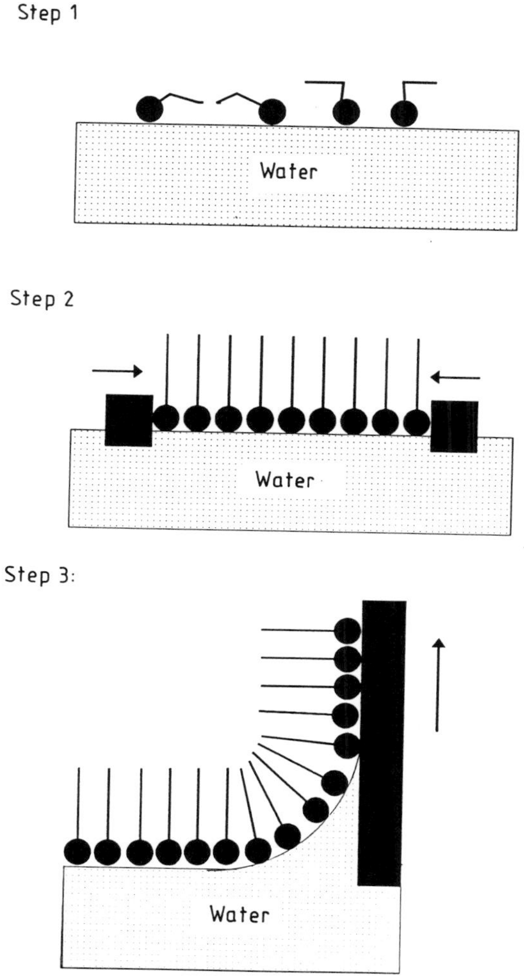

Figure 6.13 Deposition of a Langmuir–Blodgett film. Step 1, the amphiphilic molecules are spread on the air–water interface. Step 2, the molecules are compressed by movable barriers to form a close-packed monolayer. Step 3, the monolayer is transferred to a solid surface by dipping through the air–water interface. Multilayer films can be built up by repeated dipping through the interface.

expanding area of interest because the resulting modified surfaces resemble, to some extent, the surfaces of biological membranes [86]. Phospholipid monolayers can be readily assembled on mercury surfaces where they alter the interfacial properties and thus the observed electrochemistry [87]. Similar films on platinum electrodes have been used as interfaces to redox proteins [88], as permselective coatings in electrochemical detection [89], and on mass-sensitive devices for measurements in the gas phase [90–92] and in solution [93, 94]. The ability to mimic biological interfaces can be a significant advantage for sensor surfaces designed to function with biological samples, such as whole blood, or *in vivo* where surface fouling can be a problem. The work of Chapman has shown that by modifying surfaces with the same phospholipid head group as found on the outside of blood cells (phosphorylcholine) hemocompatible surfaces can be created [95]. Thus Chapman and colleagues have used Langmuir–Blodgett and dip coating techniques to coat surfaces with mixtures of lecithin and diacetylene phospholipids with the phosphorylcholine head group [96], prepared adsorbed coatings of phosphorylcholine containing methacrylate-based polymers on PVC, polyethylene cellulose and stainless steel [97], and used phosphorylcholine derivatives as plasticizers in polymers such as PVC and polyurethane [98]. Phospholipid monolayers supported on a variety of surfaces have also been used to provide a biocompatible environment within which to immobilize, or entrap, a variety of biologically active molecules. Examples include the immobilization of peptide ionophores on mercury surfaces [78, 79], binding of monoclonal antibodies [99], immobilization of glucose oxidase to biotinylated phospholipid layers [100], and immobilization of enzymes and mediators at phospholipid-coated aluminum oxide surfaces [101, 102].

6.4 POLYMER-COATED SURFACES

Another widely used approach to the chemical design of sensor surfaces is the use of multilayer or polymer films. These can be prepared in a variety of ways, either by the direct polymerization of the film onto the surface, or by first preparing the polymer and then coating it onto the surface in a subsequent step (see chapters 13 and 18). These polymeric films can be either covalently attached to the underlying surface or, more frequently, adsorbed onto the surface as the result of non-covalent interactions between the large polymeric molecules and the surface. These include contributions from electrostatic [103] and acid–base interactions [104].

One advantage of using polymeric films to modify sensor surfaces is that they can contain many monolayers of active sites and this can be helpful in increasing the response, or in prolonging the useful lifetime, of the interface. In addition the thicker films available in this way can form useful diffusional barriers to interferent species.

The methods available for the polymer modification of surfaces are quite varied and we now briefly review each in turn, starting with those methods suitable for use with preformed polymeric materials. One advantage of using such preformed materials is that they can be purified and well characterized, in terms of linkage isomerism, molecular weight distribution etc, before they are applied to the sensor surface.

6.4.1 Spin coating

This is a commonly applied technique in the microelectronics industry used to deposit thin uniform polymeric films. In this method the polymer, dissolved in a suitable solvent, is dropped onto the substrate whilst it is being spun at high speed. As a result the solution spreads out to form a thin uniform film across the substrate surface which, on evaporation of the solvent, leaves a polymer film over the surface of the substrate. The thickness of the film is determined by the rotation speed and by the viscosity of the polymer solution. The technique is only suitable for coating relatively flat substrates and for the deposition of reasonably thick (1 μm) pinhole-free films. Once deposited the polymer films can be cross-linked by suitable reactive reagents, heat, or light to produce robust, adherent films.

6.4.2 Drop coating

In this approach the polymer is dissolved in a suitable solvent which is then applied as a droplet (of volume several microliters) to the surface. Spreading and evaporation produce a thin film. With this technique the loading of the polymer can be readily calculated from the solution concentration and droplet volume. However, although simple to use, the technique is only really suitable for modification of small areas (up to about 1 cm^2) and there can be problems in achieving uniform film coverage. This approach has also been used with organosilanes to prepare copolymer films held together by –Si–O–Si– bonds formed by hydrolysis [105–111].

6.4.3 Dip coating

This procedure relies on the adsorption of the polymer onto the surface from a dilute solution of the polymer dissolved in a suitable solvent. To some extent the coverage of material can be controlled by control of the time and polymer concentration. This approach has been used with a variety of different polymer types. Again the technique can be combined with the use of hydrolytically unstable polymers able to form cross-links once deposited onto the surface.

6.4.4 Applications of polymer-coated surfaces

Polymer films have been very widely applied to modify the surfaces of chemical sensors both for solution and gas phase measurements. Solution coating has been used to coat quartz crystal oscillators [91–93, 112] and SAW devices [113, 114] with various polymeric adsorbates and polymer/lipid mixtures [90, 94] to prepare arrays of sensors for use in gas sensing and odor evaluation and discrimination.

Various modified polymers and functional polymers have been used to coat electrode surfaces either to promote electrocatalysis or to act as permselective barriers [115, 116] and thus increase selectivity. The work in this area has been

Redox polymer:

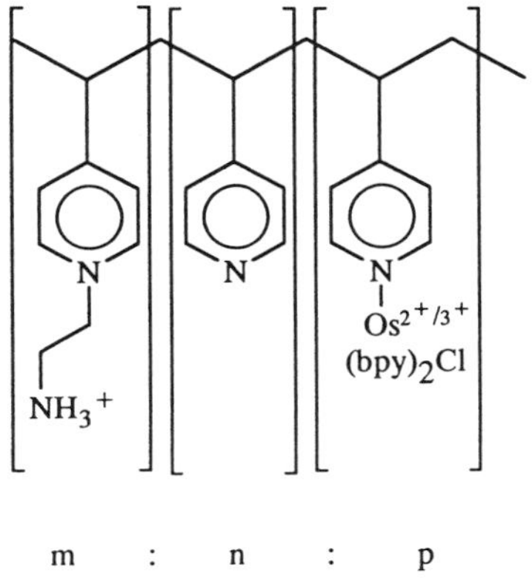

Cross linking agent:

Cross linking reaction:

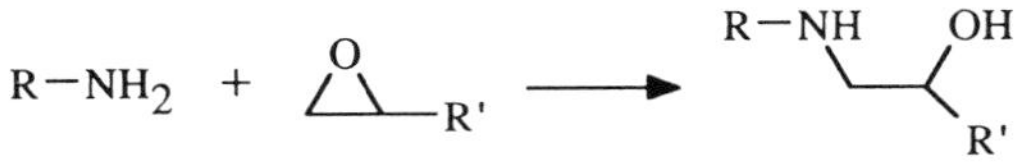

Figure 6.14 Components of an osmium containing redox hydrogel used to immobilize enzymes at electrode surfaces. Typically $m = 1.2$, $n = 4$, $p = 1$, and $q = 10$. Cross-links between the redox polymer chains are formed by reaction between the water-soluble diepoxide and amine groups on the polymer chains. (Adapted from [118] and [123].)

extensively reviewed [5, 117]. Preformed polymers used in this way can be divided into two sorts: those containing covalently bound redox-active, or other catalytic, or binding, sites; and those ion exchange polymers, such as Nafion and quaternized poly(vinylpyridine), into which ions of opposite charge can be electrostatically entrapped.

Hydrogels, cross-linked networks of polymers swollen with water, are a group of redox polymers which are now being widely applied in sensors following the results of recent studies showing that they can be used to mediate electron transfer to entrapped redox enzymes such as glucose oxidase. Work by Heller and colleagues has shown that osmium redox hydrogels can be widely applied as mediators for immobilized flavoproteins and quinoproteins [118–125]. In these systems a water-soluble diepoxide is used to cross-link a film of the polymeric redox mediator containing the enzymes after application to the electrode surface (figure 6.14). In addition to osmium redox centres other redox mediators can be used in a similar way [126].

Polymeric films of immobilized or entrapped ionophores have been widely applied in the fabrication of small, coated wire ion selective electrodes [127] and in the fabrication of ISFETs [128–130] where the ionophore is immobilized on the gate of the FET structure to give an ion selective response.

The polymers used in the work described above have, in general, been simple homopolymers. Recent developments in polymer synthesis, and in particular the use of ring opening metathesis polymerization (ROMP) [131, 132], make it possible to develop and use more sophisticated polymeric systems in which, for example, one end group is designed to bind, or bond specifically, to a surface. The first steps in this direction have been described by Albagli *et al* [133]. They used ROMP, a technique which allows the construction of nearly monodisperse polymer chains in which the nature and order of the block making up the chain can be controlled in a living polymerization process, to prepare redox-active polymer chains which were terminated at one end with different active groups designed to bind to surfaces. These end groups included a triethoxysilane for coupling to oxide surfaces. In principle the development of this type of approach to polymer attachment at surfaces should add a further degree of control over the structure of the interface at the molecular level.

6.5 ELECTROCHEMICALLY GENERATED FILMS

In addition to these general techniques there are a number of more specific techniques which have been successfully applied to coat electrode surfaces with polymer films and which rely in some way on using the electrochemistry to control the polymer deposition. This has some advantages particularly when it is desired to localize the coating to the surface of the electrode or to apply different coatings to an array of closely spaced electrodes. In addition the thickness of the coating can be controlled through control of the electrochemistry.

6.5.1 Electrochemical deposition

This technique makes use of the change in polymer solubility upon change in the ionic charge on the polymer. For example, poly(vinylferrocene) films can be deposited onto the surface of an electrode from dichloromethane by oxidizing the polymer to its less soluble ferricinium form which then adsorbs at the electrode surface [134–137]. In some ways this process is similar to metal plating. Recently this approach has been extended to make use of micellar disruption brought about electrochemically. In this approach the coating material is dissolved in a micellar solution and the micelles are disrupted by oxidation, or reduction, at the electrode surface leading to the deposition of the coating material [138]. This is an interesting and potentially flexible technique for 'electrochemically painting' the surface of a conductor.

6.5.2 Electrochemical polymerization

In electrochemical polymerization a solution of monomer is oxidized or reduced at the electrode surface to generate reactive radical species which couple together and produce an adherent polymer film at the electrode. The films can be electronically conducting, as in the case of pyrroles, anilines, thiophenes, etc; redox conductors in which conduction occurs by self-exchange between discrete redox sites attached to the polymer, as in metal poly(pyridine) complexes; or insulating, as in the case of phenols, 1, 2-diaminobenzene, etc.

In the case of conducting polymers the films can be grown quite thick by continuing growth for a sufficient time. Electropolymerized films of conducting polymers have been widely studied and have found application in a variety of sensor types. Examples of polymers are varied and diverse and one of the attractions of this approach is the ability to attach different substituents to the monomers to tailor the polymer to the application. The mechanism of electropolymerization of pyrrole, shown in figure 6.15, is typical of the reactions involved in film formation. The first step in the process is the generation of a radical cation by oxidation of the monomer. These radical cations can then either react with neutral monomer or can couple with other radicals to give a dimer. The dimers are more readily oxidized than the monomer species and therefore can undergo further oxidation and coupling reactions leading to the generation of oligomers and eventually the deposition of insoluble polymer at the electrode surface. The nature and morphology of the polymer film produced depends on the choice of solvent, counter-ion, and conditions used in the electrochemical polymerization. Because the whole process is electrochemically initiated and driven, the polymer deposition is localized at the electrode surface and is controlled by the applied potential—when the current is switched off polymerization ceases. This control over rate and spatial localization of polymer deposition is attractive in many applications, and particularly where it is desirable to coat arrays of closely spaced electrodes to make a device. This approach has therefore attracted attention as a method for modifying surfaces for various applications and for the immobilization of enzymes at surfaces [139–141]. Using electropolymerization from solutions containing mixtures of monomers it is possible to use the technique to deposit heterostructures by modulation of the deposition potential [142].

By attaching suitable substituents to the electrochemically polymerizable monomers it is also possible to deposit conducting polymer films which incorporate additional functionality. For example films with attached crown ethers [143, 144] and alkyl ether chains [145, 146] have been prepared as possible ion selective materials, and conducting polymer films with pendent

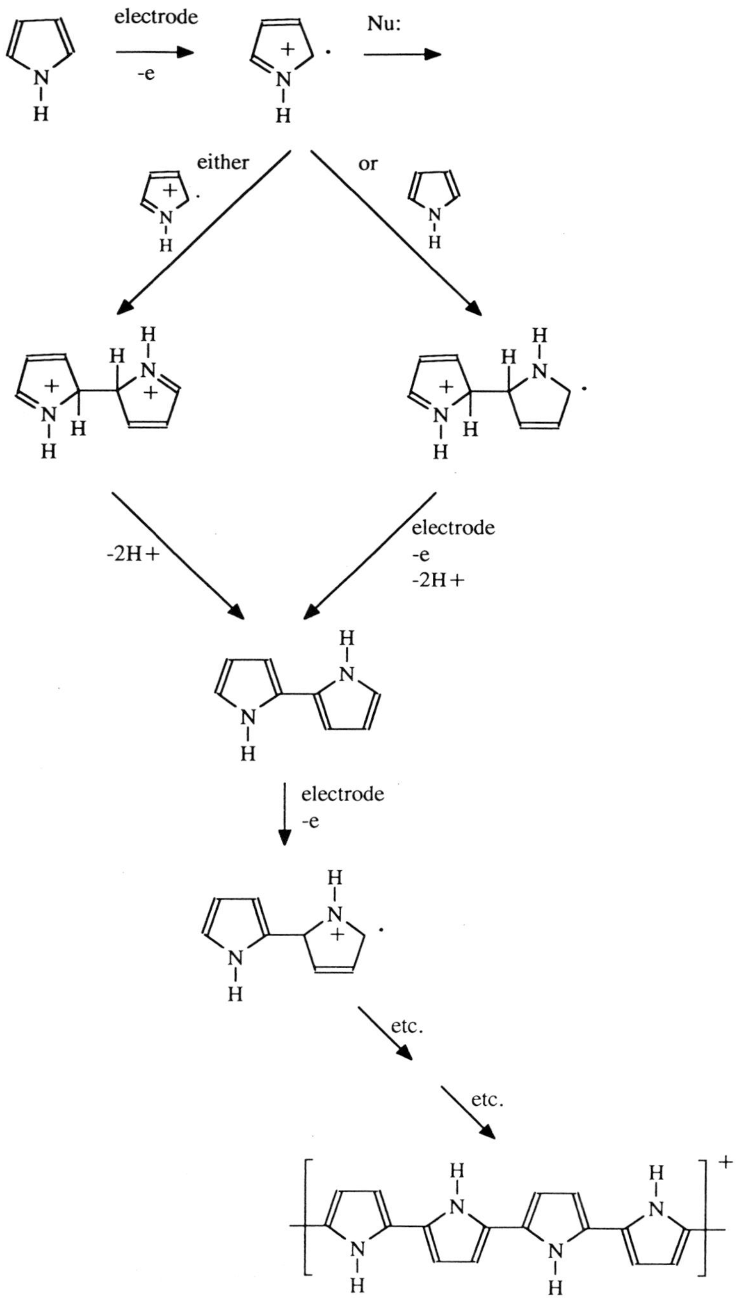

Figure 6.15 A general mechanism for the electropolymerization of pyrrole.

carboxylate groups have been shown to exhibit pH-sensitive electrochemistry [147–150]. This ability to combine conductivity with selectivity is very attractive and continues to receive attention.

Once formed it is is also possible to modify the films either chemically by covalent attachment, or by deliberate overoxidation to destroy the film conductivity. Overoxidized films have been studied as permselective barriers to reduce interference [151, 152] and for enzyme immobilization [153, 154].

Electropolymerized conducting polymer films have also been used to make gas-sensitive chemoresistors. In this application the polymer films are grown across a narrow gap between two electrodes and the resistance of the polymer measured on exposure to the gas sample. Devices of this type have been investigated as sensors for single gases such as ammonia [155], oxides of nitrogen [156], hydrogen sulfide [155], or organic vapors [157–159]. In addition arrays of conducting polymer gas sensors have been used to monitor odor and flavor [160–162]. Electropolymerized films of conducting polymers have also been used to coat the gates of suspended gate field effect transistors (SGFETs) [163, 164] and to coat quartz crystal oscillators [165, 166].

Using the ability to control the electrical conductivity of conducting polymer films Wrighton and colleagues have investigated their use as microelectrochemical transistors [167–170] and have shown that these devices can be used as simple chemical sensors for redox ions [170, 171], and, by the incorporation of catalytic metal particles, hydrogen and oxygen [172]. This has recently been extended by immobilizing enzymes onto this type of device to produce devices which are responsive to the enzyme's substrate, either as the result of direct electrochemical reaction [173, 174], or as the result of local pH changes brought about by the enzyme-catalyzed reaction [175, 176].

Electropolymerized redox polymer films have been used as modified electrodes for the electrocatalysis of solution reactions. In these cases the redox monomers used contain some functionality which undergoes polymerization on oxidation or reduction. A typical example of this is the ruthenium, osmium, and iron complexes of vinylbipyridyl and substituted vinylbipyridyl ligands. On electrochemical oxidation the corresponding poly(vinylbipyridyl) complexes are formed at the electrode surface [177–179]. This type of process has the advantage of electrochemical control over the location and thickness of the deposited film. In addition the resulting films are generally free from pinholes because any pinholes present in the films act as efficient sites for electropolymerization and thus become filled as the polymer grows. These redox polymer films act as molecular sieves differentiating between solution species on the basis of their molecular size [180, 181].

Insulating films can be formed either by using monomers such as phenol [182, 183] or 1,2-diaminobenzene which yield non-conducting polymers or by overoxidation of conducting polymer films as described above. Insulating polymers, such as poly(phenol), form as thin (0.1 μm range) pinhole-free films at the surface. This allows very thin uniform films to be deposited. Films of

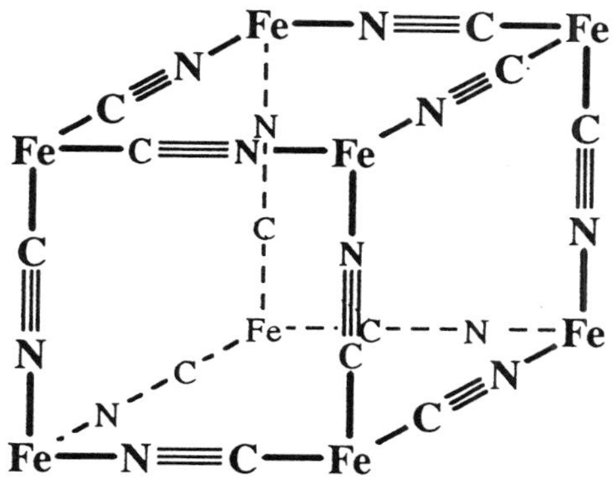

Figure 6.16 The structure of Prussian Blue. The charge balancing counter-ions are located at the centers of the cubes within the structure. Movement of ions in and out of the structure is limited by the size of the pores in the faces of the cube.

this type have been used as permselective barriers to reduce interference from solution species [181, 184–188] and to immobilize enzymes [189–193].

In addition to electrochemical polymerization, reactive monomers can be polymerized onto surfaces by using radio frequency (rf) plasma polymerization [194–197]. In this technique an electric discharge through the vapor forms a reactive plasma that chemically modifies the surface. Examples of applications of rf plasma-polymerized surfaces include the formation of $(C_2F_4)_n$ films on fiber optic sensors for detection of volatile organics [198] and the formation of alkylamine surfaces on glass fibers by plasma treatment for subsequent chemical modification [199].

6.6 OTHER SURFACE MODIFICATIONS

Inorganic lattice structures have been used as surface modifications in a number of applications. In general these materials have the advantages of good stability at elevated temperatures and under oxidizing conditions combined with the potential for size, charge, or shape selectivity based upon the pores within their structures. Four types of inorganic material for surface modification have been widely studied: the metallocyanates, zeolites, clays, and phosphonates.

The metallocyanates are a large class of compounds, the most widely studied of which are probably Prussian Blue ($Na[FeFe(CN)_6]$) in its various oxidation states and nickel hexacyanoferrate ($Na_2[NiFe(CN)_6]$). These materials can be prepared chemically or electrochemically as thin films on surfaces [200]. They differ from redox polymers in that, because of their lattice structure, there is significant interaction between the redox sites within the film and there is strong size–charge selectivity towards incorporated counter-ions. This arises because the counter-ions have to be accommodated within the lattice structure and this

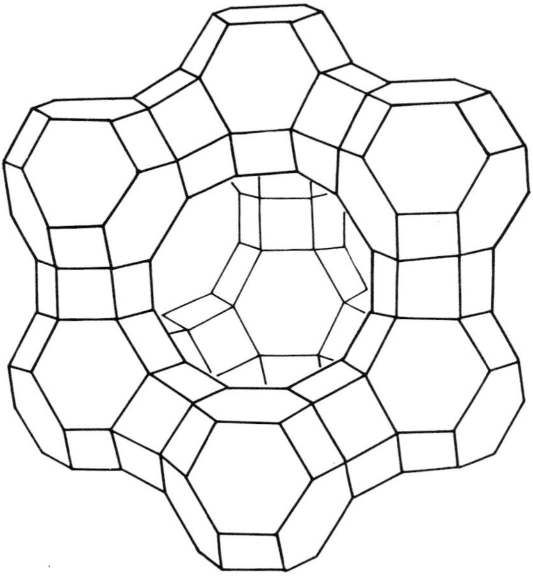

Figure 6.17 An example of a zeolite framework structure (faujasite). The aluminosilicate framework surrounds void spaces which contain the charge balancing cations. Access to these voids is restricted by the size of the apertures between pores. (Adapted from [204].)

places constraints on the size of ions that can be intercalated into the structure (figure 6.16). This size selectivity has been exploited in attempts to develop cation selective electrode surfaces based upon the use of these materials [201–203].

In some ways zeolites are similar in their properties in that they also provide size and shape selectivity because of the presence of pores within the three-dimensional structure. However in the zeolites these pores are generally much larger so that larger molecules can be incorporated into the structure. Zeolites are crystalline aluminosilicates which occur both in nature and as synthetic structures [204]. The zeolite framework is made up of aluminum and silicon atoms tetrahedrally coordinated by oxygen, giving a framework stoichiometry of MO_2. For each aluminum atom within the framework there is a formal charge of -1 which is compensated by counterbalancing cations within pores in the structure. Typically these may be cations such as Na^+, K^+, Ca^{2+}, NH_4^+, or H_3O^+. The size and shape of the void spaces within the zeolite depends on the particular material selected. Figure 6.17 shows a representative structure of this type. This structure gives the material an enormous internal surface area (typically hundreds of square meters per gram), access to which is restricted by the size of the apertures between the pores (typically 0.3–0.8 nm in diameter). Zeolites have been coated onto surfaces in a number of ways. Probably the simplest approach

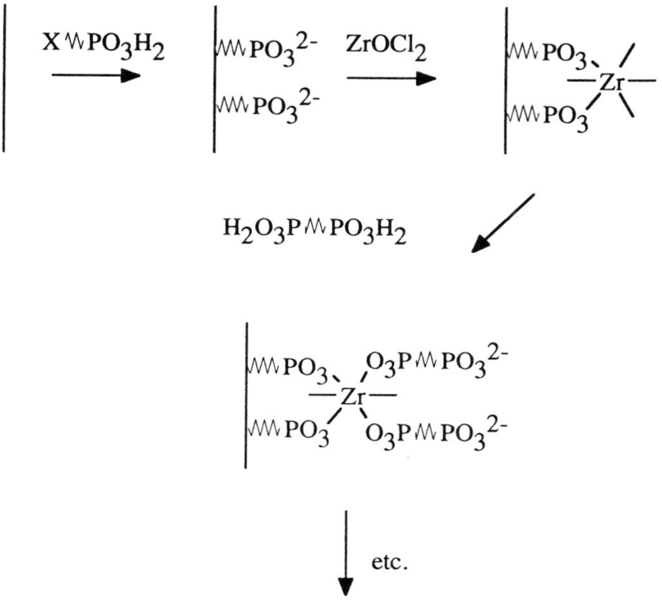

Multilayer Films

Figure 6.18 A scheme for the deposition of zirconium phosphate films by sequential reactions at a surface. (Adapted from [214].)

is to apply the zeolite as a slurry in a suitable solvent with a polymeric binder such as polystyrene [205–209]. This produces fairly thick layers (100 μm). Zeolites have also been covalently attached to surfaces using reactive silanes [210], presumably through reaction between the silane and silanol groups on the zeolite surface. Applications of zeolite-modified surfaces, in general, make use of the size and shape selectivity of the microporous structure. Thus membranes formed by embedding synthetic mordenite crystals in an epoxy membrane have been used to make cesium ion selective membranes [211], and zeolites have been used to preconcentrate ions for electroanalysis [212].

Clays are layered aluminosilicate structures made up of sheets of SiO_4 tetrahedra and AlO_6 octahedra. In the smectite clays, such as montmorillonite, the spacing between these layers can be changed and the clays swollen. In montmorillonite each layer is made up of one AlO_6 sheet sandwiched between two SiO_4 sheets with oxygens at the apices of the SiO_4 tetrahedra shared with the AlO_6 octahedra. These layers are stacked together to form sheets with the layers held together by electrostatic and van der Waals forces. The interlayer spacing depends on the extent of hydration of the exchangeable counter-ions present to balance the charge on the sheets. When exposed to high humidity the clay swells and the spacing increases from about 1 nm for the dried material to about 2 nm in the expanded form. In addition to this swelling these materials can be altered by pillaring. In the pillared clay the interlamellar cations are replaced by bulky, thermally stable, robust cations (such as $[Zr_4(OH)_{16-x}]^{x+}$ or $[Al_{13}O_4(OH)_{28}]^{3+}$) which can act as molecular pillars providing a fixed distance between the layers. These pillared clays show highly selective molecular sieving properties. Surface modification with clays has been achieved by several methods. Films can be cast from colloidal suspensions placed on the substrate surface or by spin coating on to a cleaned preheated surface. For conducting surfaces electrophoretic deposition has also been used by applying an appropriate potential to attract the charged clay particles. Clay-modified electrode surfaces have been investigated as electrocatalysts for a variety of redox reactions [213].

Recently developed techniques for the fabrication of thin-film metal phosphates and phosphonates provide another method for the preparation of layered solids at surfaces. Layers of precisely controlled thickness can be built up by alternate immersion of a suitably pretreated surface in aqueous solutions of a soluble phosphate or phosphonate followed by an appropriate metal salt. This leads to the sequential build-up of thin metal containing films at the surface [72, 214] (see figure 6.18). The method is quite flexible and can be used to build up mixed microporous films on the surface which show molecular sieving properties [215, 216]. This building up approach looks very attractive for the systematic development of thin, selective films.

6.7 CONCLUSIONS

The range of surface modification techniques available for sensors is quite large and continues to grow. While some of these approaches are well established for the modification of the surfaces of sensors others are only beginning to find application as ideas and approaches from the related fields of molecular electronics, self-assembly, and surface electrochemistry find their way into sensor and biosensor applications.

REFERENCES

[1] Israelachvili J 1992 *Intermolecular and Surface Forces* 2nd edn (London: Academic)

[2] Means G E and Feeney R E 1971 *Chemical Modification of Proteins* (San Francisco, CA: Holden-Day)

[3] Imoto T and Yamada H 1989 *Protein Function, a Practical Approach* ed T E Creighton (Oxford: IRL) p 247

[4] Woodward J 1985 *Immobilized Cells and Enzymes, a Practical Approach* (Oxford: IRL)

[5] Murray R W 1992 *Molecular Design of Electrode Surfaces, Techniques in Chemistry* vol 22 (New York: Wiley)

[6] Murray R W 1984 *Electroanalytical Chemistry* vol 13, ed A J Bard (New York: Dekker) p 191
[7] Grate J W and Abraham M H 1991 *Sensors Actuators* B **3** 85
[8] Grate J W, Klusty M, McGill R A, Abraham M H, Whiting G and Andonian-Haftvan J 1992 *Anal. Chem.* **64** 610
[9] Carey W P, Beebe K R, Kowalski B R, Illman D I and Hirschfeld T 1986 *Anal. Chem.* **58** 149
[10] Facci J and Murray R W 1980 *J. Electroanal. Chem.* **112** 221
[11] Nishizawa M, Shibuya M, Sawaguchi T, Matsue T and Uchida I 1991 *J. Phys. Chem.* **95** 9042
[12] Nishizawa M, Matsue T and Uchida I 1993 *Sensors Actuators* B **13–14** 53
[13] Nishizawa M, Miwa Y, Matsue T and Uchida I 1993 *J. Electrochem. Soc.* **140** 1650
[14] Wasserman S R, Tao Y-T and Whitesides G M 1989 *Langmuir* **5** 1074
[15] Heckl W M, Marassi F M, Kallury K M R, Stone D C and Thompson M 1990 *Anal. Chem.* **62** 32
[16] Muramatsu H, Kajiwara K, Tamiya E and Karube I 1986 *Anal. Chim. Acta* **188** 257
[17] Barton S S and Harrison B H 1975 *Carbon* **13** 283
[18] Lin A W C, Yeh P, Yacynych A M and Kuwana T 1977 *J. Electroanal. Chem.* **84** 411
[19] Ianniello R M and Yacynych A M 1981 *Anal. Chim. Acta* **131** 123
[20] Yacynych A M and Kuwana T 1978 *Anal. Chem.* **50** 640
[21] Fox M A, Nabs F J and Voynick T A 1980 *J. Am. Chem. Soc.* **102** 4029
[22] Imoto T and Yamada H 1989 *Protein Function, a Practical Approach* ed T E Creighton (Oxford: IRL) p 260
[23] Lennox J C and Murray R W 1977 *J. Electroanal. Chem.* **78** 395
[24] Lennox J C and Murray R W 1978 *J. Am. Chem. Soc.* **100** 3710
[25] Guilbault G G and de Olivera Neto G 1985 *Immobilised Cells and Enzymes a Practical Approach* ed J Woodward (Oxford: IRL) p 71
[26] Wu H M, Olier R, Jaffrezic-Renault N, Clechet P, Nyamsi A and Martelet C 1994 *Electrochim. Acta* **39** 327
[27] Pantano P and Kuhr W G 1993 *Anal. Chem.* **65** 623
[28] Pantano P and Kuhr W G 1991 *Anal. Chem.* **63** 1413
[29] Pantano P, Morton T H and Kuhr W G 1991 *J. Am. Chem. Soc.* **113** 1832
[30] Jaegfeldt H, Torstensson A B C, Gorton L G O and Johansson G 1981 *Anal. Chem.* **53** 1979
[31] Jaegfeldt H, Kuwana T and Johansson G 1983 *J. Am. Chem. Soc.* **105** 1805
[32] Gorton L 1986 *J. Chem. Soc. Faraday Trans.* I **82** 1245
[33] Persson B and Gorton L 1990 *J. Electroanal. Chem.* **292** 115
[34] Persson B 1990 *J. Electroanal. Chem.* **287** 61
[35] Armstrong F A, Hill H A O and Walton N J 1988 *Accounts Chem. Res.* **21** 407
[36] Allen P M, Hill H A O and Walton N J 1984 *J. Electroanal. Chem.* **178** 69
[37] Di Gleria K, Hill H A O, Lowe V J and Page D J 1986 *J. Electroanal. Chem.* **213** 333
[38] Armstrong F A, Cox P A, Hill H A O, Lowe V J and Oliver B N 1987 *J. Electroanal. Chem.* **217** 331
[39] Cooper J M, Greenough K R and McNeil C J 1993 *J. Electroanal. Chem.* **347** 267
[40] McNeil C J, Greenough K R, Weeks P A, Self C H and Cooper J M 1993 *Free Radical Res. Commun.* **17** 339
[41] Whitesides G M and Laibinis P E 1990 *Langmuir* **6** 87

[42] Dubois L and Nuzzo R G 1992 *Annu. Rev. Phys. Chem.* **43** 437
[43] Bain C D, Troughton E B, Yao Y-T, Evall J, Whitesides G and Nuzzo R G 1989 *J. Am. Chem. Soc.* **111** 321
[44] Mebrahtu T, Berry G M, Bravo B G, Michelhaugh S L and Soriaga M P 1988 *Langmuir* **4** 1147
[45] Nuzzo R G, Dubois L H and Allara D A 1990 *J. Am. Chem. Soc.* **112** 558
[46] Sabatani E and Rubinstein I 1987 *J. Phys. Chem.* **91** 6663
[47] Chidsey C E D and Loiacano D N 1990 *Langmuir* **6** 682
[48] Finklea H O, Snider D A and Fedyk J 1990 *Langmuir* **6** 371
[49] Chailapakul O and Crooks R M 1993 *Langmuir* **9** 884
[50] Bain C D and Whitesides G M 1989 *J. Am. Chem. Soc.* **111** 7164
[51] Bain C D and Whitesides G M 1989 *J. Am. Chem. Soc.* **111** 7155
[52] Finklea H O and Henshaw D D 1992 *J. Am. Chem. Soc.* **114** 3173
[53] Song S, Clark R A, Bowden E F and Tarlov M J 1993 *J. Phys. Chem.* **97** 6564
[54] Tarlov M J and Bowden E F 1991 *J. Am. Chem. Soc.* **113** 1847
[55] Nakashima N, Taguchi T, Takada Y, Fujio K, Kunitake M and Manabe O 1991 *J. Chem. Soc. Chem. Commun.* 232
[56] Rubinstein I, Steinberg S, Tor Y, Shanzer A and Sagiv J 1988 *Nature* **332** 426
[57] Rowe G K and Creager S E 1991 *Langmuir* **7** 2307
[58] Wollman E W, Frisbie C D and Wrighton M S 1993 *Langmuir* **9** 1517
[59] López G P, Albers M W, Schreiber S L, Carroll R, Peralta E and Whitesides G M 1993 *J. Am. Chem. Soc.* **115** 5877
[60] Sagiv J 1980 *J. Am. Chem. Soc.* **102** 92
[61] Maoz R and Sagiv J 1984 *J. Colloid Interface Sci.* **100** 465
[62] Gun J, Iscovici R and Sagiv J 1984 *J. Colloid Interface Sci.* **101** 201
[63] Netzer L, Iscovici R and Sagiv J 1983 *J. Thin Solid Films* **100** 67
[64] Cohen S R, Naaman R and Sagiv J 1986 *J. Phys. Chem.* **90** 3054
[65] Stern D A, Laguren-Davidson L, Frank D G, Gui J Y, Lin C-H, Lu F, Salaita G N, Walton N, Zapien D C and Hubbard A T 1989 *J. Am. Chem. Soc.* **111** 877
[66] Sun L and Crooks R M 1991 *J. Electrochem. Soc.* **138** L23
[67] Finklea H O, Snider D A and Fedyk J 1990 *Langmuir* **6** 371
[68] Nordyke L L and Buttry D A 1991 *Langmuir* **7** 380
[69] Netzer L and Sagiv J 1983 *J. Am. Chem. Soc.* **105** 674
[70] Tillman N, Ulman A and Penner T 1989 *Langmuir* **5** 101
[71] Lee H, Kepley K J, Hong H-G, Akhter S and Mallouk T E 1988 *J. Phys. Chem.* **92** 2597
[72] Li D, Ratner M A, Marks T J, Zhang C, Yang J and Wong G K 1990 *J. Am. Chem. Soc.* **112** 7389
[73] Allara D L and Nuzzo R G 1985 *Langmuir* **1** 45
[74] Allara D L and Nuzzo R G 1985 *Langmuir* **1** 52
[75] Facci J S 1992 *Molecular Design of Electrode Surfaces* ed R Murray (New York: Wiley) p 119
[76] Minassian-Saraga T 1955 *J. Chim. Phys.* **52** 181
[77] Fujihara M and Araki T 1986 *J. Electroanal. Chem.* **205** 329
[78] Nelson A 1991 *J. Electroanal. Chem.* **303** 221
[79] Nelson A 1991 *J. Chem. Soc. Faraday Trans.* **87** 1851
[80] Bilewicz R and Majda M 1991 *J. Am. Chem. Soc.* **113** 5464
[81] Kuritara K, Ohto K, Tanaka Y, Aoyama Y and Kunitake T 1991 *J. Am. Chem. Soc.* **113** 444
[82] Sun S, Ho-Si P-H and Harrison D J 1991 *Langmuir* **7** 727
[83] Bott B and Thorpe S C 1991 *Techniques and Mechanisms in Gas Sensing* ed P

T Moseley, J O W Norris and D E Williams (Bristol: Hilger) p 139
[84] Roberts G G, Petty M C, Baker S, Fowler M T and Thomas N J 1985 *Thin Solid Films* **132** 113
[85] Grate J W, Klusty M, Barger W R and Snow A W 1990 *Anal. Chem.* **62** 1927
[86] Stelzle M, Weissmüller G and Sackman E 1993 *J. Phys. Chem.* **97** 2974
[87] Nelson A, Auffret N and Borlakoglu J 1990 *Biochim. Biophys. Acta* **1021** 205
[88] Salamon Z and Tollin G 1991 *Bioelectrochem. Bioenerget.* **25** 447
[89] Wang J and Lu Z 1990 *Anal. Chem.* **62** 826
[90] Chang S-M, Iwasaki Y, Suzuki M, Tamiya E, Karube I and Muramatsu H 1991 *Anal. Chim. Acta* **249** 323
[91] Yokoyama K and Ebisawa F 1993 *Anal. Chem.* **65** 673
[92] Cai X X, Sun A, Cui L and Hai X L 1993 *Sensors Actuators* B **12** 15
[93] Okahata Y, En-na G and Ebato H 1990 *Anal. Chem.* **62** 1431
[94] Hayashi K, Yamanaka M, Toko K and Yamafuji K 1990 *Sensors Actuators* B **2** 205
[95] Chapman D 1993 *Langmuir* **9** 39
[96] Albrecht O, Johnston D S, Villaverde C and Chapman D 1982 *Biochim. Biophys. Acta* **687** 165
[97] Chapman D and Charles S A 1992 *Chem. Br.* **28** 253
[98] Hayward J A, Durrani A A, Lu Y, Clayton C and Chapman D 1986 *Biomaterials* **7** 252
[99] Pisarchick M L and Thompson N L 1990 *Biophys J.* **58** 1235
[100] Šnejdárková M, Rehák M and Otto M 1993 *Anal. Chem.* **65** 665
[101] Parpaleix T, Laval J M, Majda M and Bourdillon C 1992 *Anal. Chem.* **64** 641
[102] Bourdillon C and Majda M 1990 *J. Am. Chem. Soc.* **112** 1795
[103] Derjaguin B V and Toporov Yu P 1983 *Physicochemical Aspects of Polymer Surfaces* vol 2, ed K L Mittal (New York: Plenum) p 605
[104] Fowkes F M 1983 *Physicochemical Aspects of Polymer Surfaces* vol 2, ed K L Mittal (New York: Plenum) p 583
[105] Wrighton M S, Austin R G, Bocarsly A B, Bolts J M, Haas O, Legg K D, Nadjo L and Palazzotto M C 1978 *J. Electroanal. Chem.* **87** 429
[106] Wrighton M S, Palazzotto M C, Bocarsly A B, Bolts J M, Fischer A B and Nadjo L 1978 *J. Am. Chem. Soc.* **100** 7264
[107] Rocklin R D and Murray R W 1981 *J. Phys. Chem.* **85** 2104
[108] Bolts J M and Wrighton M S 1978 *J. Am. Chem. Soc.* **100** 5257
[109] Wrighton M S, Austin R G, Bocarsly A B, Bolts J M, Haas O, Legg K D, Nadjo L and Palazzotto M C 1978 *J. Am. Chem. Soc.* **100** 1602
[110] Ghosh P K and Spiro T G 1981 *J. Electrochem. Soc.* **128** 1281
[111] Ghosh P K and Spiro T G 1980 *J. Am. Chem. Soc.* **102** 5543
[112] Ema K, Yokoyama M, Nakamoto Y T and Moriizumi T 1989 *Sensors Actuators* **18** 291
[113] Ballantine D S, Rose S L, Grate J W and Wohltjen H 1986 *Anal. Chem.* **58** 3058
[114] Katritzky A R, Savage G P and Pilarska M 1991 *Talanta* **38** 201
[115] Fan Z and Harrison D J 1992 *Anal. Chem.* **64** 1304
[116] Bindra D S and Wilson G S 1989 *Anal. Chem.* **61** 2566
[117] Hillman A R 1987 *Electrochemical Science and Technology of Polymers 1* ed R G Linford (London: Elsevier) p 103
[118] Heller A 1992 *J. Phys. Chem.* **96** 3579
[119] Gregg B A and Heller A 1990 *Anal. Chem.* **62** 258
[120] Gregg B A and Heller A 1991 *J. Phys. Chem.* **95** 5970
[121] Gregg B A and Heller A 1991 *J. Phys. Chem.* **95** 5976
[122] Katakis I and Heller A 1992 *Anal. Chem.* **64** 1008

[123] Ohara T J, Rajagopalan R and Heller A 1993 *Anal. Chem.* **65** 3512
[124] Wang D L and Heller A 1993 *Anal. Chem.* **65** 1069
[125] Ye L, Hämmerle M, Olsthoorn A J J, Schuhmann W, Schmidt H-L, Duine J A and Heller A 1993 *Anal. Chem.* **65** 238
[126] Calvo E J, Danilowicz C and Diaz L 1993 *J. Chem. Soc. Faraday Trans.* **89** 377
[127] Cattrall R W and Hamilton I C 1984 *Ion Selective Electrode Rev.* **6** 125
[128] Hall E A 1990 *Biosensors* (Milton Keynes: Oxford University Press) p 86
[129] Covington A K and Sibbald A 1987 *Phil. Trans. R. Soc.* B **316** 31
[130] Blackburn F 1987 *Biosensors, Fundamentals and Applications* ed A P F Turner, I Karube and G S Wilson (Oxford: Oxford University Press) p 481
[131] Grubbs R H and Tumas W 1989 *Science* **243** 907
[132] Schrock R R 1990 *Accounts Chem. Res.* **23** 158
[133] Albagli D, Bazan G C, Schrock R R and Wrighton M S 1993 *J. Am. Chem. Soc.* **115** 7328
[134] Merz A and Bard A J 1978 *J. Am. Chem. Soc.* **100** 3222
[135] Peerce P J and Bard A J 1980 *J. Electroanal. Chem.* **114** 89
[136] Peerce P J and Bard A J 1980 *J. Electroanal. Chem.* **112** 97
[137] Itaya K and Bard A J 1978 *Anal. Chem.* **100** 1487
[138] Safi T, Hoshino K, Ishii Y and Goto M 1991 *J. Am. Chem. Soc.* **113** 450
[139] Gardner J W and Bartlett P N 1991 *Nanotechnology* **2** 19
[140] Bartlett P N and Cooper J 1993 *J. Electroanal. Chem.* **362** 1
[141] Bartlett P N and Birkin P R 1993 *Synth. Met.* **61** 15
[142] Iyoda T, Toyoda H, Fujitsuka M, Nakahara R and Tsuchiya H 1991 *J. Phys. Chem.* **95** 5215
[143] Bartlett P N, Benniston A C, Chung L-Y, Dawson D H and Moore P 1991 *Electrochim. Acta* **36** 1377
[144] Youssoufi H K, Hmyene M, Garnier F and Delabouglise D 1993 *J. Chem. Soc. Chem. Commun.* 1550
[145] Li H, Garnier F and Roncali J 1991 *Synth. Met.* **41–43** 547
[146] Delabouglise D and Garnier F 1990 *Adv. Mater.* **2** 9
[147] Bartlett P N, Dawson D H and Farrington J 1992 *J. Chem. Soc. Faraday Trans.* **88** 2685
[148] Bartlett P N and Farrington J 1993 *Bull. Electrochem.* **8** 208
[149] Pickup P G 1987 *J. Electroanal. Chem.* **225** 273
[150] Delabouglise D and Garnier F 1990 *New J. Chem.* **15** 233
[151] Witkowski A and Brajter-Toth A 1992 *Anal. Chem.* **64** 635
[152] Gao Z, Chen B and Zi M 1993 *J. Chem. Soc. Chem. Commun.* 675
[153] Schalkhammer T, Mann-Braum E, Pittner F and Urban G 1991 *Sensors Actuators* B **4** 273
[154] Schalkhammer T, Mann-Braum E, Urban G and Pittner F 1990 *J. Chromatogr.* **510** 355
[155] Miasik J J, Hooper A and Tofield B C 1986 *J. Chem. Soc. Faraday Trans.* I **82** 1117
[156] Hanawa T, Kuwabata S and Yoneyama H 1988 *J. Chem. Soc. Faraday Trans.* I **84** 1587
[157] Bartlett P N and Ling-Chung S K 1989 *Sensors Actuators* **19** 141
[158] Bartlett P N and Ling-Chung S K 1989 *Sensors Actuators* **20** 287
[159] Slater J M, Watt E J, Freeman N J, May I P and Weir D J 1992 *Analyst* **117** 1256
[160] Slater J M, Paynter J and Watt E J 1993 *Analyst* **118** 379
[161] Pearce T C, Gardner J W, Friel S, Bartlett P N and Blair N 1993 *Analyst* **118** 371

[162] Persaud K C and Pelosi P 1992 *Sensors and Sensory Systems for an Electronic Nose (NATO ASI Series 212)* ed J W Gardner and P N Bartlett (Dordrecht: Kluwer) p 237

[163] Cassidy J, Foley J, Pons S and Janata J 1986 *Anal. Chem. Symp. Ser.* **25** 309

[164] Josowicz M and Janata J 1986 *Anal. Chem.* **58** 514

[165] Topart P and Josowicz M 1992 *J. Phys. Chem.* **96** 7824

[166] Charlesworth J M, Partridge A C and Garrard N 1993 *J. Phys. Chem.* **97** 5418

[167] White H S, Kittlesen G P and Wrighton M S 1984 *J. Am. Chem. Soc.* **106** 5375

[168] Kittlesen G P, White H S and Wrighton M S 1985 *J. Am. Chem. Soc.* **107** 7373

[169] Paul E W, Ricco A J and Wrighton M S 1985 *J. Phys. Chem.* **89** 1441

[170] Wrighton M S, Thackeray J W, Natan M J, Smith D K, Lane G A and Bélanger D 1987 *Phil. Trans. R. Soc.* B **316** 13

[171] Thackeray J W and Wrighton M S 1986 *J. Phys. Chem.* **90** 6674

[172] Thackeray J W, White H S and Wrighton M S 1985 *J. Phys. Chem.* **89** 5133

[173] Matsue T, Nishizaura M, Sawaguchi T and Uchida I 1991 *J. Chem. Soc. Chem. Commun.* 1029

[174] Bartlett P N and Birkin P R 1993 *Anal. Chem.* **65** 1118

[175] Nishizawa M, Matsue T and Uchida I 1992 *Anal. Chem.* **64** 2642

[176] Hoa D T, Kumar T N, Punekar N S, Srinivasa R S, Lal R and Contractor A Q 1992 *Anal. Chem.* **64** 2645

[177] Denisevich P, Abruña H D, Leider C R, Meyer T J and Murray R W 1982 *Inorg. Chem.* **21** 2153

[178] Obeng Y S, Founta A and Bard A J 1992 *New J. Chem.* **16** 121

[179] Gould S, Strouse G F, Meyer T J and Sullivan B P 1991 *Inorg. Chem.* **30** 2942

[180] Ikeda Y, Schmerl R, Denisevich P, Willman K and Murray R W 1982 *J. Am. Chem. Soc.* **104** 2683

[181] McCarley R L, Irene E A and Murray R W 1991 *J. Phys. Chem.* **95** 2492

[182] Bruno F, Pham M C and Dubois J E 1977 *Electrochim. Acta* **22** 451

[183] Mengoli G, Bianco P, Daolio S and Munari M T 1981 *J. Electrochem. Soc.* **128** 2276

[184] Ohnuki Y, Matsuda H, Ohsaka T and Oyama N 1983 *J. Electroanal. Chem.* **158** 55

[185] Ohsaka T, Hirokawa T, Miyamoto H and Oyama N 1987 *Anal. Chem.* **59** 1758

[186] Wang J, Chen S-P and Lin M S 1989 *J. Electroanal. Chem.* **273** 231

[187] Christie I M, Vadgama P and Lloyd S 1993 *Anal. Chim. Acta* **274** 191

[188] Cheek G, Wales C P and Nowak R J 1983 *Anal. Chem.* **55** 380

[189] Bartlett P N, Tebbutt P and Tyrrell C 1992 *Anal. Chem.* **64** 138

[190] Bartlett P N and Caruana D J 1992 *Analyst* **117** 1287

[191] Bartlett P N and Caruana D J 1994 *Analyst* **119** 175

[192] Sasso S V, Pierce R J, Walla R and Yacynych A M 1990 *Anal. Chem.* **62** 1111

[193] Malitesta C, Palmisano F, Torsi L and Zambonin P G 1990 *Anal. Chem.* **62** 2735

[194] Nowak R, Schultz F A, Umaña M, Abruña H and Murray R W 1978 *J. Electroanal. Chem.* **94** 219

[195] Daum P, Lenhard J R, Rolison D R and Murray R W 1980 *J. Am. Chem. Soc.* **102** 4649

[196] Facci J and Murray R W 1982 *Anal. Chem.* **54** 772

[197] Heider G H, Gelbert M B and Yacynych A M 1982 *Anal. Chem.* **54** 322

[198] Butler M A, Ricco A J and Buss R 1990 *J. Electrochem. Soc.* **137** 1325

[199] Munkholm C, Walt D R, Milanovich F P and Klainer S M 1986 *Anal. Chem.* **58** 1427

[200] Itaya K, Uchida I and Neff V D 1986 *Accounts Chem. Res.* **19** 162

[201] McCarger J W and Neff V D 1988 *J. Phys. Chem.* **92** 3598

[202] Bocarsly A B and Sinha S 1982 *J. Electroanal. Chem.* **137** 157
[203] Bocarsly A B and Sinha S 1982 *J. Electroanal. Chem.* **140** 167
[204] Newsam J 1986 *Science* **231** 1093
[205] de Vismes B, Bedioui F, Devynck J and Bied-Charreton C J 1985 *J. Electroanal. Chem.* **187** 197
[206] Gemborys H A and Shaw B R 1986 *J. Electroanal. Chem.* **208** 95
[207] Shaw B R, Creasy K E, Lanczycki C J, Sargeant J A and Tirhado M 1988 *J. Electrochem. Soc.* **135** 869
[208] Li Z and Mallouk T E 1987 *J. Phys. Chem.* **91** 643
[209] Li Z, Wang C M, Persaud L and Mallouk T E 1988 *J. Phys. Chem.* **92** 2592
[210] Li K, Lai C and Mallouk T E 1989 *Inorg. Chem.* **28** 178
[211] Johansson G, Risinger L and Fälth L 1980 *Anal. Chim. Acta* **119** 25
[212] Wang J and Martinez T 1988 *Anal. Chim. Acta* **207** 95
[213] Shaw B R and Creasy K E 1988 *J. Electroanal. Chem.* **243** 209
[214] Lee H, Kepley L J, Hong H-G and Mallouk T E 1988 *J. Am. Chem. Soc.* **110** 618
[215] Rong D, Hong H-G, Kim Y I, Krueger J S, Mayer J E and Mallouk T E 1990 *Coord. Chem. Rev.* **97** 237
[216] Hong H-G and Mallouk T E 1991 *Langmuir* **7** 2362

7

Biological and chemical components for sensors

Jerome S Schultz

7.1 INTRODUCTION

As is evident from the preceding chapters of this Handbook, one of the essential distinctions between chemical and biological sensors is the use of biological substances in biosensor devices. The function of the biological element is to provide selective molecular recognition for the analyte(s) of interest. One of the more difficult aspects of designing and producing a biosensor is the choice of the biological material for this purpose. At the cellular level, almost all biochemical phenomena are the result of a chemical interaction between cellular materials and other chemical species. To some degree then, each of these biochemical components can be thought of as candidate for a molecular recognition element that can be utilized in a biosensor.

Traditionally the recognition and interaction of cellular components with other species are classified according to topical fields such as enzymology, pharmacology, and immunology. Since there is a certain amount of similarity in the mechanism of recognition and the quantitative description of the interactive process in these different biological phenomena, the use of various biological recognition elements will be discussed as applications of the molecular phenomena associated with these traditional fields.

Recently, with the development of the field of molecular genetics, it is becoming clear that polynucleotides (DNA and RNA) also have biorecognition capabilities, and this potential will be discussed briefly as well. Finally, a field is developing within bioorganic chemistry to synthesize compounds that mimic the recognition capabilities of natural biopolymers. These biomimetics have the potential to be designed for specific analytes, and thus provide a wider range of capabilities for biosensors.

At the outset, one distinction to recognize between the classes of biological recognition compounds is that enzymes are catalysts of reactions, and thus sensors that utilize the catalytic capability of enzymes will *consume* the analyte.

This property has important implications on the characteristics of the biosensor, particularly as related to the supply rate of the analyte to the sensor. Thus for small sample volumes, the analyte may be depleted in the region of the sensor. Further, factors that affect the delivery of the analyte to the sensor, e.g. fouling of the sensor surface, can affect the sensor performance. The bioreceptors identified with cell signaling in immunology and pharmacology are usually *non-consuming* in nature, and thus have fewer potential problems due to artifacts related to analyte consumption.

7.2 SOURCES OF BIOLOGICAL RECOGNITION ELEMENTS

7.2.1 Enzymes as recognition elements

About 2500 enzymes have been identified and characterized since urease was first crystallized in 1926 by J B Summer. An excellent review of enzyme behavior and kinetics is given in [1]. Because of the difficulties with uniquely naming so many materials for publications, a standard system of enzyme classification was developed. This compendium of enzymes and their properties is available from the International Union of Biochemistry [2].

Each enzyme has been given an EC number, and the first number indicates the primary type, for example, glucose oxidase EC 1.1.3.4 is an oxidoreductase. Enzymes are cataloged under six primary types:

1. oxidoreductases
2. transferases
3. hydrolases
4. lyases
5. isomerases
6. ligases.

The simplest mechanism for an enzyme-catalyzed reaction is

$$\text{substrate (analyte)} + \text{enzyme} \rightleftharpoons \text{enzyme–substrate} \rightarrow \text{enzyme} + \text{product complex}$$

$$S + E \rightleftharpoons \text{ES} \rightarrow \text{E} + \text{P}.$$

Typical behavior for an enzyme-catalyzed reaction (such as that for glucose oxidase) is shown in figure 7.1. Here the reaction is the oxidation of glucose:

$$\text{glucose} + O_2 \xrightarrow{\text{glucose oxidase}} \text{gluconolactone} + H_2O_2.$$

The figure shows the rate of the reaction as measured by oxygen consumption as a function of glucose (substrate) concentration. It can be seen that the catalytic

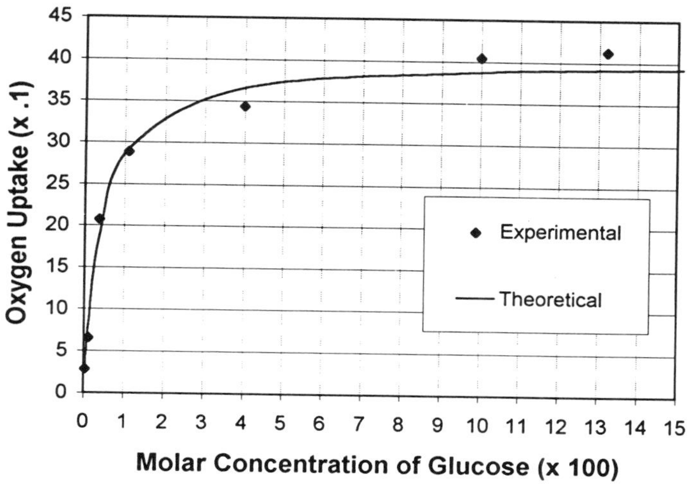

Figure 7.1 Experimental data on the oxidation of glucose catalyzed by the enzyme glucose oxidase (replotted from Keilin D and Hartree E F 1952 *Biochem. J.* **50** 331). The rate of reaction was estimated by measuring oxygen consumption. The theoretical curve was drawn based on the simple Michaelis–Menton model shown in equation (7.1).

activity of this enzyme saturates at high substrate concentrations. The leveling off of enzyme reactivity with substrate concentration is typical of most enzymes (although there are exceptions). However, enzyme activity often goes through a maximum as a function of pH or temperature at a given substrate concentraion.

The kinetics of this simple enzyme-catalyzed reaction can be represented by an equation of the form (usually called the Michaelis–Menten equation)

$$r = \frac{k_{cat} E S}{K_M + S} = \frac{V_{max} S}{K_M + S}. \tag{7.1}$$

The substrate concentration at which an enzyme reaches one-half of its maximal catalytic activity is often used as a measure of the sensitivity of an enzyme to substrate saturation. This particular substrate concentration usually has about the same numerical value as K_M, sometimes known as the Michaelis–Menten constant for the enzyme. The maximum rate of reaction per mole of enzyme is often given the symbol k_{cat}, and the maximum rate of reaction for a given enzyme concentration is often symbolized as V_{max}. Often, the kinetics of more complex enzyme-catalyzed reactions can be placed in this form under some restricted range of conditions [1].

Since the purity and molecular weight of many enzymes are not known, the amount of enzyme is usually characterized by a ‘specific activity’, i.e. units of enzyme catalytic activity per milligram of protein, where a unit of activity is something like the transformation of 1.0 μmol (10^{-6} mol) substrate/mg enzyme preparation per minute.

Table 7.1 Enzymes that have been used in commercial biosensors [4].

		K_M^a (mol l^{-1})	Substrate
Alcohol dehydrogenase	EC 1.1.3.13	1×10^{-2}	ethanol
Catalase	EC 1.11.1.6	1	sucrose, glucose
Choline dehydrogenase	EC 1.1.99.1	7×10^{-3}	phospholipids
Lactate dehydrogenase	EC 1.1.2.3	2×10^{-3}	lactate
Galactose oxidase	EC 1.1.3.9	2×10^{-1}	galactose
Glycerol kinase	EC 2.7.1.30	1×10^{-6}	glycerol
Glucose oxidase	EC 1.1.3.4	1×10^{-2}	glucose
Lactate dehydrogenase	EC 1.1.1.27	2×10^{-2}	lactate
Lactate oxidase	EC 1.1.3.2	2×10^{-2}	lactate
Mutarotase	EC 5.1.3.3	3×10^{-2}	glucose
Sulfite oxidase	EC 1.8.3.1	3×10^{-5}	sulfite
Urate oxidase	EC 1.7.3.3	2×10^{-5}	uric acid

[a] Michaelis–Menten constant (see equation (7.1)).

One convenient source of information about enzymes is the *Enzyme Handbook* [3] which provides for each enzyme the chemical reaction catalyzed, the kinetic constants for the enzyme, a definition of activity units, and references for the preparation and properties of the enzyme.

7.2.1.1 Typical enzymes used in biosensors. Several hundred enzymes have been used in the demonstration of biosensor devices. Reviews have appeared in a number of books (see, e.g. [4]) and many research papers are published in the journal *Biosensors and Bioelectronics* (published by Elsevier). A biannual comprehensive review is published in *Analytical Chemistry* (see, e.g. [5]).

However, these reports of multitudinous enzyme-based biosensors should be viewed with some caution, as it is much easier to demonstrate the possibility of using an enzyme in a laboratory prototype than to convert these observations into a reliable, and reproducible, device that can meet commercial product requirements. This is illustrated by table 7.1, which lists the few enzymes that have been reported to have been used in commercial biosensors; only about two dozen enzymes have been used commercially. Most of the enzymes are oxidases partly because of the stability of this class of enzyme, and partly because of ease of linking this type of enzyme with a Clark-type oxygen electrode.

Glucose oxidase is one of the most studied enzyme because of the widespread interest in developing a reliable glucose sensor for use by diabetics and for glucose measurements in the fields of food processing and fermentation. As mentioned above, one reason for using a biological element in a sensor is to provide selectivity to the device. In general the selectivity of a sensor should be evaluated in the sample milieu of interest; in the case of a glucose sensor one of the fluids of interest is blood. Table 7.2 shows the response of a Clark-type glucose sensor (manufactured by YSI, Yellow Springs, OH) to some potential glucose oxidase interferants that may be present in blood and that can penetrate

Table 7.2 Examples of substances in blood interfering with glucose sensor response.

Substance	Interfering level[a] (mg dl^{-1})	Normal serum level (mg dl^{-1})
Acetone	26 000	0.3–2.0
Beta hydroxybutyric acid	14 000	—
Sorbitol	14 000	—
D-xylose	730	—
D(−)adrenaline	110	—
Ascorbic acid	280	0.4–1.5
L(+)cysteine.HCI	100	0.9
D(−)fructose	5 400	< 7.5
d-galactose	300	<20.0
Glutathione	100	28–34
d-mannose	170	—
Tyrosine	160	0.8–1.3
Uric acid	400	3–7
Acetaminophen	1.5	—
Acetylsalicylic acid	167	—
Catechol	0.3	—
Sodium oxalate	11 000	—
Heparin sodium	1 800 U ml^{-1}	—
Sodium azide	360	—
Thymol	75	—
Epinephrine	—	18–26 ng dl^{-1}
Norepinephrine	—	47–69 ng dl^{-1}

[a] Corresponds to the level of interferent which would give an error of 5 mg dl^{-1} in an apparent glucose response. Measured with a Yellow Springs Instruments model 2300 sensor. Interference data courtesy of YSI, Inc.

the protective membrane barrier of the sensor. Column 2 shows the concentration of the interferant in blood that would be required to cause a 5% error in glucose estimation (normal blood glucose is about 100 mg dl^{-1}). As can be seen in the third column, most of these substances do not have a significant effect on the sensor response in blood.

7.2.1.2 Sources of enzymes. Enzymes are commercially available from a number of suppliers, such as Sigma, Worthington, and Boehringer-Mannheim. The degree of purity of an enzyme that is to be used in a biosensor depends on the type of impurity in the preparation. A contamination with another enzyme could produce erroneous results if substrates for the interfering enzyme are in the samples to be analyzed and if the product of this extraneous reaction is detected by the monitoring system. On the other hand, inert impurities usually do not affect the catalytic activity of enzymes and can be tolerated. Typically the amounts of enzymes used in biosensors are small (table 7.3). Thus the per

Table 7.3 Typical enzyme electrodes and their characteristics.

Analyte (substrate)	Enzyme	Units of enzyme per electrode	Range of operation ($mol\,l^{-1}$)
Urea	Urease (EC 3.5.1.5)	25	10^{-2}–5×10^{-5}
Glucose	Glucose oxidase	100	10^{-1}–10^{-3}
L-tyrosine	L-Tyrosine decarboxylase (EC 1.1.25)	25	10^{-1}–10^{-4}
L-glutamine	Glutaminase (EC 3.5.1.2)	50	10^{-1}–10^{-4}
L-glutamic acid	Glutamate dehydrogenase (EC 1.4.1.3)	50	10^{-1}–10^{-4}
L-asparagine	Asparaginase (EC 3.5.1.1)	50	10^{-2}–5×10^{-5}
Penicillin	Penicillinase (EC 3.5.2.6)	400	10^{-2}–10^{-4}
Amygdalin	Glucosidase (EC 3.2.1.21)	100	10^{-2}–10^{-5}
Nitrate	Nitrate reductase/ nitrite reductase (EC 1.9.6.1/1.6.6.4)	10	10^{-2}–10^{-4}
Nitrite	Nitrate reductase (EC 1.6.6.4)	10	5×10^{-2}–10^{-4}

unit cost of enzymes for the sensor will be a rather small fraction of the cost of producing the sensor.

One of the limitations of the repertoire of naturally occurring enzymes is that they are primarily active on normal biochemical metabolites. Thus finding an enzyme with the appropriate properties for the development of a biosensor for a non-metabolite may be difficult. Recent discoveries have demonstrated catalytic activity in antibodies [6]. As discussed below, the immune system has the potential to produce thousands of antibodies with different binding specificities. By challenging the immune system with analog compounds that have structural similarities to the target analyte in the activated enzyme–substrate complex, antibodies can be isolated that have enzymatic activity; these antibodies have been called abenzymes. This methodology is in the early phases of development, but may provide an approach for the facile engineering of an enzyme for a wide diversity of analytes, for example Blackburn *et al* [7] have used a catalytic antibody has been used in developing a biosensor for phenyl acetate.

An important concern for the use of enzymes in biosensors is their stability. Part of the skill in developing an enzymatic biosensor is developing a formulation that promotes the stability of enzyme activity. Immobilization

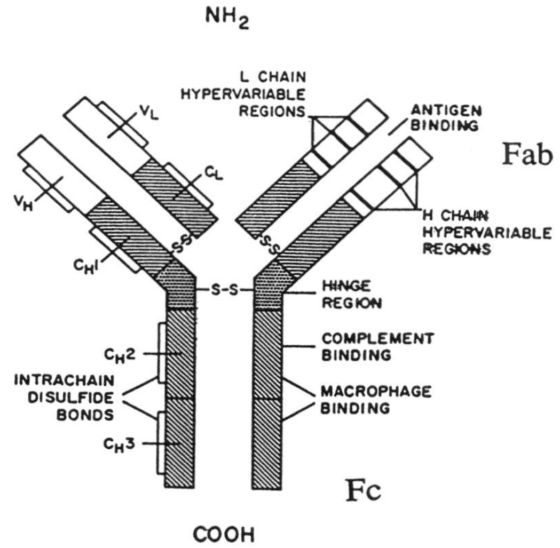

Figure 7.2 The schematic structure of the antibody IgG.

methods can be selected to stabilize enzyme activity, as discussed in the chapter on immobilization of proteins (chapter 20).

7.2.2 Recognition elements derived from the immune system

7.2.2.1 Immunological bioreceptors. One of the key proteins that is responsible for initiating an immune response is an immunoglobulin called an antibody [8]. There are a number of types of these soluble materials, IgA, IgD, IgE, IgG, and IgM, but the one that is by far best understood and most readily available for use in sensors is the IgG class. The structural features of this approximately 150 000 MW protein are shown in figure 7.2. Notable characteristics of antibodies are that they are bifunctional receptors, and that the twin recognition sites are located in the terminal hypervariable regions of the molecule. These two regions of molecule are highly variable in amino acid sequence, which accounts for the approximately 10 000 000 different specificies that can potentially be elicited by challenges to the immune system of mammals. The branches of the molecule (called the Fab fragments) can be separated while retaining their binding function. The base of the molecule also has biological functions, and for sensors, the Fc site can be used as a 'hook' to attach the antibody to other molecules or surfaces either chemically or through the use of a bridging receptor such as protein A [9].

Figure 7.3 shows the binding behavior of a typical antibody as a function of ligand concentration. The form of this hyperbolic curve is similar to figure 7.1, the pattern for enzyme kinetics. Antibodies also show saturation behavior at

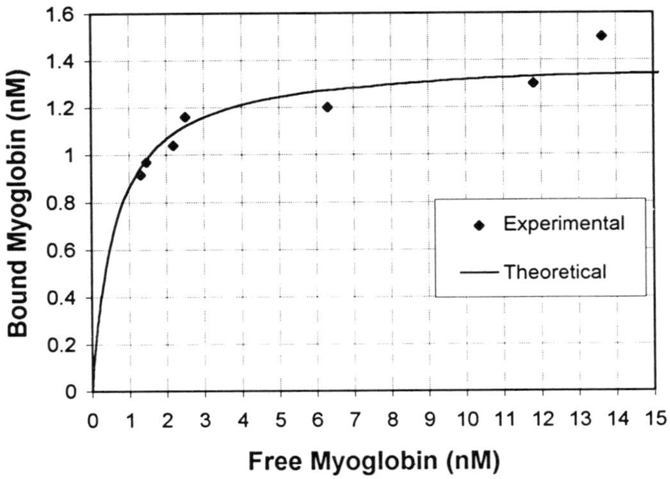

Figure 7.3 Experimental data for the binding of a ligand (myoglobin) to a monoclonal antibody (antimyoglobin) (replotted from Berzofsky J A *et al* 1980 *J. Biol. Chem.* **255** 11 188). The theoretical curve was drawn based on the mathematical model shown in equation (7.2).

high ligand concentrations. The tenacity of ligand (antigen) binding to an antibody is quantitated by the affinity constant, K_A (sometimes called the association constant or binding constant), which is the equilibrium constant for this reversible reaction

$$\text{antibody (Ab)} + \text{ligand (Lg)} \underset{k_d}{\overset{k_a}{\rightleftharpoons}} \text{antibody–ligand (Ab–Lg)}$$

$$K_A = \frac{[\text{Ab–Lg}]}{[\text{Ab}][\text{Lg}]}. \qquad (7.2)$$

The affinity constant K_A has the units of l/concentration (l mol^{-1}) and the larger the value of K_A, the less is the concentration of ligand needed to saturate the antibody. For example, the K_A for the antibody shown in figure 7.3 is about 0.8×10^{-9} mol $^{-1}$. Thus a sensor using this antimyoglobin would have an analytical range of about 0.2×10^{-9} to 2×10^{-9} mol $^{-1}$.

The kinetics of ligand binding to an antibody usually follows a simple biomolecular reaction mechanism

$$r = k_a[\text{Ab}][\text{Lg}] - k_d[\text{Ab–Lg}] \qquad (7.3)$$

where K_d is the association rate constant and k_d is the dissociation rate constant. At equilibrium the net rate of reaction is zero ($r = 0$), resulting in a simple relation between the affinity constant and the kinetic reaction rate constants:

$$K_A = \frac{k_a}{k_d}. \qquad (7.4)$$

The forward rate (k_a) is usually diffusion controlled, meaning that for most situations the antibody and ligand react as fast as these molecules collide.

The process for producing an antibody to a specific analyte consists of linking the analyte (or an analog thereof) to a protein (such as albumin) and injecting the complex into a highly susceptible and efficient antibody producing animal (such as a rabbit). The injection of this artificial foreign protein causes the stimulation of a group of B-cells that eventually synthesize a family of polyclonal antibodies: a group of antibodies with slightly different specificities and affinities. Antibodies are usually prepared by researchers for specific analytes; this is time consuming and requires specialized facilities. However, for some common specificities commercial sources are available. Some suppliers are: Sigma, Amersham, Biorad and Boehringer-Mannheim.

An alternative procedure is to isolate the B-cells, fuse them with a tumor line (myeloma), and separate these hybrid cells into individual clones, to produce antibody producing hybrids called hybridomas [10–12]. These separate cell lines can be screened to select a clone that produces an antibody with a single structure, called a monoclonal antibody. For analytical purposes, one advantage of a monoclonal antibody is that the cell line can be preserved so that a consistent source of antibody can be obtained. Again, usually researchers prepare their own monoclonals or obtain cell lines from other investigators who have published reports about the use of their monoclonals. Many researchers deposit their cultures with the American Type Culture Collection, Rockville, MD.

A new and exciting breakthrough for the selection and production of monoclonal antibodies is the preparation of antibody libraries in phage and bacteria [13]. The method starts with splicing randomized genes coding for IgGs into the chromosome of a bacteriophage or bacterium. Key requirements for this technology are (i) to ensure that a random mixture of DNA strands is used in the region of the gene that codes for the hypervariable binding sites of the antibody, (ii) millions of DNA copies are generated to produce a wide diversity of the variable elements of the antibody, and (iii) the DNA is spliced into the genome of the host so that it is expressed on the surface of the phage (figure 7.4). Then this random combinatorial library of antibodies can be screened to select a clone with the desired binding characteristics for a specific analyte of interest. One approach to utilizing this *in vitro* immune system is a process called 'biopanning' illustrated in figure 7.5. The analyte of interest (or an analog thereof) is immobilized on the surface of a test tube or Petri dish. A suspension of the phage–antibody particles is added to the tube, a sufficient amount of phage is used to be sure that there are multiple copies of all possible antibodies—about 10^{12} phage. After an incubation period, the excess (non-binding) phage are washed off the surface. The bound phage are recovered and proliferated in *E. coli*. A new preparation of phage is obtained and exposed to the bound analyte. After several cycles, the phage population is enriched with clones having high binding specificity for the target analyte. Then the DNA can be excised from the phage and placed into a bacterial expression system

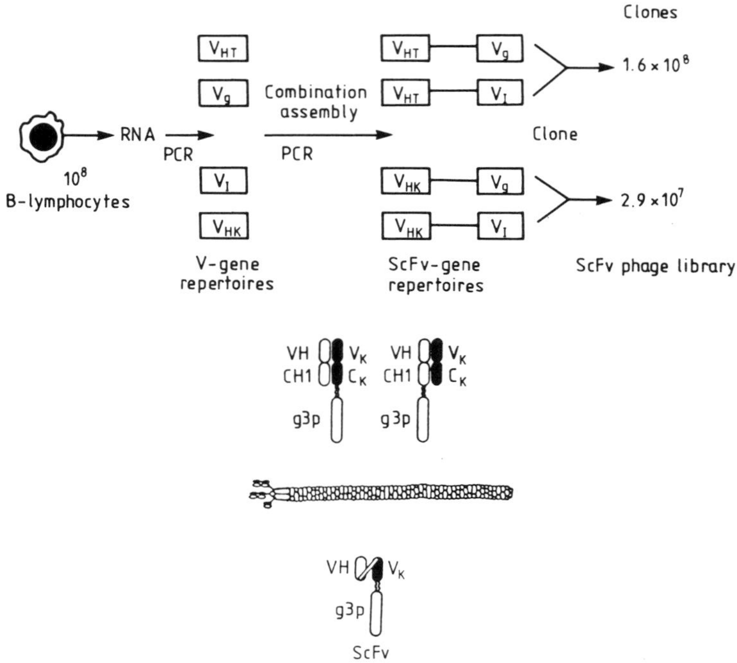

Figure 7.4 A biosynthetic antibody library expressed on a phage [13].

and a monoclonal antibody produced. Several companies have developed these phage display libraries and offer services to produce monoclonal antibodies to order (e.g. Cambridge Antibody Technology, Cambridge). Since the production of proteins by microorganisms is much less expensive than by cell culture (hybridomas), it is likely that the availability and cost of monoclonal antibodies will be much more favorable in the future.

7.2.3 Pharmacological recognition elements

Historically, much of the research effort in pharmacology has been to devise drugs that have specific pharmacologic effects on various control systems in the body, e.g. agents for hypertension. In the last few decades information has been accumulating on the sites of action of these drugs, usually proteineous receptors residing in the membranes of cells, and, by a variety of techniques, these receptors are being isolated, structure identified, and cloned. One of the most studied and understood of these structures is the acetylcholine receptor that is responsible for the control of Na^+ and K^+ transport into the postsynaptic membrane in nerve cells [14].

As bioreceptors with unique recognition properties, this class of materials are becoming likely candidates for use in biosensors. Several research papers

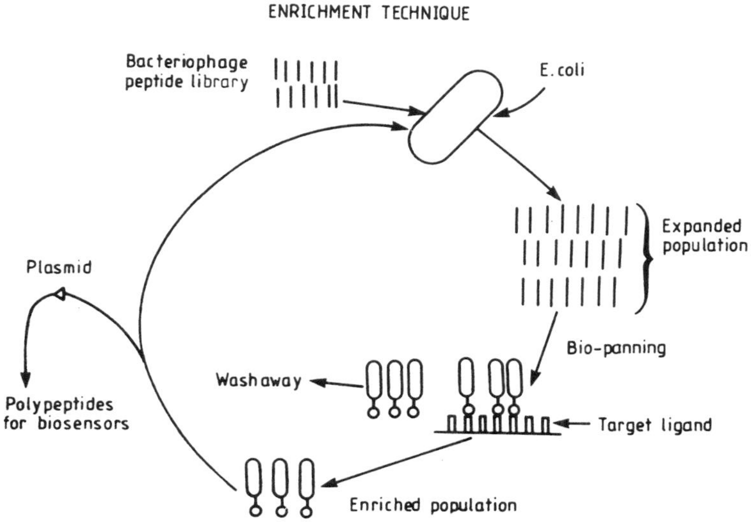

Figure 7.5 Biopanning for an antibody that binds to a target ligand [31].

have demonstrated the utility of pharmacological receptors for the quantitation of drugs, inhibitors, and toxic substances [15, 16]. Table 7.4 is a summary of compounds that might be measured using these bioreceptors.

Much of the data characterizing the binding behavior of pharmacological receptors is derived from inhibition experiments [17]. Typically the inhibition of a biological response to the natural ligand by a 'drug' is measured. The lower the concentration of drug required to interfere with the natural response, the higher the presumed affinity of the drug for the natural receptor. A typical experiment is shown in figure 7.6. The inhibition of binding of muscarine to its receptor by a series of drugs was measured (K_{bind}) as well as the inhibition of the biological response controlled by this receptor (muscle contraction, K_{pharm}). There is a direct correlation (over a millionfold range!) between the affinity of drugs for the receptor and the inhibition of bioassay.

The kinetics of binding of drugs to membrane bioreceptors is not always a simple biomolecular event as is the case with antibodies. Often these receptors have multiple binding sites and undergo a structural conformational change after a binding event, resulting in complex multiphasic sigmoidal binding isotherms. Also, the kinetics of the structural changes may be much slower than the initial binding process, so that the functional kinetics (e.g. for ion transport) may be quite complex. How these materials are used in a biosensor will result in different sensor behavior patterns. For example, Eldefrawi *et al* [16] utilized the ACh receptor to measure the concentration of several toxins. In this case, the binding dynamics of the analyte to the receptor was measured by an optical method, and the kinetics of the process could be accounted for with a simple

Table 7.4 Examples of receptor classes and their ligands [15].

	Ligand examples	
Receptor	Natural target ligand	Potential analytes
Cholinergic (nicotinic, muscarinic)	acetylcholine viscotoxins cobra/krait toxins muscarine	carbachol succinyldicholine decamethonium hexamethonium
Cholinergic (presynaptic)	black widow spider toxin botulinum toxins tetrahydrocannabinol	hemicholiniums
Adrenergic (alpha–beta)	epinephrine ergot alkaloids yohimbine	dibenamine phentolamine propranolol
Adrenergic (presynaptic)	cocaine mescaline	amphetamines imipramine
Dopaminergic	dopamine bromocriptine	apomorphine spiperone
Serotonergic	serotonin lsd psilocybin	imipramine mianserin trazodone
Opiate	codeine morphine enkephalins	heroin methadone naloxone
Histamine	histamine	cimetidine mepyramine
Aminobutyric acid (GABA)	GABA muscimol picrotoxin	barbiturates lindane benzodiazepine
Glycine	glycine strychnine	benzodiazepine chlormethiazole
Glutamate	glutamic acid kanic acid	4-fluoroglutamate methylibotenate
Adenosine	adenosine caffeine theophylline	alloxazine etazolate phenyltheophylline
Adenosine triphosphate (ATP)	ATP quinidine	antazoline dipyridylisatogen
Sodium (potassium) channels	scorpion toxin palytoxin tetrodotoxin veratridine	phencyclidines DDT ketamine amantadine

Table 7.4 (*Continued*).

	Ligand examples	
Receptor	Natural target ligand	Potential analytes
Calcium channel	batrachotoxin maitotoxin taicatoxin	phencyclidines verapamil dibenamine
Peptide hormone	angiotensins	peptide analogs (various)
Various	insulin glucagon vasopressin	
Eicosanoid	leukotrienes prostaglandins	various analogs
Immune system cell	anaphylatoxins interleukins	imidodisulfamides various analogs
Estrogen	estradiol estriol estrone	diethylstibestrol tamoxifen
Androgen	testosterone androsterone	flutamide methyltrienolone
Progesterone	progesterone pregnanediol	norethindrone dihydrogestrone
Glucocorticoid	cortisol corticosterone	dexamethasone phenylbutazone
Retinol (cellular)	retinol citral	
Retinoic acid	retinoic acid 13-*cis*-retinoic acid	various analogs
Choleragen	gangliosides raffinose	thiogalactopyranosides
Auxin	indolacetic acid phenylacetic acid	2,4-D; 2,4,5-T[a] napthylacetate
Cytokinin	kinetin benzyladenine	azidopurines aminobenzylpurines
E colicins	E colicins bacteriophage BF23 vitamin B_{12}	
Viral (glycoproteins, glycolipids)	HIV-1 polyoma virus reovirus	

[a]2,4-D, 2,4-dichlorophenoxyacetic acid; 2,4,5-T, 2,4,5-trichlorophenoxyacetic acid.

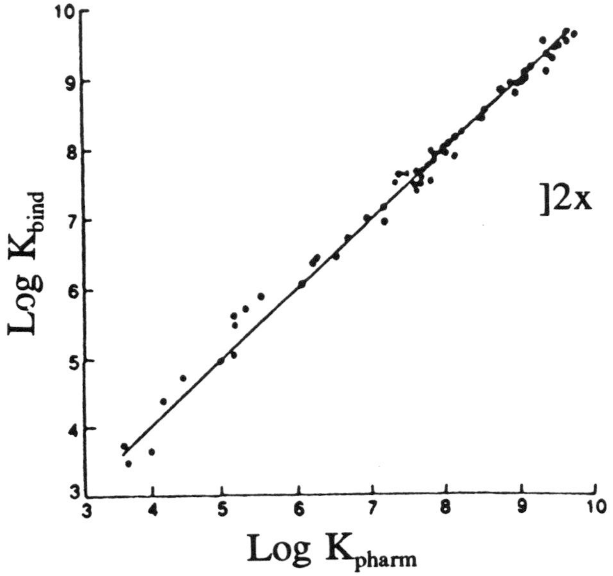

Figure 7.6 The correlation between the binding affinities of 60 muscarinic antagonists for rat cerebral cortical muscarinic receptors and their pharmacological potencies in antagonizing muscarinic contraction of the longitudinal muscle of the guinea-pig ileum. Subtype selective antagonists deviate significantly from the correlation [17].

Table 7.5 Kinetic parameters for binding of FITC-labeled toxins to nicotinic ACh receptor-coated optic fibers [16].

Toxin	K_a ($M^{-1}s^{-1}$)	K_d (s^{-1})	K_d(M) Optic fiber sensor	K_d(M) Literature
Bungarotoxin	1.4×10^5	1.3×10^{-4}	1.0×10^{-9}	0.1×10^{-9}
Naja toxin	1.0×10^5	0.8×10^{-3}	7.3×10^{-9}	4.0×10^{-9}
Conotoxin	0.2×10^5	1.3×10^{-2}	570.0×10^{-9}	200.0×10^{-9}

bimolecular scheme, such as indicated above for antibody–antigen interactions. The data summarized in table 7.5 indicate that the association rate constants (k_a) for the reaction between the receptor and the analytes (toxins) were about the same for three different toxins, and appeared to be diffusion limited as is usually the case with antibodies, as mentioned above.

On the other hand if, in a biosensor, the membrane receptor is used as an ion channel, then the response dynamics may be dominated by the speed of conformational changes. One potential of using these types of membrane receptor as ion gates in a biosensor application is the inherent amplification that

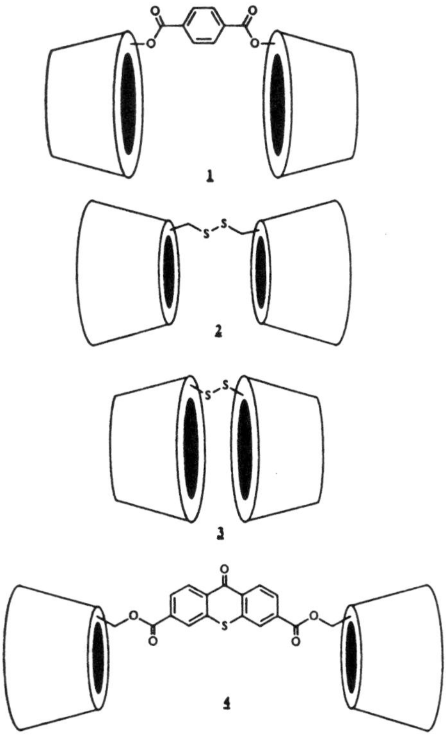

Figure 7.7 Synthetic mimics of a bioreceptor based on cyclodextrin dimer host models [18].

can be achieved in this manner. For example, for each molecule of acetylcholine that binds to its receptor, 10 000 ions will pass through the membrane channel in about 1 μs.

7.2.4 Biomimetic approaches

As the basic understanding of molecular recognition evolves, the ability to deterministically synthesize receptors with desired binding characteristics becomes more likely. Often these studies start with a chemical structure that has been found to have binding behavior, e.g. cyclodextrin, and then this structure is systematically modified while monitoring its binding properties. Some of these studies have been directed at developing catalysts that mimic enzymes [18], but the same techniques can be used for developing recognition motifs (figure 7.7).

In all these studies, the current availability of computer-based molecular modeling programs (available from Tripos, Biosym Technologies, and Molecular Simulations) facilitate the exploration of optimal molecular configurations before actually attempting to synthesize compounds of interest.

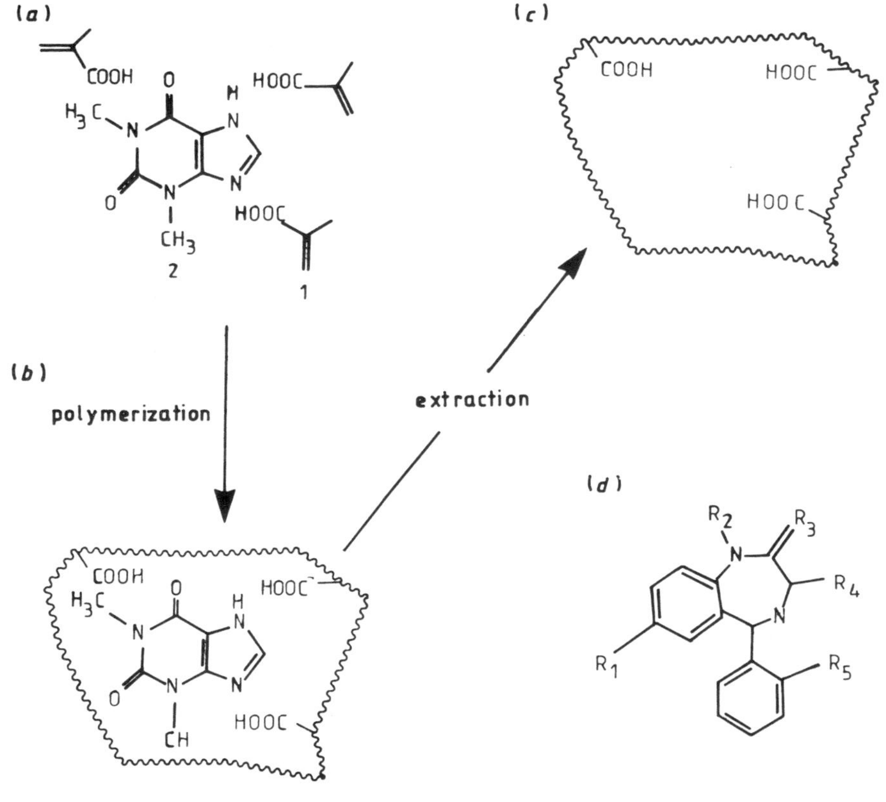

Figure 7.8 A schematic representation of the preparation of molecularly imprinted polymers [19]. (*a*) Functional monomer MAA (1) is mixed with print molecule, here theophylline (2), and EDMA, the cross-linking monomer, in suitable solvent. MAA is selected for its ability to form hydrogen bonds with a variety of chemical functionalities of the print molecule. (*b*) The polymerization reaction is started by addition of initiator (2, 2′-azobis(2-methylpropionitrile), AIBN). A rigid insoluble polymer is formed. 'Imprints', which are complementary to the print molecule in both shape and chemical functionality, are now present within the polymeric network. (*c*) The print molecule is removed by solvent extraction. The wavy line represents an idealized polymer structure but does not take into account the accessibility of the substrate to the recognition site. (*d*) Structures of the benzodiazepine derivatives used as competitive ligands in table 7.6.

Table 7.6 Diazepam antibodies. Ligands were added to drug-free serum and assayed. Cross-reactivities are expressed as the molar ratio (of theophylline and diazepam, respectively, to ligand) giving 50% inhibition of radiolabeled ligand binding. Abm, antibody combining site mimic; ab^+, natural antibody [19].

						Cross-reaction (%)	
Competitive ligand	R_1	R_2	R_3	R_4	R_5	Abm	ab^+
Diazepam	Cl	Me	O	H	H	100	100
Desmethyldiazepam	Cl	H	O	H	H	27	32
Clonazepam	NO_2	H	O	H	Cl	9	5
Lorazepam	Cl	H	O	OH	Cl	4	1

An alternative non-deterministic approach, called molecular imprinting, has been devised by Mossbach and coworkers [19]. They have shown that by selecting a polymerization system with the potential for a high degree of hydrogen bonding (carboxyl groups), and a rigid structure for the final cross-linked polymer, that a matrix with receptor recognition characteristics can be formed (figure 7.8). The receptor cavities are formed by carrying out the polymerization in the presence of the putative ligand, e.g. theophylline. A comparison of the selectivity of 'antibody' mimics to true antibodies for the drugs theophylline and diazepam is given in table 7.6. The apparent dominant binding constants and density of sites were $1/65\ \mu M$ and $1\ \mu mol\ g^{-1}$ for the theophylline polymer. For comparative purposes the site density for a pure IgG molecule is about $13\ \mu mol\ g^{-1}$.

These antibody mimics, which were most effective in apolar solvents, provide another powerful technique to engineer molecular receptors for analytical purposes, as in biosensors.

7.2.5 DNA/RNA sequences

Polynucleotide strands are one of the most specific biological recognition systems known. The use of DNA probes for isolating gene sequences by hybridization has become commonplace. In this application, the high-affinity pairing of adenine with thymine (A–T) and guanine with cytosine (G–C), coupled with the stiffness of polynucleotide chains, causes the binding specificity to increase exponentially with the number of residues. The application of this technique for use in biosensors has been minimal to date. One of the disadvantages of the use of hybridization for biosensors is that the kinetics of the binding step can be very slow. Keller and Manak [20] give a formula for estimating the hybridization rate:

$$t_{1/2} = \frac{\ln 2}{\left(3.5 \times 10^5 \left(L^{0.5}\right) C\right)}. \tag{7.5}$$

For example, they estimate that it would take about 10 minutes to reach complete hybridization of a 20-nucleotide probe.

Graham *et al* [21] demonstrated a DNA probe biosensor. An application of such a device could be to recognize a virus or bacteria in a food sample.

It is well known that messenger RNA has a very high recognition potential for amino acids, and this recognition potential of polynucleotides for *non-nucleotide* substances has led a number of research groups to develop another approach for the use of polynucleotides as recognition elements [22]. The concept is to synthesize random polynuleotide libraries, that contain of the order of thousands of different sequences (called aptamers), and then to screen these libraries to find a polynucleotide that has a binding specificity to a particular ligand (analyte). Using this technique researchers at Gilead Corporation discovered a small polynucleotide that was a powerful inhibitor of the enzyme thrombin. Although the primary application of this technique has been to discover new drugs, the method is equally applicable for selecting receptor candidates for biosensor applications.

7.3 DESIGN CONSIDERATIONS FOR USE OF RECOGNITION ELEMENTS IN BIOSENSORS

In the first section of this chapter we focused on specificity considerations in the selection of biological recognition elements for biosensors. An equally important consideration is the physical design of the biosensor to insure that the selectivity characteristics of the recognition element is fully utilized.

7.3.1 Enzyme-based biosensors

The kinetics of enzyme-catalyzed reactions can be very complex, and the mathematical representations for the effect of the concentrations of substrate, product, cofactors, and inhibitors are presented in a variety of textbooks in this field [1]. The exact form of this dependence of enzyme activity on these factors might have a profound effect on the behavior of an enzyme biosensor. However, one can delineate general rules of thumb concerning the properties of enzymes for the preliminary design of enzyme-based sensors.

For biosensor applications, the substrate is the analyte of interest, and the enzyme is chosen so that the product of the reaction has some characteristic that is readily and accurately measured, e.g. a color, change in pH, or electrochemical reactivity.

When an enzyme is utilized in a biosensor as illustrated in figure 7.9, another consideration that comes into play is the diffusion of the analyte into the sensor device and the diffusion of product from the sensor. In this situation, the relative dominance of diffusion or reaction has a major effect on the performance of the sensor. The interaction of diffusive and reactive processes complicates the

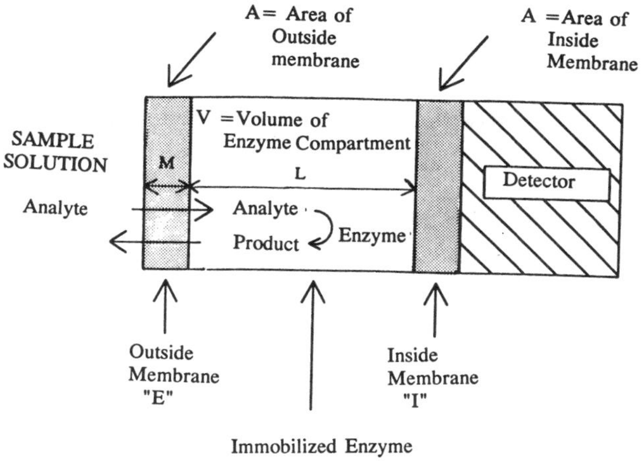

Figure 7.9 A schematic diagram of an enzyme-based biosensor.

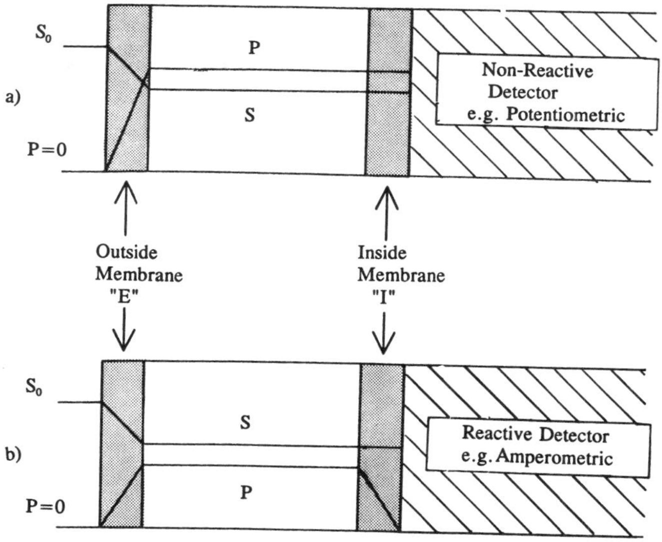

Figure 7.10 Concentration profiles in an enzyme-based biosensor where transport through the membrane layers is rate limiting.

mathematical analysis even further. However, consideration of some limiting conditions provides an accurate insight into the behavior of these devices.

Figure 7.10 shows the expected shape of the concentration profiles for analyte and product within a biosensor. Notable features are that the analyte concentration decreases from the outside surface of the sensor to the position of the detector, and the product concentration is highest inside the sensor and decreases towards both the detector element and the external surface. The shapes of these concentration profiles will depend on a variety of factors, such as the permeability of the exterior membrane (E), diffusivity of the substrate and product in the enzyme matrix, rate of enzyme reaction, and product reaction at the detector, if any. (*a*) shows typical profiles if the detector does not consume the product, e.g. a potentiometric or photometric device. (*b*) shows expected concentration profiles when the product reacts at the detector, e.g. an amperometric device.

In order to illustrate the effects of various parameters on the system, we will consider two extreme cases.

(i) The permeability of the exterior membrane (E) is rate limiting, resulting in concentration profiles as shown in figure 7.10. In this situation diffusive transport within the enzyme region is much faster than through membrane E. In practice, this can occur if the enzyme zone is much thinner than the thickness of membrane E, and/or the diffusivity of the analyte is much greater in the enzyme matrix than in the membrane material.
(ii) The diffusivity of the reactants in the enzyme phase is much lower that in the membrane matrix material. Under these circumstances the concentration gradients of substrate and products in the membrane are negligible.

7.3.1.1 The membrane-limited case. An analysis of this type of situation was given by Racine and Mindt [23]. The basic equations are as follows. For the rate of diffusion of analyte from sample phase through the membrane E

$$N_S = P_{SE}A(S_0 - S). \tag{7.6}$$

The rate of analyte conversion to product is obtained from equation (7.1)

$$r = \frac{k_{\text{cat}}ES}{K_M + S}. \tag{7.7}$$

The product will diffuse out of the enzyme phase through the exterior membrane E and also diffuse toward the amperometric electrode detector (if one is part of the sensor system). The steady state product concentration within the enzyme phase will be determined by the balance of enzymatic production of product within the enzyme region, volume V, and its disappearance by diffusion or electrode reaction

$$rV = P_{PE}A_E(P - P_0) + P_{PI}A(P - P_I). \tag{7.8}$$

If the sample solution as presented to the sensor is well mixed, e.g. by high flow rates past the sensor, or a large volume that is vigorously mixed, then

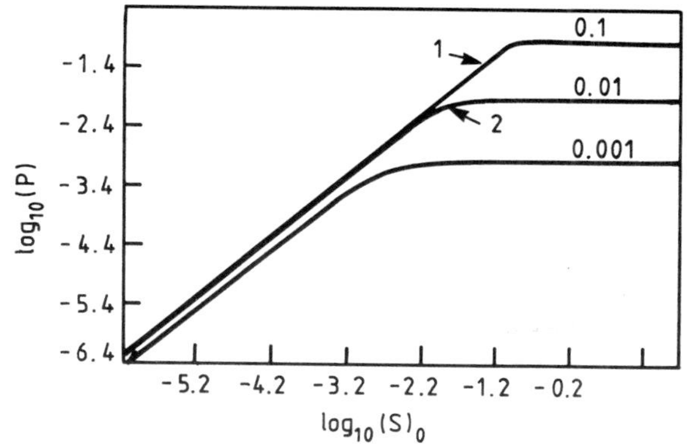

Figure 7.11 A theoretical potentiometric enzyme electrode calibration curve based on external diffusion control of the reaction: a plot of the logarithm of the product concentration in the enzyme layer versus the logarithm of the bulk substrate concentration. $K_n = 10^{-3}$; the value of $k_\alpha EV/P_{SE}A$ is given on the curve [24].

the external product concentration will be negligible; i.e. $P_0 = 0$. Also, at the electrode interface the product concentration will be zero; i.e. $P_I = 0$.

Solving these equations one obtains

$$S = -\frac{(T_S + K_M - S_0) + [(T_S + K_M - S_0)^2 + 4K_M S_0]^{1/2}}{2}$$

$$T_S = \left(\frac{k_{\text{cat}} EV}{P_{SE} A}\right) \tag{7.9a}$$

$$P = \frac{k_{\text{cat}} EV}{(P_{PE} + P_{PI})} \frac{S}{(K_M + S)}. \tag{7.9b}$$

A plot of product concentration inside the sensor (P) versus the external analyte concentrate $(S)_0$ is given in figure 7.11. One notes that the behavior of product concentration falls into two regimes: at low levels of the external analyte concentration, the internal product concentration is directly proportional to the external analyte concentration, while at high levels of analyte concentration the internal product concentration is independent of external analyte concentration. Since the detector response will be related to product concentration *in the sensor*, this result demonstrates that, above a critical range of external analyte concentrations, the sensor response will be independent of the sample analyte concentration. Clearly, the sensor must be designed so that it does not operate in the region shown on the right side of the figure.

A closer inspection of equation (7.9) provides some insight as to the dominating parameters that determine this behavior.

When the internal analyte concentration is much greater than K_M, equation (7.9b) reduces to

$$P = \frac{k_{cat} E V}{(P_{PE} + P_{PI}) A}. \tag{7.10}$$

Under these circumstances the enzyme is 'saturated' and thus the reaction rate cannot increase with increasing substrate concentrations. Thus *the sensor response is independent of external analyte concentration, and this region of operation should be avoided*.

On the other hand, when the internal analyte concentration is much less than K_M, equation (7.9*b*) reduces to

$$P = \frac{S_0}{(P_{PE} + P_{PI})(1/P_{SE} + K_M A / k_{cat} E V)}. \tag{7.11}$$

Under these conditions the sensor response is directly proportional to analyte concentration. Further, if the term in brackets (called the enzyme loading factor by Carr and Bowers [24]) is greater than unity, this equation reduces to

$$\text{if} \quad \frac{1}{P_{SE}} \gg \frac{K_M A}{k_{cat} E V} \qquad P = \left(\frac{P_{SE}}{P_{PE} + P_{PI}} \right) S_0. \tag{7.12}$$

In these circumstances of membrane permeability limitations, shown as curves 1 and 2 in figure 7.11 at external substrate concentrations below 0.01 M, the response of the sensor will be completely independent of the enzyme properties. A sensor operated in this regime would be independent of factors affecting enzyme behavior, such as denaturation, temperature, and pH. *This is the preferred regime for operating an enzyme biosensor*.

The product concentration for a detector that does not consume the product, e.g. photometric device, is accounted for by setting $P_I = P$, and $P_0 = 0$. Then, if one could select a material for the external membrane E that had the properties of a high permeability for the analyte (P_{SE}) and a low permeability for the product (P_{PE}), then one would achieve a concentration effect of the product within the sensor and an amplification of the sensor signal, given by the ratio P_{PE}/P_{SE}.

An example of a circumstance where this might be achieved is if the analyte is a neutral organic molecule and the product is a charged molecule, e.g. an acid. A membrane material such as an ion exchange polymer (polysulfate) would have a high permeability for the organic analyte, but a low permeability for the acid at neutral pH values.

7.3.1.2 The reaction-limited case. Under the circumstances that the enzyme reaction and diffusion within the immobilized enzyme zone dominate the behavior of the system, the properties of the membranes E and I can be ignored. This situation will occur when the membranes are very permeable to analyte and product, and/or the enzyme zone is much thicker than the membrane layers.

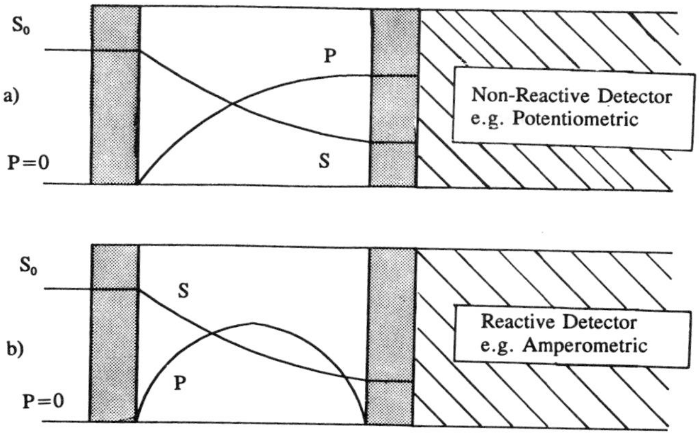

Figure 7.12 Concentration profiles in an enzyme-based biosensor when transport through the enzyme phase is rate limiting.

The governing equations for analyte and product are:

$$D_S \frac{d^2 S}{dx^2} = \frac{k_{cat} E S}{K_M + S} \qquad D_P \frac{d^2 P}{dx^2} = -\frac{k_{cat} E S}{K_M + S} \tag{7.13}$$

And for the two types of detector, non-reactive and reactive, the corresponding boundary conditions are

(i) non-reactive detector, figure 7.12(a)

$$@x = 0 \qquad dP/dx = 0$$

(ii) reactive (amperometric) detector, figure 7.12(b)

$$@x = 0 \qquad P = 0.$$

For the non-reactive detector, if the analyte concentration throughout the enzyme region is much greater than K_M, i.e. $S \gg K_M$, then the product concentration for the non-reactive detector is

$$P_{x=0} = \frac{k_{cat} E L^2}{2 D_P} \tag{7.14}$$

$$D_P \left.\frac{dP}{dx}\right|_{x=0} = \frac{k_{cat} E L}{2}. \tag{7.15}$$

Again, these conditions are equivalent to saturating the enzyme, resulting in a product concentration inside the sensor that is independent of the analyte concentration in the external sample. *This is another operating regime to be avoided in sensor design.*

For a reactive detector (amperometric) the current signal is proportional to the product gradient at the detector, i.e. $\mathrm{d}P/\mathrm{d}x @ x = 0$. Equation (7.13) yields equation (7.15), showing *another region where the sensor response would be independent of external analyte concentration*.

If the analyte concentration within the sensor is everywhere much lower than K_M, then the following expressions for product concentration are obtained:

$$\text{Non-reactive detector} \quad P_{x=0} = S_0 \frac{D_S}{D_P}\left(1 - \operatorname{sech}\left(\frac{k_{\text{cat}} E L^2}{K_M D_S}\right)^{1/2}\right) \tag{7.16}$$

$$\text{Reactive detector} \quad \left.\frac{\mathrm{d}P}{\mathrm{d}x}\right|_{x=0} = -\frac{S_0 D_S}{L^2}\left(1 - \cosh\left(\frac{k_{\text{cat}} E L^2}{K_M D_S}\right)^{1/2}\right). \tag{7.17}$$

In both these situations, the detector response will be directly related to the external analyte concentration. Further, if the second term in brackets is much smaller than unity, then it can be ignored. *Under these conditions the response of the sensor will be independent of enzyme and thus more independent of changes in enzyme activity*.

Typical values for the Michaelis–Menten constant for enzymes that have been used in making biosensors are given in table 7.1.

7.3.1.3 Time response. In most situations enzyme kinetics have very little effect on the response time of enzyme-based biosensors. From the analysis given above, it is clear that one should operate these devices under conditions where the analyte concentration within the sensor is much less than K_M. For sensors which are in the membrane diffusion limiting regime (section 7.3.1.1 above), the response characteristics of the membrane material will be governing. These depend on the thickness of the membrane and the diffusivity of the analyte in the membrane material. An approximate estimate of the membrane lag time is [25]

$$t_{\text{lag}} = 0.1 D_{SE}/M^2 \tag{7.18}$$

where D_{SE} is the diffusivity of the analyte in the external membrane (E), and M is its thickness.

For the reaction-limited regime (section 7.3.1.2 above), Carr and Bowers [24] give the time constant as

$$t_{\text{lag}} = 0.1 D_S/L^2 \tag{7.19}$$

where D_S is the diffusivity of the analyte in the enzyme phase, and L is the depth of the enzyme layer.

7.3.2 Bioreceptor binding systems

A brief review of the kinetics and equilibrium binding between ligands and macromolecular bioreceptor binding agents will provide an insight into the parameters that govern the behavior of these systems for use in biosensor applications. The recognition of the ligand (or analyte in this context) and receptor is usually confined to a limited region of the biomolecule, called the active or binding site. Some receptors have multiple binding sites, e.g. IgG antibodies have two binding sites. Also the analyte may have multiple binding sites if it is a macromolecule, e.g. a single dextran has several pendent glucose residues that can bind to concanavalin A. The analysis of multivalent ligand binding is quite complex, and still under investigation [26]; however, some of the general characteristics of ligand–receptor binding can be discerned by the consideration of simple systems.

7.3.2.1 Direct binding systems. Many biosensors that utilize receptors as a mode of obtaining selectivity depend on the detection of the binding event, using devices such as surface plasmon resonance, surface acoustic waves, and fluorescence quenching [27].

The reaction that occurs can be symbolized as

$$S + R \rightleftharpoons SR \tag{7.20}$$

where S is the analyte and R is the receptor.

In a typical sensor system the amount of receptor is fixed, e.g. bound to the detector surface or confined in a limited volume in the sensor. Thus the amount of receptor in free and bound state provides the indication of analyte concentration.

$$R_t = R + SR \tag{7.21}$$

where R_t is the total amount of receptor in the sensor.

The form of the equation that describes the degree of occupancy of receptor sites as a function of analyte concentration is the same, whether the receptor is bound to a surface (Langmuir isotherm) or in solution (reversible reaction)

$$\frac{SR}{R_t} = \frac{S}{K_R + S}. \tag{7.22}$$

Binding of analyte to receptor shows saturation behavior (figure 7.3). From this figure, one can readily appreciate that the choice of receptor to be used in a sensor for a particular analyte will depend on the range of analyte concentrations of interest. If the analyte concentration is much higher than the numerical value of K_R, there is little change in the degree of occupancy of the receptor with variations in analyte concentration. Thus one of the considerations in the selection of a receptor is that the K_R is greater than S_{avg} (the average expected analyte concentration). Another characteristic of bioreceptor methods for this type of analytical application is that the dynamic range of the sensor is usually of the order of 10, i.e. $0.15K_R < S < 1.5K_R$. For example, the operational analytical range for antimyoglobin antibody (figure 7.3) is about 0.2 to 2.0 monomolar.

7.3.2.2 Indirect binding methods. The range of applications of bioreceptors in biosensor design is expanded by the use of analyte analogs that can be detected at a much higher level of sensitivity than the direct alteration in receptor properties when the analyte is bound to the receptor. Analog or surrogate analytes can be engineered to optimize the detection process. For example, highly fluorescent dyes, or highly charged ionic species, can be used to measure the extent of occupancy of the receptor by optical or capacitive methods.

These systems can be modeled as follows

$$S + R \rightleftharpoons SR \qquad K_1 = [S][R]/[SR] \tag{7.23}$$

$$S^* + R \rightleftharpoons S^*R \qquad K_2 = [S][R]/[S^*R] \tag{7.24}$$

$$R_t = R + SR + S^*R \tag{7.25}$$

$$S_t^* = S^* + S^*R. \tag{7.26}$$

The corresponding expression to (7.22) for the extent of receptor occupancy for this system is [30]

$$\left(\frac{S^*}{S_t^*}\right)^2 + \frac{S^*}{S_t^*}\left[\left(\frac{R_t}{S_t^*} - 1\right) + \left(\frac{K_1 S + 1}{S_t^* K_2}\right)\right] - \left(\frac{K_1 S + 1}{S_t^* K_2}\right) = 0 \tag{7.27}$$

This expression is plotted in figure 7.13, and shows some of the same characteristics as the direct binding method (figure 7.3), namely that the system displays saturation behavior. The introduction of the analog analyte (S^*) provides additional opportunities to engineer the sensor to meet particular application requirements. Both the total amount of the analog-analyte (S_t^*) and its binding constant (K_2) can be selected to provide a response in the range of analyte concentration of interest.

The parametric curves in figure 7.13 show that the sensitivity range for a given bioreceptor (i.e. for a given K_1) in the indirect method can be additionally modulated by a factor of 10 by modifying K_2. An example of this approach is provided by the data shown in figure 7.14 [28]. Here the binding of dinitrophenol derivatives to an antibody was increased up to 15-fold by synthesizing multivalent derivatives of the analyte.

Based on these considerations Schultz [30] suggested a rational approach as a starting point for the design of a biosensor using the indirect method.

(i) Estimate the mid-range of expected analyte concentrations to be analyzed, S.
(ii) Select a bioreceptor with a binding constant numerically of the order of $10/S$.
(iii) Estimate the minimum concentration of analog-analyte, S_{m}, that is reliably measured by the detector. Load the sensor with analog-analyte at a concentration about 50 times this value, i.e. $S_t = 50\,S_m$.
(iv) Select, synthesize, or modify the analog–analyte so that the criteria $K_2 = S_m$ is satisfied.

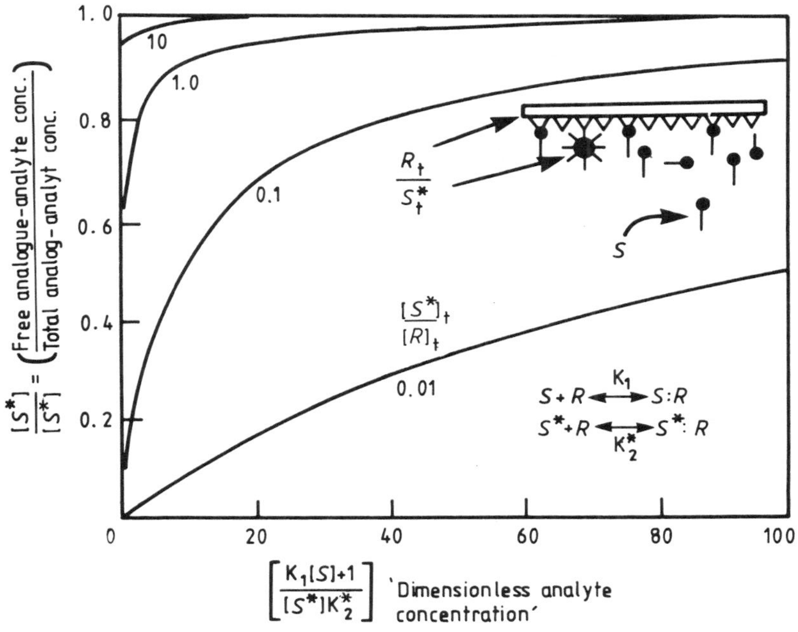

Figure 7.13 The response of a biosensor system utilizing a competitive analog analyte S, e.g. a fluorescently labeled compound.

(v) Develop methodologies to load the sensor transducer compartment with the bioreceptor so that the total concentration of receptor sites is of the order of $100S_t$.

Of course this recipe would be only the starting point of a sensor development plan.

7.3.2.3 Response time of bioreceptor-based sensors. As in enzyme-based sensors the two main factors that determine the responsiveness of bioreceptor-based sensors are diffusive and kinetic phenomena.

Diffusive limitations due to membranes in the biosensor can be estimated as discussed above, i.e. diffusion lag times are of the order of L^2/D.

The kinetic behavior of monovalent binding is less complex than enzyme reactions, and can be represented simply as shown in equations (7.2)–(7.4) [29].

One approach to estimating the time response of this type of reaction is to find the effect of a small perturbation in analyte concentration when the system is in equilibrium. The result of such an analysis Schultz [30] showed that the response time is given approximately by

$$t_{1/2} = -(\ln 2)/(2k_d). \tag{7.28}$$

That is, the response time is dominated by the dissociation rate of the reaction, k_d.

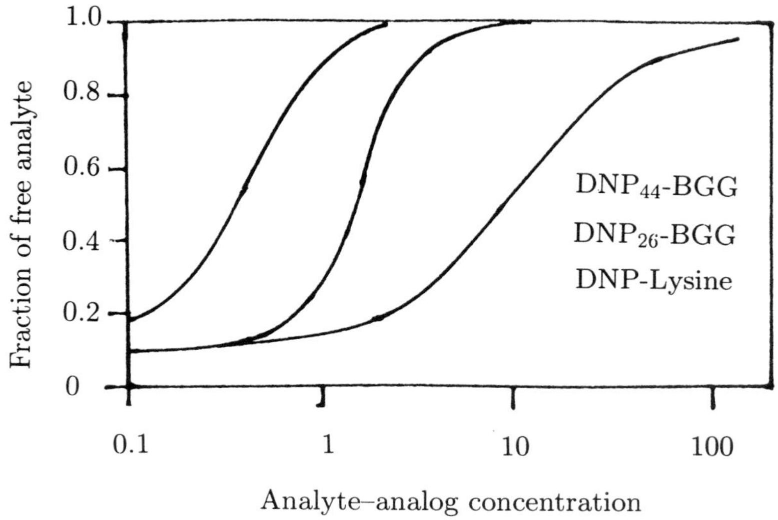

Figure 7.14 The effect of multivalency of analog analyte on the response of a biosensor.

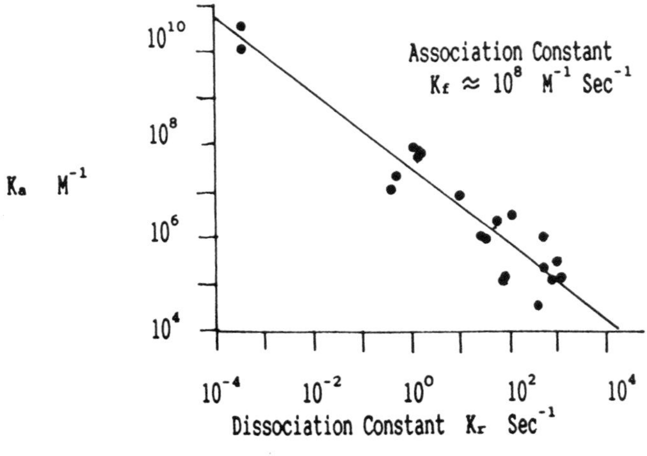

Figure 7.15 The relationship between the association equilibrium constants, K_a, and the dissociation rate constants, k_r, for a series of antibodies to dinitrophenol (DNP).

Some values of association and dissociation reaction constants for bimolecular bioreceptors are given in figure 7.15, where data are shown for a single analyte, dinitrophenol (DNP), and a variety of monoclonal antibodies. Note that although the equilibrium binding constants range over one millionfold,

Table 7.7 Association and dissociation rate constants for enzyme– ligand complexes.

Enzyme	Ligand	k_a ($M^{-1}\,s^{-1}$)	k_d (s^{-1})	K_A (M)
Old yellow enzyme	FMN	10^6	10^{-4}	10^{-10}
Malate dehydrogenase	NADH	10^8	10	10^{-7}
Lactate dehydrogenase	NADH	10^7	10	10^{-6}
Chymotrypsin	Proflavin	10^8	10^{-3}	10^5
Creatine kinase	ADP	10^7	10^4	10^{-3}
Aspartate aminotransferase	Glutamate	10^8	10^6	10^{-2}

the dissociation constants are very similar in magnitude, shown by their clustering about the line drawn in the figure. One explanation of this behavior is that the association reaction is diffusion controlled, that is the rate of binding is less dependent on chemical interactions than how fast the two molecules, analyte and receptor, can collide. An estimate of the theoretical maximum rate of collision of two molecules given by the Smoluchowski approximation is about $10^{-9}\,M^{-1}\,s^{-1}$.

Table 7.7 gives some typical kinetic data for enzyme–ligand complexes. Again, although the affinity constants (K_A) vary over 100 000 000-fold, the association rate constants (k_a) vary only by a factor of 100 and are very similar to those found with antibodies. Thus the forward rates of these reactions are probably diffusion controlled as well.

In the discussion given above it was noted that in designing a biosensor, one would want to select a bioreceptor with an affinity constant of about the same order of magnitude as the concentration of analyte in the samples to be assayed. If the analyte is at nanomolar concentration levels, then the affinity constant should be of the order of 10^{-9} M. Then as seen in figure 7.15 the dissociation rate constant will be about $10^{-2}\,s^{-1}$. Substituting this value in the expression above for the response time of such a sensor, one estimates a response time of of the order of minutes. *One can appreciate that there is an inverse trade-off on sensitivity versus responsivity; higher sensitivities (higher affinities) give longer response times*.

Sometimes statements appear in the literature that suggest that the specificity of antibodies is much narrower than the broad range of specificities found with bioreceptors. Yet if one compares figure 7.15 with figure 7.6, it can be seen that the range of affinities for a single type of receptor and a wide variety of ligands is of the order of 10^6 whether it be a membrane bioreceptor (figure 7.6) or an antibody (figure 7.15).

REFERENCES

[1] Fersht A 1985 *Enzyme Structure and Mechanism* 2nd edn (New York: Freeman) p 475

[2] Webb E C (ed) 1984 *Enzyme Nomenclature. Recommendations of the Nomenclature Committee of the International Union of Biochemistry* (Academic) p 646

[3] Schornburg D and Salzmann M (ed) 1991 *Enzyme Handbook* (Berlin: Springer)

[4] Turner A P F, Karube I and Wilson G S (ed) 1987 *Biosensors: Fundamentals and Applications* (Oxford: Oxford University Press) p 770

[5] Janata J, Josowicz M and DeVaney D M 1994 Chemical sensors *Anal. Chem.* **66** 207R–28R

[6] Schultz P G 1988 The interplay between chemistry and biology in the design of enzymatic catalysts *Science* **240** 426–33

[7] Blackburn G F *et al* 1990 Potentiometric biosensor employing catalytic antibodies as the molecular recognition element *Anal. Chem.* **62** 2211–6

[8] Edelman G M 1970 The structure and function of antibodies *Sci. Am.* **123** 3

[9] Ibrahim S 1993 Immunoglobulin binding specificities of the homology regions (Domains) of Protein-A *Scand. J. Immunol.* **38** 368–74

[10] Kohler G and Milstein C 1975 Continuous cultures of fused cells secreting antibody of predefined specificity *Nature* **256** 495–7

[11] Hurred J 1982 *Monoclonal Hybridoma Antibodies, Techniques and Applications* (Boca Raton, FL: Chemical Rubber Company) p 231

[12] Seaver S S 1987 *Commercial Production of Monoclonal Antibodies: A Guide for Scale-up* (New York: Dekker) p 327

[13] Hogenboom H R, Marks J D, Griffiths A D and Winter G 1992 Building antibodies from their genes *Immunol. Rev.* 13 041–68

[14] Changeux J P, Devillers-Thiery A and Chemouille P 1984 Acetylcholine receptor: an allosteric protein *Science* **225** 1335–45

[15] Taylor R F 1991 Immobilized antibody and receptor-based biosensors *Protein Immobilization* ed R Taylor (New York: Dekker) pp 263–309

[16] Eldefrawi M, Eldefrawi A, Rogers K and Valdes J 1992 Pharmacological Biosensors *Immunochemical Assays and Biosensor Technology for the 1990's* ed R Nakamura, Y Kasahar and G Rechnitz (Washington, DC: American Society of Microbiology) pp 391–406

[17] Hulme E C and Birdsall N J M 1992 Strategy and Tactics in receptor-binding studies *Receptor–Ligand Interactions* ed E C Hulme (Oxford: IRL) pp 63–176.

[18] Breslow R, Greenspoon N, Guo T and Zarzycki R 1989 Very strong binding of appropriate substrates by cyclodextrin dimers *J. Am. Chem. Soc.* **1111** 8297–9

[19] Vlatakis G, Anderson L I, Miller R and Mosbach K 1993 Drug assay using antibody mimics made by molecular imprinting *Nature* **331** 645–7

[20] Keller G H and Manak M M 1989 *DNA Probes* (New York: Stockton) p 259

[21] Graham C R, Leslie D and Squirrell D J 1992 Gene probe assays on a fiber-optic evanescent wave biosensor *Biosensors Bioelectron.* **7** 487–93

[22] Griffin L C, Toole J J and Lewing L L K 1993 The discovery and characterization of a novel nucleotide-based thrombin inhibitor *Gene* **137** 25–31

[23] Racine P and Mindt W 1971 On the role of substrate diffusion in enzyme electrodes *Experentia Suppl.* **18** 525–34

[24] Carr P W and Bowers L D 1980 *Immobilized Enzymes in Analytical and Chemical Chemistry* (New York: Wiley) pp 460

[25] Park G S 1983 Transport principles—solution, diffusion and permeation in polymer

and membranes *Synthetic Membranes: Science, Engineering and Applications* ed P M Bungay, H D Londsdale and M N de Pinko (Dordecht: Reidel) pp 57–107

[26] Perelson A S 1984 Some mathematical models of receptor clustering by multivalent ligands *Cell Surface Dynamics: Concepts and Models* ed A S Perelson, C Delisi and F W Wiegel (New York: Dekker) pp 223–76

[27] Meadows D and Schultz J S 1988 Fiber optic biosensors based on fluorescence energy transfer *Talanta* **35** 145–50

[28] Liu B L and Schultz J S 1986 Equilibrium binding in immunosensors *IEEE Trans. Biomed. Eng.* **BME-33** 133–8

[29] Pecht I and Lancet D 1977 Kinetics of antibody–hapten interactions *Mol.Biol.Biochem. Biophys.* **24** 306–38

[30] Schultz J S 1987 Sensitivity and dynamics of bioreceptor-based biosensors *Biochemical Engineering V; Ann. NY Acad. Sci.* **506** 406–11

[31] Cram D J 1988 The design of molecular hosts, guests and their employees *Angew. Chem. Inst. Ed. Engl.* **27** 1009–112

8

Immobilization methods

Richard F Taylor

8.1 INTRODUCTION

The key functional and distinguishing feature of both chemical sensors and biosensors is the surface layer which interacts with the analyte to be measured. This layer may be as simple as a polymer which differentially reacts with a series of analytes, or may be more complex, with multiple layers of chemical and biological components.

Chapter 6 of this text has presented means to chemically modify transducer surfaces for active layer production, and practical examples of sensor fabrication using polymerics and other chemical layers are presented starting with chapter 13. The purpose of this chapter will be to review methods for immobilizing chemical and biological molecules onto activated and polymeric sensor surfaces with an emphasis on biomolecule immobilization.

8.2 IMMOBILIZATION TECHNOLOGY

Immobilization is the key to the development of sensors which rely on the use of functional molecules where the molecules are labile, are rare (expensive), are at a low concentration in comparison to a base (inert) layer, and/or must be oriented in a specific direction. Such sensors include many polymer-based chemical sensors and most (if not all) biosensors.

There are five major immobilization methods used in the preparation of chemical sensors and biosensors.

(i) *Covalent binding*: the attachment of the active component to the transducer surface using a chemical reaction such as peptide bond formation or linkage to activated surface groups (thiol, epoxy, amino, carboxylic, etc).
(ii) *Entrapment*: physical trapping of the active component into a film or coating.

(iii) *Cross-linking*: similar to entrapment, only a polymerization agent (such as glutaraldehyde) is used to provide additional chemical linkages between the active, entrapped component and the film or coating.
(iv) *Adsorption*: association of the active component with a film or coating through hydrophobic, hydrophilic, and/or ionic interactions.
(v) *Biological binding*: association of an active biomolecule to a film or coating through specific, biochemical binding.

Immobilization may take place as part of the actual formation of the sensor active surface, or may be done after a base layer is established on the transducer. For example, entrapment and cross-linking immobilization methods involve the mixture of active component with carriers and polymerizing agent(s) and application of the mixture directly to the transducer to form the active surface. Covalent, adsorption, and biological immobilization methods attach the active component to a previously prepared surface, such as activated silica.

8.2.1 Covalent immobilization

The detection polymer or molecules of a chemical or biosensor can be covalently bonded directly to the sensor transducer or to a membrane or film coating the transducer. Covalently-bonded sensor surfaces have the highest resistance to changes in pH, ionic strength, and temperature, and are among the most stable to reuse and recycling. Disadvantages to covalent bonding include potential losses in the activity of the immobilized molecule, especially if it is a labile biomolecule such as an enzyme or receptor. Since most covalent coupling methods must be carried out at pH ranges which may be detrimental to biomolecules, it is necessary to carefully evaluate the effect of coupling conditions on biological activity during sensor development.

Covalent linkages in sensor active surface preparation utilize active functional groups in the molecule to be immobilized. The most used include primary amine, thiol, hydroxyl, and carboxylic acid groups. In the case of biosensors, proteins also can be covalently linked utilizing not only the latter groups, but also to the active groups found in arginine (guanidino), tyrosine (phenolic), histidine (imidazole), cysteine and cysteine(sulfhydryl), tryptophan (indole), and methionine (thioether) amino acid residues.

Table 8.1 lists the more commonly used methods for covalent immobilization on sensor surfaces. Table 8.2 lists examples of sensors utilizing covalent immobilization methods. More detailed information is available in other review articles [14, 15].

Peptide (amide) bond formation is one of the most used immobilization methods for sensor preparation. In peptide bond formation, nucleophilic groups on a protein, peptide, or other molecule including amino, thiol, and phenolic groups attack activated groups on the sensor transducer or the film/membrane coating the transducer. These activated groups are prepared using a variety of methods including those listed in table 8.1 and in chapter 6 of this text. The

Table 8.1 Commonly used methods for immobilization of active sensor components. (i) Peptide bond formation. (ii) Other immobilization methods.

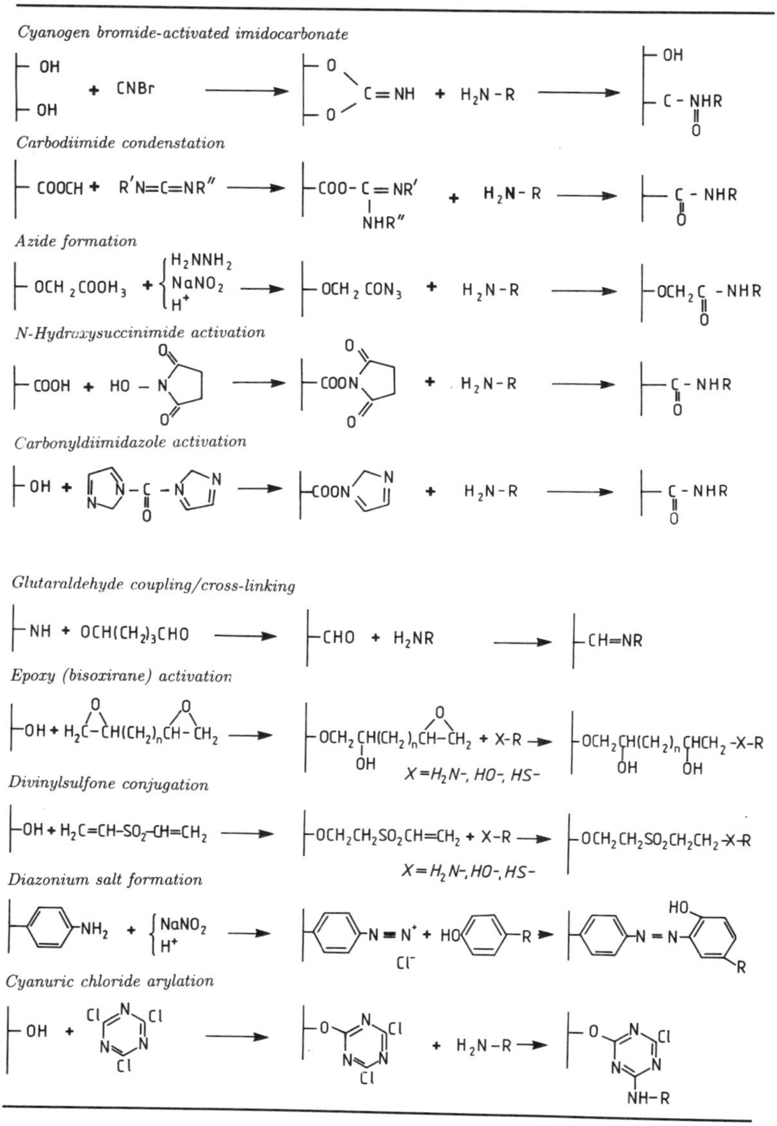

methods illustrated in table 8.1 are all for the preparation of activated surfaces: the compound to be immobilized reacts directly with the activated functional group on the surface.

Table 8.1 also illustrates other, non-peptide bond methods for sensor surface immobilization. Such methods are needed if groups other than amino groups are available for covalent linkage in the molecule to be immobilized. Of these methods, glutaraldehyde is among the most used due to the stability of the resulting matrix. This may also be due to glutaraldehyde-induced cross-linking, which further stabilizes the active surface (see below).

Hydrazine activation can also be used as the basis for *site-directed immobilization* or *directional immobilization* [16]. Directional immobilization is the immobilization of the sensor detector molecules in a highly ordered and reproducible manner. For example, in antibody-based biosensors requiring a monolayer coating of antibody on the transducer, it is critical to sensor functionality and sensitivity that as many of the antibodies as possible are immobilized such that their antigen binding sites are directed outward and easily accessible by antigen (analyte).

Such directional immobilization can be accomplished by reacting a hydrazine-activated transducer surface or film with antibody which has been oxidized using sodium periodate. The periodate acts on vicinyl hydroxyl groups of sugars located in the heavy chains (non-antigen binding C_H2 domain) of the antibody to result in formyl groups. The antibody (Ab) can then be directionally immobilized by the following reaction:

$$\text{–CONHNH}_2 + \text{OHC–sugar–Ab} \Rightarrow \text{–CONH–N=C–sugar–Ab.} \qquad (8.1)$$

Antibodies may also be reduced with agents such as 2-mercaptoethylamine to produce two half-immunoglobulin molecules with free sulfhydryl groups in the C_H2 chain region. These sulfhydryl groups can then be used to immobilize the antibody fragments by covalent coupling using, for example, an iodoacetyl-activated surface [17]:

$$\text{–NHCOCH}_2\text{CH}_2\text{–I} + \text{HS–Ab} \Rightarrow \text{—NHCOCH}_2\text{CH}_2\text{–S–Ab.} \qquad (8.2)$$

It has long been recognized in methods such as affinity chromatography and enzyme immobilization that the distance between the immobilization surface and the immobilized active molecule can be critical to functionality [18]. This appears to be true as well for sensors. For example, gold is a common substrate for sensor transducers such as the electrodes of capacitance sensors. Gold surfaces can be activated by reaction with alkanethiols which spontaneously adsorb to the gold to form densely packed, highly ordered monolayer films according to the reaction

$$\text{Au} + \text{HS(CH}_2)_n\text{X} \Rightarrow \text{Au–S–(CH}_2)_n\text{–X.} \qquad (8.3)$$

In the case of a series of alkanethiols where X = OH or COOH, it was found that if $n \leqslant 10$, the surface properties of the gold are determined by X and are independent of chain length and the gold surface [19]. Immobilization of an active molecule to the hydroxyl or carboxyl group of such films can result in active sensors [1–3].

Polymeric films can also be considered as utilizing covalent immobilization methods since such polymers are formed from the covalent linkage of monomers. Essentially, solutions of monomers are ether cast or electrochemically deposited onto the transducer surface, or precast films are grafted onto the transducer (see chapter 6 of this text).

Conductive or *electroactive* polymers represent a specialized category of polymeric coatings and are increasingly being used for sensor fabrication. Such polymers have good (semirigid) mechanical properties, have good electron transport properties, can be oxidized or reduced using electrolysis, and can be fabricated as films on sensor transducers [20]. These include conductive polymers such as polyacetylene, polythiophene, polyfuran, polyphenylene, polyquinoline, polyaniline, polypyrrole, and polyisothianaphthene.

8.2.2 Immobilization by entrapment

Entrapment has been applied primarily to biomolecules for biosensor fabrication (table 8.3). Essentially, the molecule to be immobilized is formulated with the components of a membrane or film prior to laying down the membrane or film, or is added after the membrane or film is in place on the transducer. In the latter case, the molecule to be entrapped diffuses into the membrane or film. Typical films/matrices used for entrapment include starch, polyacrylamide, silicone rubber, polyvinyl alcohol, and polyvinyl chloride.

Entrapment is a gentle method utilizing non-chemical treatment and mild reaction conditions. It is especially applicable to very labile biomolecules which may degrade and/or lose activity outside normal, physiological temperature and pH. The primary disadvantage with entrapment is that the weak bonding between the detector molecule and the matrix or transducer can be easily disrupted, resulting in leakage of the molecule from the sensor and leading to sensor performance degradation.

One of the primary applications of entrapment immobilization has been to prepare enzyme-electrode-based biosensors [27], and one of the first functional enzyme electrodes utilized urease entrapped in an acrylamide film to detect urea using an ammonium ion selective electrode [28]. Highly hydrophobic bilayer lipid membranes and liposomes have also been used to entrap highly labile biomolecules (see chapter 9). Such films and layers are, however, inherently unstable themselves and are useful primarily as research tools.

Electrochemical polymerization is another entrapment method which can be used to immobilize a detector molecule into a conductive, redox, or insulating film on a transducer (see chapter 6). For example, glucose oxidase (GOD) was immobilized into a poly(*m*-phenylenediamine) film formed by potentiostatic electropolymerization of the monomer on gold-coated glass slides [21]. Monomer and GOD in buffer were polymerized on the gold surfaces using a working electrode potential of 0.662 V for 15 min.

Table 8.2 Examples of sensors using covalent immobilization methods.

Immobilized material/method	Analyte(s) or measurement	Ref.
Synthetic lipid (dioctadecyldimethyl ammonium polystyrene and dimyristoylphosphatidylethanolamine) cast from solution onto piezoelectric quartz crystals (5μ, ordered layers); IgG covalently linked to thiolated, carbodiimide-activated gold electrodes on piezoelectric quartz crystals	Bitter/sweet compounds; anti-IgG	[1]
Galabiose and galabiose-BSA covalently linked to thioalkylcarboxy-activated gold surfaces	P-fimbriated strains of *Escherichia coli*	[2]
Photoisomerizable dinitrophenyl spiropyran linked to a gold electrode by an aminoethanethiol spacer	Antibody to dinitrophenol	[3]
Linkage of peptide cysteine groups to silicon surfaces using long-chain trichlorosilanes	Peptide antibodies	[4]
Coupling of antibody to T_2 toxin to amino-activated quartz fibers using cyanogen bromide	T2 toxin from *Fusarium tricinctum*	[5]
Coupling of F(ab′) fragments to amino-groups on bovine serum albumin–cellulose acetate membranes using succinimidyl 4-(N-maleimidomethyl)-cyclohexane-1-carboxylate, and attachment to platinum electrodes	human IgG	[6]
Linkage of recombinant HIV proteins to a capacitance transducer by glutaraldehyde coupling to amino-terminated silanes	HIV antibody	[7]
Directed IgG immobilization on gold by covalently linking sodium periodate-oxidized antibody to polylysine and linking the IgG–polylysine complex to alkanethiols on the gold surface	Antigen	[8]
Immobilization of carbonic anhydrase on radio-frequency-oxidized graphite electrodes using N-hydroxysuccinimide and carbodiimides	Nitrophenylacetate	[9]
Ionophore poly(vinyl chloride), hydroxylated PVC, and aminated PVC membranes on polyimide (Kapton) wafer electrodes	pH, pK, and pCa	[10]
Coupling of zeolite crystals to SAW devices using 3-mercaptopropyl-trimethoxysilane	Organic vapors	[11]
Aminopropyltrimethoxysilane coupling of phthalamic acid to silica particles on a quartz crystal microbalance	Uranyl nitrate	[12]
Immobilization of fluoresceinamine in an acrylamide–methylenebis(acrylamide)copolymer covalently linked to an activated glass optical fiber	pH	[13]

Table 8.3 Examples of sensors using entrapment immobilization.

Immobilized material/method	Analyte(s)	Ref.
Potentiostatic electropolymerization of glucose oxidase into poly(*m*-phenylenediamine) films onto gold-coated glass slides	Glucose	[21]
Entrapment of urease within poly[di(methoxyethoxyethoxy)phos-phazene] hydrogels	Urea	[22]
Immobilization of glucose oxidase on planar electrode arrays in photopolymerized poly-2-hydroxyethylmethacrylate membranes	Glucose	[23]
Entrapment of nicotinic acid, tryptophan, picolinic acid, and 6-aminohexanoate into poly-$[Ru(v\text{-}bpy)_3]^{2+}$ coatings on platinum microelectrodes	Mercury	[24]
Immobilization of choline oxidase and glucose oxidase on platinum electrodes by entrapment into amorphous polyester cation exchange films	Choline, glucose	[25]
Valinomycin entrapped in photopolymerized poly-(vinylchloride) membranes on silica and ion selective electrodes	K^+	[26]

8.2.3 Immobilization by cross-linking

Cross-linking combines features of both covalent bonding and entrapment. Cross-linking agents such as glutaraldehyde, hexamethylene diisocyanate, difluoro-dinitrobenzene, bismaleimidohexane, disuccinyl suberate, and dimethyl suberimidate are used both to polymerize a base layer or film, and to anchor the entrapped detection molecule in the layer or film by forming intermolecular linkages between the membrane and the detector molecule. The resulting active layer is more stable toward leakage of the detector molecule and to changes in temperature and pH. By choosing the cross-linker and its concentration carefully (often by trial and error), the degree of cross-linking should not interfere significantly with the activity of the detector molecule.

For example, we have used cross-linking to immobilize very labile receptor molecules onto interdigitated electrode transducers [29]. Essentially, opiate and acetylcholine receptors were immobilized in a bovine serum albumin base containing phospholipid, detergent, antioxidant(s), and cholesteryl ester using 2–5 wt% glutaraldehyde [30]. Concentrations of glutaraldehyde less than 2% resulted in incomplete polymerization and soft, gel-like layers while concentrations above 5% led to hard, rigid layers and a progressive loss of receptor activity with increasing glutaraldehyde concentrations.

Table 8.4 presents additional examples of cross-linking used in the preparation of chemical sensors and biosensors.

Table 8.4 Examples of sensors using cross-linking immobilization.

Immobilized material/method	Analyte(s)	Ref.
Neurological receptors and antibodies entrapped into glutaraldehyde-polymerized protein films on gold electrodes	Organophosphates, opiates, drugs, blood immunoglobulins	[29, 31]
Acetylcholinesterase or parathion antibody immobilized onto piezoelectric crystals in a glutaraldehyde-polymerized bovine serum albumin coating	Organophosphate pesticides	[32]
Tyrosinase entrapped in poly(ester sulfonic acid) and carbon-fiber-based coatings for organic phase FIA analysis	Phenols	[33]
Penicillinase cross-linked with bovine serum albumin using glutaraldehyde cast on a pH-sensitive field effect transistor	Penicillin	[34]
Xanthine oxidase cross-linked with glutaraldehyde into cellulose triacetate membranes and attached to a Teflon membrane on an oxygen electrode	Hypoxanthine as an indicator of fish freshness	[35]
Glucose oxidase cross-linked with glutaraldehyde into bovine serum album bovine serum albumin membranes on carbon fiber electrodes	Glucose	[36]
Glucose oxidase cross-linked with poly(ethylene glycol) diglycidyl ether into an osmium polyvinylpyridine polymeric base on interdigitated electrode arrays	Glucose	[37]
Electrodeposition of glucose oxidase and bovine serum albumin on a platinum black electrode followed by glutaraldehyde cross-linking and covered with a polyethylene glycol-polyurethane outer membrane	Blood glucose	[38]
Immobilization of the respiratory chain from *Escherichia coli* by co-immobilization with gelatin, cross-linking with glutaraldehyde and attachment of the resulting film to an oxygen electrode	NAD(P)H, lactate, succinate, malate, and 3-glycerophosphate	[39]
Immobilization of glycerol phosphate oxidase, glucose oxidase, and lactate oxidase onto glassy carbon electrodes using an $Os(bpy)_2Cl$-bromoethylamine-derivitized polyvinyl pyridine polymer and cross-linking with polyethylene glycol diglycidyl ether	Glycerol phosphate, glucose, and lactic acid, respectively	[40]
Fluoresceinamine immobilized into poly(vinyl alcohol) membranes using cyanuric chloride and glutaraldehyde on fiber optic bundles	pH	[41]

8.2.4 Immobilization by adsorption

Adsorption is perhaps the oldest and easiest method used for immobilization of active molecules on sensor surfaces. For example, a polyethylene membrane with adsorbed glucose oxidase was proposed in 1962 as the basis for a glucose

Table 8.5 Examples of sensors using adsorption immobilization.

Immobilized material/method	Analyte(s)	Ref.
Antiserum to atrazine adsorbed onto (i) polyacetylene films doped with iodine, (ii) iodine-doped polythiophene films on gold electrodes	Atrazine	[43]
Pseudomonas aminovirans cells immobilized onto a nylon membrane and attached to a polypropylene-coated oxygen electrode	Trimethylamine	[44]
Adsorption of adenosine deaminase and bovine serum albumin onto the gas-permeable membrane of an ammonia electrode	Adenosine	[45]
Adsorption of concanavalin A onto polycrystalline gold surfaces	Glycogen, dextran	[46]
Adsorption of DNA derivitized with hydroxyethyl disulfide onto gold disks	Quinacrine	[47]
Adsorption of metal phosphonate layers onto quartz crystal microbalance surfaces	Ammonia, butylamines	[48]
Adsorption of dyes (phthalocyanines, porphyrins, rhodamine 6G) onto superconductor $YBa_2Cu_3O_{7-\delta}$-coated MgO surfaces	Specific colors of light	[49]
Potential-controlled adsorption of fructose dehydrogenase onto a platinum electrode	Fructose	[50]
Adsorption of deoxyhemoglobin onto a cation exchange resin in a fiber optic cell	Oxygen	[51]
Adsorption of a dye (Congo Red) to cellulose acetate films on silica substrates	pH	[52]

electrode biosensor [42].

In general, adsorption is achieved by applying a solution of the molecule to be immobilized to a membrane or film on the sensor transducer and allowing the molecule to adsorb to the transducer over a specified time period. The membrane or film may be hydrophilic or hydrophobic or may contain ionic groups depending on the molecule to be immobilized. Various support/surface materials have been used for adsorption but the most used are silica, cellulose acetate membranes, and polymers such as PVC and polystyrene. As shown in table 8.5, adsorption is still used in the fabrication of many chemical sensors and biosensors.

Molecules immobilized by absorption are very susceptible to deadsorption and leakage, especially with changes in pH, temperature, and sample/media ionic content.

8.2.5 Immobilization using biological binding

Immobilization by biological binding is similar in approach and character to adsorption, but differs from adsorption in its specificity and bond strength and the opportunity for directional immobilization (see above).

A primary example of biological binding immobilization is the avidin–biotin complex. Biotin (vitamin H) is a low-molecular-weight cofactor distributed ubiquitously in cells. Biotin specifically binds to avidin, an egg white protein, or to strepavidin, a similar protein which occurs in *Streptomyces* sp.

By labeling molecules on the surface of a transducer with avidin or biotin, and labeling the detection molecule with biotin or avidin, respectively, strong binding with an association constant (K_a) of approximately 10^{15} M^{-1} can be achieved [53]. The bond formed is stable to pH extremes, salts, and chaotropic agents such as 3 M guanidine hydrochloride. In addition, since avidin only requires the bicyclic ring system for recognition and binding, the carboxylic group of the biotin side chain may be activated for reaction with the transducer or a transducer coating using covalent coupling methods (see above), or with other molecules involved in the sensor detection reaction.

Protein A, a polypeptide from *Staphylococcus aureus*, is also used in biological binding immobilization. Protein A binds specifically to the Fc region of antibodies without interfering with the antigenic binding sites and is thus also useful for directional binding. Protein A–antibody complexes typically have a K_a of 10^6–10^8 M^{-1} [53]. The complexes are thus not as stable as avidin–biotin complexes and some leakage may occur in sensors using protein A as the immobilization molecule. Protein A can itself be immobilized onto transducer surfaces or films through available primary amine groups.

Other candidates for biological binding immobilization are also available for sensor fabrication but have not been used extensively to date. Most of these have been used in the preparation of affinity chromatography supports and the methods used in preparing such supports are directly applicable to their use in sensors [17]. These include protein G, a group G streptococcal cell wall protein similar to protein A; protein A/G, a chimeric protein produced by gene fusion of the Fc binding domains of proteins A and G; lectins, such as concanavalin A, which strongly bind specific sugars, polysaccharides, and/or glycoproteins; and nucleic acids/oligonucleotides, for isolation of specific proteins or complimentary oligonucleotides.

Table 8.6 lists examples of sensors utilizing biological binding for immobilization of the detection molecules.

8.3 IMMOBILIZATION OF CELLS OR TISSUES

A number of biosensors utilize whole cells or specific tissues as the basis for detection (table 8.7). Cell immobilization utilizes methods such as adsorption, entrapment, and covalent bonding. The primary consideration in cell immobilization is to restrict (contain) the cells to the sensor while not destroying them during the process of immobilization.

Table 8.6 Examples of sensors using biological immobilization methods.

Immobilized material/method	Analyte(s)	Ref.
Biotinylated monoclonal antibody to sex hormone binding globulin (SHBG) linked to a gold film treated with (i) chromic acid, (ii) biotin–nitrophenyl ester, and (iii) avidin or strepavidin	SHBG	[54]
Biotinylated *Salmonella typhimurium* antibody bound to an avidin–polyethyleneimine membrane on piezoelectric quartz crystals	*S. typhimurium*	[55]
Biotinylated bacterial antibodies linked via an avidin–biotin–hexanediamine spacer arm to carbon felt disks and attached to a platinum electrode	*Staphylococcus aureus* and *Escherichia coli*	[56]
Antibody to progesterone immobilized by binding to immobilized protein A on a gold surface	Progesterone	[8]
Alkaline phosphatase–biotin and oligonucleotide–biotin (herpes simplex virus (HSV) DNA) immobilized onto gold and silver substrates and quartz crystal microbalances coated with avidin and streptavidin	*p*-nitrophenyl phosphate and HSV, respectively	[57]
Antibodies to human immunoglobulin G (hIgG) and benzo[*a*]pyrene tetraol (BPT) bound to protein A on silica beads and then cross-linked with dimethyl suberimidate	hIgG and BPT, respectively	[58]

The primary methods used in cell immobilization are *surface adhesion, covalent attachment and entrapment*. Surface adhesion involves adsorption of cells to microcarriers such as calcium alginate beads, collagen-coated plastic beads, diatomite silica, and dextrans [68]. Immobilization can be achieved simply by suspending the cells and microcarrier together under slow agitation. Examples of such systems include immobilized bacteria for waste water treatment and adsorbed *Azotobacter vinelandii* on Cellex E for nitrogen fixation [69]. Additional examples are presented in table 7.

Covalent attachment of cells may be necessary where the cells will not naturally adsorb to a surface. Such attachment may utilize glutaraldehyde and other methods as described above. In most cases, however, living cells are too sensitive to the chemical conditions necessary for covalent immobilization.

Entrapment provides an alternative to covalent immobilization. Typically, cells are entrapped in a porous matrix and the cells grow throughout the pores/media to result in high cell densities. Cells may be added to the matrix at the time it is formed, or may be added after matrix formation. Many matrices have been used for entrapment including agarose beads, ceramics and silica, collagen microspheres, polyacrylamide, controlled pore glass, and various membranes [68, 69]. Membrane retention of cells, which has been used for biosensors (table 8.7) is, per se, not an immobilized system since the cells are

Table 8.7 Examples of sensors utilizing immobilized cells.

Immobilized cells/method	Measurement(s)	Ref.
Entrapment/adsorption of human keratinocytes, tumor cells, and murine fibroblasts into silicone microwells	Metabolic responses to bioactive agents	[59]
Entrapment of *Pseudomonas* S-17 cells in calcium alginate on gold electrodes	CO_2 content in solutions	[60]
Immobilization of L1210 leukemia cells, human foreskin fibroblast cells, and yeast cells by entrapment in carrageenan or adsorption on filter paper for placement on electrodes	Drug screening through changes in cell viability and respiration	[61]
Adsorption of *Escherichia coli* cells onto acetylcellulose membranes and attachment of the membrane to a Teflon membrane of an oxygen electrode	Screening of mutagens	[62]
Entrapment of *Bacterium cadaveris* on an ammonia electrode using dialysis membranes	L-aspartate	[63]
Entrapment of porcine kidney tissue slices or *Sarcina flava* cells into nylon nets or hollow fiber dialysis units and placement in line with an ammonia electrode for flow-though analysis	Glutamine	[64]
Adsorption of *Alcaligenes eutrophus, Corynebacterium*, and *Pseudomonas putida* on paper attached to an oxygen electrode	Biphenyl and chlorinated derivatives (PCBs)	[65]
Dissected antennule from the blue crab (*Callinectes sapidus*) mounted in a flow apparatus and monitored by glass electrodes	Neural-active compounds	[66]
Nerve-growth-factor-differentiated rat PC-12 cells cultured (attached to) silica surfaces	Acetylcholine	[67]

physically restrained in a liquid medium between two surfaces/films. This type of entrapment is, however, a major method used in cell-based sensors.

Organelles and tissues have also been used in bathing solutions (without immobilization) to detect specific compounds (for example, see [66]). While these laboratory demonstrations have been called chemical or biosensors, they are in fact extensions of classical *in vitro* physiology experiments and do not meet the definition of a practical sensor (see chapter 23 of this text).

8.4 CONCLUSIONS

The reproducible immobilization of detection molecules onto sensor transducers remains a primary challenge to the development and commercialization of chemical sensors and biosensors. Common to the development of a new technology, many of the immobilization methods developed for sensors in the laboratory are too complex and/or contain too many steps to be successfully transferred to large-scale manufacturing. As shown above, many immobilization methods exist which are simple, one-step reactions which can be carried out under conditions which do not compromise the structure and functionality of the detector molecule. Ideally, an immobilization method should have no more than three steps; should be carried out under reasonable conditions of temperature, humidity, light, etc; should not require the use of complex (and expensive) media and reagents; and should result in product which is stable under reasonable storage conditions (e.g., 4–40 °C at humidities from 10 to 80%) for at least 1 year with retention of >90% of its analyte specificity and sensitivity.

Many of the latter requirements are being met for many chemical sensors and for enzyme-electrode-based biosensors and products based on these sensors are commercially available (see chapter 23). Other, more advanced sensors, such as antibody- and receptor-based biosensors, have been commercialized to only a limited extent, primarily due to fabrication and cost limitations. New, simpler, and more cost-effective immobilization methods now being developed will be the driving force for the commercial emergence of these sensors within the next 5–10 years.

REFERENCES

[1] Furlong D N, Gedded N J, Paschinger E, Okahata Y, Ebato H, Ebara Y and Tanaka K 1993 Surface chemical sensors for bitter/sweets, smells and antibodies in aqueous solution *Chem. Australia* **October** 552–5

[2] Nilsson K G I and Mandenius C F 1994 A carbohydrate biosensor surface for the detection of uropathogenic bacteria *BioTechnology* **12** 1376–8

[3] Willner I, Blonder R and Dagen A 1994 Application of photoisomerizable antigenic monolayer electrodes as reversible amperometric immunosensors *J. Am. Chem. Soc.* **116** 9365–6

[4] Zull J E, Mundell J R, Lee Y W, Vezenov D, Ziats N P, Anderson J M and Sukenik C N 1994 Problems and approaches in covalent attachment of peptides and proteins to inorganic surfaces for biosensor applications *J. Indust. Microbiol.* **13** 137–43

[5] Sundaram P V 1990 Waveguide for T_2 toxin detection using quartz-immobilized anti-T_2 monoclonal antibodies *Ann. NY Acad. Sci.* **613** 444–7

[6] Boitieux J L, Desmet G, Wilson G and Thomas D 1990 The specific immobilization of antibody fragments on membrane for the development of multifunctional biosensors *Ann. NY Acad. Sci.* **613** 390–5

[7] Bresler H S, Lenkevich M J, Murdock J F Jr, Newman A L and Roblin R O 1992 Application of capacitive affinity biosensors *Biosensor Design and Application* ed P R Mathewson and J W Finley (Washington, DC: American Chemical Society) pp 89–104

[8] Schramm W, Paek S H and Voss G 1993 Strategies for the immobilization of antibodies *Immunomethods* **3** 93–103

[9] Crumbliss A L, Stonehuerner J, Henkens R W, O'Daly J P and Zhao J 1994 The use of inorganic materials to control or maintain immobilized enzyme activity *New J. Chem.* **18** 327–39

[10] Lindner E, Cosofret V V, Nahir T M and Buck R P 1994 Characterization and stability of modified poly(vinyl chloride) membranes for microfabricated ion-selective electrode arrays in biomedical applications *Diagnostic Biosensor Polymers* ed A M Usmani and N Akmal (Washington, DC: American Chemical Society) pp 149–57

[11] Bein T and Yan Y 1994 Design of thin films with nanometer porosity for molecular recognition *Interfacial Design and Chemical Sensing* ed T E Mallouk and D J Harrison (Washington, DC: American Chemical Society) pp 17–37

[12] Cox R, Gomez D, Buttry D A, Bommesen P and Raymond K N 1994 High surface area silica particles as a new vehicle for ligand immobilization on the quartz crystal microbalance *Interfacial Design and Chemical Sensing* ed T E Mallouk and D J Harrison (Washington, DC: American Chemical Society) pp 71–7

[13] Munkholm C, Walt D R, Milanovich F P and Klainer S M 1986 Polymer modification of fiber optic chemical sensors as a method of enhancing fluorescence signal for pH measurements *Anal. Chem.* **58** 1427–30

[14] Cabral J M S and Kennedy J F 1991 Covalent and coordination immobilization of proteins *Protein Immobilization: Fundamentals and Applications* ed R F Taylor (New York: Dekker) pp 73–138

[15] Taylor R F 1990 Development and application of antibody- and receptor-based biosensors *Bioinstrumentation: Research, Developments and Applications* ed D L Wise (Boston, MA: Butterworths) pp 355–412

[16] Domen P L, Nevens J R, Mallia A K, Hermanson G T and Klenk D C 1990 Site-directed immobilization of proteins *J. Chromatogr.* **510** 293–302

[17] Hermanson G T, Mallia A K and Smith P K 1992 *Immobilized Affinity Ligand Techniques* (New York: Academic)

[18] Cuatrecasas P 1970 Protein purification by affinity chromatography *J. Biol. Chem.* **245** 3059–65

[19] Bain C D and Whitesides G M 1988 Molecular-level control over surface order in self-assembled monolayer films of thiols on gold *Science* **240** 62–3

[20] Chidsey C E D and Murray R W 1986 Electroactive polymers and macromolecular electronics *Science* **231** 25–31

[21] Almeida N F, Wingard L B Jr and Malmros M K 1990 Immobilization of glucose oxidase by electropolymerization of monomers *Ann. NY Acad. Sci.* **613** 448–51

[22] Allcock H R, Pucher S R and Visscher K B 1994 Activity of urea amidohydrolase within poly[di(methoxyethoxyethoxy)phosphazene] hydrogels *Biomaterials* **15** 502–6

[23] Strike D J, Berg A, Rooij N F and Hep M K 1994 Spatially controlled on-wafer and on-chip enzyme immobilization using photochemical and electrochemical techniques *Diagnostic Biosensor Polymers* ed A M Usmani and N Akmal (Washington, DC: American Chemical Society) pp 298–306

[24] Abruna H D, Pariente F, Alonso J L, Lorenzo E, Tribe K and Cha S K 1994 Electroanalytical strategies and chemically modified interfaces *Interfacial Design and Chemical Sensing* ed T E Mallouk and D J Harrison (Washington, DC: American Chemical Society) pp 231–43

[25] Fortier G, Chen J W and Belanger D 1992 Biosensors based on entrapment of enzymes in a water-dispersed anionic polymer *Biosensors and Chemical Sensors* ed P G Edelman and J Wang (Washington, DC: American Chemical Society) pp 22–30

[26] Harrison D J, Teclemariam A and Cunningham L L 1989 Photopolymerization of plasticizer in ion-sensitive membranes on solid-state sensors *Anal. Chem.* **61** 246–51

[27] Guilbault G G and Kauffmann J M 1987 Enzyme-based electrodes as analytical

tools *Biotechnol. Appl. Biochem.* **9** 95–113

[28] Guilbault G G and Montalvo J G Jr 1969 A area-specific enzyme electrode *J. Am. Chem. Soc.* **91** 2164–5

[29] Taylor R F, Marenchic I G and Cook E J 1988 An acetylcholine receptor-based biosensor for the detection of cholinergic agents *Anal. Chim. Acta* **213** 131–8

[30] Taylor R F, Marenchic I G and Cook E J 1993 Receptor-based biosensors *US Patent* 5 192 507

[31] Taylor R F, Marenchic I G and Spencer R H 1991 Antibody- and receptor-based biosensors for detection and process control *Anal. Chim. Acta* **249** 67–70

[32] Guilbault G G and Ngeh-Ngwainbi J 1988 Use of protein coatings on piezoelectric crystals for assay of gaseous pollutants *Analytical Uses of Immobilized Biological Compounds for Detection, Medical and Industrial Users* ed G G Guilbault and M Mascic (Dordrecht: Reidel) pp 187–94

[33] Wang J 1993 Organic-phase biosensors–new tools for flow analysis: a short review *Talanta* **40** 1905–9

[34] Caras S and Janata J 1980 Field effect transistor sensitive to penicillin *Anal. Chem.* **52** 1935–7

[35] Watanabe E, Ando K, Karube I, Matsuoka H and Suzuki S 1983 Determination of hypoxanthine in fish meat with an enzyme sensor *J. Food Sci.* **48** 496–500

[36] Tamiya E, Sugiura Y, Akiyama A and Karube I 1990 Ultramicro-H_2O_2 electrode for fabrication of the *in vivo* biosensor *Ann. NY Acad. Sci.* **613** 396–400

[37] Surridge N A, Diebold E R, Chang J and Neudeck G W 1994 Electron-transport rates in an enzyme electrode for glucose *Diagnostic Biosensor Polymers* ed A M Usmani and N Akmal (Washington, DC: American Chemical Society) pp 47–70

[38] Johnson K W, Allen D J, Mastrototaro J J, Morff R J and Nevin R S 1994 Reproducible electrodeposition technique for immobilizing glycose oxidase *Diagnostic Biosensor Polymers* ed A M Usmani and N Akmal (Washington, DC: American Chemical Society) pp 84–95

[39] Burstein C, Adamowicz E, Boucherit K, Rabouille C and Romette J L 1986 Immobilized respiratory chain activities from *Escherichia coli* utilized to measure D- and L-lactate, succinate, L-malate, 3-glycerophosphate, pyruvate or NAD(P)H *Appl. Biochem. Biotechnol.* **12** 1–15

[40] Katakis I and Heller A 1992 L-α-Glycerophosphate and L-lactate electrodes based on the electrochemical 'wiring' of oxidases *Anal. Chem.* **64** 1008–13

[41] Zhujun Z, Zhang Y, Wangbai M, Russell R, Shakhsher Z M, Grant C L, Seitz W L and Sundberg D C 1989 Poly(vinyl alcohol) as a substrate for indicator immobilization for fiber-optic chemical sensors *Anal. Chem.* **61** 202–5

[42] Clark L C and Lyons C 1962 Electrode systems for continuous monitoring in cardiovascular surgery *Ann. NY Acad. Sci.* **102** 29–45

[43] Sandberg R G, Van Houten L J, Schwartz J L, Bigliano R P, Dallas S M, Silvia J C, Cabelli M A and Narayanswamy V 1992 A conductive polymer-based immunosensor for the analysis of pesticide residues *Biosensor Design and Application* ed P R Mathewson and J W Finley (Washington, DC: American Chemical Society) pp 81–8

[44] Gamati S, Luong J H T and Mulchandani A 1991 A microbial biosensor for trimethylamine using *Pseudomonas aminovirans* cells *Biosens. Bioelectron.* **6** 125–31

[45] Deng I and Enke C 1980 Adenosine-selective electrode *Anal. Chem.* **52** 1937–40

[46] DeBono R F, Krull U J and Rounaghi Gh 1992 Concanavalin A and polysaccharide on gold surfaces *Biosensor Design and Application* ed P R Mathewson and J W Finley (Washington, DC: American Chemical Society) pp 121–36

[47] Maeda M, Nakano K and Takagi M 1994 Semisynthetic macromolecular conjugates for biomimetic sensors *Diagnostic Biosensor Polymers* ed A M Usmani and N Akmal (Washington, DC: American Chemical Society) pp 238–51

[48] Brousseau L C, Aoki K, Yang H C and Mallouk T E 1994 Shape-selective intercalation and chemical sensing in metal phosphanate thin films *Interfacial Design and Chemical Sensing* ed T E Mallouk and D J Harrison (Washington, DC: American Chemical Society) pp 60–70

[49] McDevitt J T, Jurbergs D C and Haupt S G 1994 Chemical sensors and devices based on molecule–superconductor structures *Interfacial Design and Chemical Sensing* ed T E Mallouk and D J Harrison (Washington, DC: American Chemical Society) pp 91–100

[50] Aizawa M, Khan G F, Kobatake E, Haruyama T and Ikariyama Y 1994 Molecular interfacing of enzymes on the electrode surface *Interfacial Design and Chemical Sensing* ed T E Mallouk and D J Harrison (Washington, DC: American Chemical Society) pp 305–13

[51] Zhujun Z and Seitz W R 1986 Optical sensor for oxygen based on immobilized hemoglobin *Anal. Chem.* **58** 220–2

[52] Jones T P and Porter M D 1988 Optical pH sensor based on the chemical modification of a porous polymer film *Anal. Chem.* **60** 404–6

[53] Taylor R F 1991 Immobilized antibody- and receptor-based biosensors *Protein Immobilization: Fundamentals and Applications* ed R F Taylor (New York: Dekker) pp 263–303

[54] Morgan H and Taylor D M 1992 A surface plasmon resonance immunosensor based on the streptavidin–biotin complex *Biosens. Bioelectron.* **7** 405–10

[55] Luong J H T, Sochaczewski E P and Guilbault G G 1990 Development of a piezoimmunosensor for the detection of *Salmonella typhimurium Ann. NY Acad. Sci.* **613** 439–43

[56] Rishpon J, Gezundhajt Y, Soussan L, Margalit I R and Hadas E 1992 Immunoelectrodes for the detection of bacteria *Biosensor Design and Application* ed P R Mathewson and J W Finley (Washington, DC: American Chemical Society) pp 59–72

[57] Ebersole R C, Miller J A, Moran J R and Ward M D 1990 Spontaneously formed functionally active avidin monolayers on metal surfaces: a strategy for immobilizing biological reagents and design of piezoelectric biosensors *J. Am. Chem. Soc.* **112** 3239–41

[58] Alarie J P, Sepaniak M and Dinh T V 1990 Evaluation of antibody immobilization techniques for fiber optic-based fluoroimmunosensing *Anal. Chim. Acta* **229** 169–76

[59] Parce J W, Owicki J C, Kersco K M, Sigal G B, Wada H G, Muir V C, Bousse L J, Ross K L, Sikic B I and McConnell H M 1989 Detection of cell-affecting agents with a silicon biosensor *Science* **246** 243–7

[60] Suzuki H, Kojima N, Sugama A, Takei F, Ikegami K, Tamiya E and Karube I 1989 Fabrication of a microbial carbon dioxide sensor using semiconductor fabrication techniques *Electroanalysis* **1** 305–9

[61] Xiang-Ming L, Liang B S and Wang H Y 1988 Computer aided analysis for biosensing and screening *Biotechnol. Bioeng.* **31** 250–6

[62] Karube I, Sode K, Suzuki M and Nakahara T 1989 Microbial sensor for preliminary screening of mutagens using a phage induction test *Anal. Chem.* **61** 2388–91

[63] Kobos P K and Rechnitz G A 1977 Regenerable bacterial membrane electrode for L-aspartate *Anal. Lett.* **10** 751–8

[64] Mascini M and Rechnitz G A 1980 Tissue- and bacteria-loaded tubular reactors

for the automatic determination of glutamine *Anal. Chim. Acta* **116** 169–73
[65] Beyersdorf-Radeck B, Riedel K, Neumann B, Scheller F and Schmid R D 1992 Development of microbial sensors for deterioration of xenobiotics *Biosensors: Fundamentals, Technologies and Applications* ed F Scheller and R D Schmid (Braunschweig: GBF) pp 55–60
[66] Buch R M and Rechnitz G A 1989 Neuronal biosensors *Anal. Chem.* **61** 533A–42A
[67] Shear J B, Fishman H A, Allbritton N L, Garigan D, Zare R N and Scheller R H 1995 Single cells as biosensors for chemical separations *Science* **267** 74–7
[68] Taylor R F 1991 Commercially available supports for protein immobilization *Protein Immobilization: Fundamentals and Applications* ed R F Taylor (New York: Dekker) pp 139–60
[69] Karkare S B 1991 Immobilized microbial and animal cells as enzyme reactors *Protein Immobilization: Fundamentals and Applications* ed R F Taylor (New York: Dekker) pp 319–37

9

Bilayer lipid membranes and other lipid-based methods

Dimitrios P Nikolelis, Ulrich J Krull, Angelica L Ottova and H Ti Tien

9.1 INTRODUCTION

9.1.1 Biosensors based on artificial chemoreception

In the past two decades the development of (bio)sensors has emerged as one of the major avenues of research, development, and training in various sciences including biomedical engineering and physics, clinical analysis, environmental analysis, agriculture, and process control monitoring. Exciting potential has been recognized for many years for use of this relatively new technology in the construction of active implantable medical devices for *in vivo* diagnostics and monitoring.

The majority of the biosensors that have been reported are based on the deposition of biologically active species such as enzymes and antibodies at the surface of an electrochemical or optical transducer. The most common principle is to identify the analyte by use of a chemically selective enzyme. The enzyme–substrate reaction produces a secondary chemical signal by means of catalysis, e.g. H^+ or H_2O_2. This signal is then recognized and quantitatively converted to an electrical signal, e.g. a potential, a current, or a change of absorption or fluorescence, by a suitable transducer.

Natural molecular recognition and transduction systems, i.e. receptors, that are associated with communication at the cellular level have been introduced for preparation of biosensors. The inherent biological recognition of molecules is realized through non-covalent reversible interactions between complementary structures, leading to a docking of the molecule to be recognized with a receptor site that is associated with a macromolecule. In nature this molecular recognition is generally accompanied by a conformational change of the receptor (protein).

This can lead to an immediate opening of an 'ion channel' and a depolarization of a membrane potential, or can activate an enzyme system, catalyzing the formation of a 'second messenger'. The second-messenger system may then initiate a cascade of enzyme modulations with the effect of amplification and distribution of the original signal.

It is interesting to note that most biosensors use biological molecular recognition systems, but do not use the principles of natural transduction. Thus, biological recognition systems which do not induce any transduction in nature have been the primary elements used in the preparation of biosensors, while the principles of natural transduction have been only tentatively investigated from the standpoint of biosensor development.

Much work has been done in recent years on the so-called 'active', 'adaptive', 'intelligent', 'magic', or 'smart' materials [1]. All of these may be termed collectively as 'biomaterials', including the experimental planar bilayer lipid membranes (BLMs) and spherical liposomes.

BLMs, especially self-assembled bilayer lipid membranes on solid supports (s-BLMs), first reported in 1989, have been used in the last several years as lipid bilayer-based biosensors.

s-BLMs, a new class of nano-dimensioned and self-assembled supramolecular structure, may be considered as an advanced and 'smart' material in that they are very suitable for carrying out fundamental investigations, as well as for developing practical device applications.

In this chapter, we will describe as well as review briefly the work on conventional BLMs, s-BLMs, and closely related systems. As such, planar BLMs are realistic models of biomembranes; they have been used to study the molecular basis of ion selectivity, membrane transport, energy transduction, electrical excitability, and redox reactions. However, one drawback with conventional BLMs is their mechanical instability. This major obstacle has now been overcome with s-BLMs which possess the requisite mechanical stability and other desired properties for biosensor development.

9.1.2 Biosensing strategies from natural chemoreception

Ligand-gated ion channels are among the few biological receptor–transducer systems that are used directly for preparation of biosensors. Membrane fragments containing ion channels have been isolated by the patch clamp technique, which preserves the inherent properties of selectivity and electrochemical response to ligands and inhibitors [2, 3]. Affinity chromatography and genetic engineering have provided the availability of channel proteins in larger quantities. These proteins can function to produce signals after integration into artificial lipid membranes. However, the widespread use of these sensing systems is limited by problems associated with the difficulty of preparation of chemically stable and physically reproducible membranes, and the control of the alignment and statistical distribution of ion channels within membranes. Results from the studies of channel proteins in BLMs serve to provide models for the establishment of more stable semisynthetic and even synthetic channel systems.

The mechanism of odor and taste perception has been extensively explored during the last few years. It is now generally accepted [4, 5] that direct adsorption of a given molecule to the lipid bilayer matrix of an olfactory or gustatory cell results in a conformational change initiating a membrane depolarization. The depolarization is electrically propagated to the synaptic region or impulse generating area of these cells.

Odor and taste perception permits recognition and discrimination between a large number of different molecules. The detection mechanism is based on the processing of signals from several neurons in an array processing system, and does not require the presence of specific receptor proteins. The same mechanism has been proposed for the action of eye irritant molecules and studies have been made to correlate the absorption behavior of eye irritant compounds in the lipid matrix not only by hydrophobicity, but through a more complicated procedure related to phase transition phenomena in the lipid matrix.

The question of whether natural chemoreception can provide biosensors, or whether the construction of biosensors can be deduced from the lessons taught by natural processes, is conditional in both circumstances. In many cases the immediate use or a direct transfer of chemoreceptive tissues, cells, or molecular receptors for sensor construction is very limited in terms of longevity and reproducibility, and often would only be possible for the preparation of one-shot devices.

A limited improvement in this context may be possible by the use of more stable proteins, e.g. from thermophilic bacteria. However, many principles demonstrated by nature could be transposed to sensor development. For example, biomimetic channel and carrier molecules could be used in conjunction with stabilized lipid membranes to prepare sensitive and selective electrochemical transducers which embodied the principle of intrinsic amplification by depolarization. The use of artificial receptor sites would probably result in a substantial reduction of the desired selectivity coefficients, but this could easily be compensated by the application of array processing for background correction.

The implementation of natural ion channel systems for biosensor construction will suffer from the complexity of the structural and regeneration requirements imposed by the biological matrix. In a practical context, a more promising immediate solution seems to be the development of biosensors having an ion channel activity without implementation of ion channel proteins from natural sources, i.e. artificial ion channels. This involves modulation of ion conductivity through lipid membranes by means of alteration of phase structure by a wide variety of different selective binding interactions with, e.g. enzymes, antibodies, lectins, and others.

9.2 EXPERIMENTAL BILAYER LIPID MEMBRANES

9.2.1 The structure of bilayer lipid membranes—uses in biology and biophysics

Artificial BLMs offer opportunities for development of chemically selective biosensors [6, 7]. The essential idea is that a receptor (such as a protein molecule) which can selectively bind to a specific organic or biochemical species (stimulant or analyte) can be incorporated into an ordered lipid membrane assembly such that selective binding events between receptor and stimulant will lead to alterations of the phase structure or electrostatic fields of the membrane (transduction). These perturbations can be monitored electrochemically as changes of transmembrane ion conductivity or as alterations in membrane capacitance.

The simplest BLM experiment involves the application of a small DC voltage (e.g. 25 mV) across a bimolecular ordered layer of lipids having the acyl chains directed towards each other, forming a hydrophobic region, while the polar head groups are towards an aqueous environment (this structure is known as a bilayer lipid membrane or BLM). The applied potential provides a driving force for moving ions across the membrane. If the membrane is in an electrolyte such as a group I metal chloride salt, then the resulting transmembrane ion current obtained is usually of the magnitude of a few picoamperes because the interior of the BLM is highly resistive to the passage of alkali metal and chloride ions. Interaction of the membrane and/or membrane-embedded receptor with any stimulant capable of altering the structure of the BLM will then result in the generation of a signal, most commonly observed as variations of the ion current driven by the constant applied potential.

The discovery of BLMs was first reported three decades ago by Mueller *et al* [8]. From the viewpoint of membrane biophysics and physiology, BLMs are essentially the basic structure of nature's sensors and devices [9]. Some examples include the thylakoid membrane of green plants which functions as an energy transducer converting sunlight into electrical/chemical energy; the photoreceptor membrane of the outer segment of a rod which detects photons as the initial step in visual perception; the plasma membranes of cells and organelles which possess the ability to detect, separate, and store ions, differentiating ions such as sodium and potassium with great specificity; and the plasma membrane which provides sites for a host of ligand–receptor interactions [10]. In view of these and numerous other vital functions associated with cell membranes, it is not surprising that the past two decades have witnessed an enormous research effort spent on membranes in which reconstituted planar BLMs and spherical liposomes have played a dominant role. In the field of biosensor research, the number of investigators involved in the examination of BLMs as a basis for an electrochemical biosensor has increased substantially during the last few years.

9.2.2 Lipid membranes as generic sensors

BLM systems have been accepted as models of natural biomembranes for applications in medicine, industry, and clinical laboratories. BLMs have therefore been studied extensively in combination with various proteins, and are an excellent choice for the basis for development of electrochemical biosensors. The principles behind the development of BLM-based biosensors are quite simple. The sensing element should be biocompatible and should have a structure similar to a biomembrane. Chemically selective proteins may then be embedded into the membranes with substantial retention of binding activity. The simplest way to test transducer function is by using ligand–receptor binding interactions [11]. The modified BLM then becomes a sensing element specific for the complement of the binding species. Formation of a complex or of reaction products can change the structure (physical or electrostatic) of the membrane providing a transduction strategy which is generic and broadly applicable to a wide variety of receptors. Changes in transmembrane electrochemical properties such as the membrane potential, capacitance, resistance, dielectric breakdown voltage, and other electrical parameters can then be readily measured. Since some modified BLMs are photoactive or can be made photoactive, a battery of spectroscopic techniques can also be used for detection of membrane changes. As a result, a number of BLM electrochemical and optical parameters can be monitored simultaneously or sequentially to provide multidimensional information about the system [12].

The operation of a BLM biosensor based on a generic electrochemical transduction scheme is shown in figure 9.1. The mechanism of the signal transduction is based on a receptor–ligand contact interaction at the BLM surface; this complexation occurs via the 'lock and key' recognition process and can be extremely selective. A modification of the acyl chain of the lipids of a membrane with a functional group, e.g. a hydroxyl group, can generate and/or enhance the analytical signal magnitude as this group can assist in ion permeation through the membrane. Increased ion transport can be obtained from BLMs having a large number of structural defects; for example, a disorder of the lipid membrane can assist in ion permeation occurring through pores [7, 44]. The perturbation zone will be localized but can be monitored since even a single small pore can allow the passage of a large number of ions through the BLM.

The construction of such biomimetic systems may provide a number of advantages. High sensitivity may result via an intrinsic amplification since a single binding event between a receptor and a target analyte can result in the movement of many ions. Therefore the ion current may be increased many-fold above the background residual current with amplification factors reaching 10^5. BLMs can be natural host matrices for maintenance of the activity of enzymes, antibodies, and receptors. A chemically selective BLM may have a thickness of the order of a few nanometers and a reaction zone even thinner. This indicates that response times could be very fast. BLM-based biosensors are not

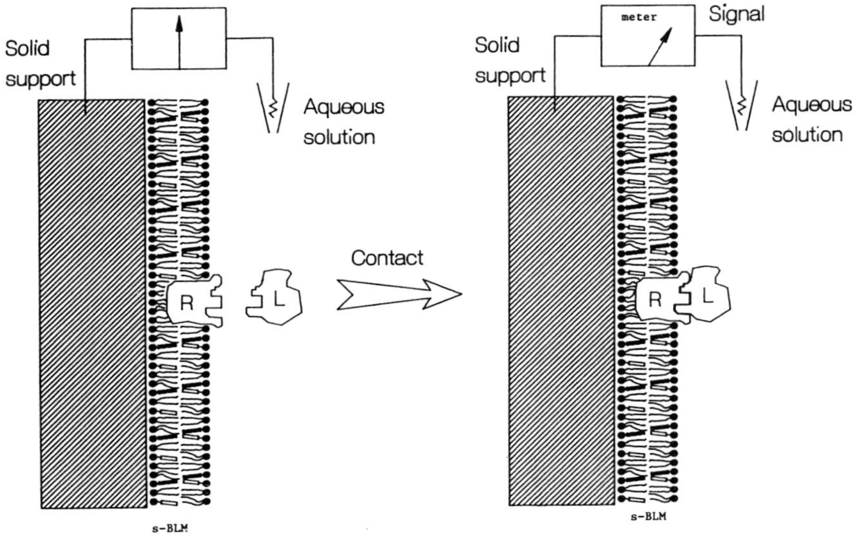

Figure 9.1 A general scheme of ligand–receptor contact interaction. The operation of an electrochemical biosensor which is based on the properties of a modified BLM or s-BLM. L, ligand (substrate, antigen, hormone, ion or electron, donor or acceptor); R, receptor (enzyme, antibody, receptor site, carrier or channel, acceptor or donor).

very sensitive to interferences which are encountered in other electrochemical sensors such as Donnan potentials. Finally, these devices may find use for *in vivo* sensing of drugs and metabolites due to the similarity of structure and transduction response between artificial BLMs and biomembranes.

There already are reports in the literature that indicate BLMs could be used to prepare inexpensive biosensors by using devices that are based on standard semiconductor manufacturing protocols. For example, BLMs have been placed on microchips such as the field effect transistor (FET) for use as sensing devices. The flow of a charge through the BLM can be altered by a voltage applied to the gate of a FET (situated between the source and sink but insulated from the substrate). In principle, if the gate is modified by an electroactive BLM, voltage changes caused by charge accumulation or by carriers in the membrane will influence the drain current [13].

A stabilized biosensor based on a phospholipid bilayer has not yet been developed commercially owing to the instability of the BLM system. This major obstacle is based on the inherent mechanical and electrical fragility of the structure of such membranes. This is a direct result of the molecular thickness of the bilayer structure. Present research is focused towards solving this problem.

9.2.3 Experimentation with bilayer lipid membranes

The species used for the formation of BLMs are primarily agents from which the bilayer structure can spontaneously self-assemble. These are phospholipids, glycerols, or other amphiphiles. Phosphatidylcholine (PC), a zwitterionic lipid, is commonly used. Other agents which modify the properties of the bilayer (e.g. membrane thickness, fluidity, head group interactions, and local internal dielectric constant) are often used as components of membranes. These agents can be sterols such as cholesterol or products of its oxidation, surfactants, hydrophobic polymers such as polyethylene or polystyrene, or polyamide containing polymers such as nylons. For example, a BLM may be formed from lipid molecules, which act as structural building blocks; cholesterol, which condenses the membrane molecular packing to increase the stability of the membrane; and various unsaturated and oxidized forms of lipids and steroids to increase conductivity of the interior of the membrane. The concentration of these components must be empirically adjusted to establish optimum conditions for the formation of stable bilayers and to control the permeability of ions through the membrane [7].

A simplified version of the experimental set-up used for the formation of planar BLMs and investigation of electrochemistry is given in figure 9.2. Figure 9.2 illustrates the two separate aqueous phases in the BLM, prepared using an electrolyte such as KCl in the concentration range of 10^{-5}–1 M. This solution acts as a source of ions for electrochemical studies and assists in the electrostatic stabilization of the membrane structure. Multivalent cations (such as Ca^{2+}) in the electrolyte solution are used to modify the head group interactions by the formation of ion–lipid complexes or larger aggregates. As a result the sensitivity of BLMs to structural perturbations caused by selective binding events can be modulated.

Planar BLMs can be prepared in electrolyte solution by two methods: the brush method [8, 15] which produces BLMs with hydrocarbon solvent remaining between the two BLM leaflets ('black lipid films') and the folding method [15–17] which produces 'solventless' or 'solvent-free' bilayers (a very low level of residual solvent may still be retained in the BLM torus). The brush method involves 'painting' of a small amount of a solution containing lipid (prepared by dissolving lipid in a non-polar solvent such as *n*-decane) across a circular aperture with a diameter of $\sim$1 mm. The aperture is located in a non-conducting plastic sheet (e.g. Teflon) which separates two aqueous phases (figure 9.2). The alkanes which are used to form painted bilayers may act as anesthetics resulting in a poor environment for ion channels, and this has prompted the development of methods for the preparation of 'solventless' BLMs. The experimental set-up that was initially used to prepare solventless BLMs involved the vertical casting of a perforated partition through a lipid monolayer located at the air/water interface of a Langmuir–Blodgett trough [16]. A plastic partition containing an aperture was held perpendicular to the planar surface of the trough, and was

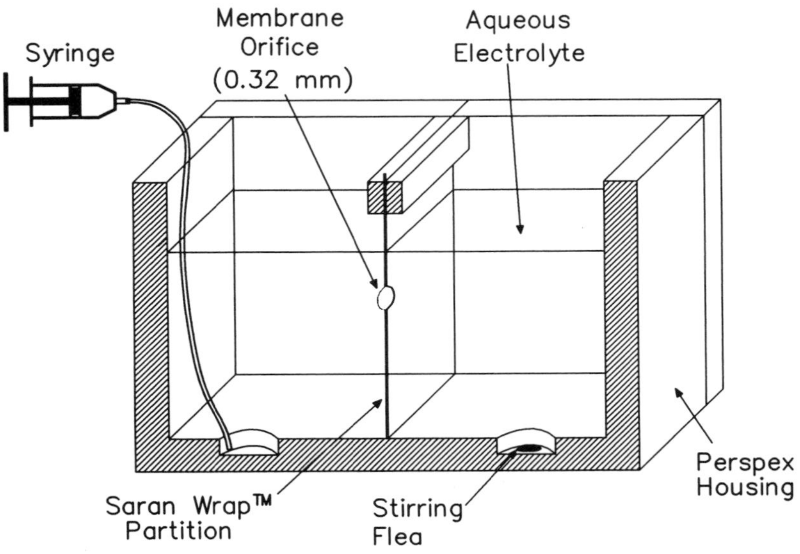

Figure 9.2 A simplified version of the experimental set-up used for the formation of BLMs. Stirring can be done by a magnetic stirring flea in a small well. The electrodes and the power supply and electrometer are not shown.

then immersed so that lipid monolayers met in opposition in the aperture. In a simplified version of this technique which was developed later, the Langmuir–Blodgett trough was simply replaced by two chambers and the electrolyte levels were raised simultaneously by the use of two syringes located with one tip inside each compartment.

The electrodes and the electrode configuration which are used for the electrochemical experiments with BLMs are chosen on the basis of the analytical parameter that is to be measured. A two-electrode system can be used when small currents are drawn across the BLM, while a four-electrode potentiostat is necessary for measurement of large currents [12]. A pair of calomel reference electrodes is useful for the measurement of transmembrane potential. Wire Ag/AgCl electrodes are generally used for the measurement of ion current because they are stable and easy to prepare [14]. Special care should be taken to ensure that one of the electrodes is grounded and that the leads to the 'hot' electrode are properly shielded. The solution cell and the sensitive electronic equipment should be isolated in a grounded Faraday cage. High-speed transient current signals can be displayed on an oscilloscope (instead of a recorder) through a low-leakage-current and fast-response *I–V* operational amplifier with a field effect transistor input. The data can be captured and displayed later with a fast-sampling and storage analog to digital converter. Recently the instrumentation associated with these electrochemical techniques

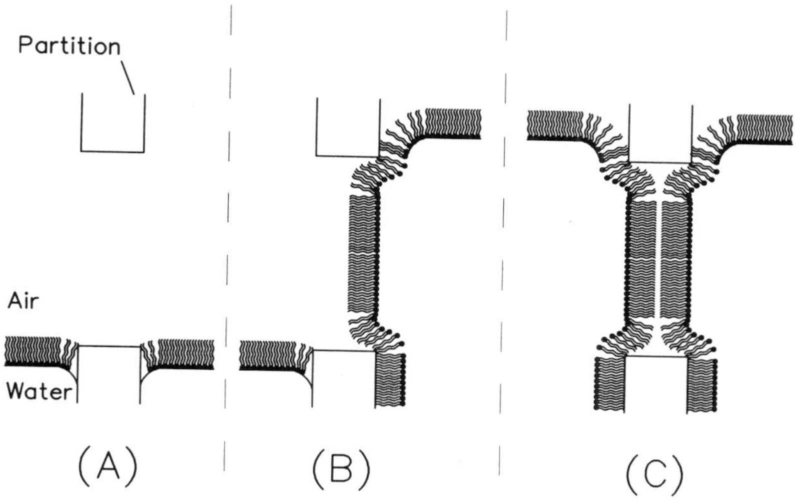

Figure 9.3 A simplified schematic overview of the formation of solventless BLMs by the monolayer folding technique: (A) both monolayers are dropped below the level of the orifice; (B) the monolayer is raised on one side of the aperture, resulting in the formation of a monolayer in the aperture; (C) the second monolayer is raised above the aperture, producing a bilayer membrane in the orifice.

has been refined to prepare a commercial prototype suitable for field studies of the electrical properties of phospholipid BLMs. This system is based in part on semiconductor components that can be switched off or on by optical radiation, so that rapid AC sampling of BLMs can be accomplished (a light addressable potentiometric sensor) [7, 14].

A simplified schematic overview of BLM formation by the folding method is presented in figure 9.3. The partitions used for the preparation of solventless BLMs require smaller thickness, $\sim$10 μm (e.g. Saran Wrap^{TM}) than those used for formation of black lipid films, and aperture areas should not exceed about 0.1 mm^2.

An improved method for the preparation of solventless BLMs that is suitable for study of BLM transduction phenomena for the development of biosensors was recently reported [18] (figure 9.3). There is no need to simultaneously raise both electrolyte levels to compensate for the surface pressure of the membrane. Instead, a small amount (10 μl) of the lipid solution in n-hexane is added dropwise onto the electrolyte surface in one cell compartment. The solvent is allowed to evaporate and then over a period of a few seconds the water level in one solution compartment is brought below the aperture and then raised again with a disposable syringe. Over 95% of the attempts to form BLMs are successful with this procedure (assuming the use of a freshly prepared dilute lipid solution) and these membranes are stable for periods of over 6 h.

These so-called solventless BLMs [15–17] can also be formed on the end of fire-polished glass pipettes with a small tip area (tip diameter ~1 μm) [19]. The lipid solution in *n*-hexane is applied to the outside of the shaft of a pipette which is oriented vertically and penetrates the surface of the aqueous solution. The drop runs down the shaft and spreads at the air/water interface. A short exposure of the pipette tip to air and subsequent reimmersion through the lipid layer leads to the formation of a BLM at the tip of the pipette. This technique has been used to obtain a digital sensor, where 'off/on' signals can be obtained when a single ion channel protein is reconstituted [20].

9.2.4 The formation of stabilized bilayer lipid membranes

The development of a portable and rugged sensing device requires that the selective recognition element be directly interfaced to the physical transducer. In the case of electrochemical transducers based on artificial BLMs, this entails stabilization of the assembly onto an electrode. The stabilization method must allow the membrane to retain characteristics of molecular mobility and fluidity which are essential for transduction and should provide sufficient ruggedness to permit use over an extended period of time (several months) without severe alteration of the response characteristics of the membrane.

Efforts to stabilize BLMs by the use of polymerizable lipids have been successful, but the electrochemical properties of these membranes were greatly compromised and ion channel phenomena could not be observed [21]. Microfiltration and polycarbonate filters, polyimide mesh, and hydrated gels have been used successfully as stabilizing supports for the formation of black lipid films [22–25] and these systems were observed to retain their electrical and permeability characteristics [24]. Poly(octadec-1-ene-maleic anhydride) (PA-18) was found to be an excellent intermediate layer for interfacing phospholipids onto solid substrates, and is sufficiently hydrophilic to retain water for unimpeded ion transfer at the electrode–PA-18 interface [26]. Hydrostatic stabilization of solventless BLMs has been achieved by the transfer of two lipid monolayers onto the aperture of a closed cell compartment; however, the use of a system for automatic digital control of the transmembrane pressure difference was necessary [27].

Stabilization of BLMs at the surface of electrodes has been reported by a number of groups [28–30]. For example, the tip of a Teflon-coated platinum microelectrode was cut *in situ* with a scalpel while immersed in a lipid solution (lipid in a hydrocarbon solvent). Upon immersion of the wire into an aqueous solution of 0.1 M KCl, the phospholipid coating adhering to the metal surface spontaneously thinned to form a BLM directly adjacent to the electrode surface [29]. These BLMs were reported to be stable for periods of 36 h. Stainless steel wire coated with a film prepared from the electrochemical polymerization of allylamine, 2-allylphenol, and 2-butoxyethanol was used recently as an

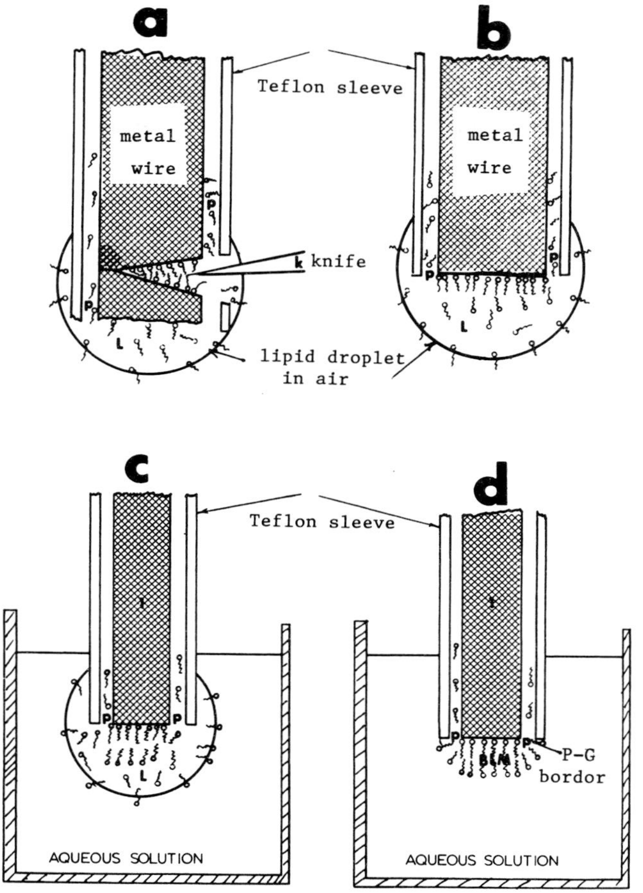

Figure 9.4 A schematic diagram showing the process of self-assembly for a lipid bilayer on a freshly cleaved metal surface: step 1, a metal wire being cut under a lipid droplet with a sharp blade forming an adsorbed monolayer of lipid; step 2, upon immersion of the lipid-coated wire tip into aqueous solution, a self-assembled BLM is formed (see [10], [11], [28], and [29] for details).

alternative to the Teflon-coated platinum wire [31]. The effects of lipid solution concentration and composition, pH, and temperature on the stability and function of these latter metal-supported BLMs were studied, and the assembly was found to be stable in urine for more than 5 h. Bilayer lipid membranes which contain voltage-dependent anion channel proteins have been immobilized onto platinum electrodes [30], and were reported to be stable and functional for periods of over 12 h. The structure of these metal-supported stabilized BLMs is shown diagrammatically in figure 9.4.

Ultrafiltration membranes composed of polycarbonate, glass microfiber,

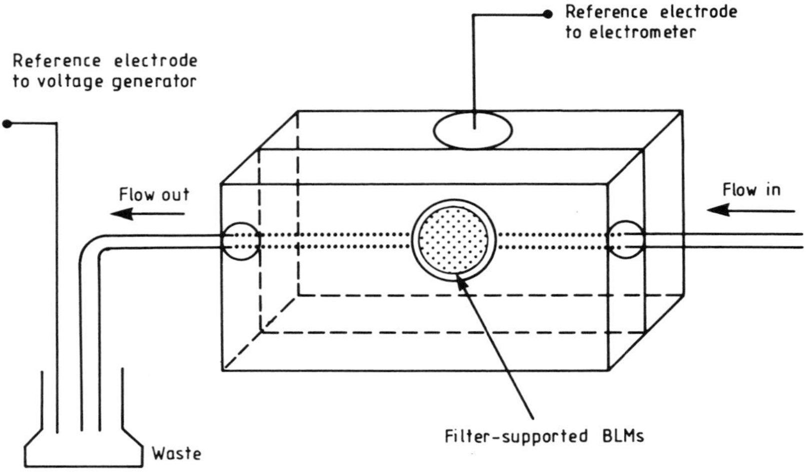

Figure 9.5 A simplified version of the apparatus used for the formation of stabilized filter-supported BLMs for flow through experiments.

and polytetrafluoroethylene (PTFE) have been explored as supports for the construction of stabilized BLMs, most recently for flow through experiments. The technique makes use of filters with nominal pore size ranging between 1 and 5 μm which are placed on an aperture separating two electrochemical cells (figure 9.5). One of these cells is used for casting of lipid membranes onto the filters while the opposing cell is used as a reaction chamber that permits the passage of a carrier stream of electrolyte solution. Injections of the analyte solution in the carrier electrolyte are made near the detection zone. The filter media support micro-BLMs in the pores of the filters, and these BLMs are stable to high flow rates of the carrier solution and can be used in flow injection analysis. Preliminary results have indicated that these systems can be used for repetitive determinations of species with a sampling rate of 220 samples h^{-1} [24].

Stabilization of a lipid membrane onto a solid support by covalent attachment also provides the physical stability necessary for the development of practical sensors. An oriented membrane can be prepared by allowing self-assembly of individual amphiphilic molecules onto a solid surface through either the reaction of terminal silane moieties with a hydroxylated surface to form a silyl ether [33, 34], or by the reaction of sulfur-terminated compounds (alkylthiols or disulfides) with gold surfaces [35, 36]. A variety of species, both with and without polar head groups, have been deposited onto surfaces such as glass, quartz, silicon, and gold [37–39]. These include phospholipids, fatty acids, and fatty amines which were synthetically altered so as to contain either a silyl chloride or a thiol moiety at the terminus of the acyl chain [40]. Both monolayers and multilayers may be formed by using either one-step or multistep synthetic strategies [39].

9.2.5 The formation of vesicular bilayer lipid membranes

A large number of studies involving lipid membranes have utilized vesicular suspensions of phospholipids. A vesicle can be unilamellar (one BLM) or multilamellar (many BLMs stacked upon one another) and may be large (over 100 nm diameter) or small (under 100 nm diameter) depending on the method used for formation of the vesicle. In general, the preparation of a vesicular suspension of lipid may be performed by evaporation of the solvent of a lipid solution in *n*-hexane and suspension of the lipid in an aqueous buffered solution followed by sonication and/or vortexing. Vesicles (liposomes) of controlled size can be prepared by incorporation of a protein or detergent into the lipid suspension followed by re-sonication. Such liposomes are dialyzed to remove the detergent or excess protein, and give vesicles of a well defined size distribution. Proteoliposomes (liposomes that contain protein) have various applications in biosensor technology; for example, they can be used as a means to transfer ion channel active proteins into BLMs [20, 41]. Liposomes with entrapped glucose oxidase have been microencapsulated on Millipore membranes adjacent to an amperometric electrode to manipulate the effective K_m value, to increase the selectivity, and to stabilize the enzyme layer [42]. The resulting glucose sensor was tested for its surface biocompatibility and did not appear to experience any biofouling phenomena, suggesting the possibility that the sensor may find use in the area of *in vivo* measurements.

9.2.6 Receptor incorporation

Membrane modification for the purpose of obtaining an analytical signal involves incorporation of receptor molecules (which in most cases is a protein macromolecule) into the lipid matrix [6, 43] as follows.

(i) The receptor molecules can be introduced directly into the lipid mixture before formation of membranes or they can be codeposited with the lipid layer at the air/electrolyte interface before the formation of the BLM.
(ii) The receptor molecules can be delivered to the bilayer structure or at the lipid film at the air/electrolyte interface by fusion of vesicles made with charged lipids and containing the receptor molecules. Vesicles are allowed to coalesce with the BLM to deliver receptor to the target membrane.
(iii) Membranes from natural sources which already carry natural receptors can be directly isolated from tissues and patch clamped into a BLM experimental configuration that is suitable for electrochemical study of BLMs.

9.2.7 Confirmation of the presence of a planar bilayer lipid membrane and characterization of structure

Any BLMs which are used for electroanalytical experimentation must be characterized to establish the existence of the BLM structure, and to establish reproducibility. Biomembranes and artificial BLMs have similar values of structural and physicochemical parameters with one major difference being that the overall charge density of biological membranes may not be as high as that of artificial lipid membranes (it seldom exceeds 0.05 m^{-2} for biological membranes, while artificial lipid membranes may carry several charged groups and can have surface charge densities of -0.4–$+0.4$ C m^{-2}) [44]. Physicochemical characterization includes measurement of membrane resistance and capacitance and the procedures for these measurements are given elsewhere [14–17].

Table 9.1 provides the values of membrane resistance (R_m), capacitance (C_m), and thickness (d) of artificial BLMs and natural cell membranes [11, 18]. The resistance of artificial membranes is much higher than that of biological membranes. This results from the presence of translocators such as peptides and proteins in the cell membranes. The resistance of artificial membranes can however be reduced to the levels of natural cell membranes when ion translocators are inserted. Specific capacitance (C_m) is the primary criterion to distinguish between solventless BLMs and black lipid films. Table 9.1 exhibits that the specific capacitance of the solventless BLMs (about 0.9 μF cm^{-2}) approaches the values measured for natural cell membranes, and is almost twice the magnitude observed for black lipid membranes. These values of specific capacitance can be used to estimate the hydrocarbon thickness, d, of membranes using the equation

$$C_m = \epsilon_0 \epsilon_m / d \tag{9.1}$$

where ϵ_0 is the permittivity of free space (8.85×10^{-2} F cm^{-2}), and E_m is the dielectric constant of the membrane, which is 2.1. The polar layers of the membrane can be represented as a large capacitor in series with a smaller one (the hydrophobic core of the membrane), and d is therefore equal to the thickness of the hydrocarbon layer of the membrane. The values of specific capacitance and apparent thickness for solvent-free membranes (see table 9.1) show that there is an exclusion of the hydrocarbon solvent. Using phosphatidylcholine which was extracted from egg yolk (egg PC) as a lipid the apparent thickness for solventless BLMs was found to be 2.5 nm [45].

Chemical characterization of the bimolecular thickness of solventless BLMs can be derived from the action of gramicidin at micromolar concentration levels. This peptide can transport cations through the membrane by forming ion channels. Gramicidin has no electrochemical effect on membranes that are thicker than one bilayer [1, 7, 11], since the length of the gramicidin molecule is not sufficient to span a distance greater than that associated with the hydrocarbon zone of a BLM.

Table 9.1 A comparison of electrochemical properties of artificial BLMS and natural cell membranes.

Property	Black lipid films	Solventless BLMs	Cell membranes
R_m (Ω cm^2)	10^6–10^8	10^6–10^8	$< 10^6$
C_m (μF cm^{-2})	0.45 ± 0.05	0.9 ± 0.14	1.0 ± 0.2
d Å obtained from			
C_m values	42	22	16–24
x-ray	—	29	28–35
Effect of 0.1 μM gramicidin on R_m (decrease)	10^3	10^3	$< 10^3$

9.3 ELECTROSTATIC PROPERTIES OF LIPID MEMBRANES

9.3.1 Overview of bilayer lipid membrane electrostatics

The electrostatic properties of lipid membranes play a significant role in ion permeation processes either directly through field effects, or indirectly by controlling the structure of a BLM. Membrane electrostatic phenomena have been extensively investigated, and it has been established that the Gouy–Chapman–Stern model describes BLM electrostatics with reasonable accuracy [44–48]. The electrostatic fields at the head group region of membranes govern the distribution of ions in the vicinity of the membrane surface and play an important role in the process of transmembrane ion transport. In addition, functional groups that are associated with amino acids of proteins incorporated into a BLM can drastically alter the electrostatic fields at the membrane surface.

The Gouy–Chapman–Stern theory relates the surface potential, Ψ_0, to the charge density of a membrane, ©, as follows (for a symmetrical electrolyte) [44, 48, 49]:

$$\Psi_0 = \begin{cases} 2kT/Ze\sinh^{-1}(Ze\sigma\lambda/2\epsilon_0 kT) \\ \sigma\lambda/\epsilon\epsilon_0 : \Psi_0 < 2kT/Ze = 25 \text{ mV} \end{cases} \tag{9.2}$$

where k, T, Z, and e have their usual meanings; the parameter ε is the dielectric constant of water and λ is the Debye screening length. The Debye screening length defines the distance from the membrane where the potential has decreased to $1/e$ of the value at the membrane surface (i.e. 1 nm for a salt concentration of 0.1 M). The Debye length can be calculated from

$$\lambda = (\epsilon\epsilon_0 kT/2 \times 10^3 Z^2 e^2 N c_i)^{1/2} \tag{9.3}$$

where N is Avogadro's number and c_i the concentration of cations in bulk electrolyte solution. The use of negatively charged lipids for BLM formation will result in an increase of the surface potential of the BLM with respect to the bulk aqueous phase and the number of cations at the membrane surface will be increased. The concentration of univalent ions at the membrane/electrolyte interface, $C_i(\chi)$, is provided by the Boltzmann relation [49]

$$C_i(\chi) = C_i \exp[-Ze\Psi_0/kT]. \tag{9.4}$$

Such a preconcentration of monovalent cations at a negatively charged BLM surface has considerable impact on transmembrane ion current values. Monovalent cations transport through BLMs to a much larger degree than anions and the ion conductivity is expected to be increased. Such an increase will affect the S/N ratio associated with selective perturbation of membrane structure.

The Gouy–Chapman double-layer theory neglects the interfacial phase structure of BLMs and hydration effects. As a result, deviations from the expected values of ion conductivity can be observed. This has resulted in the development of refined electrostatic theories which take these factors into consideration [15, 44].

9.3.2 Ion transport through bilayer lipid membranes

The transport of species through a lipid membrane can occur down a gradient of chemical or electrochemical potential (passive transport), or through the coupling of transmembrane movement of a solute with a biochemical reaction, such that the flow may be against a potential gradient (active transport). Passive transport can occur through a permanent or transient membrane pore, or by carrier-mediated diffusion. The latter diffusion processes are facilitated in the presence of a mobile carrier such as valinomycin. Active transport can be primary active such as encountered in the process of transport of sodium or calcium ions by use of ATPase with ATP hydrolysis, or secondary active which is more complex and involves the use of a biological rather than artificial membrane. Ion permeability can be described as a three-step process: ion partitioning into a membrane, transport through the membrane, and desorption of the ion into the opposing solution. Membrane permeation by an ion is energetically unfavorable due to the requirement for transfer of an ion from the polar exterior into the apolar BLM interior. Membrane pores or the formation of ion–carrier complexes can reduce this energy requirement. The activation energy required for an ion to overcome the interfacial barrier height can be set equal to the potential of mean force for the ions approaching a membrane. Therefore gradients near or across the membrane cause a net flow of material along or through the membrane. The ion flow is principally governed by electrostatic fields and structural properties (packing densities and molecular mobility) of a BLM.

The residual (background) ion current which flows through an unmodified BLM determines the background noise of the measuring system and the analytical signal is reflected as a change in the magnitude of the current value caused by a 'receptor' binding event [50]. The residual ion current depends on the surface potential as already discussed, and also involves another source of electrostatic potential known as the dipolar potential. This electrostatic field provides a significant contribution to internal electrostatic gradients within a BLM, and has substantial influence on the fluidity and phase structure of the membrane. These factors are discussed below.

(i) *Dipolar potentials.* The dipolar potential originates from the anisotropic structure of the membrane. The polar zone of the membrane may contain dipole contributions from phosphorus–nitrogen, carbonyl, and hydroxyl groups as well as water of hydration which combine to establish the dipolar or Volta potential, v, perpendicular to the surface of a BLM [13–15, 46, 51]. It has been suggested that the lipid head groups in an ordered BLM are partially hydrated, and that 10–12 water molecules may be bound to the ester groups and phosphate–nitrogen regions of lipids such as phosphatidyl choline. This water of hydration will contribute to the observed dipolar potential by virtue of the structural order of water at the membrane surface which can contribute directly to the total dipole moment, or can influence other lipid-associated dipoles through dipole–dipole interaction. The dipolar moieties have a time-averaged alignment which results in the establishment of a net electrostatic polarization [48, 51]. The average molecular dipole moment perpendicular to the plane of the membrane, $\mu_\perp$, can be represented as [52]

$$\mu_\perp = \Delta V / 4\pi\eta \tag{9.5}$$

where η is the number of molecules in a defined area. The presence of dipolar species in the head group zones of BLMs results in the development of electrostatic fields in two planes in the membrane. These fields are of a magnitude of hundreds of millivolts of a positive electrical potential. This is a static situation which partially controls the ease with which ions enter the membrane, and can be perturbed by selective complexation processes where there is

(1) alteration of the net lipid dipole alignment by headgroup perturbation,
(2) alteration of an initial receptor dipole contribution to the electrostatic field,
(3) introduction of a dipole due to the stimulant or complexation, or
(4) introduction of a dipole due to realignment of dipoles present in the receptor–stimulant complex.

These processes alter the total dipolar potential field, the latter usually being envisioned as a trapezoidal electrostatic field, as shown in figure 9.6.

(ii) *Molecular packing/fluidity.* The mobility of membrane components is usually expressed as fluidity, and is related to the concept of order and alignment. Order is the time and space average of the relative dispositions of molecules in space. Fluidity is not homogeneous in BLMs as structural imperfections can be present even when BLMs are prepared from a single type of lipid. Point defects in the hydrophobic region can be present and the resulting free volumes are important for the diffusion of small molecules and ions through

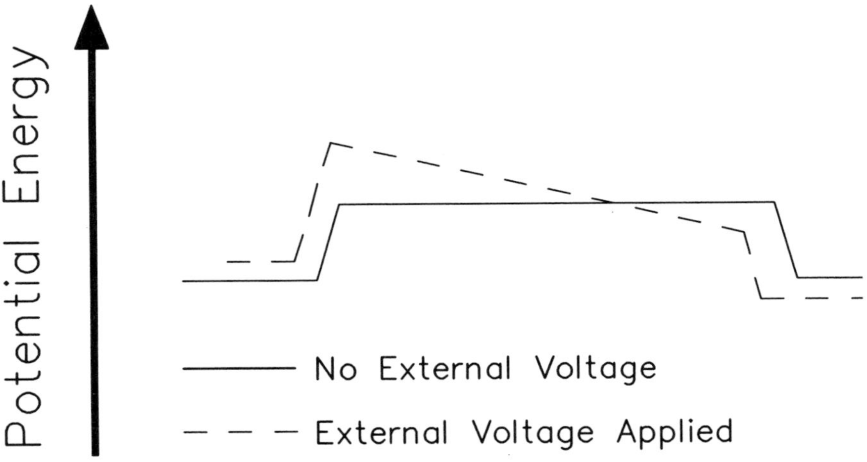

Figure 9.6 The inherent dipolar potential of the membrane as a trapezoidal potential field extending across the plane of the BLM. An external voltage can skew the electric field to act as a driving force for ion translocation.

membranes. The physical structure of a BLM is normally described in terms of distinct physical (thermodynamic) phases. Physicochemical properties are usually measured as time- and spatially averaged values, though the dynamic molecular mechanism of transient ion channel activities indicates heterogeneous distribution of membrane components at the molecular level. Bilayers generally exist in either the liquid condensed (LC) phase, in which there is loose ordering between the acyl chains of adjacent amphiphilic molecules, or the gel phase, in which there is a high density of molecules with closely packed chains and a high degree of ordering. In lipid mixtures, the coexistence of solid and fluid phases can lead to lateral phase separations, where defects are located at the boundaries between the phases. Texture defects increase near the calorimetric phase transition, when regions of disordered molecules coexist with ordered lipids. The presence of such defects has been shown to cause an increased membrane permeability at thermal phase transitions. Incorporation of exogeneous molecules into the BLMs increases the probability of the formation of physical discontinuities by stabilizing either the solid or the fluid domains. This suggests that the modulation of the chemical environment by selective chemical reactions may trigger transitions which alter the phase distribution within a lipid membrane.

Ion conduction is increased when membrane fluidity is increased and molecular packing is decreased. The insertion of polar groups into the hydrocarbon core can increase the fluidity of a membrane by preventing close packing of acyl chains [53]. In contrast, low concentrations of lipophilic molecules such as dolichyl phosphate are oriented in a membrane with their lipophilic terminus towards the interior of the bilayer, resulting in an increase of the energy barrier to ion migration through membranes [54]. This type of molecular interaction also increases the stability of a membrane. The extent of unsaturation of the phospholipid acyl chains has been related to the control of the fluidity and thickness of BLMs [28, 55]. In addition, unsaturated and oxidized hydrocarbon chains that are introduced into the interior of a BLM may act as polar defect sites in the non-polar low-dielectric region of the membrane, and can serve as electrostatic binding sites that assist permeation of ions through membranes.

9.3.3 Practical manipulation of the intrinsic ion conductivity of bilayer lipid membranes

Ion permeation through solventless BLMs composed from egg PC and the negatively charged lipid dipalmitoylphosphatidic acid (DPPA) has been investigated for the effects of surface charge and phase domain distribution on alteration of the process of ion permeation [56]. The ion translocation mechanism was found to depend on the amount and degree of ionization of the acidic lipid in BLMs. The electrostatic fields and phase structure of the BLM were indirectly determined by the latter factors. A threshold concentration of 25% w/w DPPA defined two different conduction mechanisms. At high DPPA content ion permeation occurred through a homogeneous interior, and at low DPPA content ion permeation occurred through defect sites that were associated with a heterogeneous mixed phase domain system. The introduction of calcium ions into the electrolyte solution was found to substantially affect the ion conductivity values [57]. The ion current decreased with an increase of Ca^{2+} concentration at concentrations of DPPA less than the critical amount of 25% w/w owing to the neutralization of the acidic head groups of DPPA and reduction of the surface charge of the membrane. At concentrations of DPPA larger than the transition concentration, the complexation of the acidic phospholipid with the divalent metal resulted in a phase separation involving the formation of conductive zones by the charged lipid constituent in the membrane structure. These conductivity alterations in the presence of calcium ions could be fully reversed by EDTA addition in bulk electrolyte solution. The ability to control conductivity can be used to adjust the sensitivity of transducers based on BLMs. For example, it is possible for a selective reaction to provide transients of the ion current by altering the phase structure of a BLM. The magnitude of the transients can be maximized by controlling the electrostatics and phase structure of the membrane or indirectly by manipulating the lipid composition, the pH value, and the calcium concentration in bulk electrolyte solution.

9.4 ELECTROCHEMICAL SENSORS BASED ON BILAYER LIPID MEMBRANES

9.4.1 The investigation of transduction phenomena using planar bilayer lipid membranes

Del Castillo *et al* pioneered electrochemical experiments using BLMs as transducers to obtain signals from selective biochemical interactions (enzyme–substrate and antibody–antigen pairs) [58, 59]. These observations were the basis of proposals to use BLMs as chemical transducers for various biochemical compounds by means of selective interactions with membrane-embedded receptors. Since then, a wide range of electrochemical studies using BLMs as transducers have been reported (table 9.2). The analytical signal in most cases is the relative change of transmembrane current with respect to a residual ion current, but a significant number of fundamental studies dealing with measurements of rate constants have involved capacitance, admittance, and pulsed experiments.

The mechanism of signal transduction of BLMs using various proteins can be different in each case as previously described. For example valinomycin is an ion carrier, while gramicidin transports cations across bilayers by forming channels through the membrane [15, 46, 47, 60, 61]. Unlike valinomycin, which is highly selective to potassium ions, gramicidin has a broad selectivity for cations. Alamethicin (another polypeptide) forms cation channels through the membrane, but the gating is voltage dependent [62] and the channels are induced only when the transmembrane potential is greater than a threshold value. The ion selectivity or voltage gating has offered some interesting opportunities for the development of an ammonia gas sensor based on BLMs with detection limits comparable to the conventional ammonia gas sensor, but with substantially greater selectivity [63]. The selectivity of this device was based on the use of nonactin as an ion carrier which was capable of transporting ammonium ions through the lipid membrane, but would not complex and transport any other volatile organic amines.

The interactions of concanavalin A with glycogen was studied by using BLMs and it was found that selective binding processes at the membrane surface induced large analytical signals in the form of transient ion current changes [64]. Concurrent electrochemical and fluorescence studies, in addition to pressure–area data for lipid monolayers at an air/water interface, provided structural and electrostatic information about the mechanism responsible for step current evolution. The mechanism of signal generation was associated with physical perturbations of lipid membranes by lectin–polysaccharide aggregates, resulting in the formation of localized domains of variable electrostatic potential and conductivity. The frequency of current response was used as an analytical parameter to determine analyte concentration [64]. The frequency was related to subnanomolar concentrations of concanavalin A in an aqueous solution. The

Table 9.2 Summary of electrochemical sensors based on BLMS.

Analyte	Electrical response	References
Substrates	Transient impedance signals	[32, 40, 58, 69, 70]
Antigens, hormones	Transient impedance signals	[59, 71]
Polypeptide, antibiotics	Permanent alterations in ion conductivity	[40, 60, 61, 62]
Ammonia	Permanent alterations in ion conductivity	[63]
Concanavalin A	Transient ion current signals	[64]
Ca(II)	Permanent ion-channel switching (reversible in the presence of EDTA)	[65]
L-glutamate	Transient ion current signals	[20, 66]
p_{O_2}, H_2O_2	Permanent amperometric current signals	[31]
Glucose	Permanent amperometric current signals	[68, 80]
Insecticides	Transient ion current signals	[72]

development of reproducible periodic ion current oscillations demonstrates that a digital switching process caused by a selective chemical reaction can be activated over a narrow range of concentrations at submicromolar concentration levels. It also shows that a frequency analysis of the response can be an alternative route for quantitation instead of the conventional measurements of the signal magnitude.

An extension of artificial membranes for ion selective electrochemical work was the construction of biomimetic 'ion channel sensors' [65]. These devices were based on Langmuir–Blodgett deposition of charged lipid membranes onto a glassy carbon electrode. This work indicated that a conductive zone can be opened reversibly by a stimulant–membrane interaction by surface charge alterations. This work has demonstrated how the concept of the conductivity measurement could be extended to the more common and useful technique of cyclic voltammetry.

A sensor for glutamic acid has been reported utilizing glutamate receptor as a signal amplifying element. This sensor exploits the glutamate-triggered Na^+ transmembrane ion current for multiple or single channels [20, 66]. Chemical procedures to isolate and purify ion channel components and reconstitute the proteins have made possible the examination of the function of these channels in a controlled environment. Each pulse signal from a single channel opening was the result of an active site of the receptor ion channel binding with one or more stimulant molecules. The number of sodium ions transported through an ion channel switching event was 210 000; this is the largest magnitude that has been reported for biosensors that use ion channel gating phenomena, and clearly indicates the large amplification that can be obtained during ion channel events. Analysis of the concentration dependence of the sensor responses was made on the basis of the integrated ion channel current (i.e. Coulometry), which would be a direct measure of the signal amplification by the ion channel protein. A low detection limit of the order of 3×10^{-8} M of L-glutamate and high selectivity for L-glutamate as compared with D-glutamate were reported. The detection limit of the multichannel sensor is extremely low based on the small numbers of sensory elements incorporated in the lipid membrane (about 10 ion channel proteins). The detection limit of the glutamate system is similar to that obtained from enzyme electrodes when using signal amplification based on the substrate cycling, but the response time for the BLM system can be much faster. The BLM system is superior to other conventional enzyme electrodes for glutamate which operate without the signal amplification (these enzyme electrodes exhibit a detection limit in the millimolar range). The fundamentals of utilizing biomembrane phenomena for the development of novel electrochemical biosensors including signal transduction/amplification schemes have been described [19, 67].

A new type of amperometric biosensor for glucose was fabricated using a Na^+/D-glucose cotransporter as the signal transducing element to exploit the D-glucose-triggered Na^+ ion current through BLMs [68]. The approach was based on the principle of active transport, which is substantially different from that of conventional glucose enzyme electrodes. The principle of the operation was drawn from biological processes in which active transport of glucose across the cell membrane is displayed by a transport protein and is driven by a sodium ion gradient. The Na^+/D-glucose cotransporter, isolated and purified from the small-intestinal brush border membrane of guinea pigs, was embedded into BLMs by delivery using proteoliposomes. The sensor response was measured as an ionic current through the BLM arising from cotransporting sodium ion flux under a constant applied potential and was only induced by D-glucose at concentrations above 10^{-9} M, and not by other monosaccharides except for D-galactose. This detection limit in the nanomolar range is regarded as extremely low in comparison to conventional glucose sensors based on the glucose oxidase, which exhibit a detection limit at the micromolar concentration level. The selectivity of the BLM system was comparable to that achieved using conventional enzyme-based glucose sensors (which experience interference from mannose). Finally, the effects of applied potentials, Na^+ and K^+ ion concentrations, and the addition of the competitive inhibitor phlorizin were investigated to further characterize the sensor. The approach of using a cotransport protein is expected to be extended to other cotransport and transport proteins for a wide range of substrates such as amino acids, sugars, and neurotransmitters.

A particular phenomenon which is associated with the signal transduction by BLMs that can result in a rapid response involves the storage of the electrical energy across the membrane by virtue of charging of the electrical double layer [9]. BLMs prepared from mixtures of egg PC and DPPA develop an electrical double layer at the air/water interface. Such an electrical layer is susceptible to pH alterations of the electrolyte solution. In addition, the phase structure of such BLMs is very sensitive to alterations of membrane surface charge. Both these physical parameters can be substantially altered during a hydrolytic enzyme reaction at the BLM surface. This provides the basis for the construction of 'switchable' sensors for the determination of substrates of hydrolytic enzyme reactions with rapid response times of the order of seconds to minutes. Recent experiments have shown that the immobilization of hydrolytic enzymes at the membrane surface can result in a selective transient charging signal based on the rate and extent of the biochemical interaction with the substrate [69, 70]. The interactions used in these studies were the enzyme-catalyzed reactions of acetylcholine (ACh)–acetylcholinesterase (AChE) [69], urea–urease, and penicillin–penicillinase [70]. These studies have shown that it is possible to construct BLM-based sensors which provide signals in terms of switchable conductivity states at submicromolar concentrations of the substrate without implementation of ion channel proteins from natural sources (figure 9.7). As the reactions proceed, the ionizable groups in the membrane are effectively dynamically titrated through an equivalence point with the consequent rapid reorganization of the double layer and the BLM structure. The local pH at the surface of BLMs dynamically varies during the course of the enzymatic reaction and the diffusion of the substrate and the product should control the pH at the surface of a membrane. Diffusion processes set the detection limits for determination of the substrate, which is e.g. 1 μM for ACh. The system can be reproducibly regenerated by simply washing with a buffered solution that does not contain a substrate, suggesting that the transducer may be ideal for stopped flow or flow injection analysis experiments. This offers exciting analytical sensing opportunities where signal switching is not observed until a threshold substrate concentration is established and then the signal shows a (rapid) time dependence.

Perhaps of a greater significance is the electrochemical transduction of an immunological interaction by the use of BLMs which are prepared from mixtures of egg PC and DPPA. Transient ion current signals similar to those observed for hydrolytic enzyme reactions were generated by an antibody–antigen interaction which modulated surface charge by virtue of the formation of a complex at the surface of a BLM [71]. Thyroxin (T4)–anti-rabbit T4 was used as a representative immunological reaction for these studies. An antibody–antigen complexation caused transient ion current signals due to dynamic changes of the electrostatic fields at the surface of such membranes. The mechanism of the signal generation is based on the perturbation of the electrical double layer and the surface structure of the BLMs. The transient charging signals occurred as singular or multiple events which lasted for a period of seconds. A specific lipid mixture was required in order to produce a sensor with the following response characteristics: a single, rapid ion current transient for each concentration of antigen that was determined; the magnitude of the signal logarithmically related to the concentration of the antigen in the bulk solution; determinations over a range of nanomolar to micromolar levels in a period of less than 2 min (figure 9.8). The reproducibility of the response magnitude and the speed of the response

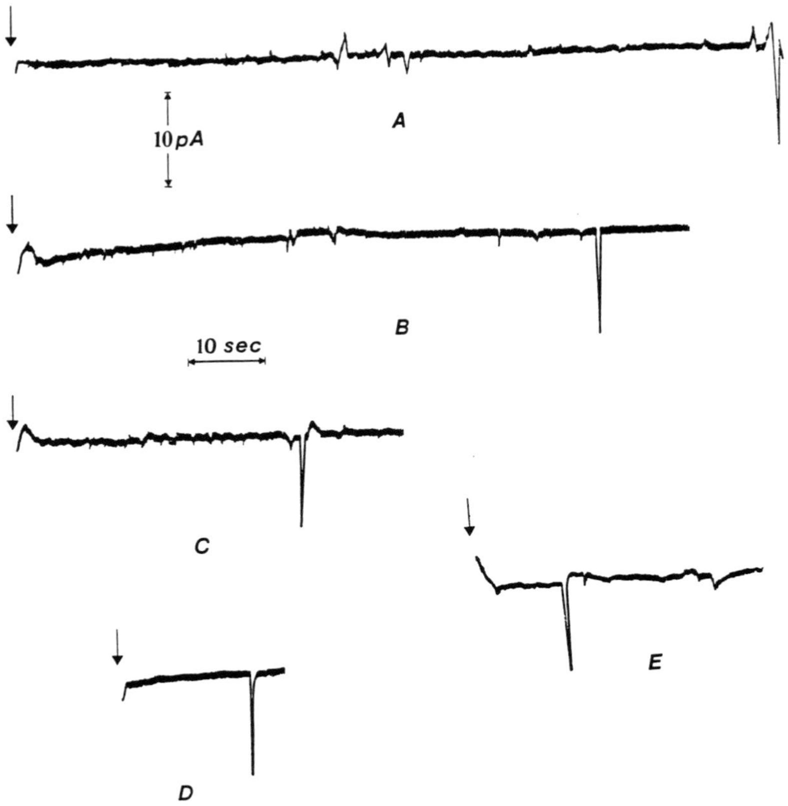

Figure 9.7 Recordings of current against time for the AChE/ACh reaction at a BLM surface at pH 8.0, with BLMs consisting of 35% (w/w) of DPPA. A volume of 3 μl of enzyme solution containing 1.0 mg ml^{-1} was codeposited with the lipid solution at the air–electrolyte interface. Substrate concentrations (μM) were (A) 2.20, (B) 5.50, (C) 11.0, (D) 38.5, and (E) 55.0.

are significant when compared to the results obtained from most chromogenic immunoassays. Conventional immunoassays often take minutes to hours to complete, and may require multiple reaction/separation steps with secondary labeling and reactants, which hampers adaptation for use in biosensors. In this connection it should be mentioned that a feasibility study was carried out using an s-BLM as a biosensor with electrical detection. The hepatitis B surface antigen was incorporated into an s-BLM, which then interacted with its corresponding monoclonal antibody in the bathing solution. This Ag–Ab interaction resulted in changes in the electrical parameters (capacitance, conductance, and potential) of s-BLMs with magnitudes directly related to the concentrations of the antibody in the bathing solution. For example, the linear

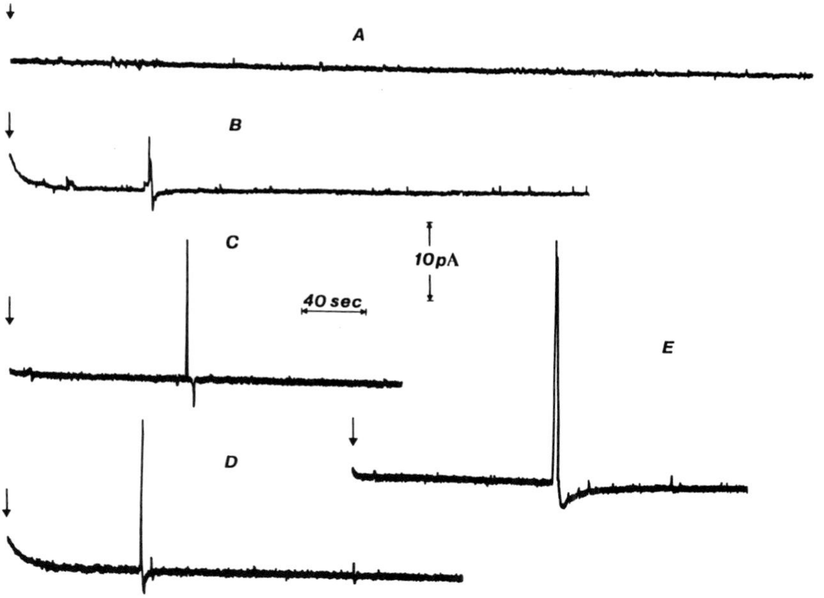

Figure 9.8 Recordings of current against time for the thyroxin–anti-rabbit T4 immunological interaction at BLMs at pH 6.0 and with BLMs composed of 15% DPPA. A volume of 3 μl of the protein solution containing 2.8 mg protein ml^{-1} was codeposited onto the air–water interface. The hormone concentrations correspond to (A) 1.12 nM, (B) 11.2 nM, (C) 33.6 nM, (D) 112 nM, and (E) 1.12 μM.

response (conductance–Ab concentrations) was very good ranging from 1 to 50 ng ml^{-1} of antibody, demonstrating the potential use of such an Ag–Ab interaction via s-BLM as a transducing device [73].

The use of BLMs for the direct electrochemical monitoring of organophosphate and carbamate insecticides such as monocrotofos and carbofuran has recently been explored [72]. Egg PC and DPPA were used for the formation of these BLMs. The interactions of the insecticides with BLMs produced a transient current signal with a duration of seconds, which reproducibly appeared within 3–5 min after an exposure of the membranes to monocrotofos and carbofuran. The sensitivity of the response was maximized by the use of high concentrations of the acidic lipid in membranes, and by the introduction of calcium ions into the bulk solution. The detection limits were of the order of nanomolar concentrations once the lipid composition of BLMs was optimized. The mechanism of the signal generation is related to the adsorption of the lipophilic insecticide molecules into the head group zone of BLMs, with a consequent rapid reorganization of the dipolar electrostatics. The magnitude of the transient current signal was linearly related to the concentration of monocrotofos or carbofuran in the bulk solution. No interferences from other insecticides such as aldicarb and methyl parathion were observed. In addition, the time responses for monocrotofos and carbofuran were substantially different (about 3 min for the former and 5 min for the latter) as controlled by the adsorption rates and partitioning of the insecticides into the BLM. This result suggests that a determination of a mixture of these two insecticides could be possible on the basis of the response time. BLM-based biosensors could replace the costly and time consuming liquid chromatographic (LC) procedures for screening applications involving insecticides. The low cost of BLM instrumentation and the ease of formation of BLMs are obvious advantages in comparison to LC instruments, and the BLM-based biosensors can have similar detection limits and do not require sample preparation.

9.4.2 Electrochemical sensors based on stabilized bilayer lipid membranes

9.4.2.1 Biosensors based on filter-supported bilayer lipid membranes. Recent work has used ultrafiltration membranes such as polycarbonate and glass microfiber filters as supports to prepare stabilized BLMs as sensors suitable for operation in flowing solution streams. The stability of the BLMs has permitted the use of these systems for the investigation of hydrolytic enzyme–substrate reactions [69, 70] in flow through cells. The response characteristics of the micro-BLMs located in the pores of the filter media were found to differ in magnitude (still observed transients of current) from those of the conventional solventless planar BLMs [22, 32]. Yoshikawa *et al* [32] have formed stable BLMs using porous filters and have obtained very valuable electron micrographs. For both of these experiments, the enzyme molecules were immobilized in the BLMs by incorporating the protein into the lipid matrix at the air/water interface before the BLM formation. In the case of the BLMs in filter supports, injections of the substrates were made into flowing streams of a carrier electrolyte solution. Hydronium ions produced by the dynamic enzyme reaction at the BLM surface caused alterations of the electrostatic fields by charging. For the filter-supported BLMs, transient signals of the order of tens of picoamperes were obtained (factors of 5 to 10 larger than for planar BLMs). The magnitudes of the transients could be correlated to the substrate concentration, which could be determined at the micromolar concentration level. The response times were of the order of about 10 s and acetylcholine, urea, and penicillin were determined in continuous flowing systems at a rate of 220 samples h^{-1}. These results highlight the advantages of BLM-based electrochemical biosensors, as the best response times for conventional enzyme electrodes have been achieved by immobilizing very thin layers of the enzyme (1–2 μm thickness) using the glutaraldehyde cross-linking immobilization technique (by spraying an enzyme–glutaraldehyde mixture) at the surface of a glass electrode. The response times of these glass enzymes electrodes (which were the fastest ever realized) were also in the range of 10 s, but sensitivity was compromised and these short response times were only observed for substrate concentrations in the millimolar range. The low detection limit obtained using the filter-supported micro-BLM systems was similar to that obtained by fluorescence methods. Therefore, it is anticipated that the analytical utility of stabilized BLMs for flow stream uses will provide new opportunities for biosensing in practical applications which demand a rapid response at a micromolar concentration level such as the analysis of fermentation broths, pharmaceutical preparations, or monitoring of industrial processes [24].

9.4.2.2 Chemical immobilization by covalent attachment on metal electrodes. Another route for the development of practical devices that are based on BLMs is the chemical stabilization of a lipid membrane on the surface of a solid metal electrode. Covalent linkages between the support and the lipid/receptor molecules may provide the highest degree of stability. This chemical stabilization may also provide a sufficient ruggedness to permit the use of a BLM-based biosensor over a period of at least several months, and could allow the membrane to retain the molecular mobility and fluidity required for the transduction of selective binding at the surface. Chemical immobilization by a covalent attachment of monolayers of lipids onto platinum or gold electrodes has produced stable lipid membranes [40]. In order to avoid the effects caused by the accumulation of ions on one side of a chemically stabilized membrane, the measurements must be made by driving the system with an AC voltage and by monitoring admittance with a phase-sensitive amplifier.

Progress was recently made in the construction of an AC admittance modulation system for surface-stabilized lipid membrane biosensors that operated on the basis of the control of the ion permeation by artificial ion channels [40]. A portable admittance modulation measurement device was designed to measure both the in-phase and out-of-phase signal components for determination of the 'effective' ion current and membrane capacitance, respectively [40]. The sensitivity and detection limit of this AC system were tested by studying the interaction of valinomycin with planar BLMs. The electrochemical phenomena were monitored through the in-phase component and measured as conductance changes of the membrane, providing a detection limit of 1 nM for valinomycin.

Metal electrodes such as platinum and gold were used as supports for the covalent attachment of amphiphiles [40]. Platinum electrodes with a surface area of 1 cm^2 were oxidized to provide a high density of surface hydroxyl sites [33, 34]. Immobilization of amphiphiles was carried out by reaction of the hydroxyl sites with silane. Gold electrodes were prepared by vacuum deposition of a 250 nm layer of the metal onto a 30 nm layer of chromium that covered borosilicate glass slides. The surface attachment of amphiphiles to this metal was achieved by sulfur–gold interactions [35, 36]. A wide variety of amphiphiles differing in hydrocarbon chain length, number of chains (one or two), chain polarity, head group charge, and head group size (e.g. acidic phosphate, carboxylic acid and ester, phosphatidylcholine) were attached to both the platinum and gold surfaces [37–39]. Two-step attachment procedures, as exemplified by the initial deposition of aminopropyltriethoxysilane (APTES) onto platinum followed by the linkage of a 10-carbon phosphatidylcholine through an amide bond to the amino group of APTES, often provided greater surface coverage than if an equivalent 16-carbon phosphatidylcholine had been deposited directly by the reaction of hydroxyl moieties with silane at the metal surface. The surface coverage for most experiments was found to be in the range 40–85% as determined by x-ray photoelectron spectroscopy, and was dependent on the deposition procedure, the chain length, and the number of chains that were associated with the deposited molecule. The electrochemical results indicated that the best blockage of the ion conductivity occurred when the amphiphiles contained long hydrocarbon chains. Species such as trichloro-octadecylsilane and octadecylthiol provided the best blockage, reducing the in-phase signal component by 95% for both types of metal electrode. Analogues of natural lipids such as dimyristoylphosphatidylcholine reduced the in-phase component to 50% of the initial value. The subsequent incubation of the derivatized electrodes in solutions containing cholesterol greatly reduced the magnitude of the in-phase component of the signal to values approaching those found for BLMs. These latter electrodes were tested for their response to valinomycin, and achieved detection limits in the range 10–100 nM [40].

A further series of experiments was performed to develop electrodes that were sensitive to pH alterations. These electrodes were used to extend the concept of control of phase domain structure to immobilized membranes. Linear 10-carbon silane and thiol carboxylic acids were immobilized onto platinum and gold surfaces, and both systems showed good sensitivity to pH (e.g. a pH change from 7 to 8 caused a conductivity change of 25%). Modification of this system to evaluate a biosensing strategy was then performed. In separate experiments, active urease and AChE have been adsorbed onto these acidic surfaces, and urease has also been covalently immobilized using the carboxylic acid functional group of the amphiphiles. The electrochemical transduction of substrates for both of these enzymes indicated the presence of time-dependent singular transients of the current, thereby confirming that covalently attached lipids on metal electrodes exhibit a similar function to planar BLM assemblies [40].

9.4.2.3 Biosensors based on metal-supported bilayer lipid membranes. BLMs, especially s-BLMs, have been used in the last three years as lipid bilayer-based biosensors [10, 11, 74–76, 82]. Hianik *et al* [75] have carried out a detailed physical study on the elasticity modulus of s-BLMs. They found that the dynamic viscosity of s-BLMs is one order of magnitude less than that of conventional BLMs [75]. It should be mentioned that in the s-BLM system, albeit attractive for certain purposes such as biosensors and molecular devices, the metallic substrate precludes ion translocation across the lipid bilayer. Therefore, the pursuit of a simple method for obtaining long-lived, planar BLMs separating two aqueous media has been an elusive one until now [81]. As reported, this much improved planar BLM system in many ways is simpler to form; it oblivates the stability problem of conventional BLMs. The new BLM system takes the advantage of frequently used agar salt bridges (sbs) in electrochemistry. The method of forming a self-assembled BLM on an sb support is quite similar to that of the s-BLM system. The formation procedure consists of three steps. In the first step, a chlorided Ag wire (Ag/AgCl) is inserted into a Teflon tubing (24–30 gages used) which has been previously filled with a mixture of agar (Sigma Chemical Co.) and KCl solution saturated with AgCl (0.3 g agar in 15 ml 3 M KCl). The AgCl electrode and the filled Teflon tubing are glued together with wax at the point of insertion. In this way an Ag/AgCl–Teflon sb is constructed. In the second step, the tip of the other end of the Teflon sb is cut *in situ* while immersed in a BLM forming solution with a scalpel. In the third and last step, the Ag/AgCl–Teflon sb with freshly lipid-solution-coated tip is immersed in 0.1 M KCl solution of the cell chamber. Alternatively, the second step described above may be carried out in air and then the freshly cut end of the sb immediately immersed in the lipid solution for a few minutes. In either case, the cell chamber filled with an appropriate aqueous solution (e.g. 0.1 M KCl) contains an Ag/AgCl reference electrode and an Ag/AgCl–Teflon sb with a self-assembed BLM at its end. The lead wires of the two electrodes are connected to the measuring instrumentation.

This newly developed sb-BLM system retains the uniqueness of conventional BLMs such as ion translocation. Its much enhanced mechanical stability permits long-term investigations [81], which is essential for membrane research and for practical applications, especially for biosensors, development should be made concerning s-BLMs.

Bilayer lipid membranes (or lipid bilayers) on solid supports have been formed by a number of methods including the two-consecutive-step technique as follows: step (i), placing a Teflon-coated metal wire (e.g. Pt, stainless steel, or Ag) to be cut in contact with a BLM forming lipid solution followed by cutting with a sharp knife and step (ii), immersing the lipid layer that has adsorbed onto the metal wire surface into an aqueous solution (see figure 9.4). For the best cutting of metal wires, a miniature guillotine has been used, where the sharp knife is moved vertically onto the wire placed on a flat surface and immersed in a lipid solution [81]. Typically we used either 1% glycerol dioleate in squalene or 1% phospholipid (lecithin) in *n*-decane. Other lipid solutions used are as follows: (i) TCOBQ (saturated) in (ii) 0.25 g cholesterol in 20 ml squalene; (iii) TCPBQ (saturated) in (ii); (iv) alpha-tocopherol (1/10 ml) in (ii). Compounds of interest for incorporation into lipid bilayers were dissolved in the lipid solution prior to cutting. As evidenced by electrical parameters (capacitance, resistance, potential, and/or current–voltage characteristics), a self-assembled BLM has been shown to form at the tip of freshly cut metal wire within 10 min or so [10, 11, 78].

Supported bilayer lipid membranes (s-BLMs) have been prepared from mixtures of natural phospholipids and synthetic amphiphatic molecules. Three classes of compounds have been immobilized into these s-BLMs as follows: (i) *Electron acceptors, donors, and mediators* including highly conjugated compounds such as mesotetraphenylporphyrins (TPP), metallophthalocyanines (PLC), TCNQ (tetracyano-*p*-quinodimethane), TTF (tetrathiafulvalene), and BEDT–TTF (bisethyldithiotetrathiafulvalene–tetraselenofulvalene). (ii) *Redox proteins and metalloproteins* such as cytochrome c, and *iron–sulfur proteins* (ferrodoxins and thioredoxins). Also included in this class are compounds that take part in ligand–receptor contact interactions: specifically, the ligand may be a substrate, an antigen, a hormone, an ion, or an electron acceptor or a donor, and the corresponding receptor embedded in s-BLM may be an enzyme, an antibody, a protein complex, a carrier, a channel, or a redox species [10, 11]. (iii) *Fine semiconductor particles* (formed *in situ*) such as CdS, CdSe, and AgCl [12]. Some of these compounds have been studied previously in conjunction with conventional BLMs. For example, methodologies developed for *in situ* formation of semiconductor particles on conventional BLMs have been applied to s-BLM systems.

The redox reactions for supported BLMs containing vinylferrocene as an electron mediator have been investigated using cyclic voltammetry. The results have shown the following. (i) Ferrocene can be very easily immobilized in the lipid bilayer on the surface of a metallic wire (s-BLM) system. This demonstrates that the s-BLM system offers a novel approach to electrode modification by simple immobilization of compounds within BLM. (ii) Ferrocene in a BLM increases the sensitivity to the potassium ferri/ferrocyanide ion by about two orders of magnitude in comparison to that of the platinum electrode [79].

The insertion of appropriate active molecules (modifiers) into the matrix of the lipid bilayer should be able to impart functional characteristics to s-BLMs [10, 11]. TCNQ (tetracyanoquinodimethane) and DP-TTF (dipyridyl-tetrathiafulvalene) were chosen as modifiers because of the properties which were representative of typical electron acceptor and donor molecules, respectively. It was found that DP-TTF could improve the stability and also the range of sensitivity of s-BLMs to hydrogen peroxide. In contrast, s-BLMs which contained TCNQ did not show much response to H_2O_2. This was not entirely unexpected since TCNQ should behave as an electron acceptor [7, 79]. The performance characteristics of metal-supported BLMs for the determination of p_{O_2} and hydrogen peroxide have been examined recently [31]. No interferences from uric and citric acid and glutathione (in millimolar concentration levels) were observed at a potential of +670 mV; ascorbic acid interfered with the hydrogen peroxide measurements. Finally, the use of the device as a sensitive oxygen sensor was reported for measuring the respiration rate of rat liver mitochondria following the addition of stimulants such as succinate, adenosine diphosphate, and methoxyphenylhydrazone. These results indicate that a metal-supported BLM-based biosensor may have performance characteristics comparable to other more conventional biosensors.

Electron conducting polypyrrole BLMs have been used for developing glucose sensors [80]. Similarly, s-BLMs have been employed for the same purpose. The same principle was recently used to prepare a simple and fast glucose-sensitive s-BLM [31]. This sensor used a biotin-modified phospholipid bilayer to which a streptavidin–glucose oxidase complex was coupled. The partitioning of a highly hydrophilic species such as glucose oxidase into a hydrophobic matrix was made possible by the high affinity of biotin to streptavidin. The assay was based on the electrochemical detection of enzymatically generated hydrogen peroxide at a potential of +670 mV. In the case of an air-saturated buffering solution the response to glucose was measured up to 50 mM, with a linear response up to 7 mM; the detection limit was 0.1 mM of glucose at conditions of air saturation, and the response time was about 45 s. The enzyme exhibited 70% of its initial activity after 2 weeks. The influence of oxygen tension, pH, temperature, and chemical interferences were investigated. No interferences from species including D-fructose, D-galactose, D-lactose, saccharose, uric acid, urea, L-glutathion, L-cysteine, and cholesterol were observed. Ascorbic acid was found to interfere, but this could be minimized by increasing the amount of negatively charged lipid in the membrane. Traces of ammonium ion increased the background current. The stability of this sensor in diluted blood, plasma, and urine was minimally 8 h. Use of the system for the measurement of glucose in blood and urine was tested, and it was determined that the sensor would function, but with poor precision.

Of all the ions crucial to the functioning of cellular processes the hydronium ion (H_3O^+) plays the leading role in enzyme catalysis and membrane transport. Thus, it is not surprising that the measurement of pH is of the utmost importance. Currently, the pH glass electrode is routinely used in chemical and clinical laboratories. However, the large size and fragility of pH glass electrodes preclude their use in many situations such as *in vivo* cell studies and in monitoring membrane boundary potentials. For example, the hydrolysis of membrane lipids by phospholipid enzymes (lipases A and C) changes the boundary potential of the BLM (or cell membrane) as a result of local pH change. Additionally, it has been known for many years that BLMs formed from chloroplast extracts exhibit Nernstian behavior as a function of pH [15, 81]. These observations suggest that s-BLMs can be used as a pH probe in membrane biophysical research and in biomedical fields where the conventional glass electrode presents many difficulties. To test this concept, a number of quinonoid compounds (chloranils) have been incorporated into s-BLMs. It was found that s-BLMs containing either TCOBQ or TCPBQ responded to pH changes (with slope responses of 55 ± 3 mV). This new pH-sensitive s-BLM offers prospects for ligand selective probe development using microelectronic technologies. In this connection mention should be made of recent experiments with sb-BLMs [81, 82]. The agar gel sb-BLM represents a new type of self-assembled lipid bilayer. It is developed on the basis of the s-BLM concept. Electrical measurements have shown that the response of the sb-BLM modified by *o*-TCBQ to hydrogen ions obeys the Nernst equation. Further, there is a linear dependence between the membrane potential of the sb-BLM containing TCNQ and the ascorbic acid concentration in the range of 10^{-4} M. Additionally, sb-BLMs modified by DP-TTF and I_2 can significantly reduce the membrane resistance towards H_2O_2 and KI. Electrical properties of agar gel sb-BLMs are consistent with those found for conventional BLMs. One major advantage of gel-supported BLMs is their superior mechanical stability [81].

More recently, a polypyrrole-modified s-BLM has been found to be sensitive to hydrogen ions [80]. Concerning pH probes, they could be the basis for monitoring blood gases (oxygen, carbon dioxide) and related variables (pH, Na^+, K^+, Ca^{2+}, Cl^-) as well as metabolites (glucose, urea, lactate, creatinine) after suitable modification. It is envisioned that all these could be monitored simultaneously and/or sequentially in a small sample of undiluted whole blood or in plasma [78].

Thin-film technology is one of the most advantageous methods in the preparation of bio/chemical sensors, which is compatible with microelectronic IC technologies. Studies have indicated that the thin-film microsystem (TF-μS) may be linked with s-BLMs for the development of 'smart sensors'. Recently, Tvarozek, Tien and their colleagues [83] have reported s-BLMs on TF-μSs arranged in the form of interdigitated electrodes or as a continuous layer. Two parameters characterizing the lipid bilayer were monitored simultaneously: the elasticity and electrical capacitance, with results similar to those obtained with s-BLMs [75]. It was concluded that TF-μSs combined with s-BLMs may represent a novel class of sensors. The technology of TF-μSs allows us to integrate an s-BLM sensoric element into an electronic circuitry necessary for signal processing [83].

Recent research has suggested a broad spectrum of possible s-BLM applications [10, 11]. These possibilities are based upon the fact that a lipid bilayer structure can be deposited on a solid substrate. This novel manner of lipid bilayer formation overcomes two basic obstacles in the way of the practical utilization of the BLM structure: (i) stability and (ii) compatibility with standard microelectronic technology. As has been repeatedly demonstrated [77, 78], the s-BLMs not only possess the advantages of the conventional BLM structure but additionally provide important new properties such as an anisotropic, highly ordered, yet very dynamic liquid-like structure, and two asymmetric interfaces. Additionally, recent work has been done on interdigitated structures (IDSs). IDSs are finger-like electrodes, in our case made by microelectronic technologies and used in microchip applications [83]. The following results were obtained on forming s-BLMs on IDSs made of platinum with a window of 0.5 mm $\times$ 0.5 mm. First, when an IDS was coated with a BLM formed from asolectin, the device responded to pH changes with a slope of 15 ± 2 mV. The conductance of an s-BLM on an IDS was about 50 times higher than that of an s-BLM on cut Pt. Second, when an IDS was coated with a BLM formed from asolectin plus TCOBQ (or TCPBQ), the pH response was linear with a 50 ± 1 mV slope. These results suggest that (i) the lipid bilayer, the fundamental structure of all biomembranes, can be attached to an IDS with responses not unlike those found in an s-BLM, (ii) this type of structure (i.e. s-BLMs on interdigitated electrodes) can be used to investigate ligand–receptor contact interactions, and (iii) s-BLMs on IDSs can be manufactured using microelectronic technologies which already exist. In concluding it should be mentioned that the experiment on an IDS chip modified with a BLM was based on a common basic aspiration, that is, to self-assemble a lipid bilayer containing membrane receptors so that a host of physiological activities, such as ion/molecular recognition, can be investigated. At the molecular level, most of these activities may be termed collectively as receptor–ligand contact interaction [9, 84]. The structures thus constituted are inherently dynamic. Receptors and ligands in such close contact normally will vary as a function of time, frequently resulting in non-linear behavior [85].

9.5 SUMMARY/TRENDS

Bilayer lipid membranes (BLMs) have been described as the basis for the construction of selective and sensitive electrochemical biosensors. Natural biological sensing has provided the principles that are necessary for the development of this relatively new technology. The electrostatic and structural properties of BLMs, and ion transport mechanisms, provide the basis for signal generation. Operation of a BLM-based transducer hinges on a selective interaction between analyte in aqueous solution and a membrane-embedded receptor; as a result of selective interactions, the electrostatic fields and/or the phase structure of the membrane change, leading to an analytical signal based on alterations of transmembrane ion current, potential, or capacitance. Examples of electrochemical sensors include those using enzymes, antibodies, lectins, and approaches that use active ion pumps for the selective determination of biochemicals such as glucose. Detection limits better than 10^{-9} M have been reported, with fast response times that can be of the order of tens of seconds. Recent research has shown that there may be a number of ways to achieve stabilization of lipid membranes, and this is a prerequisite for practical utilization of these sensors. Research will continue in the area of balancing structural stability of BLMs and covalent attachment or irreversible adsorption of selective receptors, with the intrinsic requirements for internal fluidity. Recent advances in gene technology have made possible the cloning of entire ion channel sequences, and have permitted structure–function studies utilizing site-directed mutagenesis for the construction of 'designer channels'. These advances in preparation of ion transport proteins will permit research on the function and regulation of practical membrane-bound proteins for future applications of biosensors that are based on lipid membranes.

ACKNOWLEDGMENTS

This work was supported in part by a USARO grant No DAALO3-91-G-0062 to HTT.

REFERENCES

[1] Leitmannova-Ottova A, Liu W, Zhou T-A and Tien H T 1994 *Proc. Mater. Res. Soc. Fall Meeting Symp. S (Boston, MA, 1993)* ed M Alper, H Bayley, D Kaplaan and M Navia (Pittsburgh, PA: Materials Research Society) at press

[2] Vassilev P and Tien H T 1989 *Artificial and Reconstituted Membrane Systems, Subcellular Biochemistry* vol 14 (New York: Plenum) p 97

[3] Neher E and Sackmann B 1992 *Spectrum Wissenschaft* **5** 48

[4] Tripathy A, Zviman M and Tien H T 1989 *J. Ind. Chem. Soc.* **66** 651–74

[5] Okahata Y and Ebato H 1991 *Anal. Chem* **63** 203

[6] Nikolelis D P and Krull U J 1993 *Electroanalysis* **5** 539

[7] Tien H T, Salamon Z, Kutnik J, Krysinski P, Kotowski J, Ledermann D and Janas T 1988 *J. Mol. Electron.* **4** s1–s30

[8] Mueller P, Rudin D O, Tien H T and Wescott W C 1962 *Nature* **94** 979; 1963 *J. Phys. Chem.* **67** 534

[9] Tien H T 1988 *Redox Chemistry and Interfacial Behavior of Biological Molecules* (New York: Plenum) pp 529–56

[10] Tien H T 1990 *Adv. Mater.* **2** 316–8

[11] Ottova-Leitmannova A and Tien H T 1992 *Prog. Surf. Sci.* **41** 337–445

[12] Tien H T 1985 *Prog. Surf. Sci.* **19** 169; 1989 *Prog. Surf. Sci.* **30** 1

[13] Hong F T 1989 *Molecular Electronics: Biosensors and Biocomputers* (New York: Plenum) p 259

[14] Krysinski P and Tien H T 1986 *Prog. Surf. Sci.* **23** 317

[15] Tien H T 1974 *Bilayer Lipid Membranes (BLM): Theory and Practice* (New York: Dekker)

[16] Takagi M 1967 *Experimental Techniques in Biomembrane Research* (Tokyo: Nankodo) p 385

[17] Tancrede P, Paquin P, Houle A and LeBlanc R M 1983 *J. Biochem. Biophys. Methods* **7** 299

[18] Nikolelis D P and Krull U J 1992 *Talanta* **39** 1045

[19] Tien H T 1988 *J. Surf. Sci. Technol.* **4** 1–21; 1988 *J. Clin. Lab. Anal.* **2** 256

[20] Minami H, Sugawara M, Odashima K, Umazawa Y, Uto M, Michaelis E K and Kuwana T 1991 *Anal. Chem.* **63** 2787

[21] Benz R, Elbert R, Prass W and Ringsdorf H 1986 *Eur. Biophys. J.* **14** 83

[22] Mountz J and Tien H T 1978 *Photochem. Photobiol.* **28** 395

[23] Thompson M, Lennox R B and McClelland R A 1982 *Anal. Chem.* **54** 76

[24] Nikolelis D P, Siontorou C, Andreou V and Krull U J 1994 *Electroanalysis* **7** 531

[25] Arya A, Krull U J, Thomspon M and Wong H E 1985 *Anal. Chim. Acta* **173** 331

[26] Bruckner-Lea C, Petelenz D and Janata J 1990 *Mikrochim. Acta* **1** 169

[27] Vodyanoy V, Halverson P and Murphy R B 1982 *J. Colloid Interface Sci.* **88** 247

[28] Mittal K L 1989 *Surfactants Solutions* **8** 133–78

[29] Tien H T and Salamon Z 1989 *Bioelectrochem. Bioenerget.* **22** 211

[30] Stenger D A, Fare T L, Cribbs D H and Rusin K M 1992 *Biosensors Bioelectron.* **7** 11

[31] Otto M, Snejdarkova M and Rehak M 1992 *Anal. Lett.* **25** 653
Snejdarkova M, Rehak M and Otto M 1993 *Anal. Chem.* **65** 665

[32] Yoshikawa K, Hayashi H, Shimooka T, Terada H and Ishii T 1987 *Biochem. Biophys. Res. Commun.* **145** 1092

[33] Masoom M and Townshend A 1984 *Anal. Chim. Acta* **166** 111

[34] Moody G L, Sanghera G S and Thomas G D R 1986 *Analyst* **111** 1235

[35] Bain C D, Troughton E B, Tao Y-T, Evall J, Whitesides G M and Nuzzo R G

1989 *J. Am. Chem. Soc.* **111** 321
[36] Porter M D, Bright T B, Allara D L and Chidsey C E D 1987 *J. Am. Chem. Soc.* **109** 3559
[37] Heckl W M, Marassi F M, Kallury K M R, Stone D C and Thompson M 1990 *Anal. Chem.* **62** 32
[38] Netzer L, Iscovici R and Sagiv J 1983 *Thin Solid Films* **99** 235
[39] Troughton E B, Bain C D, Whitesides G M, Nuzzo R G, Allara D L and Porter M D 1988 *Langmuir* **4** 365
[40] Brennan J D, Brown R S, Ghaemagghami V, Kallury K M, Thompson M and Krull U J 1992 *Chemically Modified Surfaces* (Amsterdam: Elsevier) p 275
[41] Wilmsen U, Methfessel C, Hanke W and Boheim G 1983 *Physical Chemistry of Transmembrane Ion Motions* (Amsterdam: Elsevier) p 479
[42] Treloar P H, Higson S P J, Desai M A, Christie I M, Ghosh S, Rosenberg M F, Reddy S M, Jones M N and Vadgama P M 1993 *Uses of Immobilized Biological Compounds* (Dordrecht: Kluwer) p 131
[43] Benga G 1985 *Structure and Properties of Membranes* (Boca Raton, FL: Chemical Rubber Company)
[44] Gallez D, Costa Pinto N M and Bisch P M 1993 *J. Colloid. Interface Sci.* **160** 141
[45] Nikolelis D P, Brennan J D, Brown R S, McGibbon G and Krull U J 1990 *Analyst* **116** 1221
[46] Atwood J L and Osa T 1991 *Inclusion Aspects of Membrane Chemistry* (Dordrecht: Reidel) p 191
[47] Chernomordik L V, Melikyan G B and Chizmadzhev Y A 1987 *Biochim. Biophys. Acta* **906** 309
[48] Blank M 1987 *Mechanistic Approaches to Interactions of Electromagnetic Fields with Living Systems* (New York: Plenum) pp 301–24
[49] Blank M 1986 *Proc. Symp. Electrochem. Soc. (Toronto, 1985)* (New York: Plenum) p 149
[50] Krull U J 1987 *Anal. Chim. Acta* **192** 321
[51] Thompson M and Dorn W H 1987 *Chemical Sensors* (Glasgow: Blackie) p 168
[52] Krull U J 1987 *Anal. Chim. Acta* **197** 203
[53] Thompson M and Krull U J 1982 *Anal. Chim. Acta* **141** 33
[54] Janas T and Tien H T 1988 *Biochim. Biophys. Acta* **939** 624
[55] Krull U J and Thompson M 1985 *Trends Anal. Chem.* **4** 90
[56] Nikolelis D P, Brennan J D, Brown R S and Krull U J 1992 *Anal. Chim. Acta* **257** 49
[57] Nikolelis D P and Krull U J 1992 *Anal. Chim. Acta* **257** 239
[58] Del Castillo J, Rodriquez A, Romero C A and Sanchez V 1966 *Science* **153** 185
[59] O'Boyle K, Siddiqi F A and Tien H T 1984 *Immunol. Commun.* **13** 85
[60] Vassilev P M, Kanazirska M P and Tien H T 1987 *Biochim. Biophys. Acta* **897** 324–30
[61] Marino A A 1988 *Modern Bioelectricity* (New York: Dekker) p 181
[62] Mittler-Neher S and Knoll W 1993 *Biochim. Biophys. Acta* **1152** 259
[63] Thompson M, Krull U J and Bendell-Young L I 1983 *Talanta* **30** 919
[64] Krull U J, Brown R S, Koilpillai R N, Nespolo R, Safarzadeh-Amiri A and Vandenberg E T 1989 *Analyst* **114** 33
[65] Sugawara M, Kojima K, Sazawa H and Umezawa Y 1987 *Anal. Chem.* **59** 2842
[66] Uto M, Michaelis E K, Hu I F, Umezawa Y and Kuwana T 1990 *Anal. Sci.* **6** 221
[67] Odashima K, Sugawara M and Umezawa Y 1991 *Trends Anal. Chem.* **10** 207
[68] Sugao N, Sugawara M, Minami H, Uto M and Umezawa Y 1993 *Anal. Chem.*

65 363
[69] Nikolelis D P, Tzanelis M G and Krull U J 1993 *Anal. Chim. Acta* **281** 569
[70] Nikolelis D P, Tzanelis M G and Krull U J 1994 *Biosensors Bioelectron.* **9** 179
[71] Nikolelis D P, Tzanelis M G and Krull U J 1993 *Anal. Chim. Acta* **282** 527
[72] Nikolelis D P, and Krull U J 1994 *Anal. Chim. Acta* **288** 187i
[73] Wang L G, Li Y-H and Tien H T 1994 *Bioelectrochem. Bioeng.* at press
[74] Schulmann W, Heyn S-P and Gaub H E 1991 *Adv. Mater.* **3** 388
[75] Hianik T, Dlugopolsky J and Gyepessova M 1993 *Bioelectrochem. Bioeng.* **31** 99
[76] Kinnear K T and Monbouquette H G 1993 *Langmuir* **9** 2255
[77] Tien H T 1994 *Biomembrane Electrochemistry (Advances in Chemistry Series 235)* (Washington, DC: American Chemical Society) ch 24
[78] Liu W, Ottova-Leitmannova A and Tien H T 1993 *Polymer/Inorganic Interfaces* vol 304 (Pittsburgh, PA: Materials Research Society)
[79] Tien H T, Salamon Z, Liu W and Ottova A 1993 *Analyt. Lett.* **26** 819
[80] Kotowski J, Janas T and Tien H T 1988 *J. Electroanal. Chem.* **253** 227
[81] Lu X-D, Leitmannova-Ottova A and Tien H T 1995 *Bioelectrochem Bioenerget.* **5** at press
[82] Ziegler W, Remis D, Brunovska A and Tien H T 1993 *Proc. 7th C-S Conf. on Thin Films (Lipt. Mikulas-Slovakia, 1993)* p 304
Ziegler W, Remis D, Brunovska A, Jakabovic J and Tien H T 1993 *Proc. 7th C-S Conf. on Thin Films (Lipt. Mikulas-Slovakia, 1993)* p 308
[83] Tvarozek V, Tien H T, Novotny I, Hianik T, Dlugoposky J, Ziegler W, Ottova A L, Jakabovic J, Rehacek V and Uhlar M 1994 *Sensors Actuators* B **19** 597
[84] Ottova-Leitmannova A, Martynski T, Wardak A and Tien H T 1994 *Molecular and Biomolecular Electronics (Advances in Chemistry Series 240)* (Washington, DC: American Chemical Society) ch 17
[85] Li Jia-Lin and Tien H T 1995 *Mater. Sci. Eng.* C at press

10

Biomolecular electronics

Felix T Hong

10.1 INTRODUCTION

A multicellular living organism is frequently compared to a modern digital computer. Superficially, this analogy is a valid one. Both digital computers and living organisms are equipped with (i) sensors for receiving input information from the outside world; (ii) a central processor for storing new information in their memory for future reference, for integrating all input information with other information retrieved from their memory and, thus, enabling them to make decisions and (iii) actuators for implementing the actions that are based on the outcome of the decision making of the central processor. There are notable differences, of course [1–3]. The most obvious and the most intriguing difference is the possession of intuition and creativity by higher organisms and the digital computers' apparent lack of it. In this chapter, rather than dealing with this analogy from the point of view of cognitive science, it will be addressed from the practical point of view of using molecular materials for the construction of functional and computing devices. In the proposed construction of biomolecular electronic devices, naturally occurring biomolecular materials or synthetic organic materials replace the conventional silicon-based materials and other inorganic materials. Biomolecular electronics research also proposes to construct devices with unconventional device architectures [4]. Readers interested in novel device architectures are referred to review articles by Conrad [5, 6] and references contained therein.

The use of conventional materials or synthetic materials in the construction of devices based on design principles inspired by Nature will also be included in this chapter (biomimetic science). It will be demonstrated that prototype electronic devices constructed with biomaterials obey the same physical laws as do conventional devices. The problem of interfacing biomolecular devices with the outside world will also be addressed. Whenever appropriate, synthetic organic materials and biomaterials are treated as if there were no differences, since many of the principles discussed apply equally well to both classes of molecular materials (for justification see [7]). This chapter will address the general features of biomolecular electronics and the problems arising from them.

10.2 ADVANTAGES OF USING MOLECULAR AND BIOMOLECULAR MATERIALS

A major motivation for pursuing biomolecular electronics is the realization by scientists and engineers that the trend of miniaturization in microelectronic integrated circuits may soon reach a limit where thermal dissipation and quantum mechanical cross-talk will make integrated circuits unreliable as digital information processing components. A living organism knows no such limits. By definition, molecular machines are of molecular dimensions and most function on the type of quantum mechanical cross-talk known as chemical reactions. Of course, a living organism is not a 'von Neumann-type' digital sequential machine. At the molecular level, informational processing is more or less analog in nature, but at the mesoscopic or the macroscopic level some biological information processing acquires digital nature, as exemplified by neural impulses in the form of action potentials. In addition, a living organism utilizes massively parallel distributed processing rather than sequential processing.

Biological molecular machines are constructed with proteins, which are copolymers made of some 20 different kinds of monomer, amino acids. A protein molecule can perform a rudimentary type of information processing as will be illustrated below. It is also well known that proteins derive their functionality from the nature of their folded shape, known as conformation to biochemists. A protein loses its function when its native conformation is altered and undergoes denaturation. Conrad and his co-workers [4–6] repeatedly point out that conformation-based intermolecular recognition, the 'lock–key' paradigm, is the predominant type of biological information processing. This proposal is supported by many examples including enzyme–substrate, antigen–antibody, and hormone–receptor interactions.

It is evident that conformation-based molecular machines constructed with proteins possess important differences from conventional digital machines. The astronomical number of monomeric unit substitutions in a medium-sized protein allows gradual changes of its conformation, known as gradualism to computer scientists. As a consequence, a molecular machine can degrade gracefully whereas a digital computer degrades catastrophically even in the presence of a single 'bug' in the 'software'. However, a molecular machine is much harder to 'debug', because the exact nature of the structure–function relationships of most proteins remains an unsolved problem.

Another feature of protein-based information processing is the relatively large number of moving parts in a molecular machine. Unlike a digital circuit with electrons and holes as moving parts, protein-based machines consist of electrons, protons, and ions, as well as the atomic nuclei of the protein polymer chain. The movement of atomic nuclei arises from conformational changes. Strictly speaking, the structural components of a molecular machine do not consist of hardware. They are made of 'wetware' instead, since most proteins need at least a partially aqueous environment to function properly. Furthermore, protons instead of electrons are the predominant moving parts for its function because proteins contain proton binding groups such as carboxylate and quarternary ammonium groups. The protonation or deprotonation (acid–base reaction) of such groups is crucial for conformational changes. As a result, the maintenance of optimum pH (proton concentration) for blood and other body fluids is of greater importance than the maintenance of optimal redox potentials for proteins. Redox reactions, though less frequently occurring than acid–base reactions in the body, exist in organelles such as mitochondria and in the photosynthetic apparatus of plants and photosynthetic bacteria. Redox reactions are usually carried out in special metal-based redox centers of protein complexes in molecular machines. This peculiarity of the relative abundance of acid–base reactions poses a problem in interfacing a protein-based device to conventional circuitry such as a metal electrode (see below).

10.3 ELECTRICAL BEHAVIOR OF MOLECULAR OPTOELECTRONIC DEVICES: THE ROLE OF CHEMISTRY IN SIGNAL GENERATION

Many prototypes of molecular electronic devices have been designed in layered thin-film configurations in which electrical events reflect charge movements in the direction perpendicular to the plane of the film (vectorial charge movements) [8, 9] (figure 10.1). This type of configuration is also evident in naturally occurring electronic devices such as the photosynthetic apparatus and mitochondria as well as visual photoreceptor membranes. It has been demonstrated that the electrical behavior of these thin-film devices is similar to that of conventional electronic devices [10, 11]. Specifically, their electrical behavior can be quantitatively modeled with equivalent circuits. In recent years, numerous attempts to develop thin-film devices have been made. Some of them utilized organic dyes or biological pigments to construct the thin films. Organic thin films constructed from bacteriorhodopsin, a popular biopigment, are used as examples here for three reasons. First, successful prototypes using bacteriorhodopsin have appeared. Second, bacteriorhodopsin thin films or membranes are suitable for modeling both the visual and the photosynthetic processes and for gaining insights into Nature's design principles via reverse engineering. Third, data demonstrating rigorous mathematical modeling in terms of an electrical equivalent circuit are available for bacteriorhodopsin.

Bacteriorhodopsin (BR) is the only membrane-bound protein in the purple membrane of *Halobacterium halobium* [12–16]. It is a single-chain polypeptide

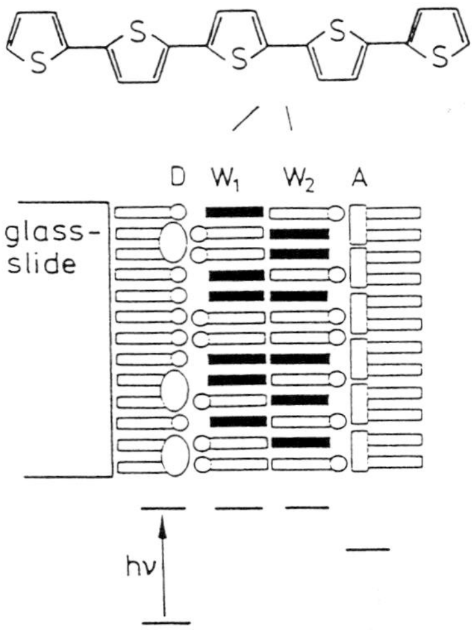

Figure 10.1 A prototype multilayered organic thin film as a molecular electronic device. The layers were deposited on a glass substrate (which may be coated with a transparent metal electrode). The first layer contains an electron donor, carbocyanine (D), dispersed in a fatty acid monolayer matrix. The second and third layers contain quinquethienyl as molecular wires (W_1 and W_2). The fourth layer (A) consists of a monolayer of bipyridinum as the electron acceptor. The energy diagram is shown below. A device like this imitates the photosynthetic reaction center (biomimetic device). (Reproduced from [8].)

which forms seven transmembrane loops of α-helices, with a covalently bond chromophore, retinal (vitamin A aldehyde). Bacteriorhodopsin resembles the visual pigment rhodopsin in chemistry and in its characteristic photoelectric signals, but functionally it comprises photosynthesis in the bacterial cell. It also possesses an exceptional chemical and mechanical stability.

Thin films or membranes reconstituted from oriented BR exhibit electrical phenomena known collectively as the *photoelectric effects* [17–19]. The photoelectric effects can be further classified into two main categories. The *DC photoelectric effect* reflects the electrical phenomenon associated with net vectorial transport of protons across the membrane from the cytoplasmic side to the extracellular side of the membrane. For example, the light-induced charge transport in BR membranes and in chlorophyll-based photosynthetic membranes begins as light-induced charge separation. A fraction of these separated charges will eventually span the entire thickness of the membranes and will constitute the net transported charges across the membrane. The other fraction of separated charges recombine without resulting in net charge transport. This latter fraction of charge movement manifests itself as a capacitance photocurrent, which is known as the *AC photoelectric current*. Detailed analyses of these phenomena have been published [20, 21]. In summary, the AC photoelectric effect is prominently evident as a transient signal when a brief light pulse is used to illuminate the membrane or thin film while the DC photoelectric effect is more readily observed if continuous light is used as the light source. Both the DC and the AC photoelectric effects are relevant to device construction.

Basically, there are two types of charge separation and recombination in BR membranes [21]. In the first type, the separated charges remain confined inside the membrane (or inside the protein itself), and the subsequent recombination is a first-order chemical process. This is referred to as the *oriented dipole mechanism* because the separated charges form a transient array of oriented electric dipoles. In the second type, the charge separation takes place across the membrane–water interface when a proton is bound to BR or is released from a proton binding site of BR into the adjacent aqueous phase. This mechanism is referred to as the *interfacial proton transfer mechanism*. The charge recombination in the latter mechanism is the back (reverse) reaction at the same interface, and is a second-order chemical process. In BR, there is a process of major intramembrane charge separation and recombination that can be detected, the *B1 component*. There are two additional components caused by interfacial charge (protons, Cl^-, etc) uptake and release at the cytoplasmic surface and the extracellular surface, respectively. They are known as the *B2* and the *B2′ components*.

The separation of these components can be implemented with several different reconstitution techniques [22]. Using the Trissl–Montal method [23] in which oriented purple membrane sheets were deposited on a thin Teflon film, we observed a composite signal comprised of the B1 and the B2 components. Using another method called the multilayered thin-film technique, we were able to isolate the B1 component in pure form without the contribution from other components that normally exist. Unlike most studies of the AC photoelectric effect, we did not decompose the signal simply into several exponential decays. Instead, we separated the components on the basis of their underlying molecular mechanisms and their sites of generation. For example, the B1 component is generated at the hydrophobic interior of BR through the oriented dipole mechanism whereas the B2 and B2′ components are generated at the two hydrophilic surface domains through the interfacial proton transfer mechanism. The objection to exponential analysis is rooted in the mode of signal generation and has been documented elsewhere [17–19]. The experimental support for our approach has also been published [24–27]. Here, we shall use the data of the B1 component to illustrate the electrical behavior of a bioelectric signal.

Based on the oriented dipole and the interfacial proton transfer mechanism described above, an equivalent circuit was established which described the relaxation time course of a photoelectric current generated by a single chemical reaction step of charge separation and recombination regardless of whether the charge separation is confined within the membrane or takes place across a membrane–water interface [21]. This equivalent-circuit analysis is notable for the absence of any adjustable parameters: each and every parameter used for the computation can be measured experimentally (figure 10.2). An example of the

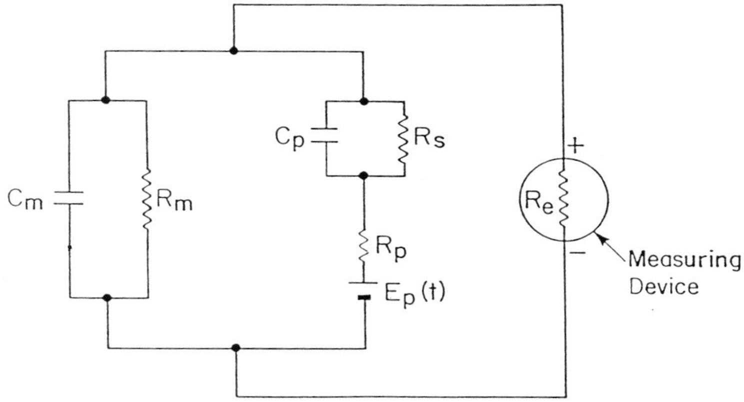

Figure 10.2 A universal equivalent circuit for the photoelectric effect. The photochemical event is represented by an RC network including (i) the photoemf ($E_p(t)$), (ii) the internal resistance (R_p) of the photocurrent source, (iii) the chemical capacitance (C_p), and (iv) the transmembrane resistance (R_p). With the exception of a strictly short-circuit measurement, the time course of the photocurrent so generated is further shaped via interaction with the RC network formed by (i) the membrane resistance (R_m), (ii) the membrane capacitance (C_m), and (iii) the access resistance (R_e). The access resistance (impedance) includes the input impedance of the amplifier, the electrode impedance, and the impedance of the intervening electrolyte solution. (Reproduced from [17].)

curve fitting is shown in figure 10.3(A). The thin-film device in this example consisted of a solid thin Teflon film support (thickness 6.35 μm) on which several layers of oriented BR membranes were deposited.

It can be instructive to compare the time courses of the photocurrent under different measurement conditions, e.g. open-circuit or short-circuit conditions. A major source of variation of the measurement condition is the access impedance (R_e), the impedance appearing at the signal input between the membrane or the thin film and the measuring device. The measurement condition is considered open circuit if the access impedance is much greater than the source impedance whereas it is considered short circuit if the access impedance is much smaller than the source impedance. Figure 10.3(B) shows two signal records measured at R_e = 20 kΩ and at 40 kΩ, neither of which is strictly short circuit nor open circuit [22]. It is evident that the apparent relaxation time constants are lengthened with increasing R_e. Also shown in figure 10.3(B) are cross-predictions of the time course of one signal based solely on information obtained by deconvolution of the other. In brief, by deconvolution, it can be shown that the time constant τ_p, extracted from both signals are identical (12.3 ± 0.7 μs). The significance of this result is better appreciated if R_e is treated as the external load to the photosignal source. The result indicates that the observed photosignal relaxation varies with the external load (R_e) and the variation is fully predictable by equivalent-circuit analysis.

The role of the *RC* (resistive–capacitative) characteristics of the inert support elements in the thin film can be appreciated by the following experiment. Similar oriented BR thin films were deposited on Teflon films of various thicknesses (6.35, 12.7, 25.4 μm). The relaxation time course varies with the thickness of the Teflon support, but the effect is predictable by the equivalent circuit. Figure 10.3(*C*) shows that the time course for the thin films of 12.7 and 25.4 μm thickness can be predicted by the equivalent-circuit analysis based solely on data obtained at the 6.35 μm thin film [10]. As expected, deconvolution of either the 6.35 or the 12.7 μm films gives essentially the same intrinsic relaxation time.

The data shown in figure 10.3(*C*) indicate that the observed signal time course can be altered by varying the *RC* characteristics of inert support materials. Thus, it appears that the electrical behavior of thin-film devices constructed with electroactive biomaterials may be fundamentally similar to that of conventional microelectronic devices made of inorganic materials. In particular, equivalent-circuit analysis is useful as a design tool in biomolecular electronics.

What sets the biomolecular electronic devices apart from microelectronic devices is the contribution of chemistry in the generation of bioelectric signals. The equivalent circuit shown in figure 10.2 was derived from combining conventional chemical kinetic analysis with the electrostatic calculation based on the Gouy–Chapman diffuse double-layer theory [20, 21]. The possibility of manipulating the photosignal by varying the underlying chemistry is shown in figure 10.4.

Evidence indicates that the B2 component is generated at or near the membrane–water interface [22]. This is most likely due to interfacial proton transfer (uptake and rerelease). While the B1 component is insensitive to changes of conditions in the (adjacent) aqueous phase, the B2 component is highly sensitive to these changes. For example, the sensitivity of B2 to aqueous pH and to the replacement of water with D_2O is expected from the participation of protons in the interfacial reaction (figures 10.4(*A*) and 10.4(*B*)). The separation of the B1 and the B2 components is most likely based on their spatial origin in different functional domains of BR, as demonstrated by the complete absence of the B2 component in the medium- to high-pH range in a mutant of BR (D212N), in which the aspartate residue at position 212 in the peptide chain is replaced by asparagine [26] (figure 10.4(*C*)). As shown in figure 10.4(*D*), the B2 component appears at low-pH range and has a polarity opposite to that in the medium- to high-pH range [27]. More recent experiments suggest that the B2 component itself may be decomposed into two subcomponents: B2-a and B2-c [28]. The B2-a subcomponent appears at low pH and is sensitive to pH as well as the Cl^- concentration (figure 10.4(*D*)). This subcomponent probably reflects a pH-modulated interfacial Cl^- transfer rather than a Cl^--modulated interfacial proton transfer. This conclusion is supported by its absence in a Cl^--free medium, and by the complete inhibition of this component

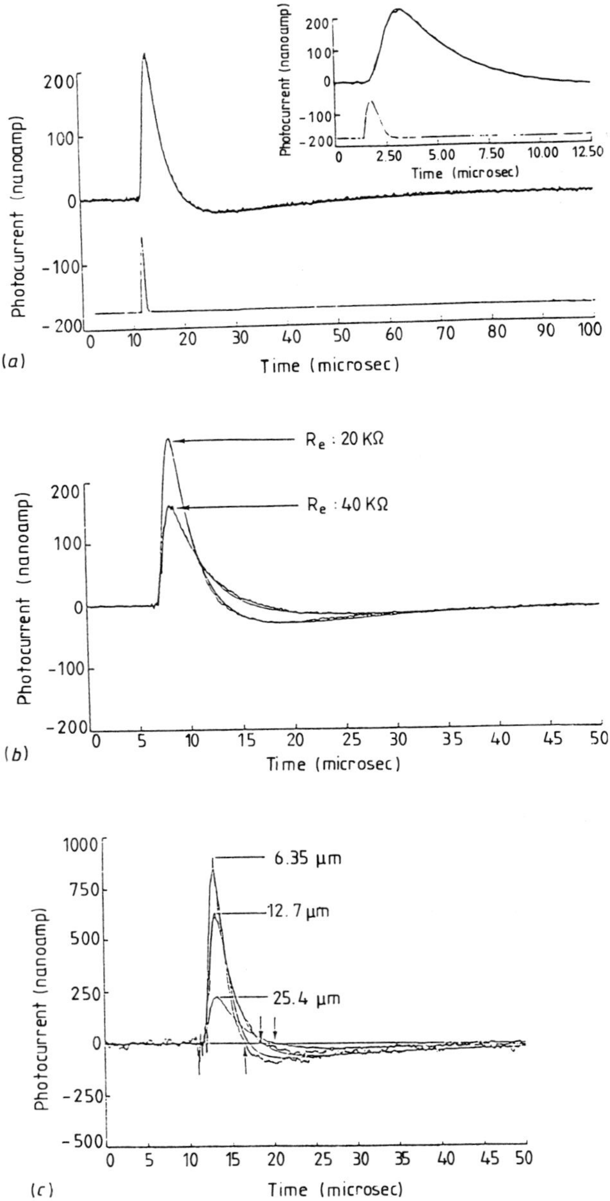

Figure 10.3 (*A*) Equivalent-circuit analysis of the B1 component. The multilayered method was used for the reconstitution. The temperature was 25 °C. The bathing electrolyte solutions contained 0.1 M KCl and 0.01 M L-histidine buffered at pH 2. The measurement was made at an access impedance of 39.2 kΩ and an instrumental time constant of 0.355 μs.

Figure 10.3 (*Continued.*) The light source was a pulsed dye laser (590 nm). The light pulse was simulated with a triangular waveform with a half-height duration of 0.8 μs. The experimentally determined input parameters were: $\tau_s = 3.4$ μs, $\tau_l = 24.0$ μs, $\tau_m = 6.42$ μs, $R_m = R_s = \infty$. These parameters together with R_e provide sufficient information to compute the following parameters: $R_p = 60.3\,\mathrm{k}\Omega$, $C_p = 211$ pF, and $\tau_p = 12.7$ μs. The computed curves (smooth traces) are superimposed on the measured curves (noisy traces) after normalization with respect to the positive peak. Normalization yields the peak photoemf of 155 mV. The same data are shown on two different scales (see the inset). The measured laser pulse is also shown in the inset (bottom trace). (*B*) The effect of varying the access impedance on the measured time course of the B1 component. Two separate measurements were made on the same membrane preparation under near-identical conditions except that the values of the access impedance were 20 kΩ and 40 kΩ, respectively. The temperature was 25 °C. The aqueous phases contained 3 M KCl and 0.05 M L-histidine buffered at pH 2. Equivalent-circuit analysis of the data measured at 20 kΩ generated a complete set of circuit parameters, which were then used to compute the theoretical time course for 40 kΩ (smooth curve). Similarly, data analysis at 20 kΩ cross-predicts a theoretical curve for the measurement at 40 kΩ. The experimental (noisy) curves are superimposed for comparison. Although the time courses are different for the two measurements, deconvolution yielded nearly identical intrinsic relaxation data: $R_p = 61.0$ kΩ, $C_p = 207$ pF, and $\tau_p = 12.6$ μs for the measurement at $R_e = 20$ kΩ; $R_p = 65.3$ kΩ, $C_p = 192$ pF, and $\tau_p = 12.5$ μs for the measurement at $R_e = 40$ kΩ. (*C*) The effect of varying the thickness of Teflon films on the measured time course of the B1 component. The photosignals were obtained from three separate multilayered dry films of purple membranes on Teflon films with thicknesses of 6.35, 12.7, and 25.4 μm, respectively. The temperature was 24 ± 1 °C. The bathing solution contained 0.1 M KCl and 0.01 M L-histidine buffered at pH 2. The access impedance was 40 kΩ. The instrumental time constant was 0.4 μs. The experimental data are shown as noisy curves. The smooth curves are predictions based on the input parameters determined by deconvolution of the experimental data of the preparation with a film thickness of 6.35 μm. ((*A*) and (*B*) reproduced from [22]; (*C*) from [10].)

by chloride ion channel blockers such as DIDS (4-acetamido-4′-isothiocyano-2, 2′-stilbenesulfonate) and SITS (4, 4′-diisocyano-2, 2′stilbenesulfonate) [28]. In contrast, the B2-c subcomponent was not sensitive to Cl^-. Instead, it was sensitive to certain divalent cations, such as Ca^{2+} and Mg^{2+}. The enhancement effect of Ca^{2+} or Mg^{2+} can be reduced or inhibited by EGTA or EDTA.

It appears that the cation sensitivity and anion sensitivity of BR are mediated by separate domains of the BR molecule. The generation of AC photoelectric signals from localized domains of BR is demonstrated in an experiment in which the pH values of the two aqueous phases are varied independently [29, 30] (figure 10.5). The membrane is reconstituted by means of a variant type of black lipid membrane according to the method originally developed by Drachev *et al* [31]. The AC photoelectric signals exhibit a positive peak and a prominent

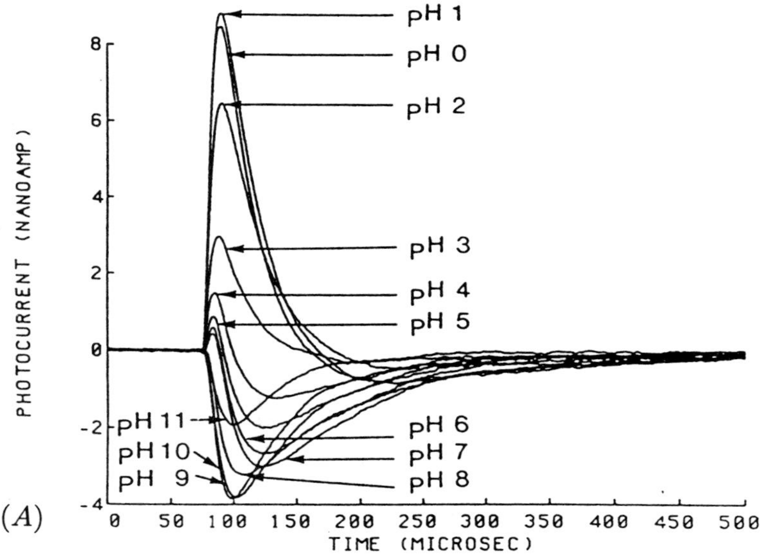

(*A*)

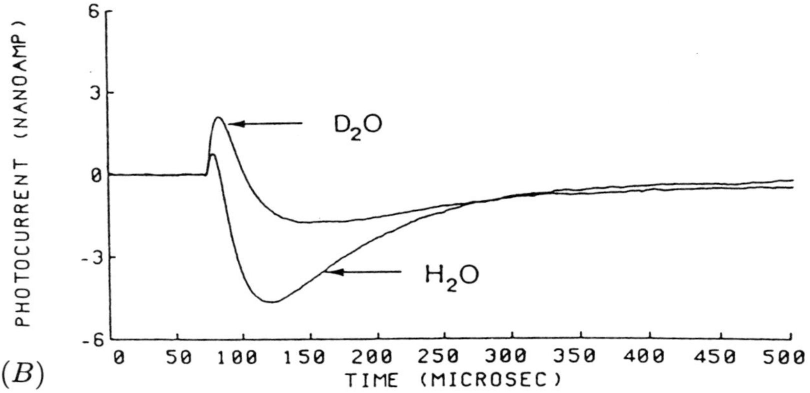

(*B*)

Figure 10.4 The effect of changing the chemical environment on the time course of the B2 component. In (*A*), the pH effect was reversible except above pH 11. The pH in (*B*) was kept at pH 7, and the electrolyte solution was exchanged with a D_2O solution of comparable electrolyte composition. The effect of D_2O–water exchange was found to be reversible. In (*C*), the photoelectric signal of the mutant D212B exhibited no B2 component in the medium- to high-pH range. The inset shows the superposition, after peak normalization, of the photosignals at low pH, demonstrating the lack of a pH dependence at low pH. In (*D*), the contribution of B1 was subtracted from the composite signal, leaving a pure B2 signal, which exhibited a polarity reversal at pH 2.7. ((*A*) and (*B*) reproduced from [22]; (*C*) from [26]; (*D*) from [27].)

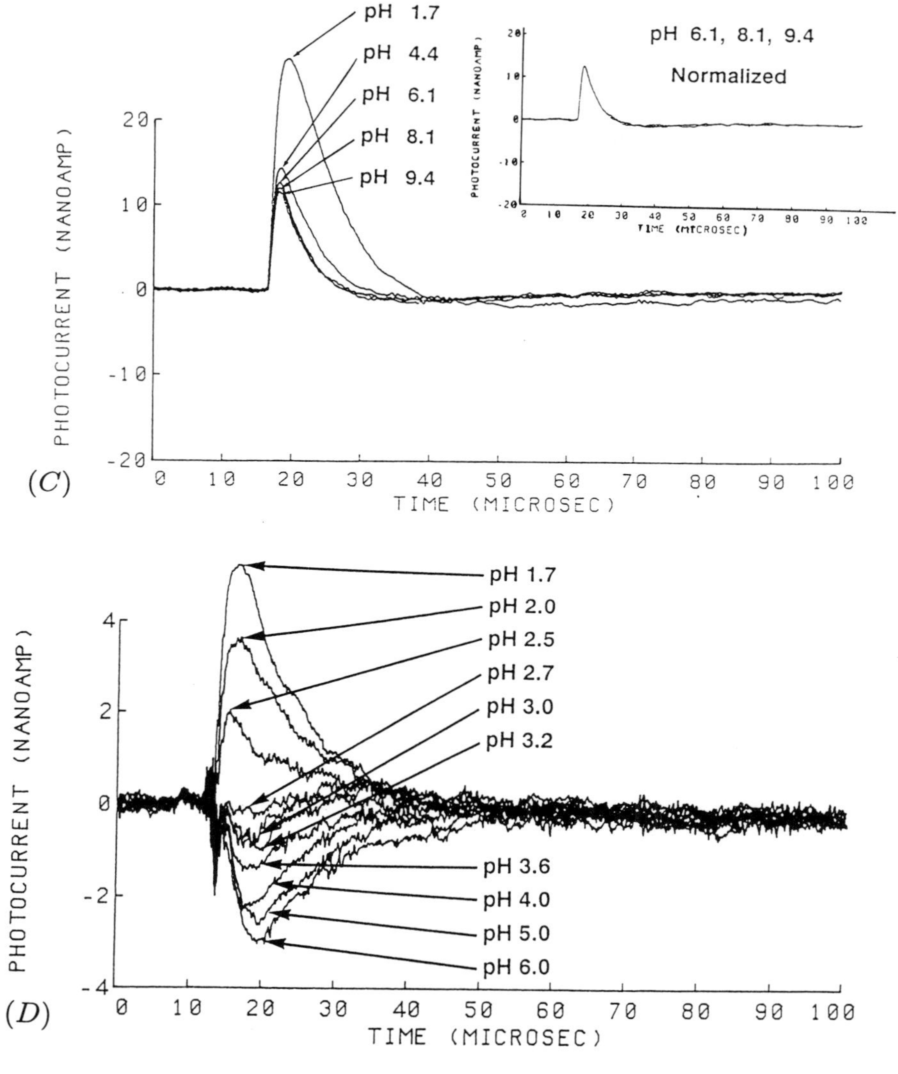

Figure 10.4 (*Continued.*)

negative peak reflecting the presence of both the B1 and the B2 components at neutral pH. When the cytoplasmic pH was lowered, the negative peak (B2 component) was reversibly inhibited. However, when the extracellular pH was also lowered a negative peak reappeared. This negative peak had a different relaxation time and also a pH dependence opposite to that of the regular B2 component. This behavior is expected from the hypothetical B2′ component. Apparently, the proton uptake at the cytoplasmic surface and the proton release

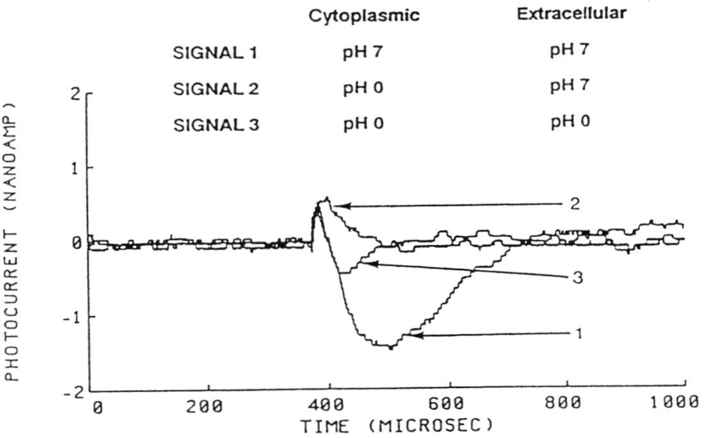

Figure 10.5 The 'differential experiment' showing the existence of two components of photosignals due to proton uptake and release, respectively, at the two membrane–water interfaces. (Reproduced from [29].)

at the extracellular surface can be treated as two independent chemical reactions although BR participates as a reactant in both reactions.

10.4 THE PHYSIOLOGICAL ROLE OF THE AC PHOTOELECTRIC SIGNAL: THE REVERSE ENGINEERING VISUAL SENSORY TRANSDUCTION PROCESS

An electrochemical approach permits correlation of the three major components of the AC photoelectric signal from a reconstituted BR membrane with three separate molecular domains of BR. The B1 component has an ultrafast risetime and has been associated with the K to L transition of the photocycle. Our interpretation, which is also shared by Birge and his coworkers [32–34], is the following. B1 is caused by the appearance of a transient electric dipole moment during photoisomerization. In other words, the B1 component reflects the necessity of the retinal protein to store the photon energy temporarily as an electric dipole array (oriented dipole mechanism) which, when dissipated, is used to drive slower conformational changes of the protein for various purposes. In this regard, it is of interest to note that membranes containing rhodopsin, BR, and halorhodopsin all share this component, which is designated R1, B1, and H1, respectively [18].

As for the remaining components, B2 and B2′, in BR, they are associated with two mandatory steps of proton transport: proton uptake and release at the two membrane–water interfaces, respectively. The role of the analogous signal in the visual photoreceptor membrane, known as the early receptor potential (ERP), is more puzzling because the photoreceptor is not primarily a photon energy converter.

The ERP was discovered in monkey retina about 30 years ago [35]. Although extensive research has been carried out to elucidate the mechanism and the physiological significance of this signal, there is no consensus on the ERP. Most investigators regard it as an epiphenomenon, playing no significant role in visual transduction. However, the reverse engineering approach suggests otherwise.

There are several lines of experimental evidence indicating that the R2 component develops at the stage of metarhodopsin II during the sequential photochemical reactions (photobleaching sequence), and that reflects a proton binding at the cytoplasmic side of the visual photoreceptor membrane (evidence is listed and reviewed in [36]). Metarhodopsin II happens to be the photointermediate that binds transducin (a G protein), activates it, and thus triggers a biochemical process of visual transduction known as the cyclic-GMP cascade [37]. This suggests that the R2 component and the associated proton binding may serve as the mechanistic trigger that leads to the binding and activation of transducin. It is well known that a surface potential is associated with an electrified interface. The obvious and immediate effect caused by the binding of a proton to rhodopsin is the formation of a positive surface potential at the cytoplasmic interface. This surface potential has been demonstrated by Cafiso and Hubbell [38]. In order to illustrate the mechanistic principle of a trigger underlying the action of a surface potential, we shall cite an experimental prototype.

Drain *et al* [39] studied an artificial lipid bilayer system in which ionic current is carried by tetraphenylboride ions and driven by an applied transmembrane potential (figure 10.6). The lipid bilayer membrane was doped with a hydrophobic pigment (magnesium octaethylporphyrin) and the two aqueous phases were loaded with equal concentrations of an aqueous electron acceptor, methylviologen. Illumination of the membrane generated two surface potentials at the two membrane surfaces with equal magnitude. No macroscopic photovoltaic effect was observed because the two interfacial vectorial electron transfers were equal in magnitude but opposite in polarity (otherwise, an AC photoelectric signal of the interfacial charge transfer type would be observable). However, the effect of these surface potentials was detectable indirectly through their effect on the surface concentration of tetraphenylboride ions. Since the tetraphenylboride ion is negatively charged, its surface concentration increases with increasing positive surface potential. As a result, the potential-driven ionic current increased accordingly upon illumination. The artificial system thus formed a prototype, photogatable ion conductor. To further test this interpretation, the tetraphenylboride ions were replaced with positively charged hydrophobic ions (tetraphenylphosphonium) and illumination caused a decrease in the ionic current as expected. This example also demonstrates that a biology-inspired design principle can be implemented in an artificial system.

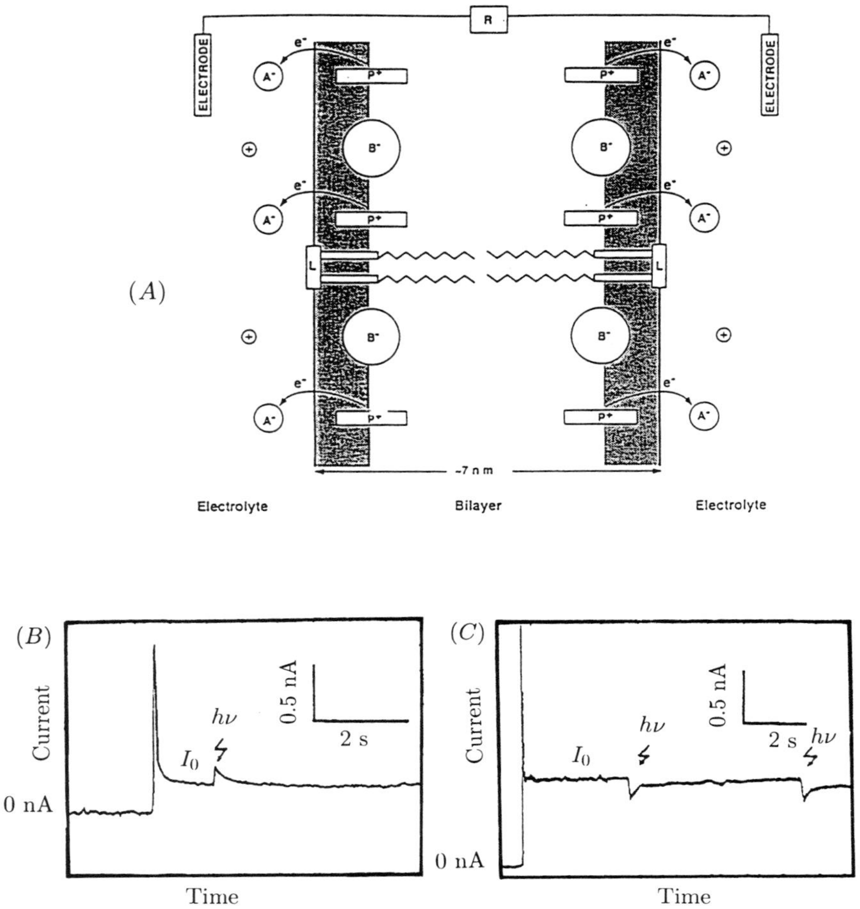

Figure 10.6 An experimental prototype showing switching of ionic currents based on light-induced surface potentials. (*A*) The lipid bilayer contains magnesium octaethylporphyrin which is lipid soluble (3.6 mM). The aqueous phases contain the electron acceptor methylviologen in equal concentrations on both sides (20 mM). Tetraphenylboride ions, B^-, are partitioned into the membrane at the region of the polar head groups of the lipid. Photoactivation of magnesium octaethylporphyrin generates two symmetrical positive surface potentials, which increase the surface concentration of B^- but decrease the concentration of tetraphenylphosphonium ions. (*B*) The ionic current of 1 μM tetraphenylboride ions increases by 50% upon illumination with a 1 μs laser pulse. (*C*) The ionic current of 5 mM tetraphenylphosphonium ions decreases by 25% upon illumination. In both (*B*) and (*C*), the spike on the left is the capacitative transient upon imposition of an applied potential of +50 mV. Light pulses are indicated by arrows. (Reproduced from [39].)

Furthermore, it may also be implemented with conventional materials. This approach of device construction is the characteristic of biomimetic science.

10.5 BACTERIORHODOPSIN AS AN ADVANCED BIOELECTRONIC MATERIAL: A BIFUNCTIONAL SENSOR

While reverse engineering may supply biology-inspired design principles, some biomaterials possess remarkable stability and may be used for device construction. Biomaterials are essential for construction of enzyme electrodes which detect biologically important solutes on the basis of specific enzymatic reactions. A common example is a glucose sensor using an electrode with immobilized glucose oxidase (GOD). Such biomaterials are regarded as intelligent materials because of their extraordinary versatility. The versatility of biomaterials is the consequence of evolution through which materials are selected for performing intricate biological functions. Bacteriorhodopsin has been successfully used to construct several photonic devices [33, 40–54] and photoelectric devices [34, 55–61] (see table 1).

Miyasaka and co-workers [55, 56] have successfully constructed a motion detector using wild-type BR (figure 10.7). The detector was constructed as a square array of eight by eight patches of oriented BR thin films deposited on transparent metal electrodes using conventional microphotolithography (figures 10.7(*A*) and 10.7(*B*)). Each BR electrode is connected by a pair of metal circuit lines which allow the sensor array to be interfaced to conventional microelectronic circuitry for further signal processing. A typical photocurrent response from a patch of BR electrode is shown in figure 10.7(*C*). A transient photocurrent develops upon the onset of illumination and another transient photocurrent of opposite polarity develops upon the cessation of illumination. The photocurrent subsides during steady illumination and during the dark after the initial transient. Such a waveform is the 'signature' of a linear high-pass *RC* filter. The photocurrent response is therefore the manifestation of the AC photoelectric effect of BR (see [17] for a detailed explanation). No DC photocurrent is detected for two reasons. First, the arrangement would not allow a DC photocurrent to be detected by external circuitry (see below). Second, the chemical capacitance acts as an internal amplifier which allows for AC signals to be detected with a disproportionately large amplitude [18, 19]. In other words, low-frequency signals and DC signals are preferentially suppressed by the high-pass filter. The linearity of the response is demonstrated in figure 10.7(*D*), in which the transients are proportional to the *change* of the level of input illumination rather than the level *per se* (differential responsivity).

Both Trissl [58] and Rayfield [59] have constructed ultrafast biological photodiodes and Rayfield has demonstrated a rise time of the photoelectric of the order of a few picoseconds. That an oriented BR thin film constitutes a photocell is apparent from our equivalent-circuit analysis. However, the claim

Table 10.1 Prototype biomolecular electronic devices constructed from bacteriorhodopsin/halorhodopsin.

Device	Mechanism of action	Biomaterial(s)	References
Biochrom-BR film Biochrom-4KBR film (Erasable optical image storage)	photochromic effect (ground state–M state transition)	BR (wildtype and modified chromophore)	[40–42]
Reversible hologram Spatial light modulator Optical pattern recognition Dynamic time averaging interferometry	photochromic effect (ground state–M state transition)	BR (genetic mutants)	[43–50]
Optical associative memory	photochromic effect (ground state-M state transition)	BR (wild type)	[33, 51, 52]
3D random access optical memory	photochromic effect (two-photon absorptivity)	BR (wild type)	[33, 52–54]
2D random access optical memory Artificial retina	AC photoelectric effect (reversal potential)	BR (wild type)	[34]
Motion detector sensor array	AC photoelectric effect	BR (wild type)	[55, 56]
Artificial neural network	AC photoelectric effect (reversal potential)	BR (wild type)	[57]
Photodiode	AC photoelectric effect	BR (wild type)	[58, 59]
ISFET	DC photoelectric effect	BR (wild type)	[60]
ISFET	DC photoelectric effect	Halorhodopsin (wild type)	[61]

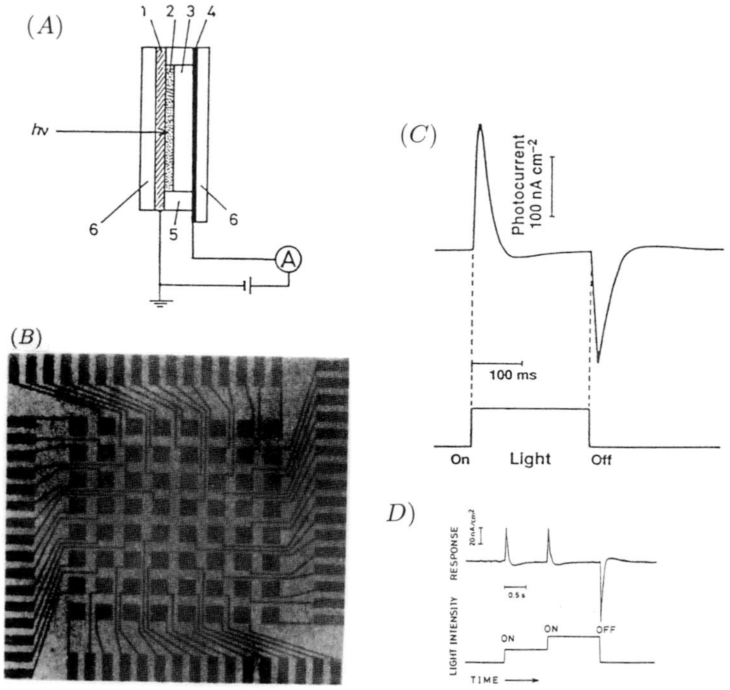

Figure 10.7 A motion-sensitive sensor array based on the AC photoelectric effect of bacteriorhodopsin. (*A*) A cross-section of a photocell, formed by immobilizing BR on a transparent electrode: 1, SnO_2 transparent conductive layer; 2, purple membrane LB film (typically six to 10 layers); 3, aqueous electrolyte gel layer (200 μm thick); 4, Au layer ($\sim$ 1000 Å) as counter-electrode; 5, Teflon ring space; 6, glass substrate. (*B*) An ITO (indium tin oxide) electrode patterned with 64 pixels used for image-sensing. Pixels of ITO (2.5 mm × 2.5 mm, 1000 Å thick transparent layer) are two dimensionally arrayed on a glass plate; each pixel has a separate wire leading to the four edge terminals along the sides for interfacing with an amplifier circuit. (*C*) The photocurrent generated in each pixel by a square light pulse of about 200 ms duration. The positive spike of the 'on' response and the negative spike of the 'off' response are characteristic of a response of a high-pass filter to a long rectangular pulse of applied current. (*D*) Photoresponses to step illumination showing the differential responsivity. The amplitude of the photosignal transient is proportional to the magnitude of step illumination. ((*A*)–(*C*) reproduced from [55]; (*D*) from [56].)

that it is also a photodiode is not evident. Using a null-current method, we have investigated the photoconductivity of an oriented BR membrane [62]. We found no rectification of the photoconductivity as exhibited by a silicon photodiode. Instead, we found that the proton conductivity of BR appears only during illumination and disappears upon termination of illumination. This latter feature turns out to be crucial for preventing photoinduced charge separation from recombining *in the dark* and thus alleviating a premature dissipation of the converted light energy—Nature's design for efficient photon energy converters that are superior to a silicon photodiode [62].

Two additional prototypes utilized the AC photoelectric effect: an artificial retina developed by Birge's group [34] and an artificial neural network developed by Lewis' group [57]. It is worth noting that both designs exploit a phenomenon known as the reversal potential in vision literature [63]. Bacteriorhodopsin, which was converted by a prior light pulse to the blue absorbing M state, responds to a second light pulse by generating an AC photoelectric signal with a reverse polarity. Just as the photochromic property of BR allows for a bistable optical transition, the phenomenon of reversal potentials allows for a bistable photoelectric transition.

From the above discussion, it is evident that BR is a bifunctional electronic material [64]: it is sensitive to light as well as to ions such as H^+, Cl^-, and Ca^{2+}. In the motion detector developed by Miyasaka *et al*, BR is configured as a photon sensor. In the cyclic-GMP cascade, a photon, via its action on rhodopsin, triggers the hydrolysis of cyclic-GMP, and thus in turn regulates the release of energy stored as a Na^+ gradient. The hypothetical trigger mechanism based on the surface potential thus works like a field effect transistor (FET), or more precisely, a phototransistor.

Alternatively, BR can be explored for its ion sensitivity. Tanabe *et al* [60] configured BR as an ISFET (ion-sensitive field effect transistor) for sensing of H^+. More recently, Seki *et al* [61] configured halorhodopsin, the second most abundant pigment of *Halobacterium halobium* and a Cl^- ion pump, as an ISFET sensor for detection of Cl^- concentration and demonstrated a linear relationship of the gate output voltage and the Cl^- concentration. The operation of these ISFET sensors depend on light for their response, since no ion sensitivity can be demonstrated in the absence of light. In other words, these sensors are addressable with light. It is thus possible to implement 'multiplexing' in a sensor array by sequentially interrogating individual sensors with light. In these prototypes, the sensors are addressed with steady light. We suggest here the use of pulsed light to interrogate these sensors so as to gain the advantage of speed as well as sensitivity: the AC photoelectric effect of BR and halorhodopsin possesses a picosecond rise time and the AC signal amplitudes are at least two orders of magnitude greater than the DC photosignal.

10.6 BIOELECTRONIC INTERFACES

In the example just cited, the photoelectric signal is capacitance coupled (AC coupled) to the metal electrode. The system possesses no interfacing problem. However, a number of investigators have observed that no DC photoelectric current can be observed from BR-coated metal electrode sensor systems. In

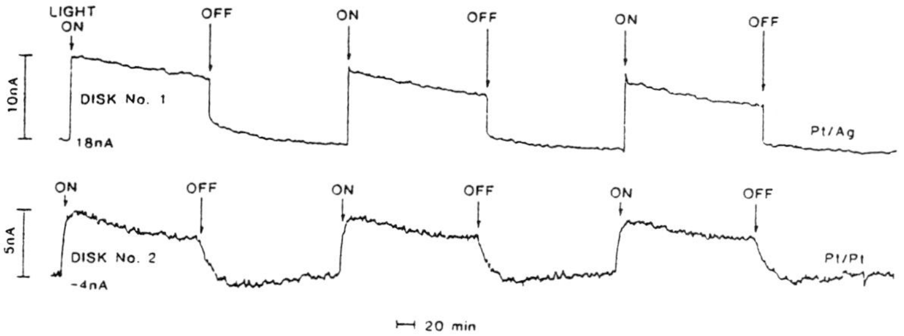

Figure 10.8 DC photocurrent measured from entrapped platinized chloroplasts. Colloid platinum was precipitated onto the external surface of photosynthetic membranes entrapped on KCl-impregnated fiber glass filter paper. The chloroplasts randomly oriented but only those with correct orientation exhibited the DC photoelectric effect. (Reproduced from [65].)

fact, no DC current should become observable in this system since there are no appropriate electrodic reactions available in the system to convert a current carried by protons in the purple membrane to a current carried by electrons in the conduction band of the metal electrode. With the exception of some redox proteins, this interfacing problem will arise in systems that generate DC photocurrents. Even with redox proteins, interfacing that permits the DC current to go through the electrode requires special bridging molecules that link the protein to the metal electrode.

Greenbaum [65] has studied the photoelectric signal in isolated thylakoid (chloroplast) membranes containing photosystem I. Photosystem II transfers electrons from plastocyanine at the intrastromal space across the membrane to $NADP^+$–ferredoxin–oxidoreductase. The net result is an interfacial proton binding, which is similar to the event at the cytoplasmic surface of the purple membrane. Charge movements start as electron transfers but are converted into proton movements in the opposite direction by the enzyme oxidoreductase. A DC photocurrent generated in the system was made observable by using a special preparation of thylakoid membrane in which colloidal platinum is directly deposited on the reducing end (stromal surface) of photosystem I. Apparently, electrons were intercepted by the colloidal platinum before they were processed by the enzyme (figure 10.8).

Aizawa has studied an enzyme (metal) electrode based on immobilized GOD, which is a redox enzyme [66]. Interfacing in this type of biosensor requires facilitation of electron transfer from the active site of the enzyme protein to the metal electrode. Various methods have been used, such as electron mediators, electron promoters, and molecular wires, to accomplish this purpose. For example, a conducting polymer, polypyrrole, formed by electropolymerization on the electrode surface, was used to form electron wires which can facilitate electron transfers to and from the metal electrode.

Interfacing technology developed for redox enzymes is unsuitable for detecting the DC photocurrent generated by the purple membrane primarily because BR is not a redox protein. A solution for this problem has been proposed on the basis of reverse engineering of the photoreactions in the photosynthetic reaction centers of the thylakoid membrane [36]. In chlorophyll-based photosynthetic systems, the primary reaction is a charge separation involving the production of electrons and holes (electron vacancies). Electron transfers dominate the process of energy conversion except at the membrane–water interfaces, where the current carrier is switched from electrons to protons. The net result is a transmembrane proton transport in the opposite direction. This conversion is implemented with the intervention of a special class of mediators known as quinoid compounds. They form the cofactors for some redox enzymes: $NADP^+$ (nicotinamide adenine dinucleotide phosphate) and NAD^+ (nicotinamide adenine dinucleotide) with oxidoreductase in the thylakoid membrane and mitochondrial inner membrane, respectively. Two additional quinoid compounds, plastoquinone and ubiquinone, serve as membrane-bound mobile electron carriers in the thylakoid and the mitochondrial inner membrane, respectively. In fact, plastoquinone and other quinoid compounds have a dual nature. They are both redox and acid/base compounds. Plastoquinone serves concurrently as a proton carrier and an electron carrier inside the thylakoid membrane and is crucial to the conversion of vectorial electron transport into vectorial proton transport in the opposite direction. Therefore, we have suggested a solution for DC current interfacing of an oriented purple membrane thin film to a metal electrode by choosing a suitable quinoid compound to make the required conversion.

10.7 IMMOBILIZATION OF PROTEIN: THE IMPORTANCE OF MEMBRANE FLUIDITY

As shown in the example of GOD-based enzyme electrodes, immobilization of proteins on a solid electrode surface is a key step in constructing a biosensor. Many approaches have been used for such immobilization. One of the oldest methods makes use of the self-assembling property of a class of organic compounds known as amphiphiles. Typical examples include soap and phospholipids. The latter compounds are the main ingredients of biological membranes. Amphiphiles contain both hydrophilic and hydrophobic parts at two separate ends of the molecules. Amphiphiles can be oriented at the air–water interface using the Langmuir–Blodgett (LB) technique. After appropriate compression which reduces the area occupied by these molecules, the thin and compact film (LB film) can be deposited onto a solid surface to form single or multiple layers of supported organic thin films [8, 9]. Thin films can be deposited on a solid support by electropolymerization, or by covalent binding using techniques such as silinization. Since a glass or quartz surface contains silicon hydroxide bonds, many silane compounds which are themselves amphiphiles can be covalently bound to the silicon atoms [67]. Functional organic thin films can then be deposited on top of the silane film via hydrophobic interactions. The silane compounds can be further functionalized with specific reactive groups that allow for coupling with the matching functional group of organic thin films to be deposited. For example, the coupling can be an immune reaction involving monoclonal antibodies and matching antigens. Another popular method makes use of specific binding between avidin and biotin [68]. Proteins are readily biotinized, as is a solid surface. The coupling of two biotinized films can then be implemented by avidin, which has four binding sites for biotin. (See Chapter 6 of this text for an overview of immobilization methods.)

Immobilization can also make use of phospholipid bilayers instead of the LB monolayers. A biological membrane *in vivo* consists of two phospholipid layers in which the hydrophobic hydrocarbon tails are facing each other and are buried inside the membrane whereas the hydrophilic polar head groups are exposed and facing the aqueous phases. In the early 1960s, Mueller *et al* [69] revolutionalized the field of membrane biophysics by introducing the technique for forming artificial lipid bilayers *in vitro*. Modifications that allow for protein incorporation have been made by other investigators [70, 71]. However, phospholipid bilayers are notorious for their lack of mechanical stability. To overcome this problem, Fare *et al* [72] used a photopolymerizable phospholipid to form bilayers which were then cross-linked by photopolymerization of the hydrocarbon chains. Tien *et al* [73] formed phospholipid bilayer membranes on a nascent surface of platinum to enhance stability. This latter system is suitable for immobilizing proteins for generating AC electric signals but not for generating DC electric signals.

In considering these efforts for immobilization, reverse engineering raises a basic question. If investigators need to find ways of immobilizing proteins on a solid surface and of stabilizing phospholipid bilayer membranes, why did Nature choose to utilize a fluidy and fragile membrane to support many molecular machines (membrane-bound proteins)? The answer lies in the fluidity of biological membranes (see [7] for discussion). For example, in the thylakoid membrane of green plants, the membrane fluidity is vital to the function of plastoquinone which shuttles electrons from photosystem II to photosystem I. In fact, the portion of the thylakoid membranes containing plastoquinones (known as the non-appressed region) has a greater fluidity than the appressed region. Fluidity is also crucial to light-induced regulation of coupling between photosystem I and photosystem II. The two systems are physically pushed apart as a result of photophosphorylation of the light harvesting protein complex II (which also serves as the antenna pigment protein for funneling absorbed photon energy into the reaction center pigment–protein complex). It appears that immobilization is merely a tentative measure in the development of biomolecular electronics. It is anticipated that membrane fluidity may be utilized in future designs of molecular devices, whereas the stability problem may be solved by alternative techniques.

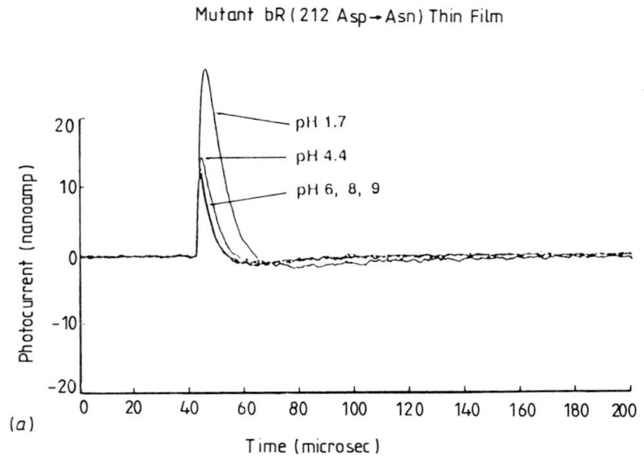

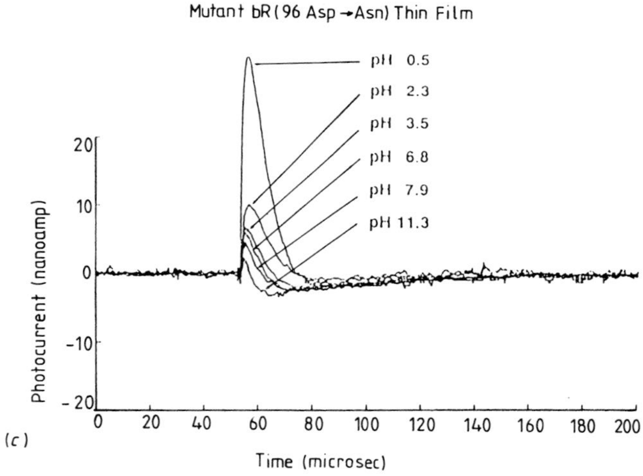

Figure 10.9 pH dependence of photosignals from TM films reconstituted from various mutant bacteriorhodopsins. (*A*) D212N; (*B*) D115N; (*C*) D96N; (*D*) D85N. (Reproduced from [80].)

Critics of biomolecular electronics often cite instability of biomaterials as a fatal flaw. However, not all biomaterials are fragile. For example, BR is notable for its legendary stability. It can withstand adverse conditions such as temperatures as high as 80 °C, and acidic conditions close to pH 0 for a limited period [13]. Furthermore, when configured as thin films of oriented BR, stability is further enhanced. Shen *et al* [74] found that BR thin films remain functionally active at 140 °C. We found that a thin film constructed from a mutant D212N

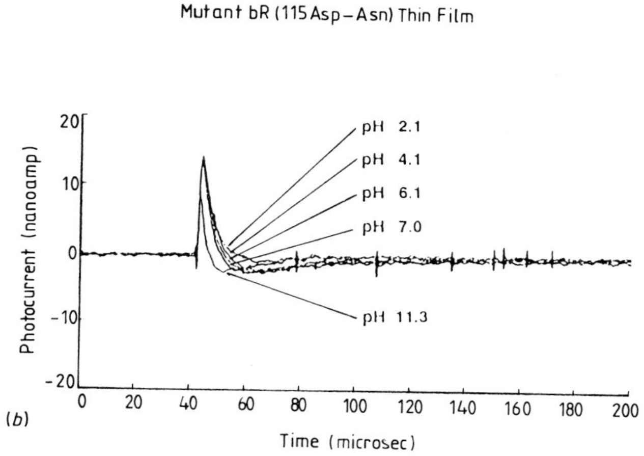

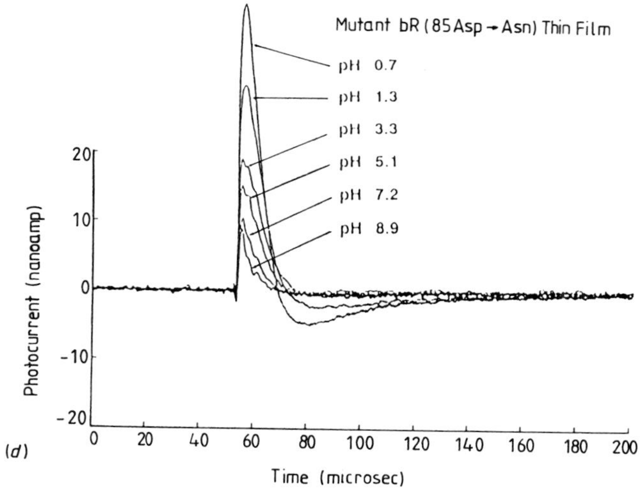

Figure 10.9 (*Continued.*)

continues to exhibit the photoelectric signal for nine consecutive days at pH 0.9 [75].

The stability of native proteins can also be enhanced by genetic engineering. Oshima has constructed a chimeric protein by fusing an enzyme with proteins from a thermophilic bacterium [76]. The catalytic activity is slightly diminished but the chemical stability is vastly enhanced.

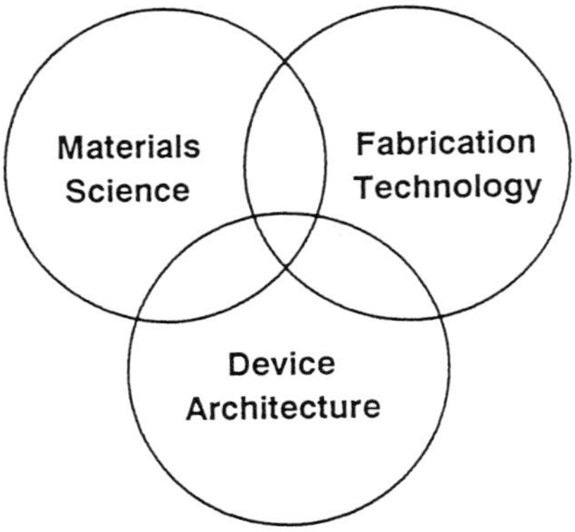

Figure 10.10 Three overlapping areas of biomolecular electronics research. (Reproduced from [81].)

10.8 THE CONCEPT OF INTELLIGENT MATERIALS

The ability of proteins to undergo conformational changes offers considerable advantages in functional applications. Among them, proteins perform fairly elaborate functions with sophisticated self-regulation. For example, the oxygen carrying protein hemoglobin has a sigmoid-shaped oxygen binding curve. The binding 'constant' actually increases with increasing oxygen concentration within most of its operating range. Such a peculiar property is made possible by interactions between its four subunits. In addition, the oxygen binding curve shifts with pH in such a way as to allow hemoglobin to bind more oxygen in the lung and to release more oxygen in the tissue than otherwise possible with a pH-independent binding curve. Such a sophisticated self-regulatory process is usually referred to as a 'cooperative effect', which is unique to molecules capable of conformational changes. From the point of view of information processing, such behavior has the flavor of parallel distributed processing without central intervention from the brain. The difference between a functional material with and without this additional self-regulatory ability thus is similar to the difference between a 'dumb' printer and a 'smart' printer, in digital computer terminology. Thus, biomolecular electronics entertains the idea of 'smart' or intelligent materials. The concept of smart/intelligent material has its origin in (non-biological) materials science. Initially, it was touted as an effort to embed functional materials (such as sensors and actuators) in structural materials [77]. Bioelectronics adopted this idea and further enhanced it. For example, the definition for bioelectronics advocated by Japan's Science and Technology Agency requires the sensor, the actuator and the processor elements to be packaged into a single molecule [78]. Furthermore, an intelligent molecule can adjust its functional properties in response to environmental changes. We have used BR as an example of an intelligent molecule, functionally equivalent to the more complex supramolecular assembly constituting the photosynthetic reaction center [79].

Biologists attribute molecular intelligence to the process of evolution. However, molecular intelligence might better be evaluated with intended function in mind since Nature may have optimized a particular molecular function on the basis of an entirely different set of criteria. Consider BR as an example. The photonic properties of BR are a byproduct of evolution rather than a primary feature essential for survival of *Halobacteria*. Before these properties can be optimally exploited for construction of a photonic device, it is often necessary to modify the structure of BR. For example, Vsevolodov and Dyukova utilized BR with a synthetic chromophore (5-keto BR) [41] and Hampp *et al* [43] utilized a genetic mutant (D96N) of BR. Modification of photoelectric properties of BR also appears to be feasible with site-directed mutagenesis [80] (figure 10.9). However, the changes of photoelectric properties caused by point mutation are often unpredictable and become known only after experimental tests. While molecular dynamics simulation can be applied to the design of small molecules and is extensively used in drug designs, it is still trial and error for the design of macromolecules such as proteins. The advent of site-directed mutagenesis and techniques such as random mutation has alleviated the problem by making the approach less time consuming and more efficient.

10.9 CONCLUDING SUMMARY AND FUTURE PERSPECTIVE

The electrical properties of devices constructed with biomaterials follow the same physical laws as do conventional microelectronic devices and thus equivalent-circuit analysis can be used as a design tool for bioelectronics. What sets biomolecular (and synthetic molecular) electronic devices apart from conventional electronic devices is the exploitation of the vast repertoire of organic and biological chemical reactions and the fact that molecular devices can be constructed with a dimension in the nanometer scale (nanotechnology and nanobiology).

Biomolecular electronics is a thriving discipline [81]. It is the culmination of the advances made in the past few decades in neuroscience, immunology, molecular biology, and computer science. Biomolecular electronics is still in its infancy, but with the convergence of various disciplines, a synergistic effect is beginning to emerge. We characterize the research and development of biomolecular electronics (or, more generally, molecular electronics) with three overlapping endeavors: materials science, fabrication technology, and device architecture (figure 10.10). The synergism between, for example, materials science and fabrication technology is exemplified by many studies in which immobilizing proteins on a solid metal electrode surface is the key to the study of biomaterial properties. On the other hand, new biomaterial capabilities will create the demand for new fabrication techniques that can fully exploit their novel properties. While biology-like device architecture projects a promising vision for the development of intelligent molecular machines, it is the research in materials science and fabrication technology that has advanced biomolecular electronics the most in recent years. At the present time, it is still too difficult to implement novel device architectures in the construction of molecular devices. Nevertheless, the implementation of novel device architectures with conventional hardware and software has led to tremendous growth in neural network research. The ultimate goal should be the molecular implementation of neural networks, but this requires removing the bottleneck of fabrication technology. This requirement is being addressed by the advent of such techniques as scanning tunneling microscopy and atomic force microscopy, which allow for characterization of molecular devices at the nanometer scale and for manipulation of single atoms or molecules. A radical approach for manipulating molecules and for constructing highly organized supramolecular structures may depend on the inherent property of molecules to form interlocking self-organizing self-assembling systems [8, 9].

ACKNOWLEDGMENTS

This research was supported by a contract from the Office of Naval Research (N00014-87-0047), a contract from the Naval Surface Warfare Center (N60921-91-M-G761), and a research stimulation fund from Wayne State University. The author also thanks his collaborators, Janos Lanyi, Lowell McCoy, Mauricio Montal, and Richard Needleman. The experimental work that forms the basis of this chapter was performed by Man Chang, Albert Duschle, Brian Fuller, Filbert Hong, Sherie Michaile, Baofu Ni, Ting Okajima, and Michelle Petrak.

REFERENCES

[1] Kaminuma T and Matsumoto G (eds) 1991 *Biocomputers* (London: Chapman and Hall)
[2] Conrad M (ed) 1992 Special issue on molecular computing *IEEE Comput.* **25** 6–67
[3] Hong F T 1995 Biomolecular computing *Encyclopedia of Molecular Biology and Biotechnology* ed R A Meyers (VCH) pp 194–7
[4] Conrad M 1972 Information processing in molecular systems *BioSystems (Curr. Mod. Biol.)* **5** 1–14
[5] Conrad M 1990 Molecular computing *Advances in Computers* vol 31 ed M C Yovits (New York: Academic) pp 235–324
[6] Conrad M 1992 Molecular computing: the lock–key paradigm *IEEE Comput.* **25** 11–20
[7] Hong F T 1992 Do biomolecules process information differently than synthetic organic molecules? *BioSystems* **27** 189–94
[8] Kuhn H 1989 Forrest L Carter Lecture: Organized monolayers—building blocks in constructing supramolecular devices *Molecular Electronics: Biosensors and Biocomputers* ed F T Hong (New York: Plenum) pp 3–24

[9] Kuhn H 1994 Organized monolayer assemblies: their role in constructing supramolecular devices and in modeling evolution of early life *IEEE Eng. Med. Biol.* **13** 33–44

[10] Hong F T 1986 The bacteriorhodopsin model membrane system as a prototype molecular computing element *BioSystems* **19** 223–36

[11] Hong F T 1994 Retinal proteins in photovoltaic devices *Molecular and Biomolecular Electronics* (Advances in Chemistry Series 240) ed R R Birge (Washington, DC: American Chemical Society) pp 527–59

[12] Stoeckenius W 1976 The purple membrane of salt-loving bacteria *Sci. Am.* **234** 38–46

[13] Stoeckenius W 1989 Light energy transducing and signal transducing rhodopsin of *Halobacteria*, *Molecular Electronics: Biosensors and Biocomputers* ed F T Hong (New York: Plenum) pp 159–63

[14] Henderson R, Baldwin J M, Ceska T A, Zemlin F, Beckmann E and Downing K H 1990 Model for the structure of bacteriorhodopsin based on high-resolution electron cryo-microscopy *J. Mol. Biol.* **213** 899–929

[15] Mathies R A, Lin S W, Ames J B and Pollard W T 1991 From femtoseconds to biology: mechanism of bacteriorhodopsin's light-driven proton pump *Annu. Rev. Biophys. Biophys. Chem.* **20** 491–518

[16] Lanyi J K 1993 Proton translocation mechanism and energetics in the light-driven pump bacteriorhodopsin *Biochim. Biophys. Acta* **1183** 241–61

[17] Hong F T 1980 Displacement photocurrents in pigment-containing biomembranes: artificial and natural systems *Bioelectrochemistry: Ions, Surfaces, Membranes* (Advances in Chemistry Series 188) ed M Blank (Washington, DC: American Chemical Society) pp 211–37

[18] Hong F T 1994 Electrochemical processes in membranes that contain bacteriorhodopsin *Biomembrane Electrochemistry* (Advances in Chemistry Series 235) ed M Blank and I Vodyanoy (Washington, DC: American Chemical Society) pp 531–60

[19] Hong F T 1995 Interfacial photochemistry of retinal proteins *Prog. Surf. Sci.* at press

[20] Hong F T 1976 Charge transfer across pigmented bilayer lipid membrane and its interfaces *Photochem. Photobiol.* **24** 155–89

[21] Hong F T 1978 Mechanisms of generation of the early receptor potential revisited *Bioelectrochem. Bioenerget.* **5** 425–55

[22] Okajima T L and Hong F T 1986 Kinetic analysis of displacement photocurrents elicited in two types of bacteriorhodopsin model membranes *Biophys. J.* **50** 901–12

[23] Trissl H-W and Montal M 1977 Electrical demonstration of rapid light-induced conformational changes in bacteriorhodopsin *Nature* **266** 655–7

[24] Michaile S and Hong F T 1994 Component analysis of the fast photoelectric signal from model bacteriorhodopsin membranes: part I. Effect of multilayer stacking and prolonged drying *Bioelectrochem. Bioenerget.* **33** 135–42

[25] Okajima T L, Michaile S, McCoy L E and Hong F T 1994 Component analysis of the fast photoelectric signal from model bacteriorhodopsin membranes: part II. Effect of fluorescamine treatment *Bioelectrochem. Bioenerget.* **33** 143–9

[26] Hong F H, Chang M, Ni B, Needleman R B and Hong F T 1994 *Bioelectrochem. Bioenerget.* **33** 151–8

[27] Hong F H and Hong F T 1995 Component analysis of the fast photoelectric signal from model bacteriorhodopsin membranes: part IV. A method for isolating the B2 component and the evidence for its polarity reversal at low pH *Bioelectrochem. Bioenerget.* **37** 91–9

[28] Petrak M R and Hong F T 1995 Component analysis of the fast photoelectric signal from model bacteriorhodopsin membranes: part V. Effects of chloride ion transport blockers and cation chelators *Bioelectrochem. Bioenerget.* submitted

[29] Hong F T and Okajima T L 1987 Rapid light-induced charge displacements in bacteriorhodopsin membranes: an electrochemical and electrophysiological study *Biophysical Studies of Retinal Proteins* ed T G Ebrey, H Frauenfelder, B Honig and K Nakanishi (Urbana-Champaign, IL: University of Illinois Press) pp 188–98

[30] Hong F T 1987 Effect of local reaction conditions on heterogeneous reactions in bacteriorhodopsin membrane: an electrochemical view *J. Electrochem. Soc.* **134** 3044–52

[31] Drachev L A, Kaulen A D and Skulachev V P 1978 Time resolution of the intermediate steps in the bacteriorhodopsin-linked electrogenesis *FEBS Lett.* **87** 161–7

[32] Birge R R 1990 Nature of the primary photochemical events in rhodopsin and bacteriorhodopsin *Biochim. Biophys. Acta* **1016** 293–327

[33] Birge R R 1992 Protein-based optical computing and memories *IEEE Comput.* **25** 56–67

[34] Chen Z and Birge R R 1993 Protein-based artificial retinas *Trend. Biotechnol.* **11** 292–300

[35] Brown K T and Murakami M 1964 A new receptor potential of the monkey retina with no detectable latency *Nature* **201** 626–8

[36] Hong F T 1989 Relevance of light-induced charge displacements in molecular electronics: design principles at the supramoleuclar level *J. Mol. Electron.* **5** 163–85

[37] Stryer L 1986 Cyclic GMP cascade of vision *Annu. Rev. Neurosci.* **9** 87–119

[38] Cafiso D S and Hubbell W L 1980 Light-induced interfacial potentials in photoreceptor membrane *Biophys. J.* **30** 243–63

[39] Drain C M, Christensen B and Mauzerall D 1989 Photogating of ionic currents across a lipid bilayer *Proc. Natl Acad. Sci. USA* **86** 6959–62

[40] Vsevolodov N N, Druzhko A B and Djukova T V 1989 Actual possibilities of bacteriorhodopsin application *Molecular Electronics: Biosensors and Biocomputers* ed F T Hong (New York: Plenum) pp 381–4

[41] Vsevolodov N N and Dyukova T V 1994 Retinal-protein complexes as optoelectronic components *Trend. Biotechnol.* **12** 81–103

[42] Vsevolodov N N, Djukova T V and Druzhko A B 1989 Some methods for irreversible write-once recording in 'Biochrom' films *Proc. Annu. Int. Conf. IEEE Eng. Med. Biol. Soc.* **11** 1327

[43] Hampp N, Bräuchle C and Oesterhelt D 1990 Bacteriorhodopsin wildtype and variant aspartate 96 $\rightarrow$ asparagine as reversible holographic media *Biophys. J.* **58** 83–93

[44] Thoma R, Hampp N, Bräuchle C and Oesterhelt D 1991 Bacteriorhodopsin films as spatial light modulators for nonlinear-optical filtering *Opt. Lett.* **16** 651–3

[45] Oesterhelt D, Bräuchle C and Hampp N 1991 Bacteriorhodopsin: a biological material for information processing *Q. Rev. Biophys.* **24** 425–78

[46] Bräuchle C, Hampp N and Oesterhelt D 1991 Optical applications of bacteriorhodopsin and its mutated variants *Adv. Mater.* **3** 420–8

[47] Hampp N, Popp A, Bräuchle C and Oesterhelt D 1992 Diffraction efficiency of bacteriorhodopsin films for holography containing bacteriorhodopsin wildtype BR_{WT} and its variants BR_{D85E} and BR_{D96N} *J. Phys. Chem.* **96** 4679–85

[48] Hampp N, Thoma R, Oesterhelt D and Bräuchle C 1992 Biological photochrome bacteriorhodopsin and its genetic variant Asp 96 $\rightarrow$ Asn as media for optical

pattern recognition *Appl. Opt.* **31** 1834–41

[49] Renner T and Hampp N 1993 Bacteriorhodopsin-films for dynamic time average interferometry *Opt. Commun.* **96** 142–9

[50] Hampp N and Zeisel D 1994 Mutated bacteriorhodopsins *IEEE Eng. Med. Biol.* **13** 67–74

[51] Birge R R, Fleitz P A, Gross R B, Izgi J C, Lawrence A F, Stuart J A and Tallent J R Spatial light modulators and optical associative memories based on bacteriorhodopsin *Proc. Annu. Int. Conf. IEEE Eng. Med. Biol. Soc.* **12** 1788–9

[52] Birge R R 1995 Protein-based computers *Sci. Am.* **272** 90–5

[53] Birge R R, Gross R B, Masthay M B, Stuart J A, Tallent J R and Zhang C-F 1992 Nonlinear optical properties of bacteriorhodopsin and protein-based two-photon three-dimensional memories *Mol. Cryst. Liq. Cryst. Sci. Technol.* B **3** 133–47

[54] Birge R R 1994 Protein-based three-dimensional memory *Am. Sci.* **82** 348–55

[55] Miyasaka T, Koyama K and Itoh I, Quantum conversion and image detection by a bacteriorhodopsin-based artificial photoreceptor *Science* **255** 342–4

[56] Miyasaka T and Koyama K 1993 Image sensing and processing by a bacteriorhodopsin-based artificial photoreceptor *Appl. Opt.* **32** 6371–9

[57] Haronian D and Lewis A 1991 Elements of a unique bacteriorhodopsin neural network architecture *Appl. Opt.* **30** 597–608

[58] Trissl H-W 1987 Eine biologische Photodiode mit höchster Zeitauflösung *Optoelecktron. Mag.* **3** 105–7

[59] Rayfield G W 1994 Photodiodes based on bacteriorhodopsin *Molecular and Biomolecular Electronics* (Advances in Chemistry Series 240) ed R R Birge (Washington, DC: American Chemical Society) pp 561–75

[60] Tanabe K, Hikuma M, SooMi L, Iwasaki Y, Tamiya E and Karube I 1989 Photoresponse of a reconstituted membrane containing bacteriorhodopsin observed by using an ion-selective field effect transistor *J. Biotechnol.* **10** 127–34

[61] Seki A, Kubo I, Sasabe H and Tomioka H 1994 A new anion-sensitive biosensor using an ion-sensitive field effect transistor and a light-driven chloride pump, halorhodopsin *Appl. Biochem. Biotechnol.* **48** 205–11

[62] Fuller B E, Okajima T L and Hong F T 1995 Analysis of the d.c. photoelectric signal from model bacteriorhodopsin membranes: d.c. photoconductivity determination by means of the null current method and the effect of proton ionophores *Bioelectrochem. Bioenerget.* **37** 109–24

[63] Cone R A 1967 Early receptor potential: photoreversible charge displacement in rhodopsin *Science* **155** 1128–31

[64] Hong F T 1990 Pigment-containing membrane as a prototype biosensor *Proc. 1990 Int. Congr. on Membrane and Membrane Processes (Chicago, 1990)* (North American Membrane Society) pp 215–7

[65] Greenbaum E 1989 Biomolecular electronics: observation of oriented photocurrents by entrapped platinized chloroplasts *Bioelectrochem. Bioenerget.* **21** 171–7

[66] Aizawa M 1994 Molecular interfacing for protein molecular devices and neurodevices *IEEE Eng. Med. Biol.* **13** 94–102

[67] Weetall H H 1976 Covalent coupling methods for inorganic support materials *Methods Enzymol.* **44** 134–48

[68] Wilchek M and Bayer E A (eds) 1990 *Methods in Enzymology* vol 184 (New York: Academic)

[69] Mueller P, Rudin D O, Tien H T and Wescott W C 1962 Reconstitution of cell membrane structure *in vitro* and its transformation into an excitable system *Nature* **194** 979–80

[70] Takagi M, Azuma K and Kishimoto U 1965 A new method for the formation of bilayer membranes in aqueous solution *Annu. Rep. Biol. Works Fac. Sci. Osaka Univ.* **13** 107–10

[71] Montal M and Mueller P 1972 Formation of bimolecular membranes from lipoid monolayers and a study of their electrical properties *Proc. Natl Acad. Sci. USA* **69** 3561–6

[72] Fare T L, Singh A, Seib K D, Smuda J W, Ahl P L, Ligler F S and Schnur J M 1989 Incorporation of ion channels in polymerized membranes and fabrication of a biosensor *Molecular Electronics: Biosensors and Biocomputers* ed F T Hong (New York: Plenum) pp 305–15

[73] Tien H T, Salamon Z and Ottova A 1991 Lipid bilayer-based sensors and biomolecular electronics *Crit. Rev. Biomed. Eng.* **18** 323–40

[74] Shen Y, Safinya S R, Liang K S, Ruppert A F and Rothschild K J 1993 Stabilization of the membrane protein bacteriorhodopsin to 140 °C in two-dimensional films *Nature* **366** 48–50

[75] Hong F H, Chang M, Ni B, Needleman R B and Hong F T 1994 Genetically modified bacteriorhodopsin as a bioelectronic material *Biomolecular Materials by Design* ed M Alper, H Bayley, D Kaplan and M Navia *Mater. Res. Soc. Symp. Proc.* vol 330 (Pittsburgh, PA: Materials Research Society) pp 257–62

[76] Oshima T 1990 A chimera of 3-isopropylamalate dehydrogenase with a mesophile head and a thermophile tail *Protein Engineering* ed M Ikehara, T Oshima and K Titani (Tokyo: Japan Scientific and Springer) pp 127–32

[77] Ahmad I and Rogers C (eds) 1988 *Smart Materials, Structures and Mathematical Issues; US Army Research Office Workshop (Blacksburg, VA: 1988)*

[78] Takagi T (ed) 1989 *The Concept of Intelligent Materials and the Guidelines on R&D Promotion* (Tokyo: Government of Japan, Science and Technology Agency)

[79] Hong F T 1992 Intelligent materials and intelligent microstructures in photobiology *Nanobiology* **1** 39–60

[80] Hong F T, Hong F H, Needleman R B, Ni B and Chang M 1992 Modifying the photoelectric behavior of bacteriorhodopsin by site-directed mutagenesis: electrochemical and genetic engineering approaches to molecular devices *Molecular Electronics—Science and Technology* ed A Aviram (New York: American Institute of Physics) pp 204–17

[81] Hong F T 1994 Molecular electronics: science and technology for the future *IEEE Eng. Med. Biol.* **13** 25–32

11

Sensor and sensor array calibration

W Patrick Carey and Bruce R Kowalski

11.1 INTRODUCTION

Sensors play an extremely important role in routine analytical measurements. Sensors, both physical and chemical, are used in laboratories and as process monitors throughout industry. However, a critical aspect of any analysis, especially one involving sensors, is the calibration. Calibration involves the manipulation of signals in order to obtain information about a sample and provide a correlation for further studies. Principles for sensor calibration are independent of the type of sensor; there are several techniques available for every type of sensor signal. Understanding the sensor signal and choice of the proper calibration technique to use is as important as selection of the appropriate sensor. Additionally, the information gathering capability of sensor systems can be enhanced by knowledge of how the calibration techniques operate.

This chapter primarily deals with the issue of correlation between sensor signal and analyte concentration in chemical measurements, i.e. the creation of a model based on standards, and the estimation of unknown samples based on that model. Sensor calibration is dependent upon the type of sensor signal such as linear versus non-linear response and sensor format involving one or many sensors simultaneously. The advantages of moving from one sensor to several sensors and from several sensors to several sensors coupled with analyte concentration modulation can yield remarkable information about a sample [1]. Each level of sensor system complexity, when coupled with the proper analysis tools, creates an unique situation which yields information about each component in a mixture and potential interferences.

Analytical instrumentation in general can be defined by its output signal according to tensorial algebra. This algebraic approach characterizes different levels of data complexity and relates one level to another. A single output (scalar) is in the zero-order level. A vector output is in the first order and a matrix output is in the second order. Different mathematical advantages exist with each order

Table 11.1 Characteristics of multidimensional instrumentation.

	Data output		
Classification	zero order	first order	second order
Format	single datum	vector of data	matrix of data
Sensor types	ISE, QCM, SAW, single-λ spectroscopy	sensor arrays (QCM, SAW, ISE) multi-λ spectroscopy, mass spectrometry, chromatography	GC–MS excitation–emission–fluorescence
Methods	least squares, polynomial regression, non-linear regression	PLS, PCR, MARS, PPR, ANN	GRAM, TLD

as listed in table 11.1. This approach has recently been reviewed as a theory of analytical chemistry [2].

Calibration of individual zero-order sensors is usually based on standard least-squares fitting routines with which almost everyone is familiar. However, this type of calibration model may be in error when an interferent is present or if the response is non-linear, causing deviations from predicted values. Sensor arrays address the interfering component problem by calibrating it into the model. The sensor array (first-order sensor) uses several sensors to obtain an identifiable response pattern for each analyte in the sample. The resolution of each component only depends on the array's selectivity. Individual sensor selectivity no longer has to be 100%, but each sensor must have a differential selectivity so that each analyte's response pattern is different. The two-dimensional sensor array (second-order sensor) uses the response from a sensor array to measure the modulation of a sample's concentration. For example, if an analyte eluting from a chromatographic column is measured as a function of time, its concentration is modulated (as the band of analyte elutes), and a unique first-order array response pattern for each analyte can be obtained without knowing about other compounds in the sample. This chromatograph effectively removes the interference problem of detecting multiple analytes simultaneously.

The application of chemical sensors now and in the future will be oriented around real time monitoring of chemical and environmental processes [3, 4]. These processes are complex systems where a variety of chemicals are present. From a calibration point of view, sensor systems for these applications will either have to be fully selective or be able to handle multicomponent samples. Since the perfectly selective sensor is difficult to almost impossible to develop, sensors in the array or two-dimensional array format have certain advantages.

11.2 ZERO-ORDER SENSOR CALIBRATION (INDIVIDUAL SENSORS)

11.2.1 Linear sensors

As with most of the sensor development currently being pursued, the individual sensor (zero-order sensor) is the most common researched and developed sensor type. The zero order is defined as an element that gives one analytical datum for each sample. An example of a zero-order sensor is the pH electrode, which gives one response value in millivolts for a given sample. The advantage of the individual zero-order sensor is that signal gathering and processing is fairly easy and straightforward. Implementation of the sensor in remote areas is simplified.

If one examines the calibration techniques for correlating a single sensor response to a single analyte, several techniques are available. Linear techniques such as ordinary linear regression, linear regression based on multiple parameters such as polynomial fitting, and non-linear regression based on non-linear estimation of parameters are all commonly used for both modeling and prediction [5, 6]. Ordinary linear regression is the simplest and most recognized technique used. Its basis comes from the linear correlation of responses to analyte concentration by the simple function

$$y = \beta_1 x + \beta_0 + \varepsilon \tag{11.1}$$

where y is the vector of sensor responses and x is the vector of known analyte concentrations. The parameters of the regression β_1 and β_0 represent the slope of the correlation and the y axis intercept, respectively. ε is the error of the model or the information in y not represented by the expression $\beta_1 x + \beta_0$. This error is the deviation of the points about the regression line, and it is assumed that the mean error is zero with Gaussian distribution, $E(\varepsilon) = 0$ and $\mathrm{VAR}(\varepsilon) = \sigma^2$. An examination of ε characterizes the quality of the model. Does the magnitude of ε change from small to large x axis values, or is ε large or small when compared to y? The assumptions of this model are that the errors in the measurements are contained only in the independent variable y, and that the concentrations of the calibration samples, x, are known exactly. However, usually there is error in both variables. Standard samples with known concentrations for x still have some measurement and preparation error associated with them. The primary problems with the ordinary least-squares approach are that matrix effects (the component does not respond to the sensor but affects the response of other analytes) and interferences (background response) cause unknown errors in sensor responses. If one knows about these factors, they can be dealt with in a number of ways. For matrix effects, one should use standard addition techniques [7]. To correct for background interferences, prior sample preparation such as the use of separation columns or masking agents is desirable. Weighted regression is also used when error varies with magnitude of signal [8].

Non-linear sensor responses can be modeled using linear equations with the same assumptions as the ordinary least-squares approach. Using a polynomial to fit curves is better in most cases than using transformations to linearize the data. For example, the following equation can be used to estimate a curvilinear calibration curve:

$$y = \beta_n x^n + \cdots + b_2 x^2 + b_1 x + \beta_0 + \varepsilon \tag{11.2}$$

where the constants, β, are linear parameters and can be estimated in the same fashion as in equation (11.1). Polynomial regression tends to overfit data, and it has difficulty predicting future samples without solving a quadratic equation. However, polynomial regression is still a mathematical linear technique with linear parameters which simplify its solution.

An advantage of using regression models that are linear in the parameters is that confidence intervals can be estimated on the parameters and on the future predicted concentration values. Confidence values are a key aspect of calibration and almost never used. Being able to predict future values from a model depends on the quality of the original model. Simply stating the error, ε, above does not state how the lack of fit affects future use of the model. If one measures a contaminant in a ground water well with a sensor, the stated value should have upper and lower 90% or 95% confidence limits on the value so that a measure of the accuracy is known and that two different methods can be compared to see how well they are measuring the sample. When moving up in sensor order, as seen later in this chapter, confidence values on predicted values are difficult to calculate. This is an area that needs future theoretical developments.

When using linear calibration models, the calculations of signal to noise ratio, selectivity and sensitivity are easily obtained in accordance with IUPAC standard definitions. The calculation of confidence limits is also achievable using the following equation [5]:

$$\left.\begin{matrix} x_U \\ x_L \end{matrix}\right\} = \hat{x}_0 + \frac{(\hat{x}_0 - \bar{x})g \pm (ts/\beta_1)\left((\hat{x}_0 - \bar{x})^2/S_{xx} + (1-g)/n\right)^{1/2}}{1-g} \tag{11.3a}$$

where

$$g = \frac{t^2 s^2}{S_{xx}\beta_1^2}. \tag{11.3b}$$

In this equation the confidence interval (x_{upper}, x_{lower}) for a predicted value of x_0 is calculated. The parameter β_1 is the slope of the regression line, t is the Student t statistic, s is the residual mean square error, S_{xx} is the sum of squares of the x values from the mean of the calibration x values, and n is the total number of calibration samples. The values from these confidence limits place upper and lower bounds for the analyte concentration, x_0, at all points along the calibration range. The best confidence is found near the mean of the data, whereas lower confidence is found at the extremes.

11.2.2 Non-linear sensors

Non-linear regression with models that are non-linear in the parameters is often used with solid state sensors and correlations that are based on natural decay or exponential relationships. Sensors such as gas-sensitive metal oxide semiconductors and some humidity sensors are non-linear. Examples of models with non-linear parameters are as follows:

$$y = \exp(\theta_1 + \theta_2 t^2 + \varepsilon) \tag{11.4}$$

$$y = [\theta_1/(\theta_1 - \theta_2)][e^{-\theta_2 t} - e^{-\theta_1 t}] + \varepsilon \tag{11.5}$$

where the θ are the non-linear regression coefficients. The determination of θ is most often performed by iteratively (trial and error approach) seeking the best values and then examining the error in the model [5, 9]. Methods such as simplex optimization and steepest descent are used to search a surface contour for the best values of the regression coefficients; however, there may be many local minima and only one global minimum. Distinguishing between the two is difficult.

Non-linear regression can be approached using least-squares methods by formulating normal equations for the model and examining the derivatives of the normal equations for each parameter to be estimated. When the derivatives of the normal equations for ordinary linear regression are performed, the resulting equation is a function of y and x only. When the equations are non-linear, the derivatives also include the parameter θ. Solutions for the non-linear parameters are not unique, and they require iterative methods to find the best set of parameters. There are methods of obtaining estimates for the non-linear parameter-based techniques such as Taylor series expansions and Marquardt's compromise. These methods go beyond the scope of this chapter, but should be investigated by users of non-linear sensors.

When confronted with non-linearly responding chemical sensors, forms of linearization are exploited to make the regression easier. The most common is the technique of log transforms. If a non-linear equation can be transformed into a linear one, it is said to be intrinsically linear. Using the response characteristics of a metal oxide semiconductor gas sensor as an example, an examination of transformation, linear regression, and confidence intervals can be performed. The response equation for metal oxide sensors has been theoretically determined as the following [10]:

$$\frac{R}{R_0} = P_{O_2}^{-\beta}(1 + P_i^{n_i})^{\beta} \tag{11.6}$$

where R and R_0 are the response and initial resistance of the metal oxide, respectively, with the gas analyte partial pressure P_i. P_{O_2} is the partial pressure of oxygen and can be assumed to be constant, n is a parameter based on surface dissociation and reaction pathways of the analyte, and β is a surface energy term which is temperature dependent. In figure 11.1, experimental data of a

Table 11.2 Response, residual, and 95% confidence limit data for the TGS response in figures 11.1 and 11.2. Regression line, $y = -0.49 \log(x) + 0.50$. $S_{xx} = 3.72$ ppm^2; $g = 0.179$; mean of $x = 204$ ppm; $n = 39$; $s = 0.03$; t at 95% confidence = 0.68.

x (ppm)	y $\log(R/R_0)$	Residuals $\log(R/R_0)$	x_L (ppm)	x_U (ppm)
50	−0.40	0.05	31.9	78.4
100	−0.49	0.00	75.1	133.2
150	−0.55	−0.03	120.7	186.4
200	−0.59	−0.05	164.3	243.4
250	−0.64	−0.05	204.1	306.2
350	−0.72	−0.04	274.1	447.0
50	−0.35	0.01		
100	−0.48	−0.02		
150	−0.57	−0.01		
200	−0.64	0.00		
250	−0.67	−0.02		
300	−0.68	−0.04	240.4	374.3
350	−0.75	−0.01		
50	−0.35	0.02		
100	−0.47	−0.02		
200	−0.63	−0.01		
250	−0.67	−0.02		
300	−0.70	−0.03		
350	−0.74	−0.02		
50	−0.35	0.01		
100	−0.48	−0.02		
150	−0.59	0.01		
200	−0.64	0.01		
300	−0.71	−0.02		
350	−0.75	0.00		
50	−0.36	0.02		
150	−0.60	0.02		
200	−0.65	0.01		
250	−0.67	−0.02		
300	−0.72	−0.01		
350	−0.73	−0.03		
400	−0.82	0.04	305.7	523.4
450	−0.87	0.05	335.8	603.1
500	−0.87	0.04	364.7	685.5
450	−0.84	0.03		
500	−0.88	0.05		
400	−0.79	0.00		
450	−0.84	0.02		
500	−0.88	0.04		

TGS 825 sensor (Figaro Inc.) in the presence of toluene vapor is presented along with a regression model determined by linear least squares using a log

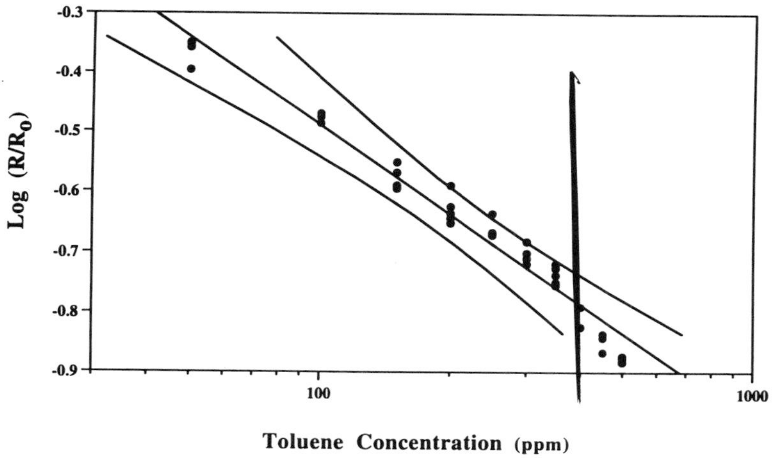

Figure 11.1 A log–log regression between the response of a Taguchi 825 sensor and toluene concentration. The 95% confidence limits for the prediction of unknown toluene concentrations are shown.

transformation of both the sensor response, R/R_0, and analyte concentration, P_i. Due to the parameter n in equation (11.6), the model is intrinsically non-linear. However, using log transformations of both x and y data yields a fairly linear correlation. There are some non-linear effects not modeled as the residuals in figure 11.2 reveal. The residuals are a good way of examining the regression model visually. The assumptions made in fitting a least-squares regression state that the residuals, ε, should be normally distributed around a mean of zero. In this example, there is a non-linear trend in the data which is not accounted for by the log transformation and the model. 95% confidence intervals are calculated based on equation (11.3). The upper and lower x value limits aid in determining the bounds for accuracy (figure 11.1). If a response value is obtained, the prediction of x is found by projecting the signal across the regression line. Where the signal intercepts the upper and lower bounds gives the 95% confidence limits for that estimation. The original data for this example are found in table 11.2.

One of the problems of transformations of the variables for calibration models is the relationship the transformations play in respect to calculating future concentrations. To estimate the concentration in an unknown sample, the logarithms of the responses are calculated, fitted into the regression model, and the antilogarithm of the resulting value taken to find the concentration datum. This procedure introduces error in the final prediction since the log–antilog procedure amplifies noise. Linear sensors are usually more statistically sound when trying to calibrate and predict values since less error amplification is imparted by the regression coefficients themselves. The alternative calibration procedures for non-linear sensors are transformations, polynomial estimates, or non-linear regression. All three have their drawbacks, with transformations the

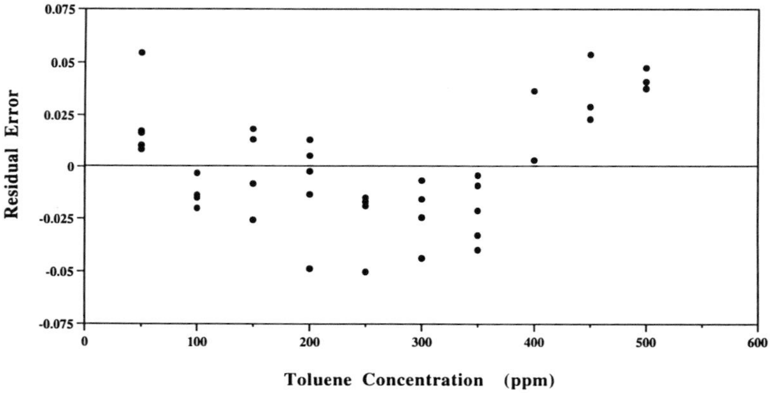

Figure 11.2 The residuals of the linear log–log model developed in figure 11.1 as a function of toluene concentration.

easiest of the three to implement if a linear correlation of the data can be obtained.

Another problem with the use of an individual sensor is selectivity. Most sensors currently under development are single sensors for a single analyte. The big question is what happens if there is an interferent present in the sample. From the above examination, the single sensor is powerful statistically speaking, but it has severe disadvantages if the sample is not strictly controlled. In the laboratory, samples can be prepared in a manner to separate out unwanted interferences that give rise to a signal, but the future application of sensors will be concentrated in real time monitoring of work spaces, industrial processes, and the environment. Development of new and novel sensors is necessary, but the amount of work being spent on making the 'perfectly' selective sensor compromises the impact that sensor would have if employed now. The ability of using arrays of partially selective sensors overcomes the disadvantage of selectivity. This leads to the world of the first-order sensor.

11.3 FIRST-ORDER SENSORS (SENSOR ARRAYS)

11.3.1 Introduction

Sensor arrays have several advantages over single sensors due to the higher degree of information carried in their signals. Arrays can carry information about several responding analytes simultaneously. This in turn allows the researcher to perform a variety of analyses on the sample. First of all, arrays can perform multicomponent qualitative studies using pattern recognition techniques. They also can perform multicomponent quantitative analysis using multivariate regression models. Lastly, they can detect interfering species which are known by calibrating them into the model, and unknown species by detecting them as outliers from the calibration models. Understanding how to properly calibrate sensor arrays requires a familiarity with linear algebra and geometry in higher-dimensional spaces.

The definition of the first order is that the instrument response gives a vector of data for each sample. This vector of data is a grouping of responses from several different or similar sensors. The most common example of a sensor array is in spectroscopy where a group of several wavelengths is used to probe a sample. The resulting spectrum gives a response that is unique for each component in the sample. The infrared spectrum, consisting of several hundred wavelengths, can identify specific characteristics of an organic compound and identify it from all others. Arrays of chemical sensors can be used in a similar fashion as long as each sensor gives some unique information about each component in a sample.

11.3.2 Qualitative analysis using sensor arrays

The linear algebra approaches used in first-order methods for pattern recognition are simple mathematical distance measurements in a multidimensional space. By using some examples of data plotted in a two-dimensional space resulting from an array of two sensors, simple relationships can be established that are identical in higher-dimensional spaces. In figure 11.3, the uppermost plot contains the responses of two sensors for pure samples with constant concentrations of three analytes denoted by circles, squares, and triangles. The two sensors have differential selectivity to the three analytes. The locations in this two-dimensional space of the analytes are physically separated from each other. The distribution of each cluster is caused by the combination of the measurement error of the two sensors.

To analyse a simple plot of this type in order to distinguish between 'circles' and 'squares', one would calculate an Euclidean distance measurement between clusters. Known samples would be a training set (calibration set) of samples in order to locate each cluster. An unknown sample can be identified based on its response to the two sensors and the relative distance to each of the known clusters. Already, with just the addition of one sensor, several analytes can be characterized using a bivariate plot.

A useful technique to explore multisensor data is principal components analysis (PCA) [11]. PCA performs an eigenvector analysis of the data and projects the samples into a new coordinate system set up by the eigenvectors. The first step is autoscaling the data (figure 11.3) by subtracting the sensor response mean and dividing by the sensor response standard deviation. This places the center of mass of all the samples at the (0, 0) coordinate. Scaling of each sensor is performed when different types of sensors are used such as temperature and voltage. Although each sensor has a different scale, it gives each sensor equal weight in the analysis. Next, the eigenvector analysis of the

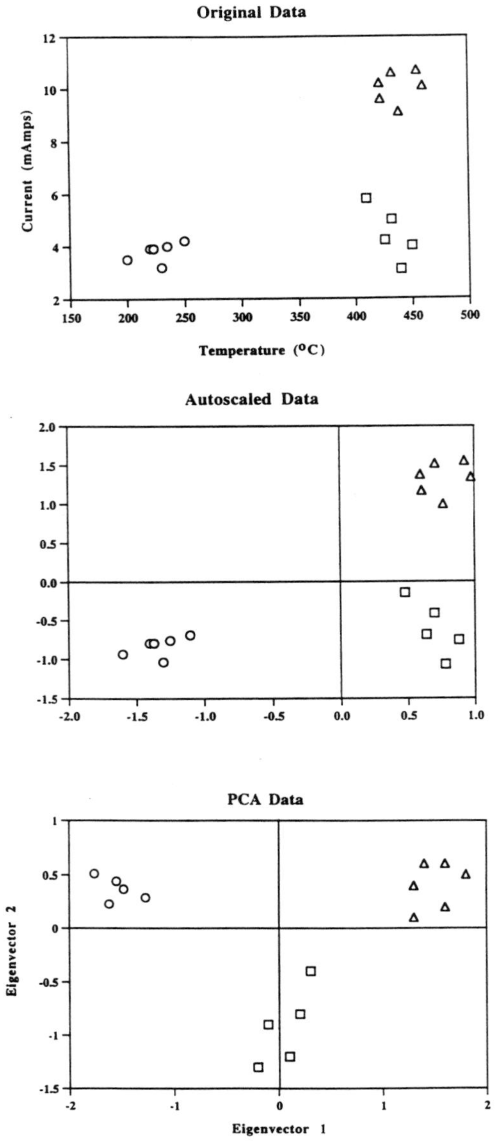

Figure 11.3 A bivariate plot of theoretical sample responses to a two-sensor array (top). The middle plot is the autoscaled data. The bottom plot is the sample data plotted in the PCA eigenvector space. As can be seen, the orientation of the data with respect to each other is maintained. Only scale and axis rotations occur with the PCA procedure.

data is performed and is shown as the bottom plot in figure 11.3. The definition of an eigenvector is that it represents the direction of maximum variance in the data. Therefore the first eigenvector is oriented with the maximum spread of the data. The second eigenvector is orthogonal to the first and represents the second direction of maximum variation.

The most useful feature of PCA is that one can examine data in a higher-dimensional space such as samples plotted as points in a ten-dimensional space if ten sensors are used. If the same three groups of samples above were plotted in this ten-dimensional space, a plane described by two eigenvectors could represent spatially how the samples were related. This two-dimensional space determined by the eigenvectors from PCA can be viewed and analysed. Other information can be obtained by comparing the angles between the eigenvectors and the original sensor response axis. The eigenvector analysis is a rotation procedure of the axis system. Finding which sensor is the most correlated with an eigenvector gives information on how well that sensor separates out information. In figure 11.3, we can observe that eigenvector 1 effectively separates the circles and triangles. The sensor with the smallest angle of rotation is the sensor that has the best selectivity between those two analytes. PCA in the form of pattern recognition has been used in many examples with arrays of chemical sensors for the detection and early warning of hazardous vapors [12–15].

The effects of changing concentrations as a sample progresses from pure A to pure B can be mapped through space. The trend can look linear or non-linear in this space depending on the data. Figure 11.4 shows an example of this type of analysis using an array of Taguchi gas sensors with two-component gas mixtures of methane and carbon monoxide [16]. The data are projected onto a three-axis space where the variation in sensor responses can be visually examined. In this example the progression of sample responses from one pure component to another is a non-linear curve in three dimensions.

11.3.3 Sensor array characterization

As described in the above paragraphs, the organization of an array needs to be an objective process with figures of merit describing each array and comparing two or more arrays in their ability to resolve different analytes. Array design is important in understanding how an array will function in a given application. One wants to maximize the resolving power of the array of sensors given a set number of analytes in the samples. Therefore figures of merit such as selectivity, sensitivity, limit of determination, and signal to noise ratios can be calculated for each array [17, 18]. An example of the selectivity of two arrays is shown in figure 11.5. The arrays consist of eight sensors with partial selectivity for two components. The top array has more discriminating power than the lower one. The mathematical term for this is collinearity. The higher the collinearity, the more similar the response patterns of the components to the array. The degree of collinearity is directly correlated with how well the array can resolve a mixture of two or more components for both identification and quantitation. The parameters of an array that collinearity influences are the figures of merit and the

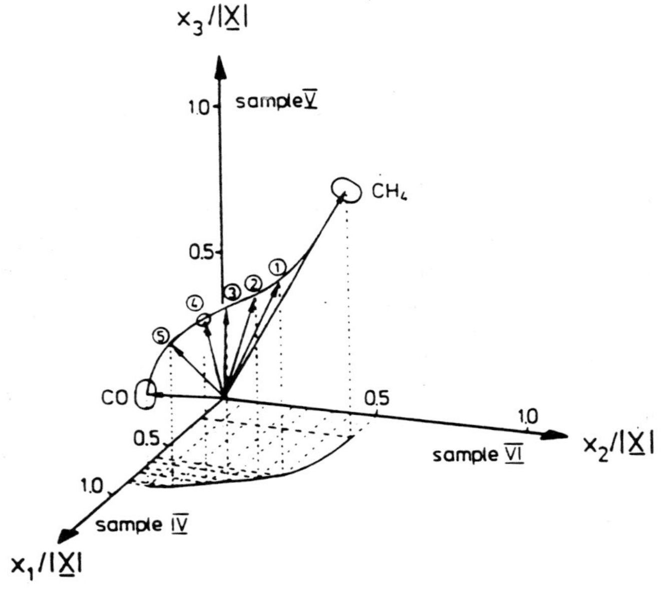

	p_{CO} (ppm)	p_{CH_4} (ppm)
①	50	8000
②	50	4000
③	100	4000
④	100	2000
⑤	200	1000

Figure 11.4 Plotted results from a multicomponent analysis to determine CO and CH_4 in humid air (50% RH) with three different SnO_2-based sensors. The three-dimensional non-linear trend is denoted. (Reprinted with permission from Elsevier Science SA from [16].)

propagation of error in calibration. As one might imagine, the use of perfectly selective sensors in the array would be ideal; the degree of collinearity would be extremely small. Therefore, the sensor array advantage is a compromise in allowing less selective sensors to perform the measurement and degrading the above figures of merit. To counteract the collinearity problem, more sensors can be added to the array in the hope of decreasing the collinearity. However, if more analytes are added, collinearity increases.

A geometric approach to looking at figures of merits for sensor arrays is presented in figure 11.6. If two analytes, A and B, are plotted as points resulting from the responses of two sensors (each sensor is a perpendicular axis), a vector can be placed to represent each response. (In figure 11.5, each analyte's response

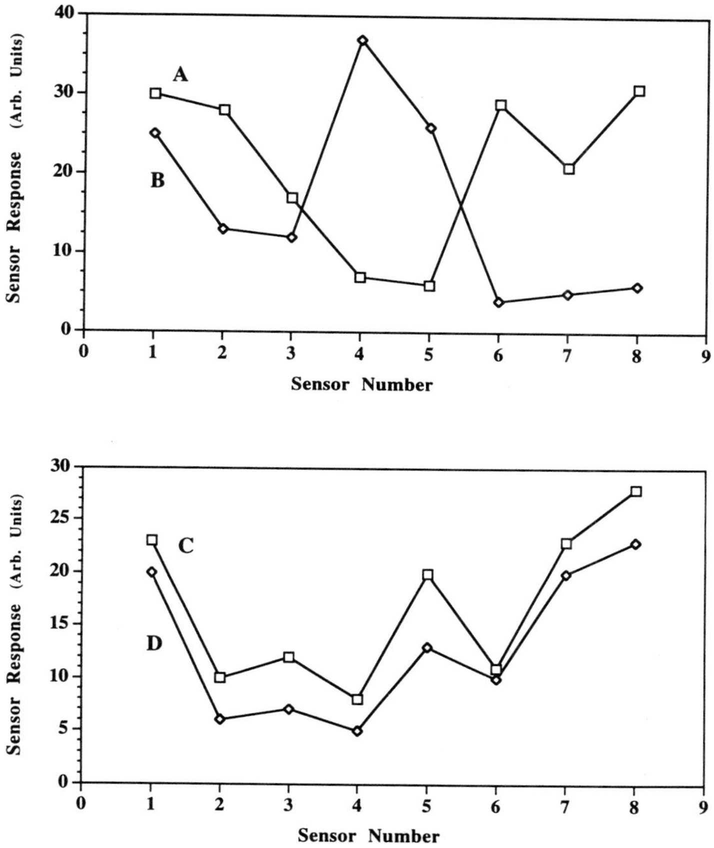

Figure 11.5 A hypothetical sensor array response patterns to two analytes, A and B and C and D. The top panel shows an array with good selectivity and low collinearity. The bottom plot represents an array with low selectivity and high collinearity.

pattern can be represented as a point in an eight-dimensional space.) If each sensor is linear in its response, all pure sample responses would fall on that vector; only the length of the vector would change with respect to concentration. Collinearity is a measure of the similarity of the response patterns of an array. In this example, collinearity would be the correlation of the two vectors. If each sensor were purely selective for one of the samples, A or B, then the two vectors would be at right angles, and the collinearity and correlation would be zero. Conversely, if the vectors were on top of each other this would result in high collinearity and correlation. In calibration, a regression coefficient must be obtained in order to translate signals into concentrations; in equation (11.1) it is the slope of the regression line. In multivariate regression, it is a vector of coefficients. This vector for analyte A can be obtained by projecting the response

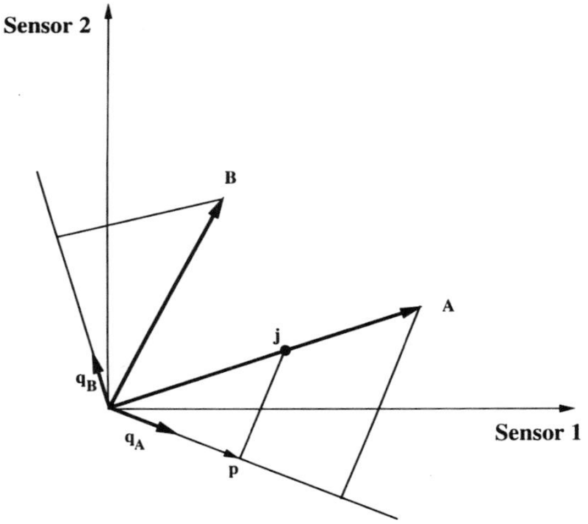

Figure 11.6 Response vectors of two samples, A and B, to an array of two sensors. The response vectors are projected onto orthogonal vectors called the net analyte signal, q_A and q_B.

vector onto a line which is perpendicular to the response vector for B. The length of the regression vector is inverted from the projection due to the inversion when solving the regression equation for the coefficients (see subsection 11.3.4). This new vector, $\boldsymbol{q}_A$, or the regression vector, is called the net analyte signal.

To calculate the figures of merit for an array to a given set of analytes, examination of the net analyte signal is performed. The net analyte signal is the unique portion of the array response pattern. To find figures of merit for the array, the net analyte signal for each component is examined with respect to the original response vector. The sensitivity of the array to component A, in the presence of B, would be the unique part of A's response divided by its concentration. Therefore a sample of known concentration of A would plot its response on figure 11.6 as the vector $\boldsymbol{j}$. Its projection onto the vector $\boldsymbol{q}_A$ would be the vector $\boldsymbol{p}$. Therefore $\boldsymbol{p}$ is the unique part of the response vector and is the net analyte signal. The sensitivity would then be calculated as:

$$\text{sensitivity}_A = \frac{\|\boldsymbol{p}\|}{C_A}. \tag{11.7}$$

$\|\boldsymbol{p}\|$ denotes the norm or length of the vector $\boldsymbol{p}$ and C_A is the concentration of A. The selectivity of the array for analyte A uses the same approach.

The selectivity is the ratio of the net analyte signal of A to the original response vector. Using the known vector $\boldsymbol{j}$ and its projection onto the net analyte signal, $\boldsymbol{p}$, one can calculate selectivity as:

$$\text{selectivity}_A = \frac{\|\boldsymbol{p}\|}{\|\boldsymbol{j}\|} \tag{11.8}$$

Table 11.3 Figures of merit for the response patterns found in figure 11.5.

	Sensitivity (with respect to unit concentration)	Selectivity
A	53.0	0.81
B	44.7	0.81
C	6.7	0.13
D	5.3	0.13

where this ratio varies from zero to one. The value of zero selectivity would denote that the response vectors A and B would be collinear. The projection of A onto an axis perpendicular to B would also be perpendicular to A. The length of the projection would be extremely small compared to the length of A. A value of one for selectivity would occur if the two response vectors were orthogonal. The net analyte signal is the same vector as the response vector. The projection of A would be itself and the ratio of the lengths would be one. The selectivity of an array for only two components is the same value for each component. That is easily seen in figure 11.6 since the angles between the net analyte signals and the response vectors for both A and B are the same. However, when more than two components are calibrated, the values can take on dissimilar values.

Table 11.3 provides the calculations of the figures of merit for the two examples given in figure 11.5. The degree of collinearity influences the magnitudes of both sensitivity and selectivity of each array.

Additional figures of merit, which can be calculated using the net analyte signal approach, are not as common in practice but can be useful. These two figures of merit are signal to noise ratio and limit of determination. To calculate these two parameters, an estimation of the noise of the array is needed. In single-sensor cases, the noise is best characterized by the standard deviation of the noise or variation. With an array there are k measurements if k sensors are used. Each sensor would have its own standard deviation associated with its noise or variability. The array error is estimated by summing the squares of the variabilities of each sensor and dividing by the appropriate degrees of freedom

$$\varepsilon = \left[\left(\sum_{i=1}^{k}(d_i^{obs} - d_i^{calc})^2\right) \Big/ (k-m)\right]^{1/2}. \tag{11.9}$$

In this equation m is the number of analytes and d is the sensor array response vector for a given sample, *obs*, and the mean values for replicate measurements, *calc*. By looking at the precision of array measurements with replicate samples, a good indication of the measurement noise can be obtained. Using this expression as a value of the error, the same formulations as in single-sensor cases apply.

The signal to noise ratio for a given response, j, is as follows:

$$\text{signal to noise}_A = \frac{\|\boldsymbol{p}\|}{\varepsilon}. \tag{11.10}$$

The limit of determination (LOD) is the minimum level at which a sensor array can detect an analyte in a mixture of other analytes which respond to the array. This differs from the limit of detection which is characterized for a single sensor for a pure analyte in a sample. The definition usually states that the minimum signal that can be accurately detected is three times the noise of the sensor. Therefore, for the single sensor, the limit of detection is three times the noise divided by the slope of the calibration line. In the array case with multiple analytes, the limit of determination is

$$\text{LOD}_A = \frac{3\varepsilon}{\text{sensitivity}_A}. \tag{11.11}$$

In this expression the array sensitivity is used to represent the influence of collinearity on the limit of determination.

As in the above discussion, the net analyte signal is an important concept in the characterization and the design of sensor arrays. Using the figures of merit derived from the net analyte signal, different arrays can be compared and evaluated as to the effectiveness of their qualitative and quantitative ability. The design of arrays can be optimized by using the concepts of the net analyte signal. The substitution of one or several sensors in a given array can make a big difference in improving the differential selectivity of the sensors in the array. Sensor array design or the choosing of which sensors to use for an array has been a small but active area of research in the past few years. Primarily, the approach must choose sensors that are stable, have minimum error, and are sensitive. However, choosing the sensors with the best differential selectivity is difficult when just mentally examining the various responses to a multiple set of analytes. PCA as discussed above can be used for this evaluation [19]. PCA can examine which sensors give the most differential response by calculating the correlations between the sensor responses and the orthogonal eigenvectors. The data matrix used in this PCA analysis is composed of the responses of each candidate sensor to all of the analytes that are going to be investigated. Choosing sensors that have the most orthogonal response components aids in both qualitative and quantitative analysis. The PCA method calculates angles between the eigenvectors and the original response vectors of each sensor. The best array will be composed of sensors that have high correlations with individual eigenvectors.

11.3.4 Quantitation with sensor arrays

Quantitative calibration for sensor arrays involves the same type of regression approach as found in the single-sensor case extended into multivariate statistics.

Linear algebra is used where the sensor and concentration data are assembled into vector and matrix forms. The regression coefficients for a sensor array are a matrix of coefficients, relationships between each sensor, and each analyte. Two types of modeling are currently used with sensor arrays involving multicomponent data and are based on the initial set-up of the regression equation. The first method is the classical method using the sensor array responses as the dependent variables and the analyte concentrations as the independent variables [20, 21].

$$\mathbf{R} = \mathbf{CK} + \mathbf{E} \tag{11.12}$$

where **R** is a matrix of sensor array responses where rows correspond to samples and columns correspond to each sensor of the array. **C** is the matrix of concentrations with rows matching the samples in **R**, and columns corresponding to analytes. To solve for **K**, the matrix of regression coefficients, the **C** matrix needs to be inverted and multiplied with **R**. The matrix **K** contains the regression vectors or the net analyte signals for each analyte. The generalized inverse can be used for this as follows:

$$\mathbf{K} = (\mathbf{C}^{\mathrm{T}}\mathbf{C})^{-1}\mathbf{C}^{\mathrm{T}}\mathbf{R}. \tag{11.13}$$

The generalized inverse is sensitive to noise during the inversion, but in this case the concentration data matrix is user defined and should not pose an inversion problem. This provides a good model of the system and is used commonly to examine pure component response patterns. The **K** matrix will contain the pure component response patterns for each analyte which is helpful for sensor array characterization and design. Spectroscopists use this technique to review true spectra of samples and identify features (often called the **K** matrix approach). This is a good modeling technique and has very good precision when predicting future samples. To predict unknown samples, another inversion is needed to invert the **K** matrix and multiply it against the **R** matrix. Since two matrix inversions are used, error propagation can be amplified if noise and collinearity in **R** exist. Successful use of this technique requires that all components of the system are known and calibrated. Any interfering component has to be quantified in the calibration. The number of sensors is not restricted and spectroscopists can have hundreds of wavelengths to use. In chemical sensor arrays, the number of sensors is more limited due to practical considerations.

The more common approach for sensor array calibration uses the direct regression model where the independent variables are the analyte concentrations and the dependent variables are the sensor responses.

$$\mathbf{C} = \mathbf{RK} + \mathbf{E}. \tag{11.14}$$

This rearrangement of the regression equation is advantageous for prediction of future samples since only one inversion is required. However, this inversion involves the response matrix. To solve for **K** the generalized inverse of **R** is used.

$$(\mathbf{R}^{\mathrm{T}}\mathbf{R})^{-1}\mathbf{R}^{\mathrm{T}}\mathbf{C} = \mathbf{K}. \qquad (11.15)$$

Using this approach, calibration can be performed knowing only the one component of interest in the system. Interfering compounds only have to be present, not quantified. They are implicitly modeled with this approach. The major implication of this technique is that application of the sensor array in remote environments is better facilitated. A restriction that this model imposes is that the number of sensors must be less than the number of calibration samples in order to perform the generalized inverse. This method usually has more error propagation due to the instability of the **R** matrix inversion. Collinearity plays an important role in this case.

The more collinear the sensor array response patterns, the more error during a matrix inversion. To solve this problem, several mathematical techniques have been used to reduce the collinearity of the **R** matrix. Stepwise regression finds only a couple of sensors out of the total set to use. This iterative process usually finds two or three sensors which are nearly orthogonal. This approach works if many sensors are used as in spectroscopy. A more common method involves using a better inversion technique. This is the basis of principal components regression (PCR). The matrix **R** is factor analysed using an eigenvector approach just as in PCA. The response data are projected onto the orthogonal eigenvectors, **V**, and the new matrix replaces the **R** matrix.

$$\mathbf{C} = (\mathbf{R}\mathbf{V}^{\mathrm{T}})\mathbf{K} + \mathbf{E}. \qquad (11.16)$$

This has several advantages, one of which is that the new matrix usually is of smaller dimension than the original **R** matrix. A sensor array of 10 sensors with collinearity can be projected into a five-dimensional eigenvector space while mainly noise is removed. This signal to noise enhancement aids in the calibration performance. The new matrix is orthogonal and is easily inverted. To predict future samples, the response of the sensor array has to be projected onto the same eigenvectors in order to solve for the **C** matrix.

Another technique which evolved from PCR is partial least-squares regression (PLS) [22, 23]. PLS performs the same type of function as PCR by creating an orthogonal matrix from the **R** matrix. However, the vectors on which the **R** data are projected are influenced by information from both **R** and **C**. The theory behind PLS is that a better correlation could be built between the new vectors and **C** if the direction of the new vectors were influenced by the concentration data. These new vectors, latent variables, contain more quantitative information than PCR vectors, especially when only a few variables are used.

Examples of using MLR and PCR in a hypothetical case using the response patterns in figure 11.6 are shown in table 11.4. The orientations of the concentration and response matrices are denoted as well as prediction results of an unknown sample. Noise was added to sensor responses only. The prediction

Table 11.4 Example matrix configurations used in multivariate calibration with response patterns A and B in figure 11.5.

Calibration Concentrations	Responses
$\mathbf{C} = \begin{pmatrix} 3 & 4 \\ 2 & 7 \\ 8 & 6 \end{pmatrix}$ (columns A, B)	$\mathbf{R} = \begin{pmatrix} 190 & 136 & 99 & 169 & 122 & 103 & 83 & 117 \\ 235 & 147 & 118 & 273 & 194 & 86 & 77 & 104 \\ 390 & 302 & 208 & 278 & 204 & 256 & 198 & 284 \end{pmatrix}$
Unknown sample Concentration (true)	**Measured array response**
$c_{unk} = (4 \quad 5)$ (columns A, B)	$r_{meas} = (245\ 177\ 128\ 213\ 154\ 136\ 109\ 154)$

Prediction results for examples using both response patterns sets A–B and C–D found in figure 11.5. Normal distribution of noise with standard deviation of 5.0 response units added. Prediction of the c_{unk} sample ($A = 4.0$, $B = 5.0$) and ($C = 4.0$, $D = 5, 0$) in both cases.

	Unit concentration (% error)			
Method	A	B	C	D
MLR	3.69 (7.8)	5.40 (8.0)	3.03 (24.3)	6.02 (20.4)
PCR[a]	4.00 (0.0)	5.05 (1.0)	3.81 (4.8)	5.36 (7.2)

[a] PCR uses three principal components in the model.

results show that collinearity and model type have an impact on prediction error. For the analytes A and B, there is low collinearity and the prediction results have low error. PCR does better at predicting than MLR because it filters the data by using only the relevant eigenvectors in the model (in this case three). When the collinearity is worst, as in analytes C and D, the low selectivity impacts both MLR and PCR. PCR is more useful in this case since the error reduction is more significant.

Some of the advantages of PCR and PLS are that the methods can handle the collinearity problem much better than classical approaches to multivariate regression and qualitative information can be obtained via PCA using the eigenvectors and latent variables. However, the use of multivariate regression techniques allows one additional advantage over the single-sensor case. That advantage is the ability to detect outliers and unknown interferences [24, 25]. In multivariate regression the samples are grouped into a defined space much like the clusters in figure 11.3. An outlier or unknown interference (which was not present during calibration) will place the sample outside the region of the calibration samples. By measuring the distance away from the calibration points, an estimation of its uniqueness can be made. This greatly improves the viability of array instruments in field analysis. Being able to detect these interferences enables the users not to report false signals and erroneous concentration predictions.

A variety of applications of chemical sensor arrays coupled with multivariate data analysis for quantitative measurements have been studied [26–32]. Each study investigated different types of sensors in the array, such as quartz crystals, ion-selective electrodes, metal oxide semiconductors, and chemFETs, with various types of modeling techniques as described above.

11.3.5 Quantitation with non-linear sensor arrays

One of the challenges of multivariate statistics is the ability to handle non-linear data. As mentioned above in the single-sensor case, non-linear responses are a little more difficult to handle since iterative techniques are often needed. In the multivariate case, using non-linear sensor arrays faces this same problem. The techniques used are iterative and are based more on approximation techniques such as spline fitting or transformation of the data into some type of linear-like form. Additionally, the figures of merit for non-linear sensors and arrays are more difficult to establish. Parameters such as sensitivity and selectivity are functions of the response magnitude and vary over the dynamic range of the sensor. For example, the selectivity of an array for two or more components may vary over the dynamic range of the analyte concentrations. Because of this variation, it is possible to have variable levels of accuracy corresponding to this selectivity change. The maximum accuracy for each component (if the number of analytes is greater than two) may not occur at the same concentration levels.

Calibration of non-linear sensor arrays has taken several approaches in the past several years using methods based on non-parametric regression. The techniques used most often for sensor array calibration are projection pursuit regression (PPR) [33, 34], multivariate adaptive regression splines (MARS) [35], and artificial neural networks (ANNs) [36–39]. Currently, research has been looking at combining two types of modeling procedure, placing PPR or PLS inside an ANN technique to help maximize correlations for future prediction. In this chapter, we will discuss only PPR and MARS.

PPR uses a projection technique similar to PCR and PLS which creates a smaller space into which the original response data are projected. However, these new vectors are found through an iterative technique since each newly formed vector is transformed using a non-linear transformation function, ϕ

$$\mathbf{C} = \phi(\mathbf{RA}^{\mathrm{T}})\mathbf{K} + \mathbf{E}. \tag{11.17}$$

To test the correlation, a smoothing function is fitted to the plot of each $\boldsymbol{c}$ vector versus each $\boldsymbol{r} \cdot \boldsymbol{a}^{\mathrm{T}}$ vector to minimize the squared error of $\boldsymbol{c}$ and the model's fit to $\boldsymbol{c}$. The transformation of the projected vectors is performed by a smoothing function that closely resembles a spline fit and smooth. The regression coefficients, **K**, model the response to the concentrations by a linear combination of the projected and transformed vectors. The one feature that PPR offers is that cross-terms of varying degree of the independent variables can be used. Correlations with sensor A times sensor B to the concentration data can be found.

MARS is a non-parametric technique which uses simple spline fitting to map the non-linear response regions and correlates it back to the concentration data. The spline fitting is a series of truncated, one-sided linear models which when placed together to approximate a complex function undergoes a smoothing function which removes all sharp corners. The smooth is a cubic approximation. Again, this method allows cross-terms of various degrees of the independent variables. The function for just a single independent variable, $\boldsymbol{x}$, is as follows:

$$f(\boldsymbol{x}) = \sum aB(\boldsymbol{x}) \tag{11.18a}$$

where

$$B(\boldsymbol{x}) = [s(\boldsymbol{x} - t)]_{+}. \tag{11.18b}$$

The spline function, $B(\boldsymbol{x})$, for each variable is obtained by placing a knot, t, and fitting a linear equation on each side of the knot location. In this equation, a is a constant and s is the slope of the linear fit. The plus sign denotes that only one-sided splines are considered by the algorithm. Giving the algorithm the maximum number of knots to use in the total model, it locates them at optimal locations in order to minimize the modeling error just as in any regression approach.

In [40] data have been presented using these non-parametric approaches compared to linear approaches using PLS and PCR. When the data are non-linear, the non-parametric approaches perform better at modeling the relationships between instrument response and concentration. PLS and some forms of PLS with non-linear terms substituted into the algorithm have been used to model non-linearities in response data. In spectroscopic experiments, non-linearity arises at high detection signals and when shifts in peaks occur. When these non-linearities are small compared to the entire data set, PLS can model them effectively by using a few more latent variables to represent them. However, when the entire data set of responses is non-linear as in semiconductor sensor arrays, the ability of PLS to model the non-linearity is encumbered by having to use all of the possible latent variables available. This negates all of the power of PLS, which is signal to noise enhancement by removing unneeded latent variables. When all of the latent variables are used, the regression model is no different from the multiple linear regression approaches as outlined in equation (11.14).

In the comparison of calibration methods, the results show that the non-parametric techniques decreased the regression error by approximately 50% over the parametric approaches. The sensor array used in these examples was constructed from eight Taguchi gas sensors. These sensors are tin dioxide-based semiconductors with dopants to increase selectivity to various vapors and gases. The array was used to analyse mixtures of organic solvents in two- and three-component situations. In the two-component cases, prediction errors for the PLS model were very large at the low-concentration samples (10–100 ppm of vapor) and much more improved at concentrations from 200 ppm to 500 ppm, the upper limit of the samples. The noise of semiconductor sensors is much larger at low concentrations than at high concentrations. PLS has a maximum of eight latent variables to use due to the eight sensors in the array. To model these sample sets, seven or eight latent variables are needed. The non-parametric approaches perform better since they have the ability to work harder at the lower concentration range and incorporate non-linear functions to model the sensor responses. Plots of predicted versus known concentration of each analyte in the two-component cases reflected similar prediction errors at all concentration levels of the analyte.

11.4 SECOND-ORDER CALIBRATION

11.4.1 An introduction to two-dimensional arrays

As outlined previously in this chapter, definite advantages arise when progressing from the single sensor to the sensor array. These advantages are understood and extracted by knowledge of the linear algebra and statistical approaches to regression. By examining the mathematics for the next level of complexity, the second order, even more advantages are evident. These advantages include the ability to quantitate a specific analyte without knowing any information on the rest of the sample. In addition, if an unknown interferent entered into a later prediction sample, it would not disrupt the quantitation of the analyte of interest. From these advantages, the implications for sensor applications is enormous. If a second-order sensor which monitored groundwater or industrial processes were available, it could easily be implemented into complex systems without the worry of knowing all of the interferences that might possibly come into contact with it. The crux of the problem is that second-order methods can resolve the pure response pattern of the analyte of interest in the presence of unknown interfering components.

This excitement about second-order sensor calibration has led to a search by chemometric researchers to find equivalent instrumentation that gives rise to second-order data. The definition of second-order instruments is slowly solidifying and currently a foundation has been established to classify which techniques are true second-order devices. This definition of second-order instruments is simply two sensor arrays which are independent of each other. However, in order for the arrays to be independent, one of the arrays must modulate the sample's analyte concentrations. The best known instrument currently is gas or liquid chromatography with an array detector such as mass spectrometry or light spectroscopy. The response for a single sample is a matrix of data where the columns may indicate the response of the array detector and the rows are the retention times of the chromatograph. The chromatograph modulates the analyte concentrations while the sensors monitor the sample at given time intervals. The term bilinear response matrix has been given to this type of data.

11.4.2 Second-order quantitation

There are three basic approaches to solving second-order sensing problems. The first method, called rank annihilation, was investigated in the 1970s as a technique to subtract a known response matrix of a pure component from a matrix of responses due to the component of interest and other interfering compounds [41]. With this technique, the known matrix of the pure component, which is of rank one, is subtracted from the complex matrix of rank greater than one. The rank of the complex matrix decreases as the amount (concentration) of the pure matrix is subtracted. When the rank drops one value, i.e. from five to four, the amount of the pure component has been completely removed and the concentration of that component is the amount subtracted. This is an iterative process which consumes a lot of time.

In the mid-1980s a new technique called rank annihilation factor analysis (RAFA and GRAM) was developed which performed this process non-iteratively [42, 43]. Now an exact solution can be estimated directly on the sample matrix if the pure component response matrix of the analyte of interest is known. The third technique uses a range of concentrations yielding several pure component data matrices. This approach, called trilinear decomposition (TLD), stabilizes the prediction of the unknown sample due to the range of known response matrices [44]. Figure 11.7 shows a diagram of the process in which the format of the data is assembled and the resulting information is obtained. The pure component data matrix is a two-dimensional structure in which the pure array response patterns in each dimension are resolved. The resolution of the original matrix is based on the fact that this matrix is a cross-product of two vectors (the two arrays).

The basis of RAFA is the direct eigenvalue solution of the mixture matrix of responses based on a singular-value decomposition. Let $\mathbf{N}_k$ represent a data matrix of the pure component response of analyte k from the two-dimensional sensor array and $\mathbf{M}$ represent the sample response of the same instrument to the analyte of interest and all interfering components; the solution of the rank annihilation problem can be stated as follows:

$$\mathbf{M} = \sum_k c_k \mathbf{N}_k \tag{11.19a}$$

where

$$\mathbf{N}_k = \boldsymbol{x}_k \cdot \boldsymbol{y}_k^{\mathrm{T}}. \tag{11.19b}$$

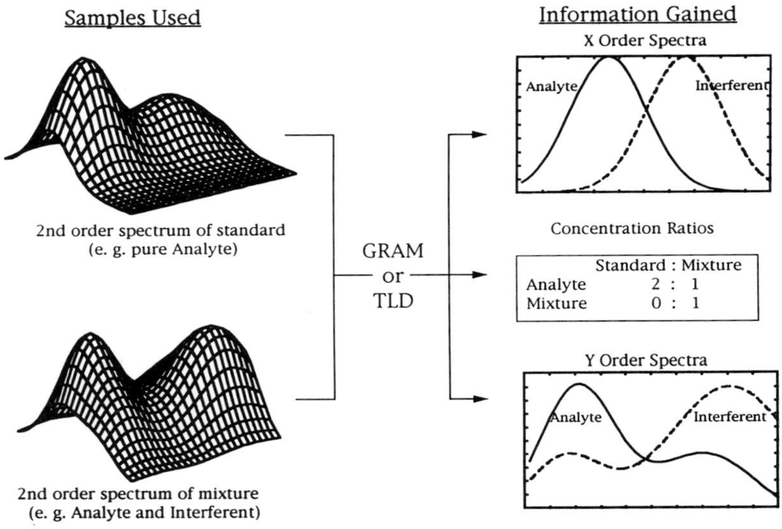

Figure 11.7 An illustration of the process used in two-dimensional sensor arrays with both GRAM and TLD. The left-hand side of the figure shows the two-dimensional response patterns used in calibration, and the right-hand side of the figure shows the resulting information from the second-order techniques. (Reprinted with permission from Elsevier Science BV.)

The vectors $\boldsymbol{x}$ and $\boldsymbol{y}$ are response patterns of the two separate arrays used. If the system is truly bilinear, these two vectors are orthogonal. c_k is the concentration of analyte k as $\mathbf{N}_k$ is defined as the bilinear response pattern for a unitary concentration level. If the matrix $\mathbf{M}$ has a rank of p, then subtracting $c_k\mathbf{N}_k$ from $\mathbf{M}$ will drop one unit of rank $(p-1)$ at the correct concentration level of c_k. Therefore the determinant of $\mathbf{M}$ will approach zero at the correct concentration

$$\det(\mathbf{M} - c_k\mathbf{N}_k) = 0. \tag{11.20}$$

To solve for c_k, the generalized eigenvalue problem is used with the singular-value decomposition technique. The results of the problem indicate both the pure component response patterns $\boldsymbol{x}$ and $\boldsymbol{y}$ and the ratio of concentrations of the pure components to the standard response concentration.

TLD uses multiple standards instead of one as in RAFA. The algorithm is similar to RAFA in that an eigenvalue problem is determined and solved through the use of singular-value decomposition. One important result of using second-order approaches such as RAFA and TLD is that the results of singular-value decomposition give estimates of the pure component response patterns which can be compared to the true response patterns. This ability allows users to check the performance of the procedure and identify components in the sample.

Problems that arise in the first-order sensor arrays can exist in the second order as well. The primary problem is collinearity of response patterns. If two components in a second-order analysis have the same response pattern for $\boldsymbol{x}$ as stated in equation (11.19*b*), then the combination of these two components contribute only one additional rank to the mixture matrix, **M**. When this occurs, non-bilinearity exists in the data matrices. Parameters to characterize the analytes in a second-order analysis have been established just as have the figures of merits for first-order sensor arrays [45]. These parameters are based on a second-order net analyte signal, **NAS**

$$\mathbf{NAS}_j = c_j \boldsymbol{x}_j \cdot \boldsymbol{y}_j^{\mathrm{T}}. \tag{11.21}$$

The vectors $\boldsymbol{x}_j$ and $\boldsymbol{y}_j$ are the resolved response patterns for component j with a standard concentration, c. These vectors do not represent the pure component response patterns, but the unique part of the original responses relative to the other analytes in the sample mixture. From the **NAS** the figures of merit for second-order instruments can be established

$$\text{Selectivity} \quad j = \| \mathbf{NAS}_j \|_F / \| \mathbf{N}_j \|_F \tag{11.22}$$

$$\text{Sensitivity} \quad j = \| \mathbf{NAS}_j \|_F / c_j \tag{11.23}$$

$$S/N \quad j = \| \mathbf{NAS}_j \|_F / \| \mathbf{E} \|_F . \tag{11.24}$$

$\| \cdot \|_F$ represents the F norm of a matrix, which is the square root of the sum of the squares of all elements. The matrix **E** is the error matrix of the second-order measurements.

11.4.3 Examples of second-order sensors

Development of second-order sensors for monitoring environmental and industrial processes is currently under way in several research groups. Two examples of second-order sensor arrays for environmental monitoring are both based on fiber optic spectrometry as one array and temporal information of analytes crossing a membrane as the second sensor array or the concentration modulator. The use of temporal information as an independent measurement axis is viewed as the time-elapsed elution profiles of analytes across a partition membrane. This approach has difficulties in resolution of different analytes since permeation rates are often similar. However, speed of measurements on the second array such as spectroscopy aid in the determination of species that may have elution profiles that differ only on the time-scale of a few seconds. In a report by Lin and Burgess [46], an optical measurement system for metal ions is coupled to a collection system based on permeation of the metal ions across a dialysis membrane (Nafion) into a collection chamber. The elution of the ions across the membrane is achieved by forming a thiosulfate complex which is negatively charged and repelled out of the membrane. Once in the collection chamber, the ions are chelated with 4-(2-pyridylazo)-resorcinol for an absorbance-based colorimetric determination. Figure 11.8 shows the matrices of the pure component response pattern of the Pb(II) ion and the matrix for the

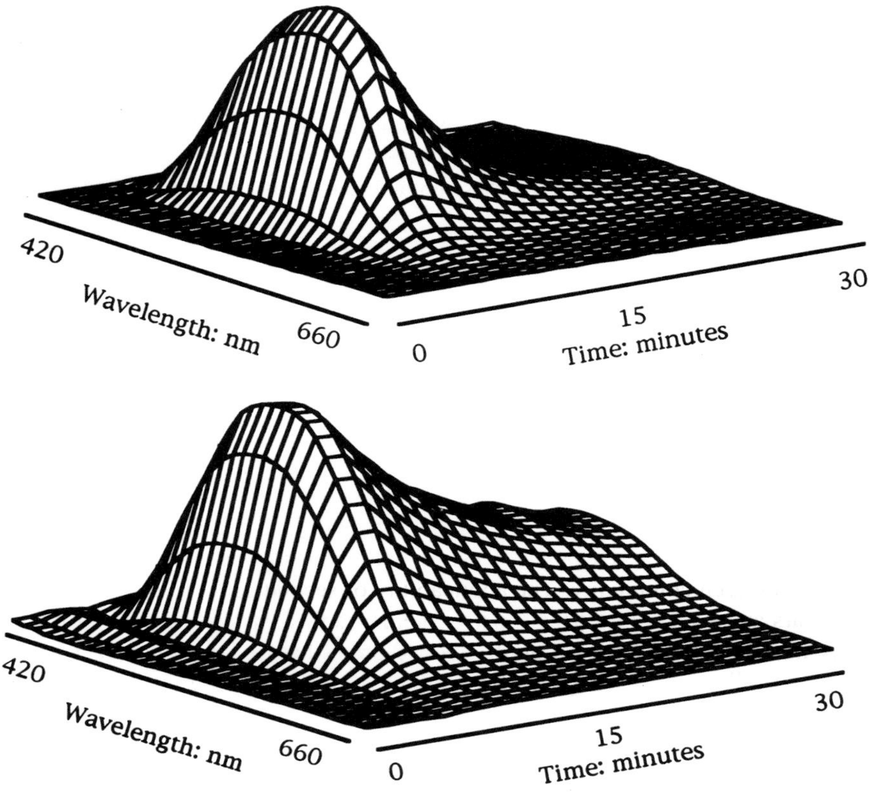

Figure 11.8 Second-order response patterns of the pure, top graph, and mixture, bottom graph, metal ion analysis using the two-dimensional fiber optic probe. The pure component is the chelated absorbance spectra of lead over time. The mixture contains lead and various other metal ions.

response of the instrument with several metal ion interferences. Results of both RAFA and TLD were compared to a single-wavelength measurement not using the temporal information for the determination of the Pb(II) and Cd(II) ions [47, 48]. In all cases the second-order methods performed better in prediction of unknown concentrations of lead and cadmium than the single-wavelength measurement. RAFA and TLD results were similar with no specific inference of better performance between the two. Error in this type of analysis is primarily focused on repeatability of the two sensor arrays between measurements. The permeation membrane approach is difficult to use due to the stability and repeatability of elutions after long periods of time and repeated samples.

The second technique is based on the detection of chlorinated hydrocarbons in groundwater applications [49, 50]. The sensor, similar to the metal ion device, allows chlorinated hydrocarbons such as trichloroethylene (TCE), trichloroethane (TCA), and chloroform present in the headspace of underground wells to permeate a Nafion membrane into a chamber filled with a colorimetric reaction-based chemistry. The Fujiwara reagent forms brightly colored complexes with reaction products of the original chlorinated hydrocarbons. In this second-order sensor, the temporal information is based on both the permeation rate of each analyte across the membrane and the reaction rate it has with the reagent. The second sensor array is again multiwavelength visible spectrometry.

11.5 CONCLUSION

The use of chemometric techniques to design, characterize, and implement sensors is a growing field due to the complexities of the measurement needs in the field as opposed to laboratory analysis. Because of these application complexities, sensors have to be more robust to meet the challenge. The design of instruments based on first- and second-order modeling techniques can greatly facilitate these measurement challenges. As shown in this chapter, the use of first-order sensor arrays is maturing rapidly with more widespread use among the world's research community. The applicability of second-order sensors is only a few years away with the current developments leading the way. We can foresee many more instruments based on the second order than the two described here. There are still many challenges in the development of techniques for calibration, array design, and characterization.

REFERENCES

[1] Carey W P, Beebe K R, Sanchez E, Geladi P and Kowaski B R 1986 Chemometric analysis of multisensor arrays *Sensors Actuators* **9** 223–34
[2] Booksh K S and Kowalski B R 1994 Theory of analytical chemistry *Anal. Chem.* **66** 782A–91A
[3] Callis J B, Illman D L and Kowalski B R 1987 Process analytical chemistry *Anal. Chem.* **59** 624A
[4] Riebe M T and Eustace D J 1990 Process analytical chemistry: an industrial perspective *Anal. Chem.* **62** 65A
[5] Draper N and Smith H 1981 *Applied Regression Analysis* (New York: Wiley)
[6] Box G E P, Hunter W G and Hunter J S 1978 *Statistics for Experimenters: An Introduction to Design, Data Analysis and Model Building* (New York: Wiley)
[7] Sharaf M A, Illman D L and Kowalski B R 1986 *Chemometrics* (New York: Wiley) p 242
[8] de Levie R 1986 When, why and how to use weighted least squares *J. Chem. Ed.* **63** 10–15
[9] Burton K W C and Nickless G 1987 Optimisation via simplex part I. Background, definitions and a simple application *Chem. Int. Lab. Syst.* **1** 135–49
[10] Clifford P K 1983 Homogeneous semiconducting gas sensors: a comprehensive model *Anal. Chem. Symp. Series 17* (Washington, DC: ACS) p 135

[11] Wold S, Esbensen K and Geladi P 1987 Principal component analysis *Chem. Int. Lab. Syst.* **2** 37–52

[12] Grate J W, Rose-Pehrsson S L, Venezky D L, Klusty M and Wohltjen H 1993 Smart sensor system for trace organophosphorus and organosulfur vapor detection employing a temperature-controlled array of surface acoustic wave sensors, automated sample preconcentration, and pattern recognition *Anal. Chem.* **65** 1868–81

[13] Rose-Pehrsson S L, Grate J W, Ballantine D S and Jurs P C 1988 Detection of hazardous vapors including mixtures using pattern recognition analysis of responses from surface acoustic wave devices *Anal. Chem.* **60** 2801–11

[14] Stetter J R, Jurs P C and Rose S L 1986 Detection of hazardous gases and vapors: pattern recognition analysis of data from an electrochemical sensor array *Anal. Chem.* **58** 860–6

[15] Nakamoto T, Fukuda A and Moriizumi T 1993 Perfume and flavour identification by odour-sensing systems using quartz-resonator sensor array and neural-network pattern recognition *Sensors Actuators* B **10** 85–90

[16] Schierbaum K D, Weimar U and Gopel W 1990 Multicomponent gas analysis: an analytical chemistry approach applied to modified SnO_2 sensors *Sensors Actuators* B **2** 71–8

[17] Lorber A 1986 Error propagation and figures of merit for quantification by solving matrix equations *Anal. Chem.* **58** 1167–72

[18] Carey W P and Kowalski B R 1986 Chemical piezoelectric sensor and sensor array characterization *Anal. Chem.* **58** 3077–84

[19] Carey W P, Beebe K R, Kowalski B R, Illman D L and Hirschfeld T 1986 Selection of adsorbates for chemical sensor arrays by pattern recognition *Anal. Chem.* **58** 149–53

[20] Haaland D M and Thomas E V 1988 Partial least-squares methods for spectral analyses. 1. Relation to other quantitative calibration methods and the extraction of qualitative information *Anal. Chem.* **60** 1193–202

[21] Beebe K R and Kowalski B R 1987 An introduction to multivariate calibration and analysis *Anal. Chem.* **59** 1007A–17A

[22] Geladi P and Kowalski B R 1986 Partial least-squares regression: a tutorial *Anal. Chim. Acta* **185** 1–17

[23] Naes T and Martens H 1984 Multivariate calibration. II. Chemometric methods *Trends Anal. Chem.* **3** 266–71

[24] Osten D W and Kowalski B R 1985 Background detection and correction in multicomponent analysis *Anal. Chem.* **57** 908–17

[25] Kalivas J H and Kowalski B R 1982 Compensation for drift and interferences in multicomponent analysis *Anal. Chem.* **54** 560–5

[26] Beebe K R, Uerz D, Sandifer J and Kowalski B R 1988 Sparingly selective ion-selective electrode arrays for multicomponent analysis *Anal. Chem.* **60** 66–71

[27] Carey W P, Beebe K R and Kowalski B R 1987 Multicomponent analysis using an array of piezoelectric crystal sensors *Anal. Chem.* **59** 1529–34

[28] Hierold C and Muller R 1989 Quantitative analysis of gas mixtures with non-selective gas sensors *Sensors Actuators* **17** 587–92

[29] Walmsley A D, Haswell S J and Metcalfe E 1991 Evaluation of chemometric techniques for the identification and quantification of solvent mixtures using a thin-film metal oxide sensor array *Anal. Chim. Acta* **250** 257–64

[30] Otto M and Thomas J D R 1985 Model studies on multiple channel analysis of free magnesium, calcium, sodium and potassium at physiological concentration levels with ion-selective electrodes *Anal. Chem.* **57** 2647–51

[31] Forster R J, Regan F and Diamond D 1991 Modeling of potentiometric electrode

arrays for multicomponent analysis *Anal. Chem.* **63** 876–82
[32] Sundgren H, Lundstrom I, Winquist F, Lukkari I, Carlsson R and Wold S 1990 Evaluation of a multiple gas mixture with a simple MOSFET gas sensor array and pattern recognition *Sensors Actuators* B **2** 115–23
[33] Friedman J H and Stuetzle W 1981 Projection pursuit regression *J. Am. Stat. Assoc.* **76** 817–23
[34] Beebe K R and Kowalski B R 1988 Nonlinear calibration using projection pursuit regression: application to an array of ion selective electrodes *Anal. Chem.* **60** 2273–8
[35] Friedman J H 1991 Multivariate adaptive regression splines *Ann. Stat.* **19** 1–141
[36] Ema K, Yokoyama M, Nakamoto T and Moriizumi T 1989 Odour-sensing system using a quartz-resonator sensor array and neural-network pattern recognition *Sensors Actuators* **18** 291–6
[37] Nakamoto T, Fukunishi K and Moriizumi T 1990 Identification capability of odor sensor using quartz-resonator array and neural-network pattern recognition *Sensors Actuators* B **1** 473–6
[38] Sundgren H, Winquist I, Lukkari I and Lundstrom I 1991 Artificial neural networks and gas sensor arrays: quantification of individual components in a gas mixture *Meas. Sci. Technol.* **2** 464–9
[39] Wang X, Fang J, Yee S S and Carey W P 1993 Mixture analysis of organic solvents using non-selective and nonlinear Taguchi gas sensors with artificial neural networks *Sensors Actuators* B **13** 455–7
[40] Carey W P and Yee S 1992 Calibration of nonlinear solid-state sensor arrays using multivariate regression techniques *Sensors Actuators* B **9** 113–22
[41] Ho C, Christian G F and Davidson E R 1978 Application of the method of rank annihilation to quantitative analyses of multicomponent fluorescence data from the video fluorometer *Anal. Chem.* **50** 1108–13
[42] Lorber A 1984 Quantifying chemical composition from two-dimensional data arrays *Anal. Chim. Acta* **164** 293–7
[43] Sanchez E and Kowalski B R 1986 Generalized rank annihilation factor analysis *Anal. Chem.* **58** 496–9
[44] Sanchez E and Kowalski B R 1990 Tensorial resolution: a direct trilinear decomposition *J. Chemometr.* **4** 29
[45] Wang Y, Borgen O S, Kowalski B R, Gu M and Turecek F 1993 Advances in second order calibration *J. Chemometr.* **7** 117–30
[46] Lin Z and Burgess L W 1994 Chemically facilitated Donnan dialysis and its application in a fiber optic heavy metal sensor *Anal. Chem.* **66** 2544–51
[47] Lin Z, Booksh K S, Burgess L W and Kowalski B R 1994 Second order fiber optic heavy metal sensor employing second order tensorial calibration *Anal. Chem.* **66** 2552–60
[48] Booksh K S, Lin Z, Burgess L W and Kowalski B R 1994 Extension of trilinear decomposition method with an application to the flow probe sensor *Anal. Chem.* **66** 2561–9
[49] Henshaw J M, Burgess L W, Kowalski B R, Smilde A and Tauler R 1994 Multicomponent determination of chlorinated hydrocarbons using a reaction-based chemical sensor. 1. Multivariate calibration of Fujiwara reaction products *Anal. Chem.* **66** 3328–36
[50] Smilde A K, Tauler R, Henshaw J M, Burgess L W and Kowalski B R 1994 Multicomponent determination of chlorinated hydrocarbons using a reaction-based chemical sensor. 2. Chemical speciation using multivariate curve resolution *Anal. Chem.* **66** 3337–44

12

Microfluidics

Jay N Zemel and Rogério Furlan

12.1 INTRODUCTION

Microfluidics is concerned with the flow of fluids in micromachined ducts whose hydraulic diameter, D_h, is defined in the conventional fashion as [1]

$$D_h = 4 \times \text{area of the duct/wetted perimeter of the duct.} \tag{12.1}$$

Microfluidics is also a cross-disciplinary subject that uses the methods and principles of microelectronics to construct very small analogs or models of such macroscopic fluidic elements as wind tunnels, valves, or fluidic amplifiers. The natural question that comes to mind is: at what dimensional scale does fluid motion depart from the extremely well understood and well established laws of fluid dynamics? There is no definitive answer to that question yet since the study of fluid motion in microscale and nanoscale structures is still at an early stage.

Because the fabrication process is such an essential part of microfluidics, an overview of the principles underlying the microfabrication technology is presented. Pressure, flow, and temperature measurements are essential variables for characterizing fluid motion in any system. An important goal is the design and construction of self-contained microfluidic systems. Because of their small size, incorporation of pressure, flow, and temperature sensors directly on the microfluid system chip is highly desirable. There are relatively few examples where microfluidic systems have been constructed with these on-board sensors. There have been so many microsensor developments in recent years that it is only a matter of time before such systems will appear. Small-scale actuators to provide either open- or closed-loop control of the flow in microchannels are needed and these efforts are addressed. While experimental work on fluid flow itself in microscale structures is rather sparse, some results will be presented that emphasize the similarity and/or differences between macroscopic and microscopic flow of liquids. Although there are not many applications of microfluidic technology as yet, the chapter will close with some speculations on possible future directions for this subject.

12.1.1 Older historical background

Fluid dynamics is an old subject. Not only has the study of fluids been of considerable practical importance since antiquity, but it has also had an important role in the development of modern science and technology. The development of the relation between the pressure, area, and average fluid velocity by Daniel Bernoulli [2] launched a period where scientists of the day described such diverse topics as combustion (phlogistic theory) and electricity in terms of the motion of a featureless, homogeneous fluid [3]. The overthrow of the phlogistic model of combustion by Lavoisier [4] led to contemporary chemistry while Poiseuille's classic paper on flow in capillaries gave rise to modern fluid dynamics [5]. Benjamin Franklin's understanding that fluids flow from high to low pressure remains with us in the electrophysical convention that electrical currents flow from a positive to a negative electrode, thus requiring electrons to be negative in charge to satisfy the condition that the electron flux moves from a negative to a positive electrode [6]. Beginning with the seminal work of Poiseuille in the 1840s, the study of fluid dynamics has been important to the evolution of scientific thought and its application to technology.

Fluid flow continues to be intensively studied. Chaotic and turbulent motion are among the more aggressively studied topics in the physical sciences [7–9]. In the macroscopic world where the smallest geometric dimension of the container or duct is large in comparison to physical length scales (such as diffusion lengths, mean free paths, or molecular dimensions), the continuum model for fluids is both valid and powerful. Laminar and creeping flows tends to be the general rule in the world of small structures and biology [10]. In order for a fluid to move at all, some form of potential gradient must act upon it. Pressure and temperature differences are among the more common and recognizable causes of fluid motion but electromagnetic and chemical potential differences can also induce motion in certain classes of fluids [11]. Inducing fluid motion in complex microstructures of the type that may eventually evolve is a challenge. The difference in physical characteristics of liquids and gases is substantial [12]. This suggests that material-specific methods for actuation may be likely. In addition, some of the measurement schemes for characterizing fluid flow in microchannels may depend on whether the fluid is a gas or a liquid. The physics embodied in the Navier–Stokes equations and energy conservation provides the basis for describing with reasonable precision topics as diverse as astrophysics and the motion of liquids in capillaries in trees.

12.1.2 Recent historical background

While silicon had been used as a structural material for pressure sensors as early as 1961, it is certain that the seminal paper by Bergveld in 1970 provided the powerful stimulus needed to consider silicon as a basis for chemical sensors [13]. In 1973, Samaun *et al* described a capacitive pressure sensor for biomedical applications which further broadened the subject [14]. By 1975, the chemical sensor effort involving silicon devices began in earnest when Moss *et al* [15] and Lundstrom *et al* [16] reported devices that yielded usable information for pH and H_2, respectively, and Zemel reviewed the emerging subject [17]. In each case, the core technology was identical to that being employed by the microelectronics industry for integrated circuit fabrication. The wealth of knowledge of materials and processes developed there has led to techniques for chemical machining of both silicon and the films grown or deposited thereon. While these techniques were important for the rapidly advancing microelectronics industry, it was the application of these methods to chemical and mechanical sensors in silicon that generated the new microsensor industry [18–20].

One of the earlier silicon-based microchannel structures to receive attention was a fluidic element, the gas chromatograph on a silicon wafer. This structure, developed by Angell and coworkers, consisted of a number of valve seats and a spiral column etched into a single silicon wafer [21]. Macroscopic valves were attached to the valve seats located strategically on the column to allow injection and extraction of the gas. While the device demonstrated feasibility, it never reached full commercial development. Nevertheless, this device structure fired the imagination of the micromachining community and led to a rapid expansion of research and development efforts on other types of device. At about the same time, the needs of the automotive and biomedical communities for small, high-precision pressure and flow meters stimulated further efforts on micromechanical systems. The maturation of the silicon micromachined, integrated pressure sensor was a major accomplishment of the late 1980s [22]. The devices, primarily piezoresistive sensors, have become commercially available in huge quantities as the automotive industry employed these elements in engine control systems [23]. Capacitive pressure sensors also found application in many areas. Howe and Muller and Guckel and Burns showed that silicon micromachining technology could be extended through the use of sacrificial layers [24, 25]. This led to a wide variety of other micromechanical elements including electrical micromotors. In the past few years, silicon micromachined accelerometers using piezoresistive sensors have become available for the airbags of automobiles [26]. These efforts not only created a new microelectromechanical systems (MEMS) industrial activity, but also demonstrated the versatility and utility of silicon as a material for combining physical and chemical sensors with microelectronics to create 'smart sensors' [27].

During this time, Tuckerman and Pease showed that micromachined silicon heat exchangers were capable of transferring extremely high heat flows [28,

29]. Silicon has a very high thermal conductivity, actually in the metallic range. The original elements used by Tuckerman employed saw slots rather than micromachining, but chemically etched structures will do as well. Research on various types of microstructure suitable for cooling electronic systems was subsequently reported along with theoretical studies on the characteristics of silicon-based heat exchange structures [30, 31].

Convective heat transfer from a heated element is an important means for monitoring mass flow. The first studies of silicon-based fluid flow measurements were reported by Huijsing *et al* [32] who employed thin silicon membranes as hot-film anemometers. Later work by Petersen *et al* on heated polycrystalline silicon filaments demonstrated their equivalence to hot-wire anemometers [33]. The silicon thermal elements have both the advantages and disadvantages of all resistive hot-wire anemometers. They are small and reasonably precise in a limited flow range but require relatively high operational temperatures ($\geqslant$100 °C) to provide adequate sensitivity. The manufacturing reproducibility of the elements is satisfactory and the flow calibration for a given wafer is good. Because of their small size and relatively high resistance, polycrystalline silicon filaments may be heated to rather high temperatures with small power level. This makes it possible to incorporate heaters onto membranes, cantilevers, or bridges that are part of a CMOS chip without raising the overall temperature of the chips appreciably. This in turn allows for the use of on-board electronics to amplify and condition the signal from the filaments.

An alternative to the resistive elements is the pyroelectric anemometer based on $LiTaO_3$ [34]. These devices provide wide rangeability and high reproducibility but, because they are fabricated separately, can only be used as hybrid elements [35]. While this is a drawback in some applications, the results of Yu *et al* established that it is possible to use even relatively large pyroelectric anemometers (active regions ~3 mm × 3 mm) to monitor flows as low as 10 mm min^{-1}, values that silicon-based devices generally cannot match.

In recent years, the characteristics of fluid flow in microchannels have been studied in some detail at the University of Pennsylvania [36]. This work will be discussed in greater detail in section 12.5. As a result of these more fundamental studies, Wilding and coworkers were led to investigate the flow of biological and mixed-phase liquids (water suspensions of small fluorescent polystyrene beads, whole blood, and washed blood cells) as well as distilled water and saline solutions through trapezoidal microchannels of different hydraulic diameters [37]. They developed a method for using silicon microstructures as chambers for polymerase chain reactions to amplify various nucleotides [38]. Their channels and associated flow structures were considerably larger than the dimensions of the devices used for the earlier Pennsylvania flow investigations. Following on with the use of microchannels for biologically important subjects, Kricka and coworkers used a microchannel maze to evaluate quantitatively sperm motility [39]. Through the use of microchannels, they demonstrated that the velocity of sperm can be determined by limiting their degrees of freedom, thereby simplifying the velocity measurements. Unpublished work showed that microchannel structures can efficiently generate *in vitro* fertilization of mice ova. Other studies by Bousse *et al* demonstrated that microstructures could be constructed that allow for the assay of drugs on single cells. They employed a combination of mechanical entrapment of the cells, injection of nutrients and drugs into the microenvironment, and chemical sensing of cell metabolism with a variant of the pH–ion-sensitive field effect transistor [40].

As sensors and MEMS advanced, the need for microstructures for controlling fluid flow became increasingly clear. The first control elements were normally open valves formed from either layers of anodically bonded silicon and Pyrex or silicon–silicon bonded wafers. The valve seats were formed from silicon using conventional micromachining methods. Piezoelectric or solenoidal actuators to effect closure were cemented onto the valve bodies. While silicon is a strong material, it is sufficiently brittle that 'hard' closure is avoided to prevent striction and damage to the valve seat. A number of investigators used variations of the silicon membrane valve as micropumps. The motive power in most cases has been piezoelectric drives [41–43]. Other methods have been used including electrostatic [44], magnetic [45], hydraulic [46], and thermal [47] actuators. These methods have the following characteristics [47]:

(i) piezoelectric drivers produce considerable displacement (>30 μm) but relatively low forces;
(ii) piezoelectric stacks produce high forces but relatively little displacement (<10 μm), electrostatic actuation is sensitive to particulates and moisture and requires very narrow gaps;
(iii) hydraulic systems can produce the required pressures to operate the valves but they too require valves, albeit larger ones can be used some distance from the operating part;
(iv) magnetic actuators can produce both force and displacement but are difficult to integrate into batch-fabricated silicon assemblies particularly at submillimeter scales; and
(v) thermal actuators are limited by the required input power, although this requirement is minimized as sizes are reduced and the quantity of material to be heated is diminished.

With the possible exception of the thermal actuator, all of these structures tend to be much larger than the valve itself and most require fairly large areas. Besides these drawbacks, the long-term reliability of the valves will depend on striction and gradual seat degradation with extended use.

The need for control elements has become quite important. In a number of papers, innovative methods for measuring and controlling flows in silicon-based structures have been presented. Among the devices reported were a fluidic separator constructed using the then new LIGA technology developed in Germany [46] and a silicon etched fluidic amplifier [48] with attractive gain characteristics. Flow rates as small as picoliters per second are needed in some long-term drug delivery systems, and at these rates, extremely small channels are required. The interaction of the liquids with the walls of such small channels can play a critical role in the reliability and utility of such devices. The interplay of basic scientific problems and real world applications is one of the more attractive aspects of this field of study.

12.2 FABRICATION OF SMALL STRUCTURES

Silicon is the 14th element in the periodic table. It has an average atomic weight of 28.086 and a density of only 2490 kg m^{-3}. For the most part, single crystals and polycrystalline films of silicon will be discussed almost exclusively in this chapter. The reason for silicon's wide acceptance and use are the enormous body of work done with it, its low cost and ready availability, the ability to passivate it through oxidation and nitridation, and the possibility of including electronics, actuation, and sensing on the same substrate. No other material has these properties. While single-crystal silicon is a strong material, comparable in mechanical strength to that of steel, unlike steel it is brittle. This brittleness is one of the important reasons why micromachining single-crystal silicon requires very different methods than those used for 'tough' materials such as steel. It also cannot be formed by dies and extrusion as may metals and green ceramics. In its sphere of applicability, it has few peers and no superiors. As will be seen, almost all of the methods employed for micromachining silicon are chemical in nature. It would be nice to be able to say that research has established a clear understanding of how various methods produce the structures in such wide use. The facts are that there is still a great deal of art in this field and even such apparently simple issues as the wet chemical etching of silicon are at best qualitatively understood. In reading the methods described below, keep this in mind since the final words on microfabrication methods remain to be written.

12.2.1 Microfabrication methods

The first and most important point to be made about microfabrication is that there is no single approach to this technology. Efforts to construct small elements range from the use of atomic force microscopes to move single atoms on a surface, to self-organization of chemical systems, to conventional methods of planar photolithography practiced in the microelectronics industry. In light of the enormous diversity of methods available, the reader is directed to the literature for details. However, the conventional methods of photolithographic microfabrication have become increasingly available in both the academic and industrial communities. For these reasons, photolithographic microfabrication methods for making microfluidic elements on silicon wafers will be emphasized here.

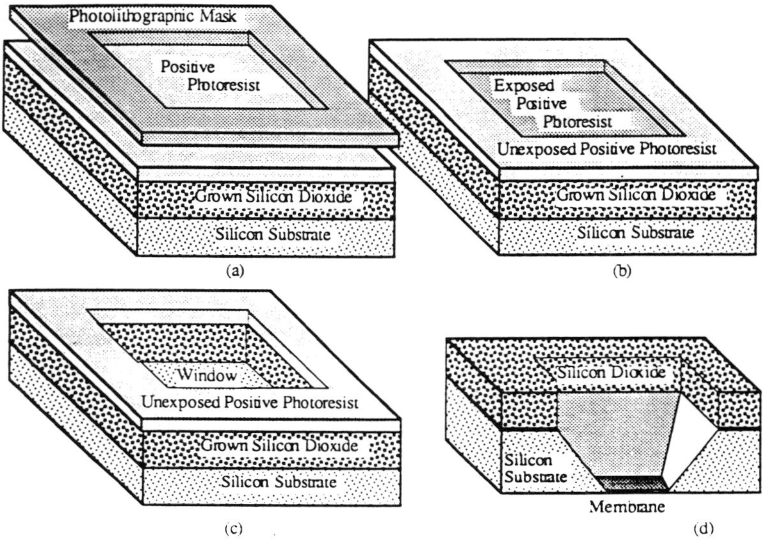

Figure 12.1 An illustration of the procedures leading up to the formation of a thin membrane used in a pressure sensor. (a) The spin-on of a 'positive' photoresist; (b) the exposed resist can then be removed by the developer leaving a window open to the SiO_2; (c) the wafer is immersed in a buffered HF solution, the SiO_2 is removed down to the silicon; (d) the region of exposed silicon can now be etched to form a cavity.

12.2.2 Masking, photolithography and photoresists

The first step in fabricating a microstructure is to selectively remove material from a wafer. Determining what remains or is removed is accomplished with a photolithographic mask on which the pattern corresponding to the desired structure is defined. The photolithographic mask may be prepared in many different ways but the most common procedure today is to use an electron beam or optical pattern generator to produce the photolithographic masks. In both cases, a beam of energy exposes the pattern in either a photographic emulsion or a photoresist optical absorbing layer. After development of the photographic plate or etching of the absorbing pattern, respectively, the mask is ready for use. To use these masks, other processing steps are required. This is partially illustrated in figure 12.1, showing procedures leading up to the formation of a thin membrane used in a pressure sensor. The first step in any micromachining operation is a thorough cleaning of the wafer. A 'field' oxide, 1–2 μm thick, is grown. Details of this procedure may be found elsewhere [49, 50].

The next procedure is the heart of the photolithographic process. A photoresist layer is spin cast over the silicon wafer from solution. It is then given a preliminary set of processing steps (such as baking at elevated temperatures) following the procedures recommended by the manufacturer. The material depicted in figure 12.1(a) is a 'positive' photoresist. Positive resists are polymers which can withstand the attack of most common inorganic acids, especially HF. Upon exposure to ultraviolet radiation, bonds in the polymer are broken (scission) and the lower-molecular-weight material is more readily dissolved. In the case of Novolak resin plus diazonaphthoquinone, one of the more common types of positive resist, the end product is indenecarboxylic acid. This compound will not inhibit the dissolution of the Novolak resin in basic solutions since it too is soluble in such solutions. The dissolution rate depends on the molecular weight of the end products of the photoscission and the polymer structure. While the details of the dissolution rate are a complicated function of the chemistry involved, it varies roughly the molecular weight to the 0.7 power for many materials. The formulation of these resists is as much art as science so that while there is a rich literature on the use of these resists, the literature on their formulation and design is singularly barren. Company secrets are the order of the day on this topic so that only broad principles are in the open literature.

The situation is not any simpler with negative resists. In this case, light exposure promotes a cross-linking reaction that makes the material resistant to the solvent for the unexposed material. Typical solvents include xylene and other aromatic solvents. A typical example is a polyisopropene matrix and a promoter that is photosensitive, for example bis-arylazide. Upon photoactivated reaction, the promoter cross-links the matrix thereby substantially increasing the molecular weight of the molecules.

As illustrated in figure 12.1(a), the photomask image is a square, typically with dimensions in the 1 mm range. The image of the transparent region is transferred to the positive photoresist by exposing the mask–photoresist combination to ultraviolet light. The exposed resist can then be removed by the developer leaving a window open to the SiO_2 as depicted in figure 12.1(b). When the wafer is immersed in a buffered HF solution, the SiO_2 is removed down to the silicon surface as indicated in figure 12.1(c). The unexposed photoresist protects the SiO_2 from attack. The region of exposed silicon can now be etched to form a cavity as shown in figure 12.1(d).

12.2.3 Etching of microchannels and microstructures

The 'bare' silicon surface (actually there is a very thin 'native' oxide that forms almost instantaneously once pure silicon is exposed) is where various structures can be generated. There are many ways this can be done once the protective oxide is removed, such as isotropic and crystallographically selective wet etching, and various types of plasma and reactive ion dry etching. Crystallographically selective wet etching and the dry etching methods are the more common means for making microfluidic structures. As indicated in figure 12.1(b), the photoresist is removed. The wafer is placed in a hot, concentrated, aqueous solution of KOH to attack the exposed silicon in the SiO_2 window [51]. The solution vigorously attacks the (100) face of silicon but etches the

(111) face quite slowly. The difference in etch rates is ~300:1. The wafer is immersed in the etch and, by carefully controlling the etch time or using an etch stop method, a deep pit is generated. Depending on the thickness of the wafer and the size of the etch window, a thin membrane of single-crystal silicon can be formed at the bottom.

The KOH-etched surfaces are generally quite smooth. However, the cross-sectional geometries of the channels are trapezoidal due to its crystallographically selective etching. While this is not an impediment for measuring flow characteristics in a simple silicon-based microchannel, it does limit the types of structure that can be made. It would be preferable to have a method that is anisotropic but not crystallographically selective. The use of plasma and reactive ion etching methods with fluoro-chloro-carbon bearing gases held out promise for this. Originally, these dry etching methods were used to make relatively shallow trenches on silicon. Typical depths were of the order of 1–3 μm. The evidence to date suggests that the anisotropy in the etching is due to polymers arising from the plasma decomposition of the C–F-based materials. These polymers adsorb on the walls of the trench and it is their presence that limits the reactions of the gas with the wall. However, as the trench deepens, the polymers also accumulate on the floor as well. It appears that the polymer layers are not uniform and they gradually interfere with the etching process to produce rough surfaces.

Consequently, the results are generally less satisfactory when these methods are extended to making deep trenches. Under ideal conditions, fairly vertical walls are obtained in channels as deep as 30–40 μm but experience suggests that the technology is not that reproducible. The advantage of this dry etching is that the walls are independent of crystallographic orientation. A wide variety of plasma or reactive ion conditions yield channels with depths of 10–15 μm and floors that are fairly smooth. With the restrictions on the use of chlorine bearing gases due to environmental concerns, obtaining these results has become even more difficult. The major problem with deep trenching using plasma and reactive ion etching is a progressive roughening of the bottom surface as depth increases. Peak to peak roughness of 10% or more of the total depth is commonly observed. Further improvements in deep trenching are needed for continued efforts in this area.

Many other materials such as very homogeneous glasses, metals, or even plastics could be used to generate microstructures. There are many recent examples of this in the literature. Ceramic materials have much to recommend them, especially their mechanical strength and the varied means developed by the microelectronics industry for hermetically sealing cavities in them. Commercial glasses quite frequently have inhomogeneous compositions. When etched in buffered HF or other similar etchants, these inhomogeneities create very rough surfaces in the etched region. Nevertheless, as shown by Harrison *et al* [52] and Kricka *et al* [53], it is possible to obtain glasses with sufficiently homogeneous compositions that excellent channels can be obtained for electroanalytical and biological studies.

Silicon can be hermetically bonded to a Pyrex glass whose coefficient of thermal expansion is matched to that of silicon to confine the fluids. An electric potential of 500 V is applied to the silicon–Pyrex combination at elevated temperatures ($\sim$300–400 °C) causing them to anodically bond. Other methods have been developed involving the bonding of silicon wafers together. As interest in more sophisticated silicon microstructures increases, the effort to find new and innovative means to seal silicon to other materials increases. The requirements for thermal expansion matching are critical because of the brittle nature of silicon. This is one reason why there is an ongoing effort to find other materials for micromachining and microfluidic research. It is fortunate that there are numerous glass formulations available with a wide range of thermal expansion coefficients. These glasses may prove useful for hermetic sealing to other micromachined materials.

12.3 SENSORS FOR USE IN MICROCHANNELS

The ongoing challenge in microfluidics is the incorporation of the necessary on-board sensors into a chip containing flow actuation. As microfluidic systems become more complex and the need for closed loop operation grows, pressure and flow measurements at different points in the system will take on increased importance. A number of reports show where pressure, flow, and temperature sensors have been combined on the same chip as a platform for use in larger systems. The three primary pieces of information that would characterize the flow (aside from the geometry of the ducts, of course) are pressure, mass flow, and temperature. Using this information, it is possible to derive a wide range of important parameters for the system.

12.3.1 Pressure sensors

All microfabricated pressure sensors derive their sensitivity from the elastic deformation of a membrane. Measuring the elastic deformation provides the desired signal information. In the broad class of microfabricated pressure sensors, the two most widely used phenomena are the change of resistance due to stress in boron-doped (p^+) resistors and the direct measurement of the membrane displacement using optical or capacitive methods.

12.3.1.1 Piezoresistive sensors. To measure the pressure, the resistance change to stress (the piezoresistance effect) may be employed. When silicon is stressed, the resulting strain breaks the cubic symmetry of the underlying crystal structure. The band structure of silicon is very sensitive to its crystal structure and, as a result, the consequent modification causes changes in the resistivity of the material (holes in the case of p^+ material). This change is

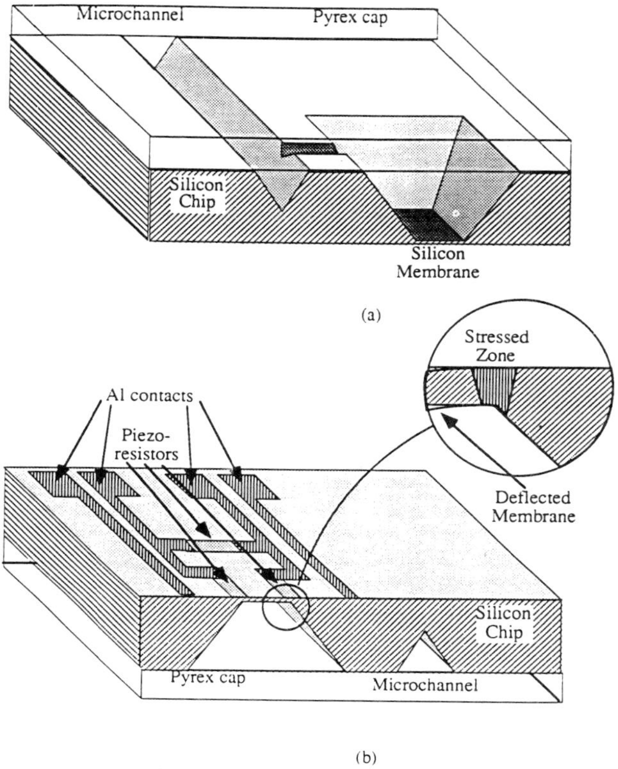

Figure 12.2 (a) A pressure sensor structure showing the essential elements connected to a microchannel; (b) the membrane is deformed by a pressure differential; this stresses the symmetrically placed pair of piezoresistors.

linear with stress. At the stress levels produced by the applied pressure, the resistance change is also linear with pressure. The interested reader can learn more about piezoresistance in monographs on solid state physics [54].

In figure 12.2(a), a structure is shown that contains the essential elements of a pressure sensor for a microchannel. The microchannel in this instance is a 'v' groove made up of (111) walls obtained by etching in a hot KOH solution as described in subsection 12.2.3. The channel is connected to the pyramidal pit (with a thin silicon membrane on the bottom) by another 'v' groove as shown in figure 12.1(d). The membrane is tethered around its perimeter by the silicon chip and bows out or in, depending on the sign of the pressure gradient. As a result, the stress in the material is a maximum in the vicinity of the membrane tether.

As this membrane is deformed by a pressure differential, it stresses the symmetrically placed pair of piezoresistors (see figure 12.2(b)). The change

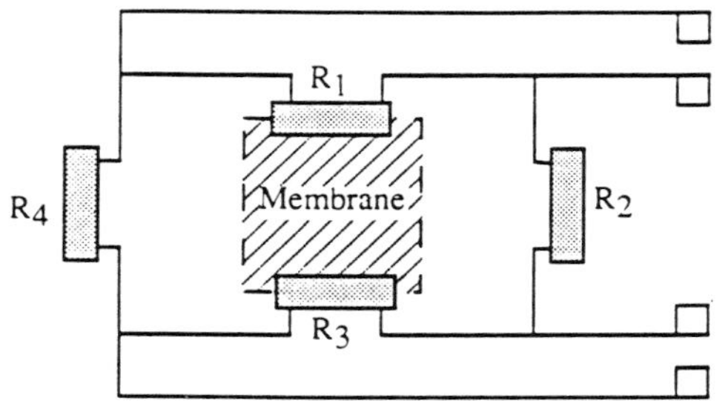

Figure 12.3 Circuit layout for a pressure sensor. Resistors R_1 and R_3 are located at the edge of the membrane and suffer the maximum stress (see the inset in figure 12.2(b)).

in the resistance is measured using a Wheatstone bridge composed of two piezoresistors that are in the region of maximum stress (R_1 and R_3 in figure 12.3) and two on the chip some distance away from the stressed region (R_2 and R_4 in figure 12.3). The piezoresistors are prepared by first growing another oxide layer for isolation on the wafer surface followed by a layer of polysilicon. The layer over the membrane is doped p^+ with a boron diffusion. Four piezoresistors are etched out of this doped layer using another photoresist–chemical etch cycle (only three are shown in figure 12.2(b) in order to depict their placement over the region of maximum stress). Aluminum leads are then generated using an Al deposition–photoresist–etching cycle to complete the sensor. As the membrane is deformed by the applied pressure, a stress field is generated at the boundary between the membrane and the body of the silicon chips as depicted by the inset of figure 12.2(b). The overall size of piezoresistive pressure sensors is typically 2–3 mm on a side but they come with circuitry for amplifying the bridge output. Some very clever elements have been built using an add-on step to fabricate the membrane in a CMOS wafer [55]. In these cases, a strong organic base etch, ethyldiaminepyrocatechol, is used rather than the KOH solutions to avoid contamination of the circuitry. The development of 'smart sensors' using CMOS-based circuitry and subsequent micromachining is a very active area that promises to supply devices that are linear, low power, and relatively inexpensive. We will refer to them again in the discussion of flow and temperature sensors.

12.3.1.2 Capacitive sensors. Essentially the same techniques as for making the piezoresistive sensor can be used to produce a capacitive sensor. An illustrative example is depicted in figure 12.4. Which one to use in a given application depends on many factors including the add-on circuitry required to obtain useful signals. Because capacitive sensors can be made using sacrificial layers, a technique where a thin layer of glass is etched away to leave a supported structure, capacitive elements may prove to be a very useful approach for pressure measurements in microfluidic systems.

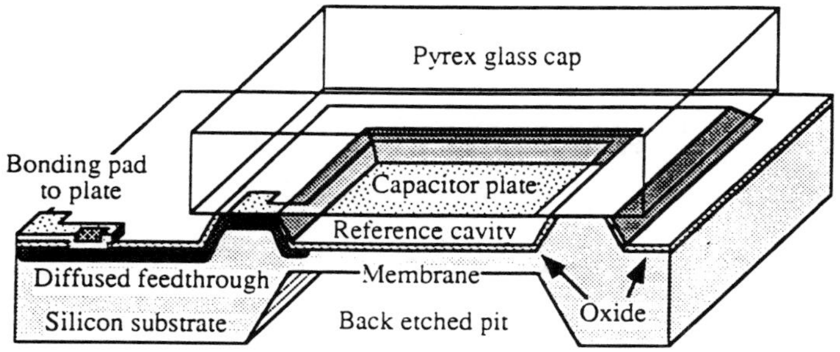

Figure 12.4 A depiction of a capacitive pressure sensor.

12.3.1.3 Fiber optic sensors. A drawback to both piezoresistive and capacitive pressure sensors is that they are subject to electromagnetic interference and occupy areas of the order of two or more square millimeters [56, 57]. If the deflection of the sensing membrane is interrogated by means of an optical technique, the electromagnetic interference problem can be avoided and smaller areas can be interrogated. Silicon-based optical fiber pressure sensors have been developed and demonstrated in the past few years [58–60]. These sensors take advantage of both silicon micromachining and fiber optics. There is a problem in maintaining precise alignment between the optical fiber and the sensing membrane and efficiently collecting the light reflected from the membrane with the fiber core. If there is misalignment of the order of one core diameter, the optical signal loss is so substantial that the device is useless. Since core diameters are of the order of micrometers, the alignment must be maintained to that degree of precision.

Recently, Tu and Zemel constructed a fiber-optic-based pressure sensor that employed an etched channel for guiding a single-mode optical fiber to a thin membrane perpendicular to the wafer surface [61]. This is depicted in figure 12.5. An etched stop located the fiber at a fixed distance from the membrane surface. The membrane is one wall of the pressure tap. These structures are micromachined in a (110) silicon wafer using a KOH crystallographically selective etch. Membranes as thin as 4 μm were fabricated. The resulting structure is anodically sealed by bonding a Pyrex glass cap to the silicon surface. A suitably prepared single-mode optical fiber is then inserted into the channel till it comes to rest at the stop. The fiber is then aligned with the maximum deflectable region of the membrane and is then fixed in place with an epoxy sealant. The interference patterns obtained are clear and stable 'Newton's rings'. The integrated intensity of the interference pattern measured as a function of applied pressure indicates that the minimum detectable optical signal is 1 μV for the 4 μm thick measuring membrane. This signal level corresponds to a minimum detectable pressure of about 5 Pa.

12.3.2 Flow sensors

The resistive type of anemometer (hot wire or hot film) described earlier relies on the temperature dependence of the hot sensing element to monitor the convective heat loss. The resistance is maintained at a constant value by adjusting the Joule heating to maintain a constant temperature. That power input is used as a measure of the convective or other heat losses. This type of anemometer is in widespread use today. However, the energy loss of the resistive anemometer depends on the temperature of the fluid ambient in which it is immersed. Changes in the fluid ambient temperature are reflected in the power required to maintain the anemometer's resistance and this is the reason why temperature compensation is needed for this type of sensor.

12.3.2.1 Hot-wire anemometry. There is enormous literature on hot-wire and hot-film anemometry. The DC heat transfer relations between a compressible laminar convective flow and a heated plate have been known for some time. Chapman and Rubesin obtained a general solution for the compressible laminar boundary layer problem, subject to an arbitrary DC temperature distribution at the plate [62]. They showed that the convective heat transfer rate between a two-dimensional surface and a flowing fluid is proportional to the square root of the flow velocity even at low flow rates provided that the plate temperature is a flow-independent boundary condition. van Oudeheusden explored a thermal model for DC differential heat transfer in considerable detail [63]. While this theoretical investigation clarifies the relationships between the sensor geometry, different heat transfer processes, and the response of silicon-based resistive structures, it is not directly applicable to the AC problem relevant to the pyroelectric anemometer (PA) structures. There have been investigations of the low-flow regime, again for the case of DC thermal distributions [64, 65]. These studies showed that the convective heat transfer process varied linearly with the flow, a process referred to in the literature as calorimetric flow sensing.

From the vantage point of microfluidics, the structures developed by Petersen *et al* [33] are the most appropriate. More recently, Baltes and coworkers combined CMOS circuitry with the microfabrication of sensors to construct a thermal mass flow system based on thin-film pyrometers [66]. As free standing mass flow sensors, they have attractive features. However, all of these silicon-based devices operate at relatively high temperatures in the 100–200 °C range. This elevated temperature limits their potential application in more complex microfluidic systems. The ideal flow sensor would be a very-low-temperature element that could be used on the walls of the microchannel.

12.3.2.2 Pyroelectric anemometry. Two years ago, Yu *et al* described a wall-mounted PA situated in a 3 mm wide, 72 μm high duct that could measure extremely low flows [35]. This device had an overall temperature rise of less than 3 °C, a value so small that it should not influence the flow in any size of

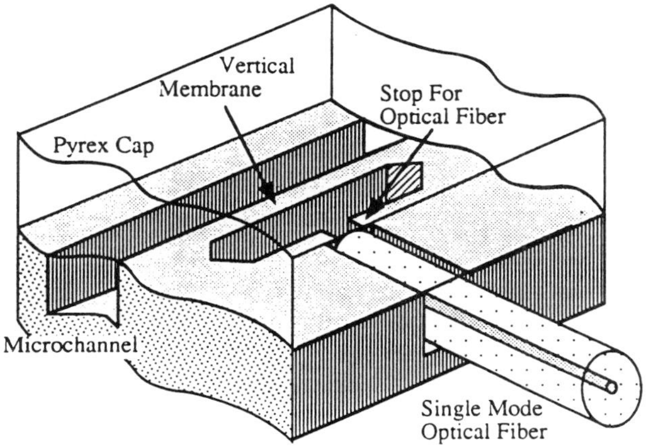

Figure 12.5 An illustration of a vertical-wall interferometric pressure sensor.

duct. Pyroelectricity arises spontaneously in ferroelectric materials. It results in the generation of surface charge when the temperature of the material is changed. The change in surface charge per degree Celsius per square millimeter along a given axis is the pyroelectric coefficient, p. Because p must be substantially temperature independent in pyroelectric anemometry, $LiTaO_3$ was selected as the sensor material. To obtain a continuous electrical output from a pyroelectric element, the temperature of the element must change continuously. This is accomplished with an AC Joule input to an on-board heater. The AC thermal signal generates an AC surface charge that follows the heat content of the pyroelectric structure. At steady state, the Joule heat in is balanced by heat losses to the surroundings, i.e. radiative, conductive, or convective heat flow. The convective heat flow modifies the *AC heat loss* which transduces into a corresponding change in the AC charge state of the pyroelectric element. When connected to an external load, the AC current generated is a direct measure of the heat loss. The PA signal is measured symmetrically from two electrodes on either side of a central heater (see figure 12.6) to obtain a measure of the differential convective heat loss. This difference is a measure of the mass flow.

Hsieh *et al* conducted extensive experimental and theoretical studies of the PA [67, 68]. They demonstrated that the PA has an extremely large flow range and excellent reproducibility. Yu's study demonstrated the possibility of using this device to measure flows as low as 0.05 sccm (or 1 $\mu l\ s^{-1}$) as shown in figure 12.6. Though the size of the channel was fairly large (a hydraulic diameter of 140 μm), it may be possible to go to even smaller dimensions. While this structure deviates from the current trend toward fully integrated systems, the results indicate that such a hybrid microstructure may be suited for the low-temperature requirements of microfluidic systems.

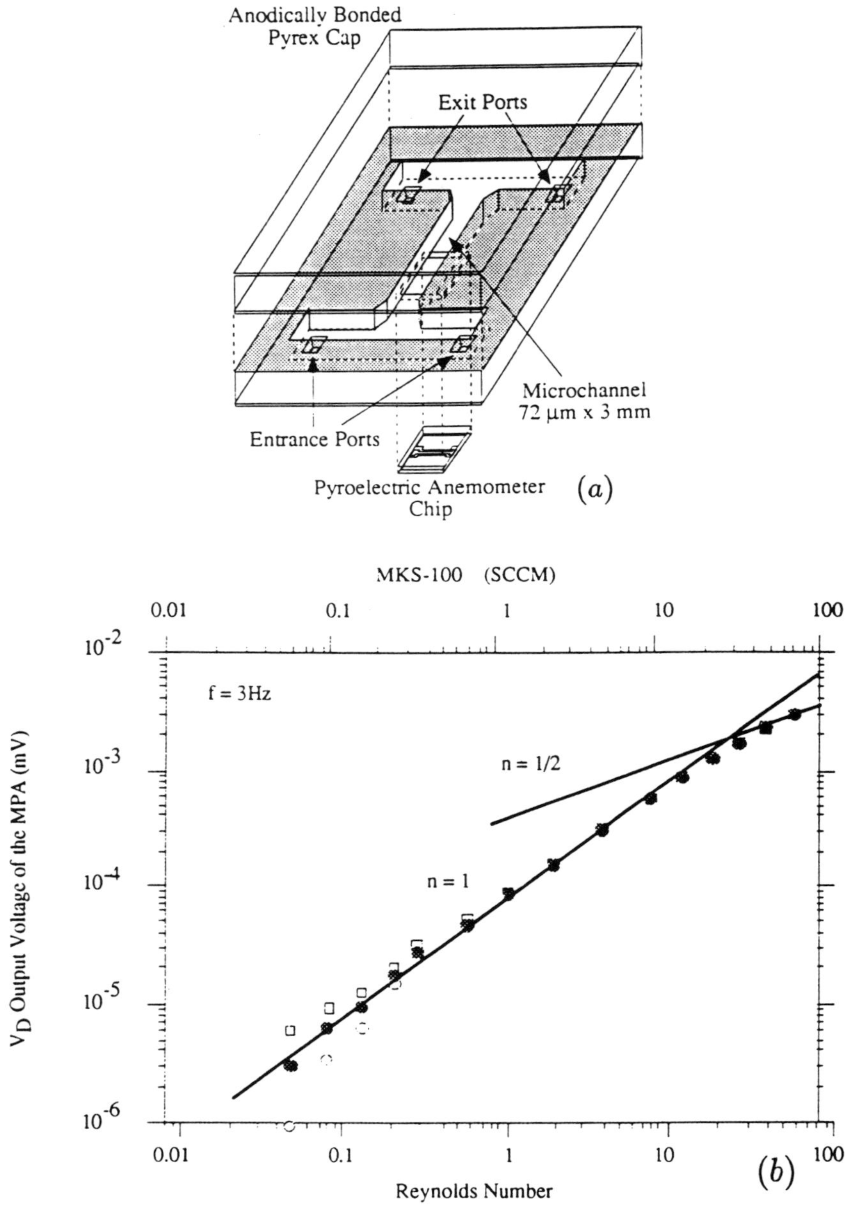

Figure 12.6 (a) An exploded view of a microchannel PA. The anemometer chip is secured in the wall of the channel with epoxy. (b) Experimental data comparing the response of the PA chip in the microchannel to flow with corresponding data taken with an MKS 100 thermal mass flow sensor.

12.4 FLOW ACTUATION AND CONTROL

As pointed out earlier, microfluidic systems have a wide range of applications, e.g. heat exchange systems for electronic devices [28–31], medical diagnostic and analytical chemical applications [69, 70], and precision dilution systems with minimal dead volume for gas chromatography [71]. More may be anticipated as the technology matures. Current research at many laboratories has shown the need to provide flow control at extremely low levels for sensor-controlled implanted drug delivery systems [72] and portable diagnostic cards for polymerase chain reaction analysis [73].

12.4.1 Microvalves and micropumps

As pointed out earlier, silicon microvalves produced with conventional microfabrication technology have been used to control flows as well as generate them. Because of the mechanisms used to drive the valves and pumps, there are limits on how small a device can be made. At present, there does not seem to be a completely satisfactory solution for mechanically operated microvalves and pumps. Interesting designs have been proposed for thermofluidic drives involving phase changes [47] and there have been recent advances with a LIGA-based micropump that shows promise in the valve area as well [46].

12.4.2 Fluidic amplifiers

There are only a few studies on alternative approaches to the conventional microvalve system based on the microfluidic amplifier [48, 74]. The major advantage of microfluidic switching elements is that they provide fluid control without moving parts. It is the fluid itself that controls the flow. Macroscopic fluidic amplifiers are a well developed area of the technology, a subject generally known as 'fluidics' [75]. Developed in 1959, these devices found widespread use in control systems, the aerospace industry, medicine, personal-use items, materials handling, etc. In general, a fluidic amplifier may be categorized by the function it performs or by the phenomenon that is the basis for its operation [75, 76]. These amplifiers are either analog (proportional) or digital (bistable). Considering the fluid phenomena, fluidic amplifiers can be classified as jet deflection (or beam deflection), wall attachment, impact modulation, flow mode control, vortex devices, etc. Jet deflection and wall attachment fluidic amplifiers are the devices that have gained attention for applications in microfluidics. These fluidic amplifiers have at least four basic functional parts: (1) a supply port; (2) one or more control ports; (3) one or more output ports; and (4) an interaction region. The jet from the supply nozzle is directed to one or another output, depending on the pressure/flow of the control inputs. Most amplifiers also contain vents (5) to isolate the effects of the output loading from control flow characteristics.

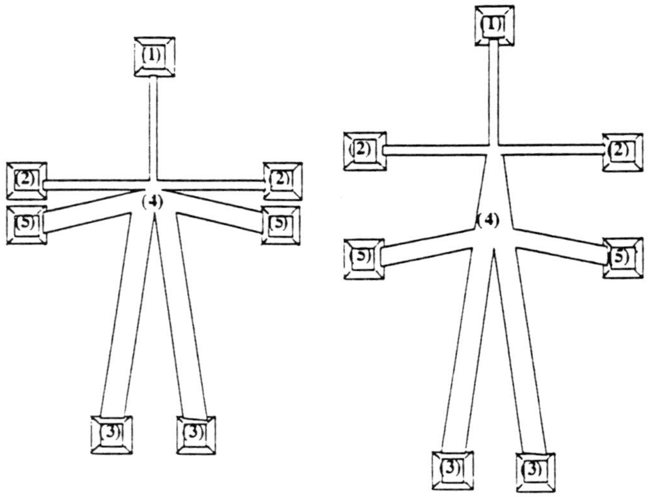

Figure 12.7 Schematic diagrams of jet deflection and wall attachment fluidic amplifiers. (1) is the source inlet; (2) are the control ports; (3) are the waste ports; (4) is the control region; (5) are the output ports.

In jet deflection amplifiers, the control input generates a proportional or analog output by continuously varying (over a limited range) the jet direction. The jet is divided by the configuration of the output ports as seen in figure 12.7. In wall attachment amplifiers, the 'Coanda effect'—a turbulent jet's property of attaching itself to a wall—causes the device to be bistable. The jet will remain attached to one of the walls downstream of the control ports until a sufficient pressure signal is applied to the control port on that side. Different geometrical configurations of wall attachment amplifier can produce most of the common digital logic functions. In addition, a fluidic oscillator can be produced by providing a feedback loop from each output receiver to its corresponding control port. Studies here have shown that jet formation can occur at very low Reynolds number in microfluidic structures [77].

The results to date indicate that the control function can be achieved fairly easily. What is not certain is whether systems can be developed that are tolerant of the waste stream that is essential for effective control. Like many aspects of microfluid research, there is much that remains to be done in the control area and it appears more and more likely that hybridized structures are likely to be the appropriate route to follow rather than the fully integrated system.

12.5 FLUID FLOW PHENOMENA

No matter what eventually transpires in the area of microfluidic devices, what will always be required is as complete an understanding of fluid flow in these small structures as can be obtained. There have been several studies on fluid flows in small simple structures that provide a valuable basis for understanding more complex flow systems. These structures are little more than linear ducts. Flow in such ducts may be characterized by the Fanning friction factor, f_F, the Reynolds number, Re, and their product, the Poiseuille number Po:

$$f_F\mathrm{Re} = \mathrm{Po} \qquad f_F = 2\bar{\tau}_w/\rho\langle U\rangle^2 \qquad \mathrm{Re} = \rho D_h\langle U\rangle/\mu \tag{12.2}$$

where ρ is the density, μ is the absolute (dynamic) viscosity, and U is the cross-sectionally averaged velocity. These relations hold for steady, isothermal, incompressible, fully developed laminar flow of Newtonian fluids, i.e. those in which the rate of shear deformation is proportional to shear rate, through a duct of arbitrary but constant cross-section. The perimeter- (P-) averaged wall shear stress, $\bar{\tau}_w$, is defined as

$$\bar{\tau}_w = \int_P \mu\left(-\frac{\mathrm{d}u}{\mathrm{d}r}\right)_w \mathrm{d}l = \frac{D_h}{4}\frac{\mathrm{d}p}{\mathrm{d}x} \tag{12.3}$$

where u is the local velocity and p is the pressure. r and x are the directions normal and parallel to the wall (and flow), respectively. Combining equations (12.2) and (12.3) yields the Poiseuille number in terms of experimental quantities

$$\mathrm{Po}_{exp} = AD_h^2\,\Delta p/2\mu QL \tag{12.4}$$

where Δp is the pressure drop across a channel of length L, A is the cross-sectional area, and $Q = UA$ is the volumetric flow rate. A schematic illustration of the type of channel employed is given in figure 12.8.

The Poiseuille number is a dimensionless constant which is independent of fluid material properties, velocity, temperature, and duct size. It is solely a function of the duct shape. Po_{theory} may be calculated directly from equations (12.2) and (12.3) for simple geometries, such as circular ducts (Po $= 16$) or infinite parallel plates (Po $= 24$). For the isosceles triangular channels, one of the geometries depicted in figure 12.8, the Poiseuille number was calculated numerically to be $\mathrm{Po}_{triangle} = 13.30$, which agrees well with the values presented by Shah and London [10]. For the trapezoidal channel (with height/width aspect ratio of 0.0277), the Poiseuille number was calculated numerically to be $\mathrm{Po}_{trap} = 22.94$. The ratio of Po_{exp} to Po_{theory} yields a parameter Po*. Determining Po* for microchannels is of interest not only to test the hydrodynamic laws for small structures but also to obtain a better understanding of the design requirements for microscale fluidic systems.

12.5.1 Earlier studies

There have been few experiments on fluid flow in microscale ducts ($\leqslant 25$ μm) because they are difficult to make and even more difficult to characterize. There

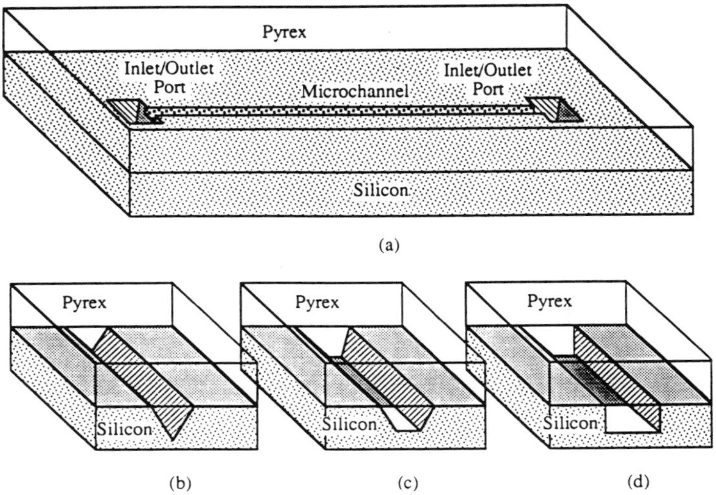

Figure 12.8 (a) An illustration of a simple straight microfabricated channel. Note that the inlet/outlet ports go through the back of the silicon chip. (b) A crystallographically etched triangular cross-sectional channel produced by the selective attack of KOH on silicon. (c) A crystallographically etched trapezoidal cross-sectional channel produced by the selective attack of KOH on silicon. (d) A rectangular cross-sectional channel produced by reactive ion etching.

were two notable, closely related studies on flow in very small channels in the 1970s. These investigations arose from the so-called 'polywater' problem. Beck and Schultz [78] and Anderson and Quinn [79] constructed nanoscale channels with hydraulic diameters, D_h, in the nanometer range (3 nm $\leqslant D_h \leqslant$ 40 nm) by etching damage tracks induced by nuclear particles in mica. Beck and Schultz measured flow employing a pressure head of only 0.78 kPa. Both diffusion and conductivity measurements were carried out in the Beck–Schultz studies but only ionic transport processes were examined by Anderson and Quinn. The Anderson–Quinn study did include a measurement of the temperature dependence of the ionic conductivity. The agreement between the temperature dependence of the ionic conductivity in the bulk and the nanoscale channels was excellent. Beck and Schultz found that the distribution of damage track channel diameters in mica was respectably narrow, with a standard deviation of the order of 13% of the mean. By averaging over this relatively narrow D_h distribution, they showed that this variation did not materially influence their results. Both groups assumed effusive flow conditions to estimate the average diameter of the pores from gas flow through their samples. They concluded that there were no anomalies in the viscosity of water for pores in the 3–40 nm range and temperatures below 40 °C. Their values for the viscosity of water were approximately the same as those obtained by Poiseuille in his original, seminal paper on distilled water flow in capillaries with dimensions down to 29 μm [5]. Interestingly, Poiseuille's values agree to within 1% of currently accepted values for the viscosity of water.

12.5.2 Microchannel studies

The earlier work on mica pores gave strong support to the hypothesis that the viscosity was a local property of liquids down to the molecular level. Supporting this result were some of Israelachvili's studies on shearing of layers several molecular distances thick. Because of the potential importance of microfluidic systems, several studies were initiated to examine fluid flow in micrometer scale ducts. Photolithographic microfabrication was used to make channels with D_h as small as 1 μm. For these dimensions, Re is quite small and so is the volumetric flow. To put the point more finely, consider the case of a triangular microchannel with $D_h = 10$ μm, $L = 10$ mm, and $\Delta p = 10$ kPa. Using equation (12.4) with $\mathrm{Po}_{theor} = 13.30$, the volume flow for water is of the order of 160 μl day^{-1}!

The studies by Harley and Pfahler employed crystallographically selective (trapezoidal cross-section, figure 12.8(c)) or reactive ion etching (rectangular cross-section, figure 12.8(d)) [80]. Urbanek used a crystallographically selective etch to form a triangular cross-sectional channel (figure 12.8(b)) [81]. Harley's work on gas flows through microchannels with 1 $\mu\mathrm{m} \leqslant D_h \leqslant 50$ μm supports the predictions of Navier and Stokes and rarefied gas dynamics theory [82]. For small D_h values, attention must be paid to the ratio of the collision mean free path of the gas molecules, λ_{mfp}, and D_h. This is the Knudsen number, Kn, defined as

$$\mathrm{Kn} = \frac{\lambda_{mfp}}{D_h}. \tag{12.5}$$

Gas flows may be treated as behaving according to hydrodynamic laws for $\mathrm{Kn} \leqslant 0.01$. At higher values of Kn, various corrections must be applied [81]. Kn never exceeded 0.36 in Harley's studies but at these levels the flow is in the slip regime. Pfahler studied the flow of 2-propanol and a light silicone oil through similar microchannels [77]. He found indications of duct-size-dependent Poiseuille numbers. Urbanek examined the temperature dependence of the Poiseuille number for five different liquids using three different values of D_h and found a small but persistent increase in the Poiseuille number with increasing temperature [81].

Since equation (12.4) is the basis for the Harley–Pfahler–Urbanek studies, the accuracy of the channel dimensions, volumetric flow, pressure, bulk viscosity, and temperature determine the uncertainty of the final results. The largest uncertainty was the measurement of channel dimensions which translates into errors of 8–16% in Po_{exp}. The actual variation from theory is somewhat better than these numbers. A meniscus method was used by Harley, Pfahler, and Urbanek to obtain reproducible data in a precision capillary. Urbanek was able to show that this system worked reasonably well over an 80 °C range of temperatures. Pressure sensors were unavailable to determine the fluid pressure in the microchannels themselves. Consequently, pressure sensors were installed upstream and downstream of the test channel. Calculations indicated that the pressure losses in the connecting tubing between the test channel and the pressure sensors were negligible. Temperature stability in all measurements was better than 1 °C and pressure variations were less than 1%. Consequently, reproducibility was typically ±4–6% in the Pfahler–Harley–Urbanek measurements. Because of the difference in the character of the fluids investigated (gas and liquid), it is convenient to divide the discussion of the experimental results accordingly.

12.5.2.1 Gas flows in microchannels. Figure 12.9 depicts Harley's data on the measured Poiseuille number normalized with the calculated one for nitrogen, helium, and argon flow in a 11.04 μm deep channel. The discharge pressure was atmospheric. The ratio between the experimental and theoretical values is close to one (within experimental error) which indicates that gas flow in this size channel behaves according to the predictions of the Navier–Stokes equations with a non-slip velocity at solid boundaries. These data were typical of the results obtained with a number of different channels with varying D_h. During the course of these measurements, it was observed that choking occurred when the inlet pressure became high enough. This is not particularly surprising since the large length to hydraulic diameter of these microchannels is expected to strongly impede the flow. This result is of particular interest for extremely-low-flow-rate wind tunnels. As an illustration, in the case of a 10.9 mm long, $D_h = 8.88$ μm channel with an inlet pressure of 1.124 MPa and an outlet pressure of 0.101 MPa, the measured flow of N_2 was 29.1 μl s^{-1}, a value some 15% less than the theoretical value.

Figure 12.10 depicts the normalized Poiseuille number as a function of the Reynolds number for nitrogen and helium flow in a 0.51 μm deep conduit. The dashed line represents data normalized with the Poiseuille number predicted by the Navier–Stokes equations using a non-slip boundary condition. Witness that at low Reynolds numbers, the experimental Poiseuille number is about 18% below the theoretical one. This deviation far exceeds our experimental error but it can be explained by wall slip. As the Reynolds number decreases so does the average pressure in the channel, which in turn leads to an increase in the Knudsen number (the ratio between the molecular mean free path and the conduit's characteristic size). In figure 12.10, the maximum Knudsen number achieves a value of 0.36. This suggests that the flow is non-continuum in nature. When a first-order slip boundary condition (the solid line in figure 12.10) is incorporated into the Navier–Stokes equation, we were able to predict most of the experimental data with reasonable accuracy. The deviations for Re $<$ 0.02 suggest that at this degree of refinement one cannot use a simple, first-order slip model.

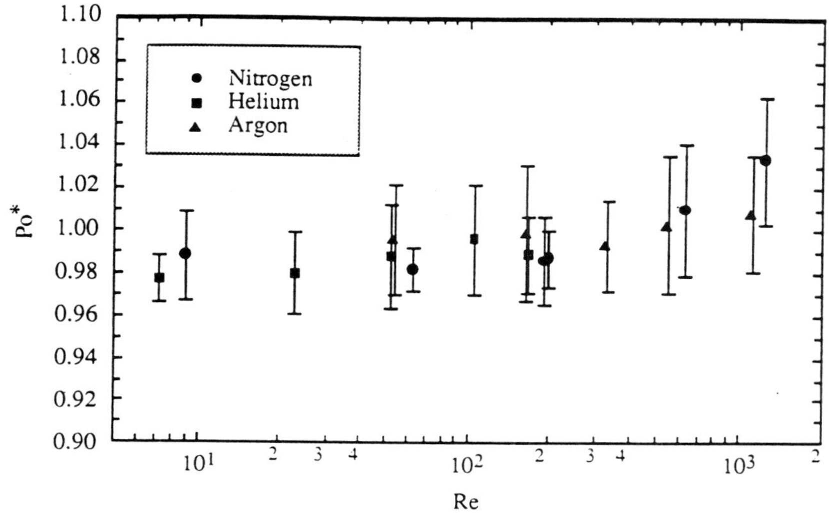

Figure 12.9 The normalized Poiseuille number (Po*) is depicted as a function of the Reynolds number (Re) for nitrogen, helium, and argon flow in a 11.04 μm deep channel. The vertical bars represent two standard deviations.

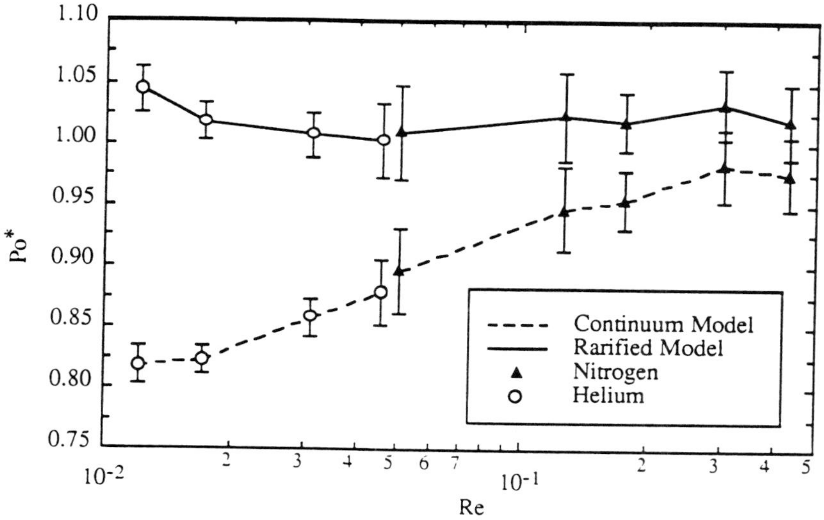

Figure 12.10 Normalized Poiseuille numbers (Po*), calculated with non-slip (dashed line) and slip (solid line) boundary conditions, are depicted as functions of the Reynolds number (Re) for nitrogen and helium flow in a 0.51 μm deep channel. The vertical bars represent two standard deviations.

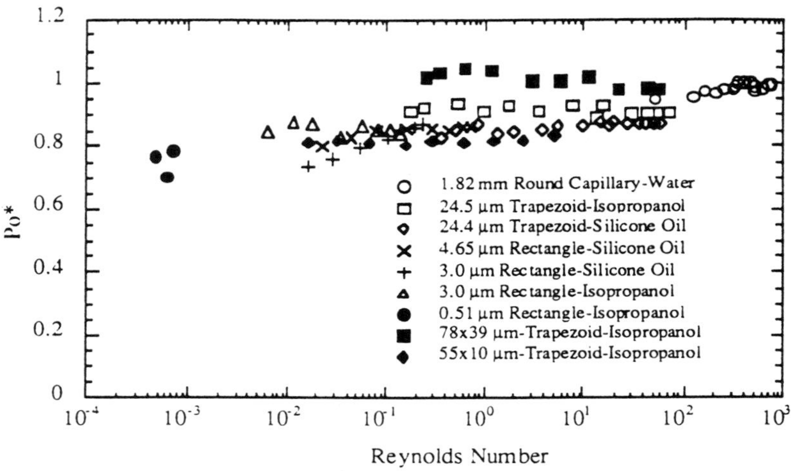

Figure 12.11 The variation of the Poiseuille number for a wide range of channel sizes and geometries as noted in the legend. Three different liquids, 1-propanol, a light silicone oil, and water, were used.

12.5.2.2 Liquid flows in microchannels. The work of Beck and Schultz and Anderson and Quinn provided strong if not compelling evidence for the validity of hydrodynamic theory for water flow in extremely small channels. The more recent studies by Israelachvili gave further substantiation to the theory [83]. Pfahler's study employed a number of different liquids such as isopropanol, water, and a light silicone oil [36, 77]. Unlike the situation with the etched pores in mica with their statistical size distribution, Pfahler prepared well defined channels in the 1–40 μm hydraulic diameter range. Po* for his liquids were independent of the Reynolds number as seen from the data in figure 12.11. However, when these data for each size channel are averaged over all values of Re, there appears to be a dependence on the channel's characteristic size. When the Re-averaged Po* values are plotted, as in figure 12.12, there appears to be a distinct downward bias for 40 μm $\geqslant D_h \geqslant$ 1 μm. The disagreement is small, corresponding to an effective viscosity decrease of no more than 25%. The problem here, unlike the situation with the gas data, is that there is no obvious mechanism to account for this behavior. Since Po* depends on the fourth power of D_h, a measurement uncertainty of 5–6% brings all the data into agreement with theory.

In the light of Pfahler's observation of a possible D_h dependence, Urbanek undertook a series of measurements to examine the temperature dependence of Po* for five different liquids, 1- and 2-propanol, 1- and 3-pentanol, and water, in three different channels [81]. The results of this series of experiments were somewhat surprising. After correcting for the temperature dependence of

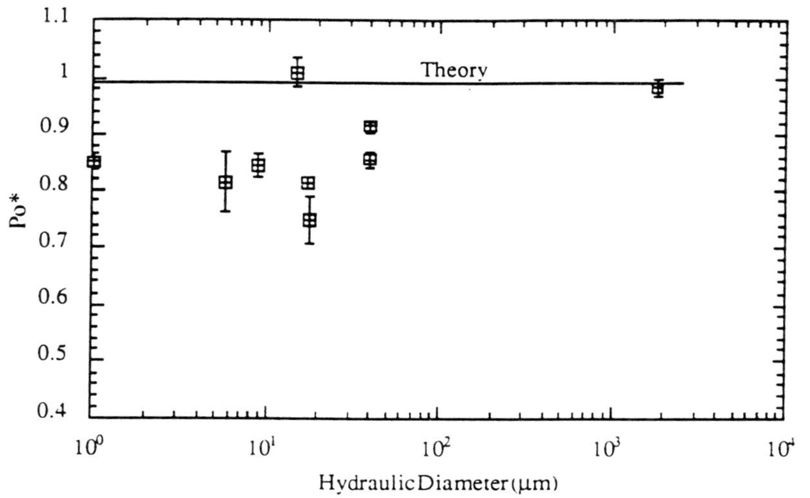

Figure 12.12 The variation of the Poiseuille number with hydraulic diameter averaged over Re for all the different liquids depicted in figure 12.10.

viscosity, it was found that Po* increased as a function of temperature. A sample of the data on two 25 μm channels is shown in figure 12.13(a) and on two 12 μm channels in figure 12.13(b). These data indicate that Po* is not less than the theoretical value, suggesting that the slight bias toward Po* < 1 observed by Pfahler actually fell within the range of experimental uncertainty for the dimensions of the channels. However, while the values of Po* for the four channels examined in figure 12.12 are greater than unity by approximately 6% (which might have been due to a size error of the order of 1–2%), the upward bias could not be so readily dismissed. No source of experimental error was found that could account for the systematic 7% increase in Po* observed over the 60 °C range studied. The increase in Po* becomes more evident at higher temperatures. In the 0–40 °C range, the variation is only 4%, a value within the experimental error of most of the research previously reported. It remains to be seen whether these relatively small deviations from theory are signals of new phenomena or a demonstration of the difficulty of carrying out these experiments.

12.6 CONCLUSION

Microsensors for general use are rapidly moving into the mainstream of contemporary microelectronic efforts. Pressure, temperature, and flow sensors based on micromachined elements are widely employed in all parts of the world economy and their use can only grow. The introduction of similar sensors into microfluidic systems has barely begun so it would be unwise to predict what

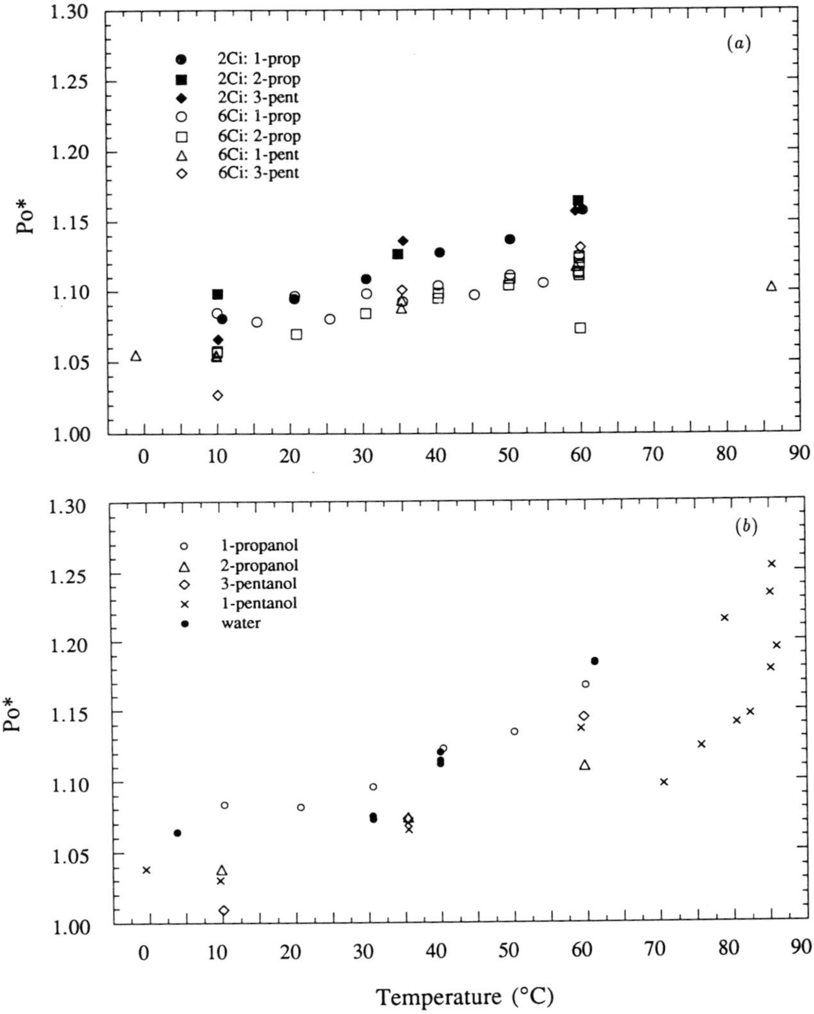

Figure 12.13 The temperature variation of Po* for two channels: (a) solid symbols, 2Ci, $D_h = 25.6\ \mu$m, and open symbols, 6Ci, $D_h = 26.0\ \mu$m; (b) 6Civ, $D_h = 11.6\ \mu$m.

might be the outcome of such efforts. Information on the basic elements of pressure, temperature, and flow are essential for any control elements that might be incorporated into future microfluidic systems. What can be said is that our knowledge of fluid fundamentals is moving along nicely and that should allow new classes of microsystems to arise.

There are opportunities for both new applications and knowledge to be derived from microfluidic investigations. The development of fluidic logic elements with hydraulic diameters in the 1–50 μm range is just beginning.

Harley has demonstrated that choked flow arises in microchannels in this dimensional range. It will be interesting to determine whether it is possible to generate choked flow in microfluidic amplifiers and whether it is possible to control flows under these or more extreme conditions. In any event, controlling flows with these small fluidic elements could be an important enabling result of research in these areas, especially for the control of injected materials into a flow stream. The major challenge confronting this area of research is to find appropriate means for rapidly altering the flow conditions in these extremely small devices.

The measurements by Harley, Pfahler, and Urbanek not only provide a solid basis for modeling fluid flows in small ducts but also raise a question about the nature of that flow at elevated temperatures. The lower-temperature data justify the use of hydrodynamic theory in simple ducts. Whether this will hold in more complex flow structures needs further study. For gas flow in ducts where the Knudsen number is 0.05 or greater, slip flow is observed. Urbanek's data suggest that there may be increased wall interactions as the temperature approaches the boiling point. A more definitive study is needed to clarify this point.

To close this chapter, it should be clear that the future of this topic is as likely to be decided by researchers who are outside this field at present as the current students of the subject. The stroke of intuition that led a clinical chemist to use a microchannel to monitor sperm motility will undoubtedly be repeated in different venues for different problems. In light of the results obtained to date, there is every expectation that completely unique applications of microfluidics will arise shortly.

REFERENCES

[1] White F M 1991 *Viscous Fluid Flow* 2nd edn (New York: McGraw-Hill)

[2] Bernoulli D 1738 *Hydrodynamica* Argentorati

[3] Wolf A 1939 *A History of Science, Technology and Philosophy in the Eighteenth Century* (New York: Macmillan)

[4] Lavoisier A L 1864–1893 *Œuvres de Lavoisier, publices par les soins du Ministre de l'Instruction Publique* (Paris)

[5] Translation by Herschel W of Poiseuille J 1940 Recherches expérimentales sur le mouvement des liquides dans les tubes de très petits diamètres *Mémoires Presentés par Divers Savants à l'Académie Royal des Sciences de l'Institut de France (Rheological Memoirs 1)* ed E Bingham (Lancaster: Lancaster Press)

[6] An interesting but short description of Benjamin Franklin's studies of electricity may be found in a biographical article in the *Encyclopedia Britannica* 1910 vol 11, pp 29–30

[7] Stanley H E and Ostrowsky N 1988 *Random Fluctuations and Pattern Growth: Experiments and Models* (Dordrecht: Kluwer)

[8] Bau H H and Wang Y-Z 1991 Chaos: a heat transfer perspective *Annual Reviews in Heat Transfer* vol IV, ed C L Tien, pp 1–50

[9] Ott E, Grebogi C and Yorke J A 1990 Controlling chaos *Phys. Rev. Lett.* **64** 1196–9

[10] Shah R K and London A L 1978 *Laminar Flow Forced Convection in Ducts* (New York: Academic)

[11] De Groot S R and Mazur P 1966 *Irreversible Thermodynamics* (Amsterdam: North-Holland)

[12] Hanson J P and MacDonald I R 1991 *Theory of Simple Liquids* (New York: Academic)

[13] Bergveld P 1970 Development, operation and application of the ion sensitive field effect transistor *IEEE Trans. Biomed. Eng.* **BME-19** 70

[14] Samaun, Wise K D and Angell J B 1973 An IC piezoresistive pressure sensor for biomedical instrumentation *IEEE Trans. Biomed. Eng.* **BME-20** 101–9

[15] Moss S D, Janata J and Johnson C C 1975 Potassium ion-sensitive field effect transistor *Anal. Chem.* **47** 2238

[16] Lundstrom I, Shivaraman S, Svenson C and Lundqvist L 1975 A hydrogen sensitive MOS field effect transistor *Appl. Phys. Lett.* **26** 55–6

[17] Zemel J N 1975 Ion-sensitive field effect transistors and related devices *Anal. Chem.* **47** 255A–268A

[18] Petersen K E 1982 Silicon as a mechanical material *Proc. IEEE* **70** 420–57

[19] The journal *Sensors and Actuators* is an excellent source of references on sensor science and technology since 1980. A great deal of material may also be found in the proceedings of a series of biennial conferences on *Solid State Sensors and Actuators*

[20] Reports on progress in the general area of microelectromechanical systems (MEMS) began at the *Microrobots and Teleoperator Workshop (Hyannis Port, MA, 1987)* and has continued as a biennial event since then

[21] Angell J B, Terry S C and Barth P W 1983 Silicon micromechanical devices *Sci. Am.* **248** 44–55

[22] Chang S-C and Ko W H 1994 Capacitive sensors *Mechanical Sensors (Sensors: A Comprehensive Survey 7)* ed H H Bau, N F de Rooij and B Kloek (Series ed W Göpel, J Hesse and J N Zemel) (Weinheim: VCH) ch 4

[23] Igarashi I 1989 Automotive: on-board sensors *Fundamentals and General Aspects (Sensors: A Comprehensive Survey 1)* ed T Grandke and W H Ko (Series ed W Göpel, J Hesse and J N Zemel) (Weinheim: VCH) ch 14

[24] Howe R and Muller R S 1984 Integrated resonant microbridge vapor sensor *Digest IEEE IEDM* (New York: IEEE) 354–7

[25] Guckel H and Burns D W 1986 Fabrication techniques for integrated sensor microstructures *Digest. IEEE IEDM* (New York: IEEE) 176–9

[26] Ristic L 1994 Sensors for the Automotive Industry *Sensor Technology and Devices* (Boston: Artech House) pp 401–2

[27] Muller R S, Howe R T, Senturia S D, Smith R L and Ehite R M (ed) 1991 *Sensors* (New York: IEEE)

[28] Tuckerman D B 1984 Heat transfer microstructures for integrated circuits *PhD Dissertation* Stanford University

[29] Tuckerman D B and Pease R F W 1981 High performance heat sinking for VLSI *IEEE Electron Device Lett.* **EDL-2** 126–9

[30] Aung W (ed) 1988 *Cooling Technology for Electronic Equipment* (Washington: Hemisphere)

[31] Weisberg A, Bau H H and Zemel J N 1992 Analysis of microchannels for integrated cooling *Int. J. Heat Mass Transfer* **35** 2465–74

[32] Huijsing J H, Schuddemat J P and VerHoef W 1982 Monolithic integrated direction-sensitive flow meter *IEEE Trans. Electron Devices* **ED-29** 133–6

[33] Petersen K, Brown J and Renken W 1985 High-precision, high-performance, mass-flow sensor with integrated laminar flow micro-channels *Proc. Int. Conf. on Solid-State Sensors and Actuators (Philadelphia, PA)* pp 361–3

[34] Hsieh H-Y 1993 Pyroelectric anemometry *PhD Dissertation* University of

Pennsylvania
[35] Yu Dun, Hsieh H Y and Zemel J N 1993 Microchannel pyroelectric anemometers for gas flow measurements *Sensors Actuators* **39** 29–35
[36] Pfahler J, Harley J C, Bau H H and Zemel J N 1990 Liquid transport in micron and submicron channels *Sensors Actuators* A **21–23** 431
[37] Wilding P, Pfahler J, Bau H H, Zemel J N and Kricka L J 1994 Manipulation and flow of biological fluids in straight channels micromachined in silicon *Clin. Chem.* **40** 43–7
[38] Wilding P, Shoffner M A and Kricka L J 1994 PCR in a silicon microstructure *Clin. Chem.* **40** 1815–18
[39] Kricka L J, Nozaki O, Heyner S, Garside W T and Wilding P 1993 Applications of a microfabricated device for evaluating sperm function *Clin. Chem.* **39** 1944–7
[40] Bousse L J, Parce J W, Owicki J C and Kercso K M 1990 Silicon micromachining in the fabrication of bio sensors using living cells *IEEE Solid-State Sensor and Actuator Workshop (Hilton Head Island, SC)* (New York: IEEE) pp 173–6
[41] van Lintel H T G, van de Pol F C M and Brouwstra A 1988 A piezoelectric micropump based in micromachining of silicon *Sensors Actuators* **15** 163–7
[42] Gass V, van der Schoot V H, Jeanneret S and de Rooij N F 1994 Integrated flow-regulated silicon micropump *Sensors Actuators* A **43** 335–8
[43] Nakagawa S *et al* 1990 Integrated fluid control systems on a silicon wafer *Micro System Technologies (Berlin, 1990)* p 793
[44] Ohnstein T, Fukiura T, Ridley J and Bonne U 1990 Micromachined silicon microvalve *Proc. IEEE Micro-Electro-Mechanical Systems Workshop (Napa Valley, CA)* (New York: IEEE) p 95
[45] Yanagisawa K, Tago A, Ohkubo T and Kuwano H 1990 Magnetic microactuator *Proc. IEEE Micro-Electro-Mechanical Systems Workshop (Napa Valley, CA)* (New York: IEEE) pp 120–4
[46] Rapp R, Schomburg W K, Maas D, Schulz J and Stark W 1994 LIGA micropump for gases and liquids *Sensors Actuators* A **40** 57–61
[47] Ji J, Chaney L J, Kaviany M, Bergstrom P L and Wise K D 1991 Microactuation based on thermally-driven phase-change *Digest IEEE Int. Conf. on Solid State Sensors and Actuators (1991)* (New York: IEEE) p 1037
[48] Zdeblick M J, Barth P W and Angell J B 1986 Microminiature fluidic amplifier *Tech. Digest Solid State Sensors Workshop (Hilton Head Island, SC, 1986)*
[49] Brodie I and Muray J J 1992 *The Physics of Micro/Nano-Fabrication* (New York: Plenum)
[50] Elliott D J 1989 *Integrated Circuit Fabrication Technology* 2nd edn (New York: McGraw-Hill)
[51] Price J B 1973 Anisotropic etching of silicon with KOH–H_2O–isopropyl alcohol *Semiconductor Silicon (Electrochemical Society Symposium Series)* ed H R Huff and R R Burgess (Pennington, NJ: ECS) pp 339–53
[52] Harrison J J, Seiler K, Manz A and Fan Z 1992 Chemical analysis and electrophoresis systems integrated on glass and silicon chips *Tech. Digest Solid-State Sensor and Actuator Workshop (Hilton Head Island, SC)* (New York: IEEE) pp 110–13
[53] Kricka L J, Ji X, Nozaki O, Heyner S, Garside W T and Wilding P 1994 Sperm testing and microfabricated glass-capped silicon microchannels *Clin. Chem.* **40** 1823–4
[54] Lang S B 1974 *Sourcebook of Pyroelectricity* (New York: Gordon and Breach)
[55] Moser D 1993 CMOS flow sensors *PhD Dissertation* ETH Zurich
[56] Bicking R E 1981 A piezoresistive integrated pressure transducer *Inst. Mech. Eng.* C **164/81** 21–6

[57] Fung C D and Ko W H 1982 Miniature capacitive pressure transducers *Sensors Actuators* **2** 321–6

[58] Uttamchandani D, Thornton K E B, Nixon J and Culshaw B 1987 Optically exited diaphragm pressure sensor *Electron. Lett.* **23** 152–3

[59] Thornton K E B, Uttamchandani D and Culshaw B 1990 A sensitive optically excited resonator pressure sensor *Sensors Actuators* **24** 15–19

[60] Valette S, Renard S, Jadot J P, Gidon P and Erbeia C 1990 Silicon-based integrated optics technology for optical sensor applications *Sensors Actuators* A **21–23** 1087–91

[61] Tu X Z and Zemel J N 1993 Vertical membrane–optical fiber pressure sensor *Sensors Actuators* **39** 49–54

[62] Chapman D R and Rubesin N W 1949 Temperature and velocity profiles in the compressible laminar boundary layer with arbitrary distribution of surface temperature *J. Aeronaut. Sci.* **16** 547–65

[63] van Oudeheusden B W 1991 The thermal modeling of a flow sensor based on differential convective heat transfer *Sensors Actuators* A **29** 93

[64] Rotem Z 1967 The effect of thermal conduction of the wall upon convection from a surface in a laminar boundary layer *Int. J. Heat Mass Transfer* **10** 461

[65] Widmer A E 1982 A calibration system for calorimetric mass flow devices *J. Phys. E: Sci. Instrum.* **15** 213

[66] Moser D, Lenggenhager R and Baltes H 1991 Silicon gas flow sensors using industrial CMOS and bipolar IC technology *Sensors Actuators* A **25–27** 577–81

[67] Hsieh H-Y and Zemel J N 1995 Pyroelectric anemometry: frequency, geometry and gas composition dependence *Sensors Actuators* at press

[68] Hsieh H-Y, Zemel J N and Bau H H 1995 The theory of the pyroelectric anemometer *Sensors Actuators* at press

[69] Dawes T, McReynolds, Modlin D and Bousse L 1994 Fluidic performance of micromachined multichannel microphysiometer systems *Proc. Symp. on Microstructures and Microfabricated Systems* 94–14 (Pennington, NJ: Electrochemical Society) p 220

[70] An example of this type of device is the multispecies chemical sensor chips developed by I-STAT Inc., Princeton, NJ, for hand-held blood analysis

[71] Klee M S 1990 *GC Inlets—An Introduction* (Avondale, PA: Hewlett–Packard)

[72] Heuberger A 1993 Silicon microsystems *Microelectron. Eng.* **21** 445

[73] Stix G 1994 Gene readers—microelectronics has begun to merge with biotechnology *Sci. Am.* **January** 149

[74] Vollmer J, Hein H, Menz W and Walter F 1994 Bistable fluidic elements in LIGA technique for flow control in fluidic microactuators *Sensors Actuators* A **43** 330–4

[75] James W J 1983 Fluidics: basic components and applications *Harry Diamond Laboratories Report* HDL-SR-83-9

[76] Foster K and Parker G A 1970 *Fluidics: Components and Circuits* (New York: Wiley–Interscience)

[77] Pfahler J 1991 Liquid transport in micron and submicron size channels *PhD Dissertation* University of Pennsylvania

[78] Beck R E and Schultz J S 1972 Hindrance of solute diffusion within membranes as measured with microporous membranes of known pore geometry *Biochim. Biophys. Acta* **255** 273–303

[79] Anderson J and Quinn J 1972 Ionic mobility in microcapillaries: a test for anomalous water structures *J. Chem. Soc. Faraday Trans.* I **68** 608–12

[80] Harley J C 1994 Compressible gas flows in microchannels and microjets *PhD*

Dissertation University of Pennsylvania
[81] Urbanek W 1994 An investigation of the temperature dependence of the Poiseuille number in microchannel flow *PhD Dissertation* University of Pennsylvania
[82] Harley J C, Bau H H, Huang Y and Zemel J N 1995 Gas flow in micro-channels *J. Fluid Mech.* **284** 257–74
[83] Israelachvili J 1992 Adhesion forces between surfaces in liquids and condensable vapours *Surf. Sci. Rep.* **14** 3–158

13

Practical examples of polymer-based chemical sensors

Michael J Tierney

13.1 INTRODUCTION

A sensor may be thought of as consisting of two parts: the transducer which converts a chemical signal to an electronic signal, and a chemically selective layer which interacts directly with the sample. A transducer itself usually does not have the selectivity, dynamic range, or sensitivity required for many measurement applications. Just as a camera uses different lenses for telephoto, macro, or wide-angle shots, a transducer requires the cooperation of a chemically selective material in the form of a coating, membrane, or electrolyte to measure different species in different applications. These materials serve as a selective receptor to mediate between the outside environment and the sensor's transducer.

During sensor operation, the chemical species to be detected is transported into the chemically selective layer from the environment; the transducer then detects the species in the layer, or detects a change in the layer itself caused by the presence of the species. The key challenge in sensor development is the selection of the proper material to act as this selective layer. Because of their wide range of chemical and physical properties, polymers are often used as chemically selective layers for many types of chemical, gas, and biosensors.

13.2 ROLES OF POLYMERS IN CHEMICAL, GAS, AND BIOSENSORS

Polymer materials have been developed with almost every imaginable combination of physical and chemical characteristics. Their diversity of properties enables polymers to play a wide range of roles in sensor technology. Several of the most important uses are listed in table 13.1 and outlined below.

Table 13.1 Uses for polymers in chemical and biosensors.

Use	Example	Analyte	Reference
Electrolyte (polyelectrolyte)	Nafion	O_2	[13]
Electrolyte (polymer electrolyte)	polyethylene glycol	CO_2	[31]
Electrolyte (hydrogel)	polyvinyl alcohol	glucose	[3]
	polyacrylamide	O_2, CO_2, pH	[23]
Electrolyte and mediator	derivatized poly(vinylpyridine)	glucose	[19]
Gas-permeable membrane	Teflon	O_2	[5]
	polysiloxane	O_2, CO_2, pH	[23]
Diffusion barrier	polyurethane	glucose	[6]
Immobilization matrix (hydrogel)	DEAE dextran	ethanol	[8]
Immobilization matrix	poly-N-methyl pyrrole	glucose	[9]
Interference removal	Nafion	glucose	[12]

13.2.1 Polymeric electrolytes

A liquid electrolyte contains ionic species which can migrate through solution and conduct electrical current. Polymers have been developed which have this same ability allowing them to be used as the electrolyte in the development of all-solid-state electrochemical sensors.

There are two types of polymeric electrolyte, based on their conduction mechanisms. The first group is the polyelectrolyte in which the polymer itself contains an anionic or cationic group, usually on a side chain. The counter-ions for these groups are typically small, inorganic ions that are mobile within the polymer matrix. Nafion, a perfluorinated sulfonated ionomer made by du Pont, is an example of this type of electrolyte [1]. Nafion has been used as the electrolyte in several amperometric gas sensors.

The second type of polymer electrolyte does not itself possess charged moieties along its chain. Rather, the polymer acts as the solvent for electrolyte ions which are able to move through the polymer matrix much as in a liquid electrolyte. Thus, the polymer serves as a solid ionic conductor. An example of this type of polymer is polyethylene oxide in which lithium and other small cations have high mobility [2].

13.2.2 Hydrogels

Hydrogels are actually a third type of polymer electrolyte. However, in hydrogels, the solvent for the ions is water; the water-soluble polymer serves only to increase the viscosity of the solution. Hydrogel electrolytes are formed when the dissolved polymer is cross-linked into a rigid three-dimensional matrix which can contain up to 98% water by volume. To this extent, the hydrogel can be considered as a 'solid', and, thus, more easily processed form of liquid water. An example of a polymer that can form hydrogels is poly(vinylalcohol). PVA can be cross-linked by a variety of agents and can form gels containing as little as 2–3 wt% PVA [3].

The consistency of hydrogels can vary widely, even when made from the same polymer. For example, PVA solutions containing 5 wt% solids form gels with varying consistencies depending on the molecular weight of the PVA and the degree of cross-linking. Typically cross-linkers for PVA are boric acid and glutaraldehyde [4]. The borate ion can bind to two sets of vicinal hydroxyl groups; when these sets are on different polymer chains, a cross-link is formed. Glutaraldehyde cross-links PVA by forming an acetal between the carbonyl group of the glutaraldehyde and two hydroxyl groups on the PVA. The rheology of the resulting gel depends on the degree of cross-linking and can vary from a brittle rubber at high cross-linking to a fluid adhesive when lightly cross-linked.

Hydrogels have found application as replacements for liquid electrolytes. Because of their viscosity, they resist leakage from sensor housings, and many sensing components can be incorporated into them, such as ionic salts, electrochemical mediators, and enzymes. For example, a glucose sensor has been fabricated by immobilizing glucose oxidase in a PVA matrix deposited on a platinized graphite electrode [3].

13.2.3 Gas-permeable membranes

Polymers are often used as gas-permeable membranes for gas sensors for both dissolved gas and atmospheric gas measurements. A typical gas-permeable membrane is hydrophobic and has a high rate of gas transport relative to other species. The membrane may be solid or microporous. Polymers that are commonly used as gas-permeable membranes are Teflon and silicon rubber [5].

Gas-permeable membranes serve a number of functions. They provide the sensor with a selective membrane covering and, thus, reduce contamination of the sensor by incompatible materials in the environment. They allow passage of gases only, preventing excess electrolyte loss in electrochemical sensors. Although the gas-permeable membrane cannot totally eliminate electrolyte loss through evaporation, the rate of diffusion of water vapor through the hydrophobic membrane is often many orders of magnitude slower than the gas transport rate. The membrane can also act as a diffusion barrier to lower the flux of gas reaching the electrode of an amperometric sensor, and thus bring the sensor into diffusion-limited operation [5].

13.2.4 Diffusion barriers

Although it is not desirable to impede diffusion of the analyte molecule to the sensor in many cases, a diffusion barrier can be useful if the analyte is at a sufficiently high concentration that the sensor fails to respond in a linear and reproducible manner. The addition of a diffusion barrier can reduce the flux of analyte to the sensor and so bring the effective analyte concentration down into a linear portion of the sensor's response curve. In this way, a diffusion barrier can increase the sensor's usable measurement range.

Diffusion barriers are typically used where the chemical reaction responsible for sensing is slow relative to the diffusion rate of the analyte to the active portion of the sensor. As shown above, a gas-permeable membrane can act as a diffusion barrier in some cases for gas sensors. Diffusion barriers are also often used in amperometric enzyme sensors which exploit the native selectivity of an enzyme–substrate reaction to measure the concentration of the substrate. If the substrate is at a high concentration, the enzyme may become saturated, especially if the enzyme turnover rate is low. In this condition, the signal will plateau and no longer be dependent upon the substrate concentration. A diffusion barrier between the enzyme layer and the sample reduces the flux of the substrate to the enzyme and, thus, prevents saturation of the enzyme and increases the linear range of the sensor.

In vivo glucose sensors commonly employ diffusion barriers [6]. A major difficulty of amperometric glucose sensors in blood is that the glucose concentration is higher than the O_2 concentration, making O_2 the limiting reagent in the glucose oxidase-catalysed oxidation of glucose. To eliminate this problem, the sensor is coated with a membrane which passes O_2 at a much higher rate than glucose. The flux of glucose to the sensor is reduced to below that of O_2, making glucose the limiting reagent for the sensing reaction.

13.2.5 Immobilization matrices

Suitable matrices for immobilization of biological reagents, especially enzymes, is an active area of biosensor research [7]. Because the activity of an enzyme is determined by its three-dimensional structure, and that structure is, in turn, related to the environment into which the enzyme is placed, the sensitivity of an enzyme-based sensor is greatly affected by this choice of immobilization matrix.

In general, an enzyme will retain its proper three-dimensional conformation when it is in an environment most like its native environment, that is, aqueous, usually close to neutral pH, and containing a moderate salt concentration. Of course, the optimum environment will vary for each enzyme. Therefore, an immobilization matrix should reflect these preferences. Although most pure polymers would be ill suited to be congenial enzyme environments, polymer membranes that are created by cross-linking in an aqueous solution, or by aqueous electrochemical polymerization, contain sufficient water to be hospitable environments.

A good example of this type of polymer is a hydrogel. Hydrogels can contain up to 98% water. Thus, they are already very much like aqueous solutions. In addition, the constituents along the polymer backbone can provide additional stability to the enzyme. For example, the hydroxyl groups in PVA

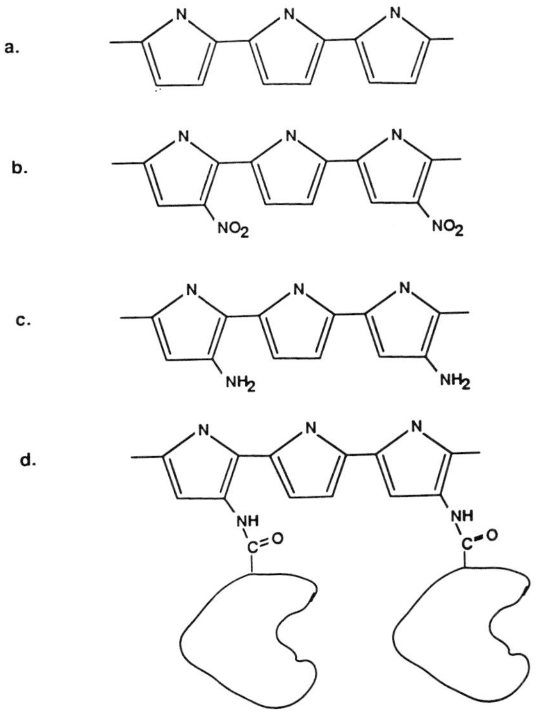

Figure 13.1 Covalent attachment of an enzyme to derivatized polypyrrole: (a) polypyrrole polymer; (b) nitration of polypyrrole; (c) electrochemical reduction of nitro groups to amine groups; (d) attachment of enzyme to amine groups with carbodiimide [32].

have been shown to provide a good degree of stability to the enzyme, probably through hydrogen bonding between the enzyme and the polymer itself. Other polyhydroxyl polymers, such as polysaccharides, are also good enzyme immobilization matrices [8].

Another type of polymeric immobilization matrix is a polymer in which the enzyme can be entrapped and cross-linked into place. For example, polypyrrole films containing glucose oxidase (GOD) can be grown by electrochemical polymerization from aqueous solution. As the polypyrrole film polymerizes on the electrode, the GOD is entrapped in the growing chains [9]. The GOD can then be cross-linked to the polymer by carbodiimide reagents. Alternately, the polypyrrole can be derivatized to provide points of attachment for enzymes (figure 13.1). Other polymers, such as polymerized 1,2-diaminobenzene, can be cross-linked to GOD with glutaraldehyde [10]. This film has the added advantage of rejecting ascorbic acid and other sensor interferences. Although immobilization of an enzyme invariably results in some loss of activity, the immobilization process can lead to greater long-term stability.

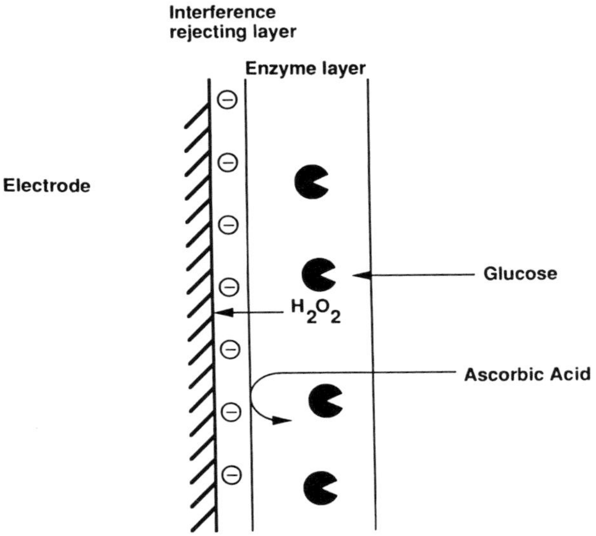

Figure 13.2 Use of an anionic membrane in a glucose sensor to prevent anionic interfering species (e.g. ascorbic acid) from being detected. The sensor detects hydrogen peroxide resulting from the reaction between glucose and glucose oxidase.

13.2.6 Permselective membrane for interference removal

Polymer membranes can also be used as permselective electrode coatings to prevent electroactive interferences from reaching the electrode and generating spurious background currents (figure 13.2). This problem was first encountered with ascorbic acid interference with *in vivo* detection of neurotransmitters [11]. Polymers with anionic or cationic charges exclude ions of similar charge by electrostatic repulsion. For example, Nafion with its anionic sulfonate side groups excludes other anions, such as ascorbic acid. Coating a sensor with a thin, pinhole-free layer of Nafion forms an efficient barrier to anionic interferences. In an analogous manner, polycationic polymers, such as polymers with amine side groups, prevent cations from reaching the electrode.

Preventing uncharged electroactive molecules from reaching a sensor is a more difficult problem. Obviously, an ionomeric polymer would have little or no effect in electrostatically screening out an uncharged interfering substance. However, there are several other screening methods which may be used, such as size exclusion, permeability differences, and diffusion barriers [12]. Eliminating an interference based on size exclusion works best when the interfering molecule has a much higher molecular weight than the molecule to be sensed. The ratio of molecular weights required to obtain exclusion is greatly dependent upon the polymer structure. A high-density polymer would be expected to selectively exclude molecules only a few times larger than the analyte, while a lower density polymer with large free volume will screen out only very large molecules.

Permeability differences between the target molecule and the interference can be used to reduce the amount of interference if a polymer can be chosen in which the interfering molecule is much less soluble than the analyte. For example, an acidic polymer, such as Nafion, would exclude CO_2, an acidic gas, to a much larger extent than O_2 because CO_2 is less soluble in Nafion than O_2 [13]. A simple diffusion barrier takes advantage of the other component of permeability, diffusion rate, by slowing the diffusion rate of the interfering molecule to a much greater extent than that of the target molecule.

13.3 PROPERTY/FUNCTION-BASED SELECTION OF POLYMERS FOR SENSORS

Chemical and biosensors are used to measure a wide variety of analytes in an equally large number of environments from glucose in blood to pH in nuclear power plant cooling water. Because of this wide range of applications, there is no one optimum sensor design for measuring each analyte, or for measuring in each environment.

Fortunately, polymers have a range of properties as diverse as the applications for sensors. Sometimes, however, the multitude of choices of polymers makes selection of the proper polymer difficult, and one resorts to trial and error methods of selection—an inefficient and time-consuming technique. However, sufficient knowledge of the physical and chemical properties of the analyte, the potential interferences, and the environment in which the sensor will operate, combined with insight into the desired function of the polymer, can make selection of polymers for sensor fabrication a much less daunting task.

Several of the more important polymer properties for sensors are reviewed below.

13.3.1 Hydrophobicity/hydrophilicity

Because water is ubiquitous both in the sensing environment and inside many sensors (especially electrochemical sensors), the hydrophobic or hydrophilic nature of a polymer used in a sensor is often crucial. For example, a polymer that is to be used as a hydrogel is by definition hydrophilic. On the other hand, gas-permeable membranes are often made of hydrophobic polymers to prevent passage of water through the membrane. These conventions are not always the case, however. An electrolyte for a sensor operating with non-aqueous electrochemistry may be less hydrophilic. Similarly, an *in situ* sensor to analyse polar degradation products in motor oil may use a hydrophilic membrane to allow passage of the analyte into the aqueous electrolyte from the non-polar hydrocarbon sample [14].

The hydrophilic or hydrophobic properties of a polymer are easily determined by looking at its solubility in various solvents. These properties can be modified by either grafting side chains of different polarities onto the main chain, or by copolymerization of different monomers. An extreme example of this process is found in Nafion. Its backbone is Teflon, a very hydrophobic polymer insoluble in all common solvents. By grafting polyether side chains which terminate in sulfonate groups onto the main chain, the polymer becomes much more hydrophilic, and is soluble in polar organic solvents.

13.3.2 pH/pK_a

The pH of a polymer is often poorly defined except in dilute aqueous solutions. However, the presence of acidic or basic groups, either along the main chain of the polymer, or, more commonly, on side groups, can give a polymer an acidic or basic nature. If the K_a or K_b of these groups is large, this effect can be the main determinant of the polymer's properties. For example, partially cross-linked poly(acrylic acid) can be dissolved in water to form a hydrogel. The pH of this gel is acidic, pH 4–5. If the pH is raised slightly, all the carboxyl groups on the chain are dissociated and the resulting electrostatic repulsion increases the viscosity of the polymer by almost two orders of magnitude.

Similarly, the acid–base properties of a polymer can determine the solubilities of various species. The sulfonate groups in Nafion are strongly acidic, especially when the polymer is hydrated. Carbon dioxide dissolves in aqueous solutions to form carbonic acid. However, the high acidity of Nafion prevents dissolution of CO_2. Therefore, Nafion could serve as a protective membrane to prevent CO_2 from entering a sensor, or as an electrolyte into which CO_2 would not dissolve.

13.3.3 Permselectivity

Permselectivity is the ability of a membrane to allow a high flux of one chemical species (typically the analyte) while reducing or eliminating the flux of other species (chemical or sensor interferences). An extreme example of permselectivity is the liquid polymer membranes used in ion selective electrodes. The ionophores contained in these membranes bind selectively to a specific ion. However, this binding is often so strong that the diffusion constant for these ions through the membrane is vanishingly small [15].

More commonly, permselective membranes create a differential in the transport rate of different species. The permeability differences may be based on physical size, solubility differences, or specific interactions between the polymer and the diffusing species. Permselective membranes are used in a variety of functions, such as diffusion barriers and membranes for interference removal. A commonly used permselective membrane is Nafion which demonstrates an extremely high affinity for organic cations, reasonable affinity for inorganic cations, and almost completely excludes both organic and inorganic anions [16].

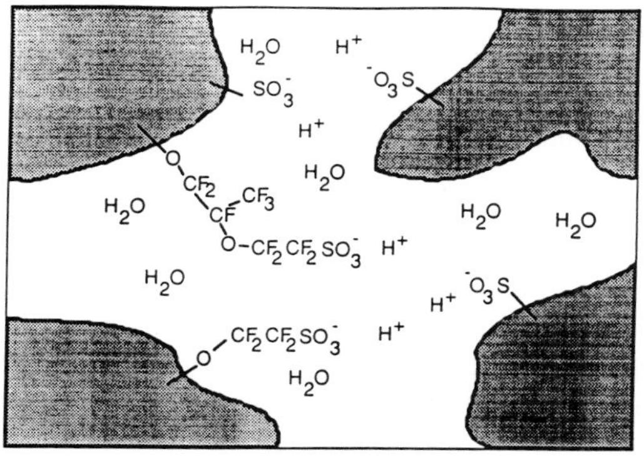

Figure 13.3 The proposed morphology of Nafion. One hydrophilic cluster is shown surrounded by hydrophobic Teflon-like domains. A mobile ion, such as H^+, must move from each hydrophilic cluster to the next.

13.3.4 Ionic conductivity

Ionic conductivity is one of the critical parameters for a polymer to serve as an electrolyte. The mechanism of ionic conduction can vary, however, from simple migration of ions in hydrogels to more complex percolation-type migration in denser polymers. Hydrogels, where up to 98% of the volume is water, act very much like free water; only a small percentage of the water molecules is tightly bound to the polymer chain. Unless the ions have some specific interaction with the polymer, they will migrate in a similar manner to their behavior in water.

At the other extreme, Nafion consists of Teflon-like perfluorinated regions and micelle-like hydrophilic regions containing pendent sulfonate groups, and their associated water molecules [1]. The protons on the sulfonate groups readily dissociate when the polymer is allowed to absorb sufficient water, and may be exchanged for other cations (figure 13.3). It is hypothesized that the hydrophilic regions are interconnected via hydrophilic pathways, and that ionic transport through the polymer occurs via these interconnections [17]. A migrating ion carrying electric current must move through these interconnected hydrophilic regions via what is probably a slow and tortuous path. It is unlikely that an ion would tunnel through the Teflon-like hydrophobic regions.

A third type of ionic conduction occurs in polymer electrolytes, such as polyethylene oxide. The mechanism of ionic conduction in polymer electrolytes is not entirely understood. However, it is believed to involve rapid polymer segmental motion which creates regions of an elastomeric nature. These elastomeric regions have relaxation times similar to liquids, and, thus, allow a higher ionic mobility than would be concluded from the polymer's macroscopic properties [2].

13.3.5 Hospitable environment for biomolecules

Enzymes, like most proteins, are unstable molecules. The stability of enzymes is of particular importance to biosensors because the rate of the enzymatically catalysed reaction is especially sensitive to the three-dimensional conformation of the enzyme. Most, if not all, enzymes are fully active only in a hydrated state. However, rarely are enzymes found in a completely aqueous medium in a sensor. Usually, the enzyme is dissolved into a hydrogel, immobilized onto a surface, or entrapped in a polymer matrix, all environments which differ in varying degrees from the natural environment of the enzyme.

In general, any sort of immobilization or entrapment of enzyme in a polymer will cause conformational changes. These distortions can result in a decrease in the affinity of the enzyme for the substrate or, if the distortion is severe, inactivation of some of the enzyme molecules [18]. Both of these situations reduce the V_{max} of the enzyme, and as a result, the signal from the sensor. Immobilization of the enzyme can effect the kinetics in more subtle ways as well. Ionic or hydrophobic groups on the polymer can alter the microenvironment around the enzyme causing changes in the solubility, hydration, or charge distribution near the active site of the enzyme. The immobilizing polymer can also reduce the diffusion rate of substrate to the enzyme and of product away. Slow transport of reactants and products inside the polymer matrix effectively changes the concentration of these species inside and outside the matrix, leading to effective enzyme kinetics that are not solely a function of the enzyme reactivity.

To minimize these effects, the enzyme can be immobilized in a polymer matrix with the following properties:

(i) hydrophilic, preferably with opportunities for hydrogen bonding with the enzyme;
(ii) containing a high percentage (40–99%) of water;
(iii) neutral pH range (pH 5–8);
(iv) the ability to contain ionic species, such as Na^+ and Cl^-; and
(v) a rapid diffusion rate of enzyme substrate and product (approaching that in solution).

Fortunately, a number of polymers fulfill these requirements. Polyhydroxyl polymers, such as poly(vinylalcohol) and polysaccharides, provide ample sites for hydrogen bonding with both carboxylic and amine groups on the surface of the enzyme. Likewise, polymers with oxygen atoms in their chain, such as poly(ethylene oxide), are also suitable enzyme immobilization matrices. Polymers carrying amide groups, such as poly(acrylamide), also have plentiful sites for hydrogen bonding. It is perhaps not surprising that all of the polymers mentioned above are also hydrogels and contain a large percentage of water as well.

13.4 POLYMER MEMBRANE DEPOSITION TECHNIQUES

There are a number of different techniques for depositing polymer layers. Many of these techniques are suitable for laboratory scale fabrication only, while others are adaptations of methods used in the silicon integrated circuit industry to fabricate reproducible membranes in large numbers.

13.4.1 Solution casting

Depositing polymer membranes from solution is probably the oldest and simplest method for polymer deposition. Typically a solution of the polymer is deposited onto the transducer surface and allowed to dry. Although this technique is simple to practice, it is a fine art to perfect. Polymer concentration, solvent composition, amount of solution deposited, and solvent evaporation rate are all crucial parameters to control in order to obtain films that adhere well to the substrate, are uniform in thickness, and are free from defects. Gregg and Heller have solution cast redox-functional hydrogels containing glucose oxidase on an electrode to fabricate a glucose sensor [19].

A variation on simple solution casting is dip coating. In this technique, the sensor is dipped into a solution of the polymer for a length of time and removed. The same solution parameters hold for dip coating as for solution casting. Dip coating is typically used to coat an entire sensor with an outer protective layer. For example, hydrophobic gas-permeable membranes or interference barriers can be coated onto transducers using this technique [20].

13.4.2 Spin coating

Another variation on solution casting is spin coating. This technique borrows from the methods developed by the semiconductor industry to deposit very thin and uniform layers of photoresist onto silicon wafers. This method has been successfully used in the sensor industry to deposit polymer electrolyte membranes onto silicon-based gas sensors [21]. Some main advantages of spin coating are that very thin and reproducible films can be produced, and that an entire array of sensors can be coated simultaneously using batch fabrication methods. In addition, spin coating equipment is readily available from the semiconductor industry.

13.4.3 Screen printing

Although screen printing has been used extensively to fabricate electrodes and circuit boards, it can also be used to deposit thin layers of polymers. In screen printing, a material is forced through a fine-gauge screen which has been partially blocked by a patterned mask. In the areas where the mask is open, the material is deposited onto the substrate; in other areas it is prevented from passing through the screen. In this way, polymers can be deposited over selected areas of a sensor. This is a clear advantage over solution casting or spin coating where the polymer is deposited over the entire substrate. For example, it may be advantageous to coat only the working electrode with a selective coating, and to leave the reference electrode of a sensor uncoated. An enzyme-based sensor for the measurement of urea levels in blood samples has been fabricated using screen printing to deposit the various layers [22].

13.4.4 Electropolymerization

Electropolymerization of polymers directly onto the surface of an electrode has been used for a number of enzyme-based biosensors. By polymerizing from a solution containing the monomer, as well as the other components of the sensor, enzymes for example, a multifunctional polymer film can be fabricated. As the polymer film grows on the electrode, the enzyme and other components are entrapped in the film [9]. GOD and other enzymes have been incorporated into sensors using electropolymerization. Advantages of electropolymerization are that the film thickness can be easily controlled by the amount of polymerization charge passed, and that the polymer film is deposited only on the sensing electrode.

13.4.5 Other techniques

A number of other polymer deposition techniques may be used in sensor fabrication. Some of the most exciting are techniques adopted from the semiconductor industry. For example, uncross-linked hydrogel precursors can be cast onto the surface of a substrate and UV cross-linked through a shadow mask. This technique allows patterning of the hydrogel to make complex sensor arrays on a single sensor substrate, for example, an integrated O_2, CO_2, and pH sensor array [23].

13.5 EXAMPLE: POLYMERS IN FAST-RESPONSE GAS SENSORS

A practical example of the challenges encountered in selecting, fabricating, and testing polymers for chemical sensors is the development of a fast-response gas sensor for respiratory monitoring. This sensor, called the Back Cell^{TM}, was designed with the express aim of reducing the response time of both amperometric and potentiometric gas sensors [24]. Fast response is achieved through the unique feature of this sensor design: the elimination of a separate gas-permeable membrane (figure 13.4). Two different gas sensors were made using the Back Cell design: an amperometric Clark cell O_2 sensor, and a potentiometric Severinghaus-type CO_2 sensor [13].

The sensor consists of a porous ceramic substrate with thin-film electrodes deposited onto it so that the electrodes are also porous. A solid layer of polymer

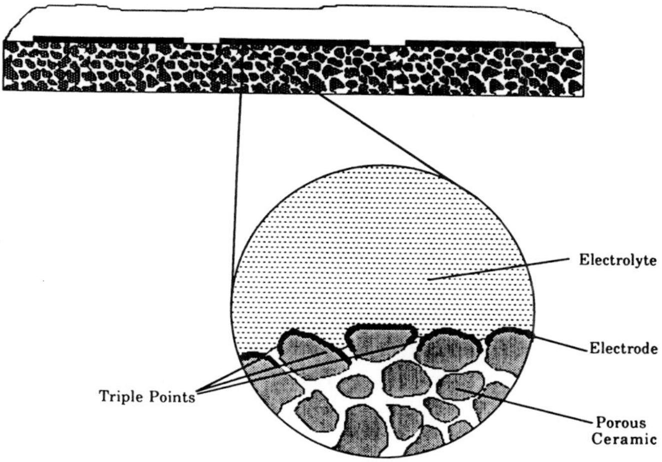

Figure 13.4 The Back Cell gas sensor design. The replacement of the gas-permeable membrane with a porous substrate enables fast response.

electrolyte covers the electrodes but does not protrude into the pores of the substrate. In this design, the gas reaches the electrodes by transport through the air-filled pores of the porous substrate instead of by diffusion or permeation through a gas-permeable membrane. Because the diffusion rate of a gas molecule through air is many orders of magnitude faster than through water, the response time is also much faster. (For example, the diffusion coefficient for CO_2 in air is $0.139\ cm^2\ s^{-1}$, and in water it is $1.96 \times 10^{-5}\ cm^2\ s^{-1}$.) When the sensor is used for amperometric sensing, the electrochemical sensing reaction begins when the gas reaches the 'triple points'—sites where the electrode, the electrolyte, and the gas phase meet.

The advantages that the Back Cell design offers amperometric sensors are also important for potentiometric sensors. To sense CO_2, Severinghaus-type sensors utilize the pH change resulting from the dissolution of CO_2 in an aqueous medium. Thus, the diffusion of CO_2 into a thin layer of aqueous electrolyte is a necessary step in CO_2 sensing. The response time when used as a potentiometric sensor is slightly slower, however, as the sensing reaction does not occur directly at the electrode–electrolyte–ambient triple points, but slightly under the surface of the electrolyte. This process and the diffusion of the resulting protons to the pH sensing electrode slow the total response of the sensor somewhat, although it is still much faster than that of conventionally designed Severinghaus sensors.

13.5.1 Sensor design and fabrication

The amperometric sensors utilize a three-electrode cell. The oxygen sensor comprises platinum working and counter electrodes, and a silver reference electrode. For potentiometric CO_2 sensing, a two-electrode cell is used: a pH-sensitive iridium oxide sensing electrode, and a silver (Ag/AgCl) reference electrode.

A major challenge in the development of these two sensors was the selection of the proper electrolyte. For the Back Cell design, the polymer electrolyte required a unique set of properties. The polymer must have relatively high viscosity upon fabrication to prevent it from entering the pores of the substrate and blocking the triple points. The polymer must have sufficient ionic conductivity to eliminate pick-up of electrical noise. In addition, the polymer must have a stable water content at the high relative humidities found in respirator circuits. An obvious requirement is that the gas must dissolve in the polymer electrolyte. For the CO_2 sensor, an enzyme, carbonic anhydrase, is used to improve response time. Therefore, the hydrogel must provide an hospitable environment for the enzyme to retain its activity.

In the amperometric O_2 sensor, Nafion was chosen as the electrolyte polymer. It is well known that Nafion can be used in this type of sensor [25]. The geometry of a planar substrate would suggest a cast film of Nafion as the electrolyte. However, the stability of the conductivity of a thin film in the varying humidity conditions that would be encountered in this application argued against it. Instead the polymer was processed into a gel which was formed by evaporating the solvent from an 10% alcoholic solution of 1100 equivalent weight Nafion (Solution Technology, Inc.) at room temperature until the Nafion reached the consistency of a rubbery gel. Both H^+ form and Na^+-exchanged Nafion were investigated. It was found that the consistency of the Na^+-exchanged Nafion was very brittle and unsuitable for gel formation. The H^+ form Nafion could be made into a clear gel that could be mechanically applied to the electrodes as a thick film. No other ingredients were added to the Nafion. Because of the lack of anions in the Nafion electrolyte (other than the sulfonated side chains), the silver electrode in the amperometric sensors is assumed to be an Ag/Ag_2O reference electrode.

Nafion was initially investigated as the electrolyte for the CO_2 sensor as well. However, preliminary tests showed no response of the sensor to CO_2. The high acidity of the Nafion effectively excluded CO_2 from dissolving into the electrolyte. In addition, the low percentage of water in Nafion probably prevents the complete hydration of the dissolved CO_2. It was determined that a neutral or basic pH hydrogel would probably have suitable properties for the polymer electrolyte. Also it is unlikely that Nafion would be a stable environment for the enzyme. A commercially available hydrogel, Promeon RG-63B (Medtronic), was used instead as the polymer electrolyte. The as-received hydrogel was swollen in a solution of 0.1 M KCl and 10^{-4} M $NaHCO_3$ until it reached its equilibrium water content. Carbonic anhydrase from bovine erythrocytes (Sigma) was mixed into the swollen gel at a concentration of 1 mg enzyme/ml gel. The swollen hydrogel was then redried to a more viscous consistency.

In both sensors, the electrolyte was coated onto the electrodes and spread out into an even layer on the surface. Both types of polymer electrolyte do not penetrate far into the pores: they are viscous enough to remain at the entrance of the pores, allowing the formation of the triple points at the electrolyte–electrode–ambient gas boundaries (figure 13.4). An additional advantage of the Back Cell design is that the electrolyte layer may be rather thick, as the gas need not diffuse through the electrolyte during sensor operation. The thick electrolyte layer results in a lower electrical impedance than that of thin-film electrolyte sensors [25] and acts as a buffer to changes in water content.

13.5.2 Amperometric oxygen sensor

Oxygen is detected amperometrically by the reduction reaction

$$O_2 + 2H_2O + 4e^- \rightarrow 4OH^-. \qquad (13.1)$$

The reaction proceeds at potentials more negative than −0.5 V against the Ag/Ag_2O reference electrode. The sensor does not exhibit the usual diffusion-limited current plateau observed in most electrochemical gas sensors. Rather, the current increases continuously past the reaction onset potential. The lack of a diffusion-limited current indicates that the transport of oxygen through the pores to the triple points is faster than the consumption of oxygen at the potentials studied. It is expected that the cell would become diffusion limited only at extremely high potentials where the rate of O_2 reduction would exceed the rate of O_2 transport to the electrode. However, this potential is likely beyond the decomposition potential of the electrolyte. For the following characterization, the oxygen sensor was operated at a potential of −0.55 V, just more negative than the onset potential, but positive enough to avoid interferences.

The response of this sensor to a step change in oxygen concentration from 0 to 100% is shown in figure 13.5. The 90% response time is of the order of 300 ms ($\tau = 150$ ms). This response is extremely fast compared to other amperometric sensors. The recovery of the sensor back to baseline upon a step change in oxygen concentration from 100% to 0% O_2 is 400 ms—slightly slower than the initial response. It is believed that recovery is slower because, during exposure to the high O_2 concentration, some O_2 dissolves into the Nafion electrolyte, and must diffuse out (a slower process) for the sensor to return to baseline. Because the recovery time constant (τ) is 150 ms—the same as the response time constant—the slower 90% figure is a result only of this slower diffusion of residual O_2 from the electrolyte.

The calibration curve for the oxygen sensor operated at −0.55 V fits a linear least-squares fit well ($R^2 = 0.993$), but exhibits a slight sublinearity. Interestingly, the calibration curve obtained at a potential of −0.65 V does not show this sublinearity. This evidence points towards the rate of the electrochemical reaction itself being the rate limiting step in the gas sensing process.

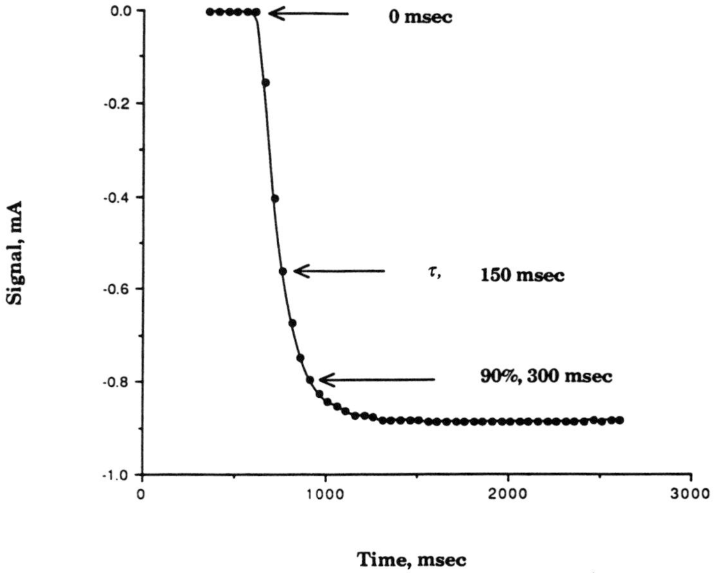

Figure 13.5 The response of the amperometric O_2 sensor to a step change from 0 to 100% O_2. Applied potential, −0.6 V against Ag/Ag_2O.

In addition to fast response, the O_2 sensors demonstrated good reproducibility and reliability. This sensor design also allows incorporation of a water reservoir on the front side (the electrode side) of the sensor to increase sensor lifetime without degrading the sensor's response time. This reservoir provides sufficient moisture to the Nafion electrolyte which requires water to be functional.

13.5.3 The potentiometric carbon dioxide sensor

The Severinghaus-type carbon dioxide sensor measures CO_2 concentration by monitoring the resultant pH change when ambient CO_2 dissolves into an internal electrolyte to form bicarbonate ions:

$$CO_{2(gas)} \rightleftharpoons CO_{2(diss)} + H_2O \rightleftharpoons H_2CO_3 \rightleftharpoons H^+ + HCO_3^-. \qquad (13.2)$$

If the internal electrolyte already contains an excess of bicarbonate ions, the pH will be related to the concentration of CO_2 in the ambient gas by the equation

$$\mathrm{pH} = A - \log P_{CO_2} \qquad (13.3)$$

where $A = \mathrm{p}K_a + \log[HCO_3^-] - \log\alpha$ (the solubility coefficient of CO_2 in water). The slope of the response then is the usual Nernstian slope of 59 mV/decade at room temperature provided a suitable pH electrode is used.

To detect CO_2 in ambient air, the sensor is utilized as a novel design Severinghaus sensor. A sputtered thin film of iridium oxide is used as the pH sensing electrode. Iridium oxide, as well as several other metal oxides, has been shown to be pH sensitive [26], although the mechanism of this effect is not understood in detail. It is believed that the pH dependence is related to one of the redox couples of iridium oxide, such as

$$2IrO_2 + 2H^+ + 2e^- \rightleftharpoons Ir_2O_3 + H_2O. \tag{13.4}$$

Iridium oxide films were produced by reactive sputtering of iridium metal in an oxygen-containing atmosphere. The exact stoichiometry of the iridium oxide film is unknown, but it is assumed that it is a mixed-valence film and contains both oxide and hydroxide species. To investigate the pH response of the iridium oxide film, the potential was monitored against a silver thin film sputtered on the same substrate. An initial rapid potential drift occurred for the first 24 h after the sensor was placed in solution as the oxide species on the film hydrated [26]. The endpoint of the drift for most electrodes was in the region of 240–260 mV. The pH response slope for six different fabrication batches of sensors was found to be Nernstian with an average of 57.452 ± 0.65. The slope is reproducible from sensor to sensor and from batch to batch. Iridium oxide has a noted redox sensitivity and is impractical for pH measurements in many samples. However, in the well defined bicarbonate buffer system, no redox agents are present to cause spurious voltage fluctuations.

In addition to slow diffusion of the CO_2 gas through the gas-permeable membrane, conventional Severinghaus-type carbon dioxide sensors suffer from an additional rate limitation, namely slow reaction kinetics. The hydration reaction of CO_2

$$CO_2 + H_2O \rightleftharpoons H_2CO_3 \tag{13.5}$$

is relatively slow compared to the other steps in the chemical reaction [27]. The rate equation of this reaction is pseudo-first order with respect to CO_2 with a rate constant of 0.03 s^{-1} ($t_{1/2} = 23.1$ s). The slow kinetics of this reaction have been known for some time and several groups have attempted to improve the response time of Severinghaus sensors through the use of carbonic anhydrase enzyme [28]. Carbonic anhydrase catalyses the hydration and dehydration reactions of CO_2 and accelerates this reaction rate approximately 5000-fold. The results of earlier attempts to use carbonic anhydrase to improve sensor response time were mixed, probably because the response times of previous sensors were limited by slow diffusion through the gas-permeable membrane, and not by reaction kinetics [29, 30].

The Back Cell design, however, does not suffer from diffusion-limited response times. Initial experiments without the enzyme gave 90% response times no faster than 2.2 s, indicating the response time of the hydration reaction. When the enzyme was added to the electrolyte at a concentration of 1 mg enzyme/ml electrolyte, the response time decreased by more than one-half to

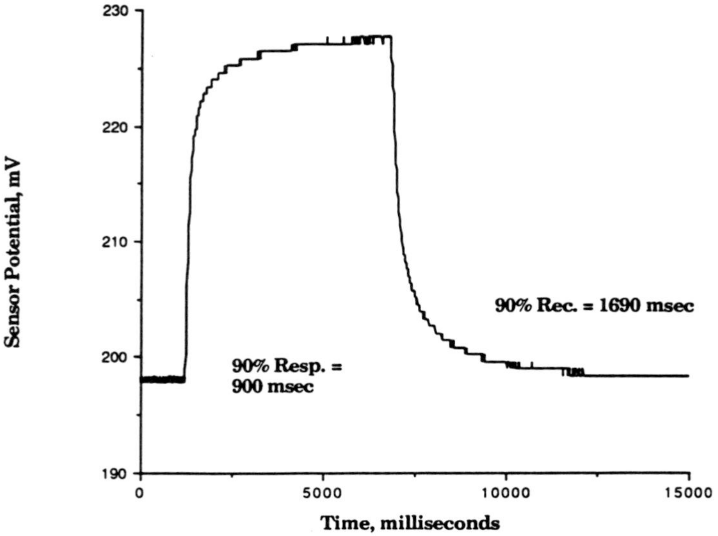

Figure 13.6 The response of the potentiometric CO_2 sensor to step changes between 1 and 9% CO_2.

900 ms (figure 13.6). The time constant, τ, of the response was 190 ms. The effect of the enzyme may be decreased somewhat by the presence of chloride ion in the electrolyte, which is a slight inhibitor of the enzyme [28]. The 90% response time of the sensor can be reduced even further to approximately 300 ms, if a smaller response magnitude is acceptable. The responses of this CO_2 sensor are at least two orders of magnitude faster than those of other conventionally designed Severinghaus-type sensors. The response time of the sensor did not degrade after several months of refrigerated storage, demonstrating the stability of the enzyme in the hydrogel matrix. The enzyme quickly lost activity, however, if the electrolyte was heated above 60 °C.

The calibration curve of the CO_2 sensor from 1 to 10% CO_2 is shown in figure 13.7. Because the sensor responds to a pH change generated as shown in equation (13.3), the potential of the sensor is controlled by the Nernst equation

$$E_{out} = K - \frac{RT}{nF}\log(\%\ CO_{2[ambient]}) \tag{13.6}$$

where K contains such constants as the standard potential of the sensor, the Ostwald coefficient for CO_2 dissolution into the electrolyte, and the constant A from equation (13.3). This logarithmic relationship may be expressed as a linear calibration curve by plotting the sensor output potential against $\log[CO_2]$. As can be seen in figure 13.7, the calibration curve is linear in the physiologically relevant range of 1–10%. The slope is approximately 50 mV/decade at room temperature, and is typical of the slope observed in sensors with fully hydrated

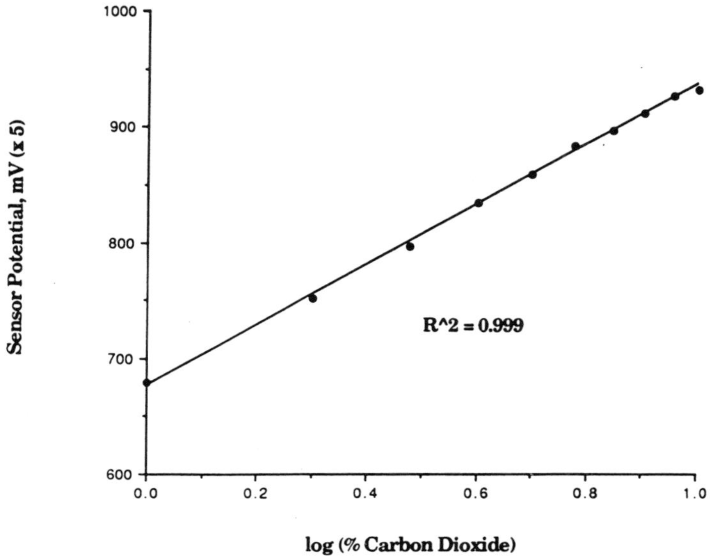

Figure 13.7 Calibration curve for the CO_2 sensor with steady state cell potential as a function of CO_2 concentration in air.

hydrogel electrolytes. The decrease from the Nernstian response is believed to be caused by a slight buffering effect of the enzyme protein. When the hydrogel was used without swelling to its equilibrium water content, the sensor exhibited lower response slopes. This reduction in slope in drier electrolytes may be a result of buffering side groups on the hydrogel polymer chain, an increase in the HCO_3^- concentration, or a decrease in the water available for CO_2 hydration. However, these partially hydrated sensors usually showed the fastest responses because the gel remained at the surface of the pores and did not flow into the pores to occlude the triple points. For some applications, such as respiratory monitoring, this trade-off of lower slope for faster response time would be advantageous.

The long-term stability of the CO_2 sensor is mainly dependent upon the humidity of the ambient gas stream. When the bare sensor was placed in a test set-up simulating a patient on a respirator, changes in both the baseline potential and the response slope occurred over the course of 8 h. The direction and magnitude of the change was related to the relative humidity of the gas supplied by the respirator circuit. High humidity tended to increase the response magnitude and decrease the baseline potential, as water condensed into the hydrogel and diluted the electrolyte somewhat. Severe electrolyte dilution decreased the response magnitude. Conversely, if the sensor operated at a low humidity, the baseline potential increased and the magnitude decreased to near zero, as the hydrogel lost water. It is believed that the dependence upon ambient humidity can be decreased or eliminated by placing the sensor in an appropriate housing containing a reservoir of electrolyte in contact with the hydrogel.

The unique property of the Back Cell sensor—fast response—is made possible through careful design, taking into account the rate limiting steps in the gas sensing properly. Implementation of the sensor design is achievable only through careful selection of polymer electrolytes with the proper physical and chemical properties. Knowledge and consideration of the types of property that can be found in the wide range of polymers possible to use in sensors can greatly reduce the time needed to develop sensors for specific applications.

REFERENCES

[1] Yeo R S 1982 Applications of perfluorosulfonated polymer membranes in fuel cells, electrolyzers, and load leveling devices *Perfluorinated Ionomer Membranes* ed A Eisenberg and H L Yeager (Washington, DC: American Chemical Society) pp 453–73

[2] Vincent C A 1987 Polymer electrolytes *Prog. Solid State Chem.* **17** 145–261

[3] Galiatsatos C, Ikariyama Y, Mark J E and Heineman W R 1990 Immobilization of glucose oxidase in a poly[vinyl alcohol] matrix on platinized graphite electrodes by gamma-irradiation *Biosensors. Bioelectr.* **5** 47–51

[4] Brinkman E, van der Does L and Bantjes A 1991 Poly(vinyl alcohol)-heparin hydrogels as sensor catheter membranes *Biomaterials* **12** 63–70

[5] Cao Z and Stetter J R 1991 Amperometric gas sensors *Chemical and Biological Sensors* ed M J Madou and J P Joseph (Gaithersburg, MD: NIST) pp 49–50

[6] Wilson G S, Zhang Y, Reach G, Moatti-Sirat D, Poitout V, Thévenot D R, Lemonnier F and Klein J-C 1992 Progress toward the development of an implantable sensor for glucose *Clin. Chem.* **38** 1613–7

[7] Coughlan M P, Kierstan M P J, Border P M and Turner A P F 1988 Analytical applications of immobilised proteins and cells *J. Microbiol. Methods* **8** 1–50

[8] Gibson T D and Woodward J R 1992 Protein stabilization in biosensor systems *Biosensors and Chemical Sensors: Optimizing Performance through Polymeric Materials* ed P G Edelman and J Wang (Washington, DC: American Chemical Society) pp 40–55

[9] Bartlett P N and Whitaker R G 1987 Electrochemical immobilisation of enzymes. Part II. Glucose oxidase immobilised in poly-N-methylpyrrole *J. Electroanal. Chem* **224** 37–48

[10] Sasso S V, Pierce R J, Walla R and Yacynych A M 1990 Electropolymerized 1,2-diaminobenzene as a means to prevent interferences and fouling and to stabilize immobilized enzyme in electrochemical biosensors *Anal. Chem.* **62** 1111–7

[11] Gerhardt G A, Oke A F, Nagy G, Moghaddam B and Adams R N 1984 Nafion-coated electrodes with high selectivity for CNS electrochemistry *Brain Res.* **290** 390–5

[12] Zhang Y, Hu Y, Wilson G S, Moatti-Sirat D, Poitout V and Reach G 1994 Elimination of the acetaminophen interference in an implantable glucose sensor *Anal. Chem.* **66** 1183–8

[13] Tierney M J and Kim H-O L 1993 Electrochemical gas sensor with extremely fast response times *Anal. Chem.* **65** 3435–40

[14] Kim H-O L, Tierney M J, Oh S and Madou M 1992 In-situ electrochemical sensor for oil quality monitoring *Proc. Sensors Expo (Chicago, IL, 1992)* (Peterborough, NH: Helmers) pp 199–203

[15] Janata J 1989 *Principles of Chemical Sensors* (New York: Plenum) pp 112–6

[16] Whiteley L D and Martin C R 1987 Perfluorosulfonate ionomer film coated electrodes as electrochemical sensors: fundamental investigations *Anal. Chem.* **59** 1746–51

[17] Moore R B and Martin C R 1988 Chemical and morphological properties of solution-cast perfluorosulfonate ionomers *Macromolecules* **21** 1334–9

[18] Schmidt H-L, Schumann W, Scheller F W and Schubert F 1989 Specific features of biosensors *Sensors: a Comprehensive Survey* ed W Göpel, J Hesse and J N Zemel (New York: VCH) p 761

[19] Gregg B A and Heller A 1990 Cross-linked redox gels containing glucose oxidase for amperometric biosensors applications *Anal. Chem.* **62** 258–63

[20] Bindra D S, Zhang Y, Wilson G S, Sternberg R, Thévenot D R, Moatti D and Reach G 1991 Design and in vitro studies of a needle-type glucose sensor for subcutaneous monitoring *Anal. Chem.* **63** 1692–6

[21] Buttner W J, Maclay G J and Stetter J R 1990 An integrated amperometric microsensor *Sensors Actuators* B **1** 303–7

[22] Schaffer B P H, Dolezal A M, Kienberger W P, Offenbacher H, Riegebauer J and Ritter C 1994 Screen printed solid state contact biosensors for the determination of urea in undiluted biological samples *Proc. 3rd World Congr. on Biosensors (New Orleans, LA, 1994)* (Oxford: Elsevier) Poster 1.0.

[23] Arquint Ph, van den Berg A, van der Schoot B H, de Rooij N F, Bühler H, Morf W E and Dürselen L F J 1993 Integrated blood-gas sensor for pO_2, pCO_2, and pH *Sensors Actuators* **13–4** 340–4

[24] Madou M J and Otagawa T Microsensors for gaseous and vaporous species *US Patent* 4 812 221

[25] Otagawa T, Madou M, Wing S, Rich-Alexander J, Kusanagi S, Fujioka T and Yasuda A 1990 Planar microelectrochemical carbon monoxide sensors *Sensors Actuators* B **1** 319–25

[26] Kinoshita K and Madou M J 1984 Electrochemical measurements on Pt, Ir, and Ti oxides as pH probes *J. Electrochem. Soc.* **131** 1089–94

[27] Kern D M 1960 The hydration of carbon dioxide *J. Chem. Educ.* **37** 14–23

[28] Lindskog S, Henderson L E, Kannan K K, Liljas A, Nyman P O and Strandberg B 1971 Carbonic anhydrase *The Enzymes* ed P D Boyer (New York: Academic) pp 587–665

[29] Constantine H P, Crow M R and Forster R E 1965 Rate of the reaction of carbon dioxide with human red blood cells *Am. J. Physiol.* **208** 801–11

[30] Lutmann A, Mückenhoff K and Loeschcke H H 1978 Fast measurement of the CO_2 partial pressure in gases and fluids *Pfluegers Arch. Ges. Physiol.* **375** 279–88

[31] Wu X-Q, Shimizu Y and Egashira M 1989 Carbon dioxide sensor consisting of K_2CO_3–polyethylene glycol solution supported on porous ceramics *J. Electrochem. Soc.* **136** 2892–5

[32] Schumann W, Lammert R, Uhe B and Schmidt H-L 1990 Polypyrrole, a new possibility for covalent binding of oxidoreductases to electrode surfaces as a base for stable biosensors *Sensors Actuators* B **1** 537–41

14

Solid state, resistive gas sensors

Barbara Hoffheins

14.1 INTRODUCTION

The phenomenon of a conductive film sensitive to the presence of gases has been known and studied for many decades [1]. Commercial products relying on this reaction have been available for well over 25 years with the development of tin and zinc oxide sensors proposed by Seiyama *et al* [2] and patented by Taguchi [3]. Uses for these types of sensor have increased dramatically. One of the first applications was an LP gas detector based upon a Taguchi-type sensor for use in Japanese buildings. Since 1968 Figaro Engineering Inc. alone has sold over 50 million sensors for domestic gas detectors in Japan [4]. The market has grown to include other gas leak detectors for domestic and commercial buildings, breath alcohol analyzers, automatic cooking controls in microwave ovens, air quality in parking garages (detectors for elevated levels of carbon monoxide), and fire alarms [5]. Although tin oxide is the most popular semiconductor gas sensor material, other metal oxides, metals, and organic materials have been successfully configured as conductive gas sensors. Of these, most are sensitive to reducing gases, but some have been developed to sense partial pressures of oxygen and carbon dioxide. Current research and development efforts are focused in many areas, including synthesis of materials that have high sensitivity and/or selectivity to gases of interest, improvements in sensor geometries and fabrication (especially in the area of microstructures), and the applications of sensor analysis techniques for solving complex, real world problems.

14.2 MATERIALS

14.2.1 Metal oxides

The primary mechanism responsible for gas reactions with metal oxide semiconductors in air at elevated temperatures (200–600 °C) is the change

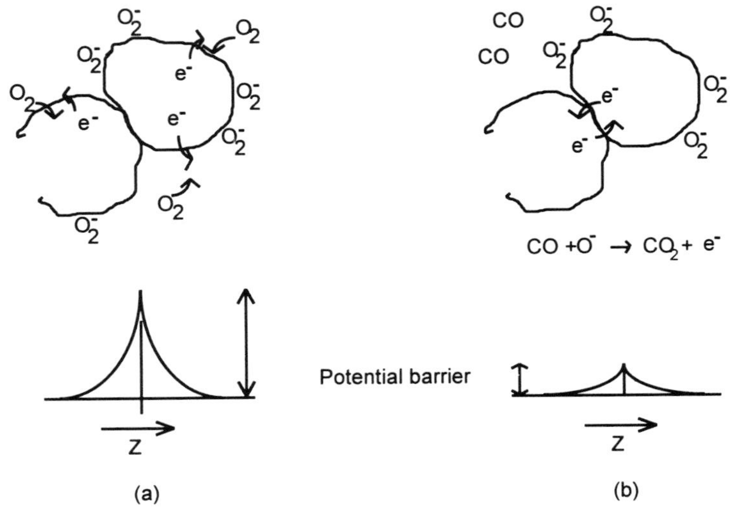

Figure 14.1 (a) Charge exchange associated with the chemisorption of oxygen at the surface of an n-type semiconductor and the potential distribution across a grain junction. (b) Charge exchange associated with a reducing gas and the effect on the potential barrier at the grain junction.

in the concentration of adsorbed oxygen at the semiconductor surface, well documented by Moseley [1] and others. Oxygen ions form at the surface by abstracting electrons from the semiconductor solid. As the charge carrier density is reduced, a potential barrier develops that, in turn, slows the oxygen adsorption rate and corresponding electron movement through the grain junctions. Conductivity of the semiconductor material is thus limited by the oxygen adsorption rate and the potential barrier effects at grain junctions. At the surface of an n-type semiconductor, reactive gas combines with adsorbed oxygen, reducing the height of the potential barrier and the semiconductor resistance. Similarly, for a p-type semiconductor, reactive gases increase the semiconductor resistance while oxidizing gases decrease it.

Semiconductor sensor materials are thus classified as n or p type based on the resistance changes to decreasing partial pressures of oxygen or to reactive gases in fixed partial pressures of oxygen. As in other semiconductor materials, solid state doping can set a metal oxide to n or p type, as desired, although many materials predictably switch behavior from n type to p type with increasing partial pressures of oxygen.

Other mechanisms affecting resistance changes in the semiconductor are adsorption of ions other than oxygen at the surface, changes in ambient humidity, or water formed by combining with adsorbed oxygen [1]. These last two related mechanisms are the underlying principles for the development of several metal oxide-based humidity sensors.

The reactions for an n-type semiconductor [6] as shown in figure 14.1 are

Adsorption of oxygen. $(SnO_2 + e^-) + \frac{1}{2}O_2 \leftrightarrow SnO_2(O^-)_{ad}$
Reducing gas. $SnO_2(O^-)_{ad} + CO \rightarrow (SnO_2 + e^-) + CO_2$.

An understanding of the fine structure of the metal oxide is an important key to optimizing sensor sensitivities. A study by Xu *et al* [7] illustrates the effects of grain size and space–charge depth at the microstructure level of tin oxide doped with aluminum and antimony. A crystallite, or grain of dimension D, will have an electron-depleted surface layer, or space–charge layer, with a depth L. The Debye length and the amount of adsorbed oxygen affect the size of L. For porous sintered elements with grain size held constant, aluminum-doped tin oxide sensors were found to have improved sensitivity over pure tin oxide sensors because the aluminum doping increased the space–charge depth. Not surprisingly, antimony-doped sensors, with decreased space–charge depth, displayed less sensitivity than the pure tin oxide sensors. For pure tin oxide sensors with grain sizes of 5–32 nm, sensitivity was greatest with $D \leqslant 6$ nm. Some generalizations were derived from this study. If $D \gg 2L$, the electron movement will be dominated by the potential barrier at the grain boundaries, independent of the magnitude of D; if $D \geqslant 2L$, electron flow is controlled by surface space charge at the necks (grain contacts with diameter approximately $0.8D$); and, lastly, if $D \ll 2L$, the sensitivity of the metal oxide is determined by the electric resistance of the grain chains.

By far the most popular metal oxide material used in resistive sensors and the most widely studied is tin oxide. As with other metal oxides, the resistance change can be large and reasonably linear over many decades of gas partial pressure. Many sensitive and inexpensive gas sensors and detectors have been created. Commercial sensors based upon tin oxide have typical sensitivities to reducing gases of 10–500 ppm [4] depending upon the target gas and the sensor type, although one sensor for the detection of methyl mercaptan and hydrogen sulfide (Figaro TGS 501) is reported to be sensitive to 0.08 ppm.

Schematic diagrams of commercial tin oxide sensors by Figaro Engineering Inc. are shown in figure 14.2. The sensor of figure 14.2(c) consists of a small ceramic tube coated with tin (IV) oxide and a suitable catalyst. A heater coil maintains the tin oxide temperature (between 200 and 500 °C). The signal is the resistance measurement between the two electrodes at opposite ends of the tube. The assembly is packaged in a protective housing with an open area (covered by wire mesh) to allow gases to enter.

Power consumption for these Figaro-type sensors ranges from 0.6 to 1.2 W. In one effort to reduce the power consumption, Figaro devised a pulse-driven sensor [8]. A microcomputer acts as a pulse generator and sensor signal processor to pulse the sensor heater at a duty ratio of 8/1000 for an effective power consumption of 14 mW. This particular sensor is sensitive to carbon monoxide and hydrogen gas.

Figaro literature [4] cautions that their TGSs (tin dioxide gas sensors) are

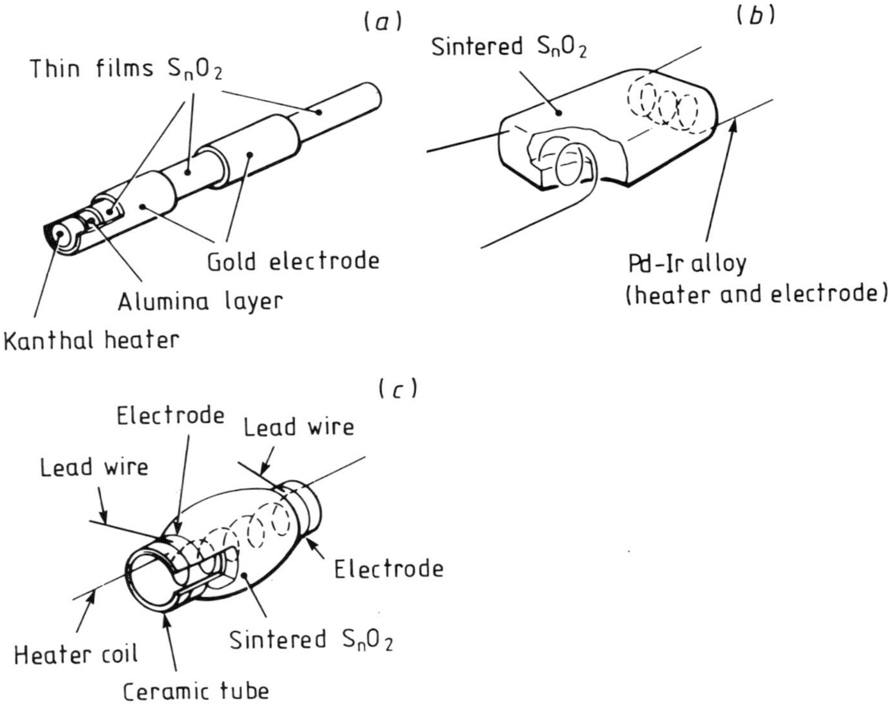

Figure 14.2 Figaro sensor types [4].

sensitive to humidity changes, have individual variation in sensitivity requiring individual calibration, and are subject to poisoning (not normally permanent) if not stored in clean, air tight containers. However, the Figaro and similar sensors have a reasonably long life of 3–5 years, are durable, and have a very low cost as compared to general purpose gas analysis instrumentation.

A serious contender for some of the Figaro market may be a sensor distributed by Sinostone, Chicago [9]. This SnO_2/Fe_2O_3 sensor boasts a lower power consumption, 0.1 W; a response time of 20 s; and a 3 year lifetime. The sensor is most sensitive to hydrogen (5 ppm) and, in decreasing order, butane, propane, carbon monoxide, ethane, and methane, all at sensitivities comparable to Figaro sensors.

Another material, indium oxide (In_2O_3), is the basis for a commercial sensor developed for the detection of hydrogen sulfide (H_2S) [10]. James Youngblood, of Gemini Detectors, Texas, has gone through an extensive process to optimize sensor performance and minimize manufacturing costs and complexity. Thin films, ultimately chosen for precise control of heating, were additionally attractive because less precious metal is used per sensor. The first sensor prototypes were modified ceramic trim potentiometers with the potentiometer used as a heater. In subsequent development, oxidized silicon wafers were chosen for the substrate material. The design was a single thin-film deposition system with integral thermistors. Input voltage was selectable at 5 or 10 V DC and power consumption was 0.235 W. In the final version, the thermistor was replaced by a platinum heater wire connected as one leg of a common Wheatstone bridge circuit from which an error signal was used to keep the heater at constant temperature. These sensors, dubbed microceristors™, are mounted on a TO-5 header, require 80 mW, and operate from three heater voltages, 1, 2, or 3.5 V DC. To counteract a documented aging problem, Youngblood patented a technique to coat the sensor surface with non-contiguous islands of gold. As a result, the sensor does not 'go to sleep' during inactivity and the response time for 10 ppm H_2S is consistently 20 s.

Some n-semiconductor materials have been studied extensively for the detection of changes in partial pressure of oxygen, carbon dioxide, and other stable gases. One example is a titania (TiO_2) sensor which has undergone much development in the United States and Japan. Resistance changes in non-stoichiometric TiO_2 provide a means to quantify oxygen partial pressure [11]. The sensor, a small, disk-shaped pellet, is simple and inexpensive; however, its rival, the zirconia sensor, based on an electrochemical mechanism, is more likely to be the oxygen sensor of choice for automotive applications because it claims greater durability.

Other metal oxides for oxygen sensors include a Ni_2O_5 thin-film sensor [12] and CeO_2 in thick- and thin-film versions [13]. Thick films of CeO_2 were deposited on alumina substrates prepared with a Pt heater and temperature resistor and fired at 1400 °C for 2 h. Thin-film sensors were made by reactive rf sputtering onto heated alumina substrates and then cured in air at 1100 °C for 2 h yielding a final thickness between 0.5 and 1.5 μm. Sensor responses to high oxygen partial pressures even after 200 test h were still reproducible, with response times of 5–10 ms. The high thermal shock resistance appears promising for combustion control in internal combustion engines.

The major advantages of metal oxides are reversibility, rapid response, longevity, low cost, and robustness. The materials and manufacturing techniques easily lend themselves to batch fabrication. Unfortunately, several disadvantages limit more widespread sensor utilization and continue to challenge sensor development. The metal oxide sensor generally requires a high temperature for measurable gas response, often leading to high power consumption (although newer fabrication techniques produce a much smaller sensor area that dramatically reduces power consumption). Metal oxide sensors are not highly selective and much effort has been involved in devising materials and methods of operation to improve specificity. These sensors are generally not appropriate for applications in vacuum or inert atmospheres and most commercial devices are not sensitive enough for sub-ppm detection. Baseline sensor drift may require periodic calibration for certain applications. Time response, even with the use of dynamic signal analysis, is often too slow for real time feedback and control applications.

Table 14.1 Examples of materials used for resistive gas sensors.

Material	Target gases	Implementation	Reference
Commercial sensors			
In_2O_3	H_2S	thin film	[10]
SnO_2	carbon monoxide, alcohols, methane, ethane, propane, ketone, ester and benzol families, isobutane, hydrogen	porous sintered tube	[4]
$SnO_2 + Fe_2O_3$	hydrogen, butane, carbon monoxide, propane, ethane, methane	sintered pellet	[9]
TiO_2	oxygen	sintered pellet	[11]
Prototype sensors			
$BiFeO_3$, $Bi_4Fe_2O_9$, $Bi_2Fe_4O_9$	ethanol, acetone, natural gas, petrol vapor	sintered pellet	[14]
CeO_2	oxygen	thick and thin films	[14]
Fe_2O_3	'smog', acetone, alcohol; H_2, C_2H_4, iso-C_4H_{10}	thin films	[15, 16]
Ga_2O_3	reducing gases	thin films	[17]
In_2O_3 + Fe_2O_3 additive	O_3	thin film	[18]
$MgAl_2O_4$	humidity	thin film	[19]
PbPc, 10 wt% RuO_2, 2 wt% Pd	Cl_2, NO_2, H_2, CO	thick film	[20, 21]
$Sr_{1-y}Ca_yFeO_{3-x}$	phosphine	thick film	[22]
$SrTiO_3$	oxygen	thin film	[23]
TiO_2	trimethylamine (fish freshness)	thick film	[24]
WO_3	2-methylpyrazine	porous sintered tube	[25]
ZnO	hydrogen, phosphine	thick film, thin film	[26]

14.2.2 Mixed oxides

Other materials, such as bismuth ferrites ($BiFeO_3$, $Bi_4Fe_2O_9$, $Bi_2Fe_4O_9$) [14] have been studied as semiconductor materials for resistive gas sensors; they respond as n-type semiconductors as with typical metal oxides. The changes in conductivity of these materials are due to a volume effect. In the absence of air, sensitivities to reducing gases such as petrol, ethanol, and acetone were at least an order of magnitude lower than under similar conditions with air present.

Perovskite compounds, deposited by thick-film methods, have successfully responded to some reducing gases and partial pressures of oxygen [27].

14.2.3 Organic sensors

The most commonly reported organic semiconductor material is phthalocyanine (Pc), although many embodiments of this material are technically classified as polymers and will not be discussed in this chapter. One sensor of this type is a p-type semiconductor sensitive to Cl_2 and NO_2 at a 150 °C operating temperature, and has the advantage of very low sensitivity to H_2, H_2S, NH_3, CH_4, SO_2, and CO [20]. Mg, Co, Ni, Cu, or Zn phthalocyanine films have also demonstrated some sensitivity to NO_2 [28]. In another study, lead phthalocyanine (PbPc) was configured as an n-type semiconductor when doped with RuO_2 and Pd and was shown to be sensitive to H_2 and CO gases [21].

Sensor fabrication has been accomplished in various ways. One method [29] combines the phthalocyanine compound with stearic acid to form a bonded substance which is then dissolved in an organic solvent. The mixture is deposited by one of several possible methods onto a substrate and subsequently fired to remove the vehicle. A thin-film model, manufactured by vacuum sublimation over interdigited electrodes on 3 mm × 3 mm alumina substrates [28], exhibited sensitivity to Cl_2, F_2, and BCl_3, but not to common reducing species. The films operating continuously did not survive longer than 6 months.

14.2.4 Metals

Resistive metal sensors rely on the change in resistance due to absorption of atomic species, such as hydrogen. The most notable examples of these sensors are platinum and palladium. As an illustration, when the surface of palladium (and many of its alloys) is exposed to gaseous hydrogen, some of the H_2 molecules will dissociate on the surface and then diffuse into the metal, reaching an equilibrium concentration that is a well characterized function of the hydrogen partial pressure [30]. The concentration of hydrogen in the palladium can be measured through several physical effects; however, the only one considered in this chapter is the change in the resistance of the palladium metal corresponding to the amount of dissolved hydrogen. For the most part, the reaction is hydrogen specific, although the reactions might be complicated by varying partial pressures of oxygen. Testing has also revealed that 100% Pd films can expand and will blister irreversibly when exposed to high concentrations of hydrogen gas [31]. One attractive feature of the palladium sensor is that it does not require operation at elevated temperatures like the metal oxides and can therefore realize significant power consumption savings.

An early resistive hydrogen sensor, developed at Bendix Corporation [32], consisted of two palladium thin films deposited on a glass slide, with gold or silver contacts at their ends. Each strip, 8 mm long and 0.5 mm wide, measured 1–2 $\Omega/\square$. One strip was covered by a mask (thin Mylar tape) to exclude hydrogen so that, by comparing the resistivities of the two strips within a bridge amplifier arrangement, some degree of temperature compensation was achieved. The inherent simplicity of the resistance bridge technique makes this approach fundamentally attractive, provided, of course, that questions of speed, sensitivity, reproducibility, and drift can be adequately resolved.

An example of the resistive platinum sensor is the 'hot-wire' sensor used for the detection of combustibles [33]. At temperatures of 1000 °C, the resistance and temperature of the platinum increase predictably in the presence of methane and other like gases. This type of sensor has been used in coal mines. Based upon the same phonomenon, a microsensor version boasting lower power consumption and use of less precious metal has been fabricated using photolithography and vacuum evaporation [34].

14.3 ENHANCING SELECTIVITY

Generally, sensors of the conductometric or resistive type cannot identify explicitly an unknown gas or mixture of gases because they are not inherently selective. Typically, these sensors, especially those developed for reducing gases, respond to many compounds. Figaro TGSs can have similar sensitivites for several gases (figure 14.3). The use of either one of these sensors would not be sufficient in a system, such as an underground parking garage, where fuel vapors, carbon monoxide, and other combustible products are likely to occur together. This dilemma of cross-sensitivity is particularly acute when dealing with hazardous industrial chemicals, CBW (chemical and biological warfare) agents, explosives, food substances, and so forth, because such mixtures are chemically complex. Rapid and accurate identification may be critical for making decisions that affect the safety of personnel and equipment, or the productivity of a manufacturing process.

14.3.1 Additives

Noble metal additives such as Pt, Pd, or Ag, are often added to metal oxide semiconductor sensor materials to enhance sensor response to a particular gas or class of gases [35]. Other metals, namely Be, Mg, Ca, Sr, Ba, have been used with tin oxide to enhance performance of an alcohol selective sensor [36], for example. Commercially produced metal oxide sensors may customarily use one or more of the above catalysts.

One effect of the catalyst is to enhance the oxidation of the target gas at the sensor surface, thereby increasing sensitivity to that gas. A study in the qualitative analysis of consommé soup illustrates this mechanism [25]. Sensors of pure WO_3 and SnO_2 exposed to 2-methylpyrazine, an important flavor component of consommé, over a temperature range between 200 and 500 °C

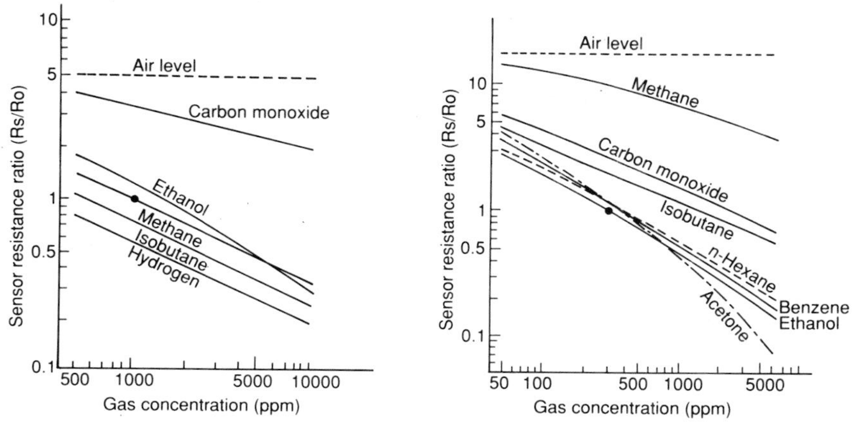

Figure 14.3 Response curves for Figaro TGS 711 and TGS 308. (Reproduced from Figaro Products Catalog, April 1994, Wilmette, Illinois.)

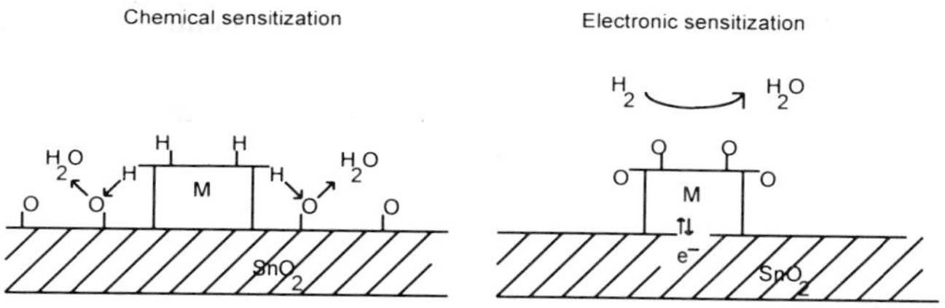

Role of additives	Activation and spillover of reactants	Electron donors or acceptors to or from SnO_2
Origin of conductivity change	Change of oxidation state of SnO_2	Change of oxidation state of additives

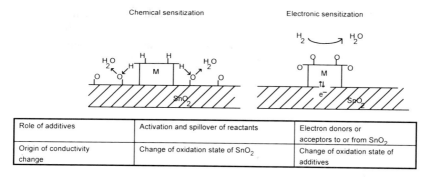

Figure 14.4 Sensitization mechanisms of SnO_2 sensors [35].

exhibited mild or little response. When the WO_3 and SnO_2 films were loaded with noble metals (0.4 wt%), not only did the study demonstrate improved sensor sensitivities but also the output of a WO_3–Rh sensor at 300 °C was highly correlated to vapors of three grades of consommé soup and could possibly be used to replace a human tester.

Two catalyst mechanisms, chemical and electronic, proposed by Yamazoe and coworkers [37, 38], are shown in figure 14.4 [38]. In chemical sensitization, gas would first adsorb to the metal ('M' in the figure), then 'spill over' to the SnO_2 semiconductor. Perhaps more likely, the electronic sensitization effect is noted by the direct exchange of electrons between the metal and the semiconductor. The oxidation state of the metal is changed upon contact with the gas; electron exchange increases and subsequently changes the conductivity of the semiconductor.

The characteristics or behavior of a gas can be a clue for enhancing sensitivity of the semiconductor material; for example, sensitivity to acidic gases, such as H_2S, or basic gases, such as NH_3, has been found to be increased by the incorporation of a corresponding acidic or basic oxide, respectively, that forms adsorption sites for the gases [39]. Similarly, sensitivity for other target gases can sometimes be increased by the addition of compounds with which the target will react.

Additives are either dispersed on the semiconductor surface or they are impregnated throughout the material, commonly in proportions between 0.4 wt% and 3.0 wt% [25]. Typically, pellets and thick films of metal oxides are impregnated and thin films have dispersed catalysts.

Other methods of enhancing selectivity for the single sensor element have included the use of molecular sieves to restrict larger gas molecules [40], temperature control [41], and filters, such as semipermeable membranes [42, 43].

14.3.2 Gas sensor arrays

One theoretical method of synthesizing selectivity or minimizing cross-sensitivity, proposed by Clifford [44], is to assemble an array of sensors having differential sensitivities to different gases and solving a series of simultaneous equations to derive the concentrations of all target species. Going beyond Clifford's first 'chemometric' approach, many workers have realized that such an array would, in principle, yield signatures for different gases that would have varying degrees of uniqueness and could be analyzed or classified by various pattern recognition approaches. The degree to which the sensor array gives a unique response for each target gas determines the success of this approach. Several strategies can be employed singly or in concert to create a sensor array: (i) by using different sensing materials, (ii) by using catalysts, (iii) by using thermal gradients or thermal cycling, and (iv) by using filters.

A number of sensor arrays consisting of an assortment of commercial metal oxide gas sensors have been reported [45–47]. For controlled tests, the sensors are mounted in an air-tight chamber fitted with gas inlets and outlets for controlled gas flow. Each sensor's heating element is controlled externally and resistance changes of the gas sensors are monitored by a computer data acquisition system. A significant effort in this area exists at the University of Warwick, Coventry, where for many years, sensor arrays, made from discrete SnO_2 sensors or miniature integrated sensors, have been studied for ultimate application to food quality and food process control [47, 48].

An early example of an integrated gas sensor array is a thick-film implementation developed by Hitachi [49], with six discrete metal oxide sensor areas on an alumina substrate (figure 14.5). The chip is heated to 400 °C with a platinum heater printed on the bottom of the substrate. More recently, other SnO_2-based arrays have been reported. A low-power, 1.5 mm^2 four-sensor array employing SnO_2 has been demonstrated for detection of toxic

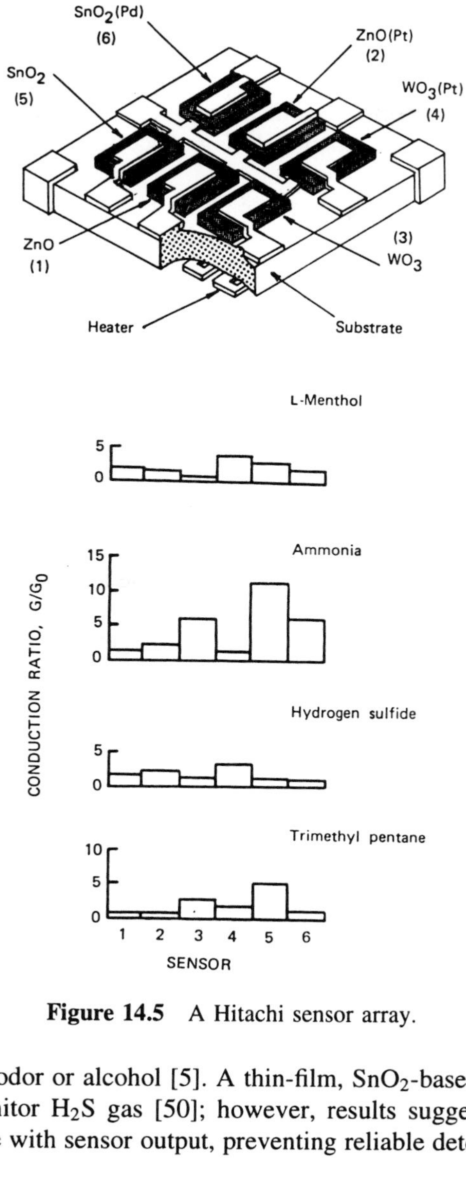

Figure 14.5 A Hitachi sensor array.

gases, and mouth odor or alcohol [5]. A thin-film, SnO_2-based array was tested in Finland to monitor H_2S gas [50]; however, results suggest that competing pollutants interfere with sensor output, preventing reliable determination of H_2S concentration.

14.3.2.1 Sensor array signal processing. As noted above, data from sensor arrays can be treated as a pattern recognition problem. Many types of traditional and novel approach have been explored for practicality and suitability.

The study and applications of 'chemometrics' embrace many of the traditional statistical approaches. One common chemometric method is the use of multivariate analysis. Some of the Warwick studies applied this approach to the outputs of 12 tin oxide sensors exposed to the headspace of three commercial coffees. Results of one study demonstrated high correlation to the blend and roasting level of the coffee [47]. If the sensor array response could be analyzed in real time, on-line quality control during processing could be dramatically improved.

Fuzzy logic was used in a Japanese study to analyze the output of a set of five sensors consisting of two Figaro TGS 203s (carbon monoxide), a Figaro TGS 109 (volatile organic compounds), a temperature sensor, and a humidity sensor [45]. The patterns of 'hot', 'smelly', 'smoky' were established by observing and coding the sensor array outputs during tests of fire, exposures to carbon monoxide, combustible gas, and cigarette smoke. Then a microprocessor-controlled system connected to the sensors was configured to distinguish disaster (fire or leaking gas) from non-disaster (cigarette smoke).

One commercially available sensor array analysis system, offered by Mosaic Industries [51], is Rhino™, a microprocessor-based instrument with an array composed of discrete, resistive gas sensors. An artificial neural network processes sensor inputs and relates them to patterns established by training the instrument with gas components and mixtures of interest for a specific application. In principle, each system is customized for an application by the choice of sensors and the gas detection needs. Potential applications for this system are limited by the availability of suitable sensors and the complexity needed for discrimination.

Recognition and quantification target gases can both be achieved as long as only one target substance is present, or if the universe of possibilities is simple enough and the sensor system is adequately sensitive. However, a more realistic and therefore more vexing dilemma is discrimination among dynamically changing mixtures of compounds. As the complexity of the problem environment increases, the ability to quantify or even identify a particular constituent becomes more difficult. The importance of selectivity and sensitivity is obvious, and many of the resistive-type sensors discussed in this chapter are neither sufficiently sensitive nor selective enough to analyze chemically complicated systems successfully, regardless of sensor combinations or analysis technique.

14.4 FABRICATION

Most of the commercial resistive sensors are processed bulk materials; however, thin-film sensors are becoming more prevalent and considerable effort in the last few years has been spent developing and optimizing methods of microsensor fabrication. Bulk material fabrication is generally much simpler and requires less sophisticated equipment; however, the emerging thin-film sensors, requiring less material, and possessing such advantages as lower power, smaller size, better reproducibility, and on-board electronics may justify initially higher production costs.

14.4.1 Bulk materials

Much of the art of the sensor is in the synthesis of the desired material. Characteristics such as homogeneity, grain size, and crystalline phase, which can be controlled to varying degrees during the synthesis process, greatly influence the ultimate sensor mechanisms [52]. Iterative experimentation with metal oxide particle size, prefired compositions, catalysts, and process variables is necessary to optimize porosity, resistance, sensitivity, and other sensor characteristics.

The metal oxide is often precipitated from chemically pure salts and other reagents; for example, to obtain tin oxide, tin chloride and stearic acid or ammonia are combined. In a similar way, an experimental compound, $ZnSnO_3$, was prepared by precipitating zinc sulphate and tin chloride with oxalic acid and ammonia [52]. This mixture was then washed with deionized water, dried at 110 °C and calcined at 400–600 °C for one to several hours. After milling the precipitated material to desired fineness, several fabrication avenues can be taken.

One option is to press the milled powders into pellets, and sinter. Electrodes or heater elements can be embedded in the pellet or applied after the sintering process. The sensor may be pretreated by exposure to gas at prescribed temperatures.

Another technique, used by Figaro Engineering Inc. and others, is to coat an alumina tube with tin oxide in an organic matrix, such as stearic acid, and sinter at 400–700 °C. A porous film results. Electrodes at either end of the tube are printed on or embedded in the sintered oxide and a platinum heater coil is positioned inside the tube.

Thick-film paste can be formulated to paint or print metal oxides onto a substrate. In formulating the paste, finely milled metal oxides or other sensor materials are combined with small amounts of comparably sized glass frit (for adhesion to the substrate), catalysts (if desired), and an organic vehicle to form a printable paste. The particle size of the constituents varies, although for screen printing powders should be 0.5 μm or less in diameter. The paste is then printed through a pattern on very fine wire mesh or applied by painting onto a ceramic substrate, typically alumina, dried, and fired at temperatures between 500 and 1000 °C for one or more hours. Standard printable thick-film materials for resistive heaters and conductor lines may be applied to the substrate before or after the sensor layer, as dictated by sensor design or firing schedules of the materials. Catalysts, often in the form of precious metal chlorides, may be applied to the surface of the processed metal oxide film and fired. Pretreatment of the sensor with target or cleaning gases may be performed to help remove residual oxygen left from the firing process or other surface contaminants that slow or impede normal sensor response.

14.4.2 Thin films and microfabrication

Thin films (generally $<1\ \mu m$) are responsible for some improvements in resistive semiconductor sensors, although most sensors based on thin films are still experimental. Much development work has focused on the pressing need to reduce the power consumption of metal oxide gas sensors by reducing the total size of the active sensing elements. Through the use of thin sensor films and associated microfabrication techniques, power consumption can be decreased by an order of magnitude and more. In addition, sensitivity and response time are often much better than that of bulk-sintered sensors. The thickness of the thin film itself may affect sensitivity. In one study of thin-film tin oxide sensors with different film thicknesses [53], a sensor with half the thickness had greater sensitivity to ethanol, although all other sensor parameters were identical. Thin-film fabrication methods can also produce sensors with good to impressive uniformity.

Many chemical and physical methods of thin-film deposition are available. For metal oxide work, the favored thin-film technique is rf sputtering from a metallic or a metal oxide target. Other methods include chemical vapor deposition, thermal evaporation, electron gun evaporation, spray pyrolysis, and DC magnetron sputtering [54]. In one example of spray pyrolysis [55], tin oxide was decomposed from $(NH_4)_2SnCl_6$ dissolved in water or $(CH_3COO)_2SnCl_2$ dissolved in ethyl acetate. The SnO_2 grain size is determined by the substrate temperature (between 70 °C and 450 °C) and the temperature used during subsequent annealing. The choice of deposition, annealing schedule, and substrate material can greatly affect sensor sensitivity and selectivity to target gases.

Desired effects of rf sputtering as with some of the other thin-film methods are the production of thin films that are stable, composed of the same crystalline phase, and highly structured. These factors increase uniformity among sensors and simplify adsorption other response characteristics. A deposition might only consist of a few atomic layers. Surface additives, in the form of noble metal catalysts, can also be applied to provide a level of selectivity to certain gases.

Other current and notable efforts in thin-film development have utilized the methods of anisotropic chemical etching and plasma etching and the sacrificial layer method [56]. One early example was a sensor array on an ultrathin, thermally isolated silicon membrane fabricated by chemical machining [57]. The device operated at 300 °C, requiring about 150 mW. Recent work at Honeywell [58] is aimed at removing chip to chip variability, through the development of a 'microbridge' structure that can be adapted to a variety of sensors, not just gas sensors. Both ZrO_2 and SnO_2 thin sensor films have been tested on the silicon microbridge structure. With a sensor prototype heated to 200 °C, NO_x was detected in the range of 0.1–100 ppm. Other parameters that can be measured by this arrangement were thermal conductivity and specific heat. An additional benefit of these fabrication methods is the ability to incorporate other stuctures

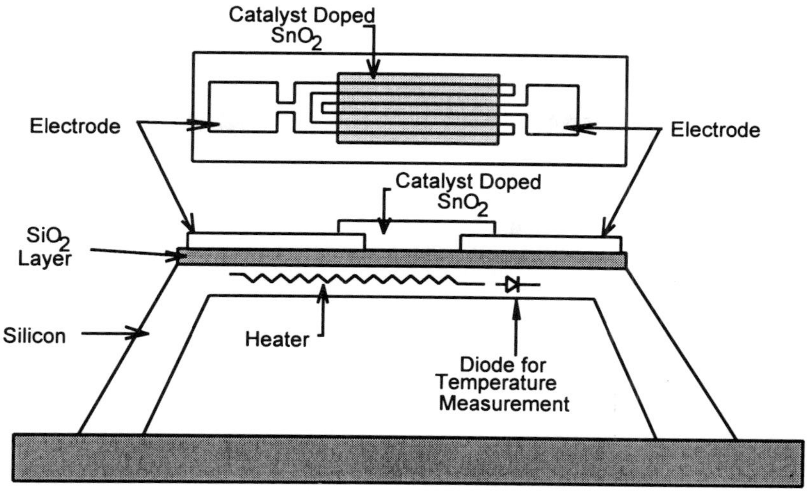

Figure 14.6 A schematic view of structures defined by micromachining techniques [71]. The SnO_2 layer is applied in a spin coating process.

for temperature control and sensing, and signal processing (figure 14.6).

Promising sensor array work [59, 60] involves the construction of a silicon heater array using standard integrated circuit (IC) fabrication techniques and one postprocess etching step to create suspended heater structures. In one version, each element of the heater array can be powered individually with 4 mA at 4 V. This 16×16 array consumes less than 100 mW. Because the heater is suspended, heat is not absorbed by the underlying silicon structure. Most sensor layers and surface additives, not being standard IC materials, would necessarily be deposited on the sensor substrate by chemical vapor deposition or some other process following the chemical etching step; although in principle, all accompanying signal processing and control circuitry can be integrated on the same silicon chip in the foundry process. In fact, a 'smart peripheral' has been demonstrated by the University of Michigan and the MOSIS foundry [61, 62]. This device consists of an eight-line interface to four independent sensors to program sensor heater and operation parameters, and read sensor output. Because the heater area is small, the sensor can be heated to over 1000 °C. At 500 °C the temperature is uniform throughout the sensor area.

One IC-based sensor array approaching commercialization is the Wide Range H_2 Sensor [63, 64], which combined CMOS-compatible metal-insulated semiconductor (MIS) diode sensors and resistive sensors (Pd–Ni alloy) with on-chip control electronics. The 'wide-range' claim relies on the MIS sensor for sensitivity to low concentrations of hydrogen gas (0.1%), and the resistive sensor for higher concentrations (1–100%). Non-standard IC processes for deposition of the Pd-based structures were devised at the Sandia foundry. Tests indicated little variability among sensors, e.g. 1–2% error. It was found that sensor exposures to low concentrations (1–100 ppm H_2) may not be reversible without a cleaning treatment (burn off with O_2), but other remedies are being evaluated.

By using IC technology, not only can size and power consumption be reduced, but sensor and on-chip control circuitry density can be dramatically increased. In designing these sometimes radically new devices, the researcher can take advantage of existing highly developed tools and methods for constructing specialized structures, although some processes must be modified to create new structures or apply sensor materials. Fabrication set-up, especially for new structures or materials, can be very costly. As a consequence, only experimental foundries are currently supplementing or modifying their processes. Significant cost savings might be realized if a sensor design is perfected to the point of justifying large-scale production.

14.5 SPECIFIC SENSOR EXAMPLES

Work performed at Oak Ridge National Laboratory (ORNL) has paralleled much of the work described previously in this chapter. Experiments have involved the construction and testing of metal oxide sensor arrays, accompanied in some cases by the application of pattern recognition techniques. More recently, new sensor work has concentrated on the development of solid state, resistive hydrogen sensors, both thin- and thick-film versions.

14.5.1 An integrated sensor array

To reduce size, power consumption, and manufacturing costs of a metal oxide gas sensor array, an integrated array (IGAS chip) was designed and constructed using conventional thick-film techniques [65]. Metal oxide semiconductors were chosen because of known sensitivity to gases and ease of fabrication. The effect of a multisensor array was achieved on a single substrate with a continuous metal oxide film whose catalytic activity varied across the chip surface. The sensor surface properties were varied by (i) establishing a thermal gradient along the length of the chip, (ii) loading the metal oxide film with different noble metals in different areas, or (iii) combining (i) and (ii). Thick-film electrodes provided access to sensing areas and the thick-film heater. The chip measured 2.5 cm × 1.0 cm, conveniently sized for use with dual in-line package (IC) sockets. The sensor heater consumed approximately 2.5 W.

Using commercial thick-film materials, electrodes and heaters were first printed and fired onto standard 96% alumina substrates. This configuration consisted of 10 opposing pairs of electrodes and a heater resistor which was printed either as a narrow strip across the end pair of electrodes to provide a thermal gradient along the substrate length, or as a large strip on the reverse side of the substrate to yield a uniform temperature. The gas-sensitive film was deposited in one of two ways, either by decomposing tin chloride as described in the Taguchi patent [66], or by screen printing fritted oxide mixtures. Metal oxides tested were SnO_2, ZnO, and Fe_2O_3.

Sensors made by the tin chloride decomposition formed crenulated surfaces, which provided large surface areas, an obvious advantage for the detection of many organic compounds. There were, however, several important shortcomings. First, it was difficult to control and reproduce uniform sensors in the laboratory, although uniformity would be improved using specialized manufacturing equipment. In addition, the strength and adhesion of the layer was quite poor, and its initial resistance varied greatly from point to point on the chip.

To overcome these difficulties, printable compositions were formulated. The printable sensor inks contained three major components: the metal oxide, glass frit for adhesion, and organic vehicles that burn off during firing. A catalyst, in the form of a precious metal chloride, was applied and fired on the sintered metal oxide layer; alternatively, precious metal resinate solutions were incorporated directly into the ink. Initial tests of these printable layers demonstrated sensor resistivities that changed rapidly and reversibly by as much as a factor of 14. The response time was a few seconds while recovery took about 1 min, although complete recovery was often longer than 16 h.

Most IGAS sensor tests were conducted using sensors fabricated by the Taguchi method of applying the metal oxide. It was noted that the tin oxide compositions were generally more responsive to acetone vapor than were the zinc oxide compositions, suggesting that SnO_2 has greater intrinsic catalytic properties than ZnO with respect to the oxidation of acetone. The relative effectivenesses of the catalysts with respect to oxidation of organic compounds studied here, in decreasing order of effectiveness, were platinum, rhodium > ruthenium > palladium > iridium, osmium. Sensor responses to three classes of compounds revealed other interesting effects. Ketones exhibited similar responses that were distinct from alcohols and normal hydrocarbons, and the hydrocarbons and alcohols had similar responses within their respective groups. The behavior of these compounds was more strongly influenced by the presence or absence of reactive groups such as CO or OH than by the hydrocarbon chain length (figure 14.7). Signatures of other substances, such as phenolic compounds, had similar features that appeared to be related to their common chemical structures.

14.5.2 A sensor array of discrete sensors

In another experiment, a sensor array was constructed with several commercial sensors to test the enhancement of sensor selectivity using pattern recognition techniques [67]. This type of sensor array was configured using two strategies: (i) operating identical sensors at different power levels (and, hence, different temperatures) and (ii) using different models of sensors, chosen for dissimilar sensitivities. The overall goal would be to choose or operate each element of

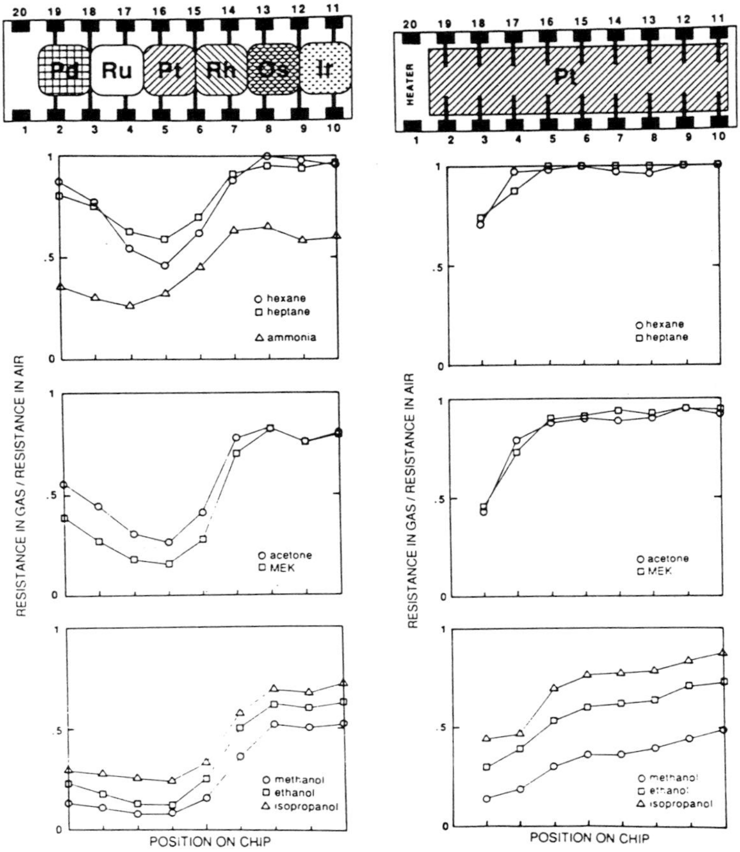

Figure 14.7 The response of two IGAS sensors to several gases. Left, note that similar compounds have similar signatures, suggesting that behavior is dominated by the most active functional group on the molecule and relatively insensitive to chain length. Right, catalyst effects, especially those of Pd, Ru, and Pt, dominate the responses of a uniformly heated tin oxide sensor chip. Note that functional groups still behave similarly.

the array to obtain response characteristics that in general were not identical for different gases. The arrangement can allow flexibility in controlling the range of sensitivities of the different sensing elements. Specifically, this array consisted of nine Taguchi-type sensors (TGS 813) operated at power levels ranging from 30 to 100% of the manufacturer's recommended operating power (900 mW) and one TGS 824 ammonia sensor operated at 100% of the recommended power. The entire array was housed in a Plexiglas chamber with a gas inlet and outlet. Vapors were introduced into the test chamber by passing air through a flask containing the test liquid.

14.5.2.1 Sensor array behavior. For many groups of substances, the sensor array exhibits different sensitivities in different temperature areas of the array. Comparing normal hydrocarbons and alcohols, the TGS 813 sensors in the warmest part of the array were more sensitive to hexane and heptane than the cooler sensors, but not as sensitive to alcohols; in contrast, the TGS sensors in the middle of the temperature gradient (40–60% of recommended power) were more sensitive to alcohols and not as sensitive to the normal hydrocarbons.

Important real world demands on a sensor array are selectivity and ability to distinguish substances of interest in a mixture. For the metal oxide sensor array the signature of a mixture appears to be the superposition of the signatures of its constituents, but, because the effect is exponential, the signature of the more reactive component will dominate. Therefore, if the substance that we seek to detect is the more reactive component, we will enjoy enhanced sensitivity (e.g. alcohol in gasoline). On the other hand, diluents such as alcohol can completely mask more subtle components such as essential oils.

Although uniqueness of signatures is considered desirable, there could be situations in which the signature of an unknown substance has some similarities to a known substance, which could provide some chemical information; for example, signatures of both oil of clove and oil of wintergreen showed an interesting increase in sensor array output at the middle-temperature zones. This response is perhaps due to the benzene ring present in both compounds. In fact, the signature of naphthalene, another aromatic compound, had similarities in the intermediate-temperature range, suggesting the possibility that part of the signature is more important or contains more chemical information in some cases.

Water had a relatively weak effect on the sensor array. This is helpful if we are trying to detect substances that are much more reactive than water. In fact, it might be possible to configure the sensor array to study flavor notes in aqueous substances such as coffee or tea. If, on the other hand, humidity measurement is desired, then the presence of more volatile constituents in the system may mask any response to humidity.

14.5.2.2 Achieving selectivity using pattern recognition. Once a sensor array is configured to yield distinctive patterns for each of the substances of interest, many types of pattern recognition may be applied to solve the analysis problem. For analysis of the sensor array described above, both cluster analysis and a simulated neural network, the Science Applications International Corporation (SAIC) ANSim™ programs were evaluated. Results obtained using the neural network program are discussed below [68].

The steady-state response of each sensor (~5 min) was recorded for each substance. A training pattern was formed by converting each set of steady-state sensor responses to a bit pattern. Bit patterns generally consisted of no more than 200 bits. The low resolution of the training pattern maximized process speed for the test; in an actual instrument, a small pattern size would also help to

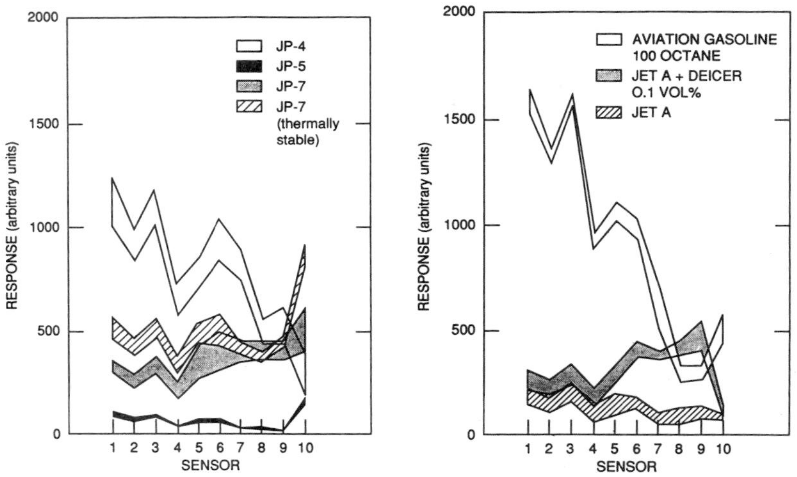

Figure 14.8 Sensor array responses for seven aviation fuels. A sensor array of three TGS 813 sensors produced a collection of outputs that could be successfully classified by neural network software.

reduce overall system complexity. After the neural network was 'trained' with the set of known patterns, other signatures, coded into patterns and treated as 'unknowns', were presented to the trained system, and matched to one of the known patterns. Training sets for aviation fuel included signatures of aviation gasoline (100 octane), Jet A (with and without deicer), JP-4, JP-5, JP-7, and JP-TS (figure 14.8). Another group of training sets was composed of signatures of three automotive gasolines of different octane numbers, and two gasoline/alcohol mixtures. Tests were of a preliminary nature and involved small data sets, but, nevertheless, support several important conclusions about the usefulness of sensor arrays for real problems.

Although the sensor arrays in this work were not optimized for any specific task, relatively simple neural network techniques were found to successfully classify aviation and motor fuels [68]. With signatures that exhibited a high degree of uniqueness, such as the aviation fuels, successful classification could be obtained by using very few carefully chosen sensors. With signatures that are much more similar to one another, such as those of the motor fuels, a greater number of sensors were needed for reliable classification. Surprisingly, a seven-sensor array performed as well as the 10-sensor array in classifying a set of motor fuels, indicating that three of the sensors were providing redundant information. The neural network easily distinguished between signatures of straight gasoline and those containing alcohol. More noteworthy is the correct classification of fuel samples of two octane grades (87 versus 92 ON), even though raw signatures of these fuels look very similar to the eye.

An implicit assumption is that the sensor array is stable enough over the long term that, once a signature is established, the array produces that signature repeatedly at any time in the future when exposed to the same substance [67]. Signatures of ethanol collected sporadically over 9 months' time seemed at first glance to have significant differences. However, examination of this series of patterns with the neural network yielded surprising results. The network consistently recognized them as ethanol after training with randomly selected signatures.

Typically, steady-state data were used to form the patterns for input to the neural network; however, a look at the dynamic responses of the sensor array for different substances revealed a potential source of information to enhance or provide needed uniqueness to the signatures [67]. In addition, it is conceivable that dynamic response could be studied as a way to increase overall speed, which is necessary in real time or on-line applications.

Using the testbed approach of independently operated sensors, optimization of the sensor array and pattern recognition methods can be performed iteratively to develop and verify an application-specific system. As the set of patterns becomes more distinctive it is likely that less resolution is required by the pattern recognition scheme; alternatively, the better the pattern recognition scheme, the more ambiguous the sensor output can be. For actual systems, many pattern recognition algorithms can conceivably be implemented in programmable or custom-designed integrated circuits. In general, it is expected that once a classification system for a sensor array is finalized and sized properly, it can be coded into suitable programmable or application-specific hardware.

14.5.3 Resistive hydrogen sensors

Both thick- and thin-film versions of a solid state, resistive hydrogen sensor were designed and fabricated at ORNL [69, 70]. Both versions of the sensors (25 mm × 25 mm × 0.6 mm) are small enough to be incorporated into hand-held leak detectors or distributed sensor systems for safety monitoring throughout large areas.

Thin-film sensors were fabricated by depositing palladium by vacuum evaporation on alumina substrates. The resistor was a sinuous pattern with a large length/width ratio for sensitivity and low power dissipation. Two layouts were tested. The first had separate metallizations for the active and reference legs and required external resistors for the other two legs of a Wheatstone bridge. In the second sensor layout, all four Pd bridge resistors were deposited as a single monolithic metallization to reduce electrical contact noise and improve consistency and reproducibility. The operating principle of this sensor requires that the reference legs be covered by a passivation that is effectively impermeable to (molecular) hydrogen. Reference legs were made by covering at least one (usually three) of the four palladium resistor legs with a passivation layer to inhibit the hydrogen reaction.

To improve adhesion of the Pd to the alumina substrate, a thin layer of Ti

(20 nm) was deposited first, followed by a layer of Au (20 nm) to prevent the Ti from removing hydrogen from the Pd. Lastly, the 200 nm Pd layer was deposited.

For passivation, ceramics were chosen rather than polymer films because ceramics are inherently less permeable and potentially better able to survive in hostile environments such as steam, solvent vapors, acids, etc. Compounds of SiO_2, Ta_2O_5, and MgF_2, typically 200 nm thick, were all evaluated. The initial films were not entirely impermeable to hydrogen, evidenced by a hysteresis effect in the output of the sensor during and after exposure to hydrogen. (The sensor thus appeared to have two 'active' elements, one with a much longer time constant than the other, rather than one 'active' and one 'passive'.) Passivation was improved by coating the SiO_2 with lacquer, increasing the SiO_2 thickness, or using either Ta_2O_5 or MgF_2 as the passivation.

The thin-film sensors were exposed to hydrogen gas from concentrations of 0.5% to 4% in matrix gases of air, argon, and nitrogen. The sensor was also tested directly in transformer oil with promising results. In all cases, the rate controlling step appears to be one or more surface processes (adsorption, dissociation, reaction with adsorbed oxygen, etc). Although solid state diffusion of atomic hydrogen in Pd is very rapid (at ambient temperatures a 200 nm film would exhibit a time constant of the order of a few milliseconds), response times for the thin-film sensor varied widely, from 20 s at 90% of maximum to as long as several minutes. In one series of tests, it was found that after an exposure of 90% NO_2/N_2, a surface 'cleaning' of 2% H_2/N_2, followed by pure nitrogen, would greatly reduce the subsequent response times [69].

A thick-film version of the hydrogen sensor was also developed [70]. Because of the difficulty of making a totally impermeable thin-film passivation, a thick-film layer was expected to be much more effective. Also, a thick-film sensing layer could be more porous and react more effectively with gas samples because of the higher surface area. Another feature of the design was the incorporation of trimmable areas on the sensor to balance the bridge circuit without external resistors. (Trimming can also be done on the thin-film sensor using laser trimming but with more difficulty.) The thick-film sensor is inherently more rugged than the thin-film sensor, as well as simpler and much less expensive to produce commercially.

The thick-film design consists of four layers, to be separately screen printed and fired on a 1 in square alumina substrate (figure 14.9). Commercial formulations were used for electrodes, bridge trimming resistors, and passivation layers. The first attempted sensor layer was a commercial silver/palladium paste modified by the addition of palladium powder. Based on the performance of the first thick-film sensors, DuPont Electronics (Research Triangle Park, NC) specifically formulated a palladium-based thick-film paste for this application.

The performance of the thick-film sensor was similar to that of the thin-film model. Although not verified, the response time may actually be faster; however, the greater surface area of the thick film may actually slow response because of

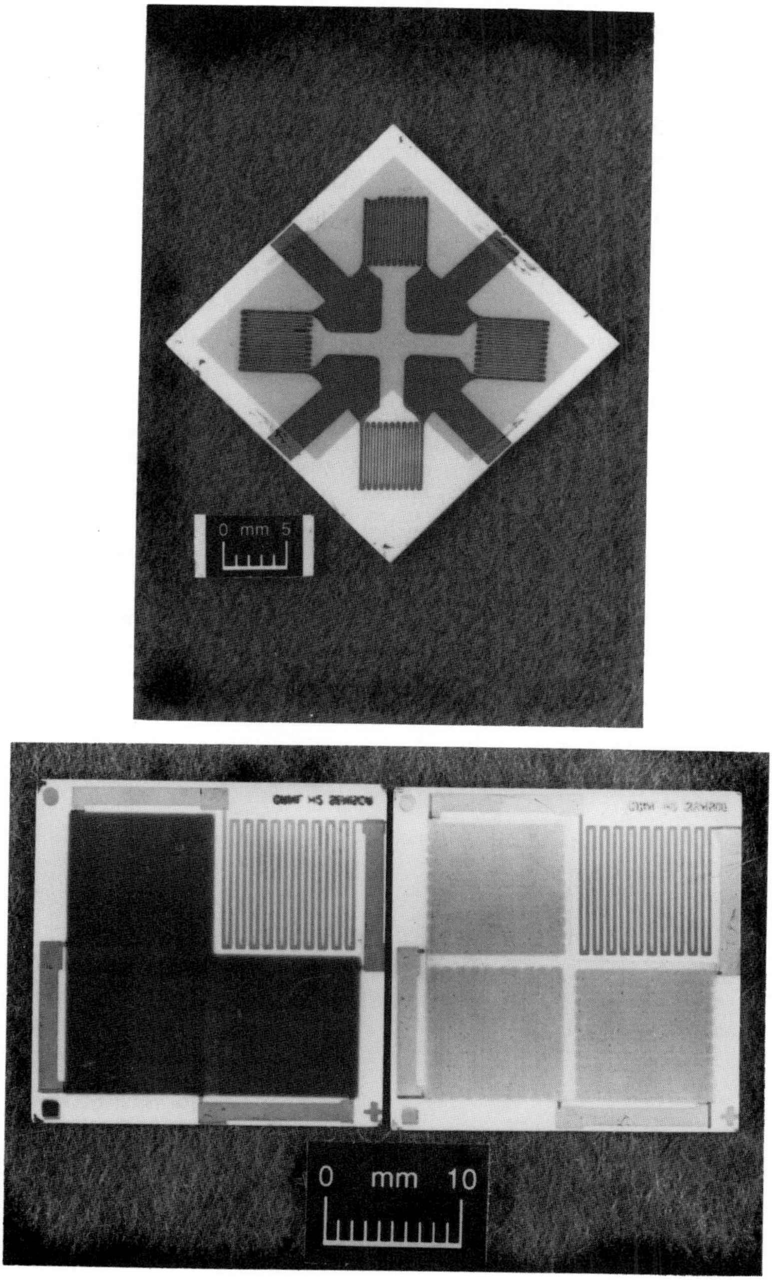

Figure 14.9 (a) A thin-film hydrogen sensor and (b) a thick-film sensor based upon the same sensing principles.

greater adsorbance of water, or other molecules, at the surface. Like the thin-film sensor, the thick-film sensor can become somewhat dormant after several days of 'inactivity' due to surface effects, although further development may correct this.

Resistive hydrogen sensors, both thin and thick film, show promise for the detection and monitoring of hydrogen, particularly near the lower explosive limit (4% in air). This sensing element showed good sensitivity, low noise, and power consumption low enough for adequate battery life. Because surface processes are rate limiting, sensor response may depend to some extent on its exposure history. The general sensor design can be easily modified for specific applications. A particular advantage of the thick-film sensor is its low cost.

Despite many limitations, resistive gas sensors continue to hold their own in the gas sensor market and appear to have a promising future. The attributes of small size and low cost have probably contributed to their popularity for specific gas sensing problems where general purpose instruments would not be practical. The desire for better and less costly solutions to problems continues to inspire innovation in this area. Significant progress is evident in that (i) sensor materials and mechanisms are better understood; (ii) new techniques of fabrication are improving sensor performance with regard to lower power consumption and increased sensitivity; and (iii) novel methods of assembling and analyzing sensor arrays are improving information content and integrity. As work continues we shall see more of these sensors applied to combustion and environmental monitoring; quality and process control; and safety devices.

REFERENCES

[1] Moseley P T 1991 New trends and future prospects of thick- and thin-film gas sensors *Sensors Actuators* B **3** 167–74

[2] Seiyama T, Kato A, Fujiishi K and Nagatani M 1962 A new detector for gaseous components using semiconductive thin films *Anal. Chem.* **34** 1502–3

[3] Taguchi N 1962 *Japanese Patent* S45-38200

[4] *Figaro Products Catalog* 1994 (Wilmette, IL: Figaro)

[5] Matsuura S 1993 New developments and applications of gas sensors in Japan *Sensors Actuators* B **13–14** 7–11

[6] Williams G and Coles G S V 1993 NO_x response of tin dioxide based gas sensors *Sensors Actuators* B **15–16** 349–53

[7] Xu C, Tamaki J, Miura N and Yamazoe N 1991 Grain size effects on gas sensitivity of porous SnO_2-based elements *Sensors Actuators* B **3** 147–55

[8] Amamoto T, Yamaguch T, Matsuura Y and Kajiyama Y 1993 Development of pulse-drive semiconductor gas sensor *Sensors Actuators* B **13–14** 587–8

[9] *RTC-01 Combustible Gas Sensor Specifications* (Schaumburg, IL: Sinostone)

[10] Youngblood J L 1986 Development of a hydrogen sulfide sensor: a case study *Sensors Expo. Proc.* (Peterborough, NH: Helmers Publishing, Inc.) pp L0782–92

[11] Igarashi I 1986 New technology of sensors for automotive applications *Sensors Actuators* **10** 181–93

[12] Baresel D, Scharmer P, Huth G and Gillert W 1980 Sintered metal oxide semiconductor having electrical conductivity highly sensitive to oxygen partial pressure *US Patent* 4 194 994

[13] Beie H-J and Gnörich A 1991 Oxygen gas sensors based on CeO_2 thick and thin films *Sensors Actuators* B **4** 393–9

[14] Poghossian A S, Abovian H V, Avakian P B, Mkrtchian S H and Haroutunian V M 1991 Bismuth ferrites: new materials for semiconductor gas sensors *Sensors Actuators* B **4** 545–9

[15] Peng J and Chai C C 1993 A study of the sensing characteristics of Fe_2O_3 gas-sensing thin film *Sensors Actuators* B **13–14** 591–3

[16] Fukazawa M, Matuzaki H and Hara K 1993 Humidity- and gas-sensing properties with an Fe_2O_3 film sputtered on a porous Al_2O_3 film *Sensors Actuators* B **13–14** 521–2

[17] Fleischer M and Meixner H 1993 Improvements in Ga_2O_3 sensors for reducing gases *Sensors Actuators* B **13–14** 259–63

[18] Takada T, Suzuki K and Nakane M 1993 Combustion monitoring sensor using tin dioxide semiconductor *Sensors Actuators* B **13–14** 404–7

[19] Gusmano G, Montesperelli G and Traversa E 1993 Humidity-sensitive electrical properties of $MgAl_2O_4$ thin films *Sensors Actuators* B **13–14** 525–7

[20] Nylander C 1985 Chemical and biological sensors *J. Phys. E: Sci. Instrum.* **18** 736–50

[21] Kanefusa S and Nitta M 1992 The detection of H_2 gas by metal phthalocyanine-based gas sensors *Sensors Actuators* B **9** 85–90

[22] Eguchi K, Kayser P, Menil F, Lucat C, Aucouturier J L and Portier J 1990 Detection of phosphine at the ppm level by $Sr_{1-y}Ca_yFeO_{3-x}$ thick layers pretreated with highly concentrated phosphine in air. Evidence of a transition of the surface conductivity from p- to n-type *Sensors Actuators* B **2** 193–7

[23] Gerblinger J and Meixner H 1991 Fast oxygen sensors based on sputtered strontium titanate *Sensors Actuators* B **4** 99–102

[24] Egashira M, Shimizu Y and Takao Y 1990 Trimethylamine sensor based on semiconductive metal oxide for detection of fish freshness *Sensors Actuators* B **1** 108–12

[25] Maekawa T, Anno Y, Tamaki J, Miura N, Yamazoe N, Asano Y and Hayashi K 1993 Development of semiconductor gas sensors to discern flavors of consommé soup *Sensors Actuators* B **13–14** 713–14

[26] Varfolomeev A E, Volkov A I, Eryshkin A V, Malyshev V V, Rasumov A S and Yakimov S S 1992 Detection of phosphine and arsine in air by sensor based on SnO_2 and ZnO *Sensors Actuators* B **7** 727–9

[27] Hafele E, Hardtl K-H, Muller A and Schonauer U 1991 Semiconductor for a resistive gas sensor having a high response speed; doped perovskite *US Patent* 4 988 970

[28] Jones T A, Bott B, Hurst N W and Mann B 1983 Solid state gas sensors: zinc oxide single crystals and metal phthalocyanine films *Proc. Int. Meeting on Chemical Sensors (Analytical Chemistry Symposia Series 17, Fukuoda 1993)* ed T Seiyama, K Fueki, J Shiokawa and S Suzuki (New York: Elsevier) pp 90–4

[29] Saeki H and Suzuki S 1992 Organic thin film semiconductor device *Japanese Patent* JPX 19881102 63-277732; *US Patent* 5 079 595

[30] Lewis A 1967 *The Palladium Hydrogen System* (New York: Academic)

[31] Barile R 1993 personal communication

[32] Michaels A 1964 *Design, Development and Prototype Fabrication of an Area Hydrogen Detector* (Southfield, MI: Bendix) (NAS8-5282)

[33] Accorsi A 1986 Les capteurs de gaz *Proc. Capteurs 86 (Paris)* pp 183–92

[34] Accorsi A, Delapierre G, Vauchier C and Charlot D 1991 A new microsensor for environmental measurements *Sensors Actuators* B **4** 539–43

[35] Matsushima S, Teraoka Y, Miura N and Yamazoe N 1988 Electronic interaction between metal additives and tin dioxide in tin dioxide-based gas sensors *Japan. J. Appl. Phys.* **27** 1798–802

[36] Fukui K 1989 Alcohol selective gas sensors *US Patent* 4 849 180

[37] Yamazoe N, Kurokawa Y and Seiyama T 1983 Effects of additives on semiconductor gas sensors *Sensors Actuators* **4** 283–9

[38] Yamazoe N, Kurokama Y and Seiyama T 1983 Catalytic sensitization of SnO_2 sensors *Proc. Int. Meeting on Chemical Sensors (Analytical Chemistry Symposia Series 17, Fukuoda 1993)* ed T Seiyama, K Fueki, J Shiokawa and S Suzuki (New York: Elsevier) pp 35–9

[39] Yamazoe N 1992 Chemical sensors R&D in Japan *Sensors Actuators* B **6** 9–15

[40] Leary D J 1982 Gas monitoring apparatus *US Patent* 4 347 732

[41] Hoffheins B S, Lauf R J and Siegel M W 1986 An intelligent thick-film gas sensor *Proc. 1986 Symp. on Microelectronics (Atlanta)* pp 154–60

[42] Klass D L 1975 Selective solid-state sensors and method *US Patent* 3 864 628

[43] Portnoff M A, Grace R, Guzman A M, Runco R D and Yannopoulos L N 1991 Enhancement of MOS gas sensor selectivity by ‘on-chip’ catalytic filtering *Sensors Actuators* B **5** 231–5

[44] Clifford P K 1985 Selective gas detection and measurement system *US Patent* 4 542 640

[45] Oyabu T 1991 A simple type of fire and gas leak prevention system using tin oxide gas sensors *Sensors Actuators* B **5** 227–9

[46] Lauf R J and Hoffheins B S 1991 Analysis of liquid fuels using a gas sensor array *Fuel* **70** 935–40

[47] Gardner J W, Shurmer H V and Tan T T 1992 Integrated tin oxide odour sensors *Sensors Actuators* B **6** 71–5

[48] Shurmer H V, Gardner J W and Chan H T 1989 The application of discrimination techniques in alcohols and tobacco using tin oxide sensors *Sensors Actuators* **18** 359–69

[49] Kaneyasu M, Ikegami A, Arima H and Iwanage S 1987 Smell identification using a thick-film hybrid gas sensor *IEEE Trans. Components, Hybrids, Manufacturing Technol.* **CHMT-10** 267–73

[50] Mizei J and Lantto V 1992 Air pollution monitoring with a semiconductor gas sensor array system *Sensors Actuators* B **6** 223–7

[51] *Product Literature* 1993 (Newark, CA: Mosaic)

[52] Shen Y-S and Zhang T-S 1993 Preparation, structure and gas-sensing properties of ultramicro $ZnSnO_3$ powder *Sensors Actuators* B **12** 5–9

[53] Pink H, Treitinger L and Vite L 1980 Preparation of fast detecting SnO_2 gas sensors *Japan. J. Appl. Phys.* **19** 513–17

[54] Sberveglieri G 1992 Classical and novel techniques for the preparation of SnO_2 thin-film gas sensors *Sensors Actuators* B **6** 239–47

[55] Pink H and Tischer P 1981 Gas detection by metal-oxide semiconductors *Siemens Forsch. Entwickl.-Ber.* **10** 78–82

[56] Wu Q, Lee K-M and Liu C-C 1993 Development of chemical sensors using microfabrication and micromachining techniques *Sensors Actuators* B **13–14** 1–6

[57] Crary S B 1986 Recent advances and future prospects for tin-oxide gas sensors *General Motors Report* GMR-5556

[58] Bonne U 1994 Thermal microsensors for environmental and industrial control *NIST Workshop on Gas Sensors: Strategies for Future Technology (1993)*

(Gaithersburg, MD: National Institute for Science and Technology) pp 31–8

[59] Semancik S and J R Whelstone 1993 NIST programs and facilities in gas sensing and related areas *NIST Workshop on Gas Sensors: Strategies for Future Technology (1993)* (Gaithersburg, MD: National Institute for Science and Technology) pp 39–45

[60] Marshall J C, Parameswaran M, Zaghoul M E and Gaitin M 1992 High-level CAD melds micromachined devices with foundries *IEEE Circuits Devices* **8** 10–15

[61] Wise K and Gland J H 1993 Microchemical sensor with temperature programmed methods *NIST Workshop on Gas Sensors: Strategies for Future Technology (1993)* (Gaithersburg, MD: National Institute for Science and Technology) pp 53–9

[62] Gland J H 1994 Microfabricated chemical sensors based on metal thin films, oral presentation at Oak Ridge National Laboratory, Oak Ridge, TN

[63] Hughes R C 1993 Gas sensor technologies at Sandia National Laboratory *NIST Workshop on Gas Sensors: Strategies for Future Technology (1993)* (Gaithersburg, MD: National Institute for Science and Technology) pp 47–51

[64] Hughes R C, Buss R J, Reynolds S W, Jenkins M W and Rodriquez J L 1992 Wide range H_2 sensor using catalytic alloys *Sandia National Laboratory Report* SAND–92-2382C DE93 007591

[65] Lauf R J, Hoffheins B S and Walls C A 1987 An intelligent thick-film gas sensor: development and preliminary tests *Oak Ridge National Laboratory Report* ORNL/TM-10402

[66] Taguchi N 1971 Method for making a gas-sensing element *US Patent* 3 625 756

[67] Hoffheins B S 1990 Using sensor arrays and pattern recognition to identify organic compounds *Oak Ridge National Laboratory Report* ORNL/TM-11310

[68] Hoffheins B S and Lauf R J 1992 Performance of simplified chemical sensor arrays in a neural network-based analytical instrument *Analysis* **20** 201–7

[69] Hoffheins B S, Lauf R J, Fleming P H and Nave S E 1993 Solid-state, resistive sensors for safety monitoring *Instrumentation, Controls and Automation in the Power Industry* vol 36 (Instrument Society of America) pp 189–203

[70] Hoffheins B S and Lauf R J 1994 A thick-film hydrogen sensor *Proc. 1994 Symp. on Microelectronics* (International Society for Hybrid Microelectronics) pp 542–7

[71] Qinghai Liu, Wei Peng, Henrique do Santos and Chung-Chiun Liu 1993 Development of Differential Mode and Solid Electrolyte Future Technology *NIST Workshop on Gas Sensors: Strategies for Future Technology (1993)* (Gaithersburg, MD: National Institute for Science and Technology) pp 71–4

15

Optical sensors for biomedical applications

Gerald G Vurek

Transforming an idea into a new product which will help provide better care for sick people is a complex challenge. There are six orders of magnitude between the first sensor demonstrating concept feasibility and the millionth sensor sold; to the developers, the amount of work needed to reach that goal may seem comparable. One must develop and document systems and processes that give assurance that the product will perform to or exceed the customers' expectations. In the United States, the Food and Drug Administration may audit a manufacturer to verify that those systems and processes are adequate and are being used consistently. The foundation for the manufacturing task is a thorough understanding of the sensor and the measurement system. Optical blood gas sensor technology is used here as an example for the tasks involved in going from concept to product.

Here an optical sensor is defined as a part of a system used to measure a change in the optical properties of a substance that reflect the concentration or activity of a local chemical of interest, such as the partial pressure of oxygen (P_{O_2}) or carbon dioxide (P_{CO_2}), or the hydrogen ion activity (pH). The sensor is located at some distance from the analytical instrument because the latter may be too large or the effort of taking samples from the site to the instrument may be too time consuming or costly. Optical chemical sensors have been a very popular topic, being the subject of over 22% of the sensor papers reviewed by Janata *et al* [1]. *In vivo* optical sensors for medical applications need to offer advantages over existing measurement methods. These advantages may include reduced cost, more convenience, more timely information, or better information. This chapter is focused on sensors to measure the variables of oxygen, carbon dioxide, and hydrogen ions in the body because these three variables play such a vital role in gaging the status of sick patients.

15.1 WHY BLOOD GAS MONITORING?

Oxygen supports the fire of life, to use Max Kleiber's [2] metaphor—without oxygen, we die. Because oxygen is carried from air to tissues by the hemoglobin in the blood's red cells and because about 15–20% of the total content is consumed per minute, interruption of the oxygen supply for less than 5 min can have severe consequences. Proper care of persons with impaired ventilation requires moment to moment oxygen monitoring. One of the byproducts of oxygen metabolism is carbon dioxide, the other 'blood gas'. Rapid changes in the blood's P_{CO_2} can affect cerebral blood flow or cause severe patient discomfort. Carbon dioxide is part of the body's acid–base buffering mechanism so that P_{CO_2} shifts can shift the body's pH away from the normal operating conditions of our myriad coupled enzyme reactions. Anesthesiologists, intensivists, respiratory therapists, nurses, and others involved with care of patients requiring ventilatory support need timely and reliable information about the status and changes in blood gas values. This need has been the driving force for the development of blood gas instrumentation as has been very well described by Astrup and Severinghaus [3]. Laboratory blood gas analysers use electrochemical sensors and early efforts to make versions that function inside the body were based on miniaturization of those sensors. Optical sensors offer advantages of better stability than electrochemical sensors for *in vivo* applications.

The performance specifications or goals for two types of blood gas sensor are listed in table 15.1. One is the fiber optic oximetry catheter (Oximeter) which incorporates optical fibers into a multifunction catheter and uses the optical properties of whole blood to measure oxygen saturation, i.e. the fraction of hemoglobin in blood that is carrying oxygen. The other is the intra-arterial blood gas (IABG) sensing system which is usually placed through an existing radial artery catheter that also allows pressure monitoring and blood sampling as well as continuous monitoring of the blood gases and pH by means of the sensors in the artery.

The oximetry catheter was developed as an addition to the flow-directed pulmonary artery catheter, a relatively large (2.5+ mm diameter) device placed in the low-pressure venous circulation. The 'sensor' of the oximetry catheter derives from the geometry of the optical fibers contacting the blood and the optical properties of the blood itself. IABG sensors for radial artery application are about 0.5 mm in diameter. IABG sensors have specific reagents that may require isolation from blood components. The size and need for special chemical sensors make the IABG system considerably more complex than the oximetry system. In order to be accepted by the market both the oximeter and IABG catheter devices must perform better, with regard to need for recalibration or standardization, than the existing bench instruments. These generally are recalibrated or standardized several times per day but the intravascular systems are expected to operate without requiring the sensors to be standardized for at least 24 h.

Table 15.1 Performance goals for intravascular blood sensors. (IABG values from the article by Soller [25]; copyright 1994 IEEE, with permission.)

	pH	P_{CO_2}	P_{O_2}
Operating range			
Oximeter			0–99% (saturation units)
IABG	6.8–7.8 pH units	20–100 mm Hg	30–600 mm Hg
Resolution			
Oximeter			1% (saturation units)
IABG	0.01 pH units	1 mm Hg	1 mm Hg (30–150 range) 5 mm Hg (150–300 range) 10 mm Hg (300–600 range)
Operating temperature	20–40 °C	20–40 °C	20–40 °C
Stability	<0.03 pH unit/72 h	<6 mm Hg/72 h	<8 mm Hg/72 h
Response time	<3 min	<3 min	<3 min
Accuracy relative to blood gas analyser	±0.02–0.04 pH units		±5–10%

15.2 OXIMETRY

The most successful *in vivo* optical sensor is the fiber optic oximetry catheter, illustrated in figure 15.1. The sensor consists of sending and receiving optical fibers which are used to measure the oxygen-sensitive spectral properties of whole blood. The instrumentation delivers light to the blood at two or more wavelengths via the sending fiber and detects the intensity of the backscattered light. Figure 15.2 shows the absorption spectra of three hemoglobin forms. Native hemoglobin (Hb) and its derivatives, oxyhemoglobin (HbO_2), and carboxyhemoglobin (HbCO) absorb very strongly below 600 nm. There is a non-linear relation between the partial pressure of oxygen in the plasma and the binding of oxygen with Hb in the cells to form HbO_2. Approximately 50% of the hemoglobin is saturated at a P_{O_2} of 27 mm Hg; the actual partial pressure at which 50% saturation occurs depends on many factors including pH and P_{CO_2}. Bench oximeters make multiwavelength spectrophotometric measurements on hemolysed, non-scattering samples. *In vivo* oximeters work in whole blood which scatters and absorbs light. Within the wavelength range shown in figure 15.2, the relative amounts of scattering loss and absorption loss are comparable so neither can be assumed to dominate the losses. Oximeters using either

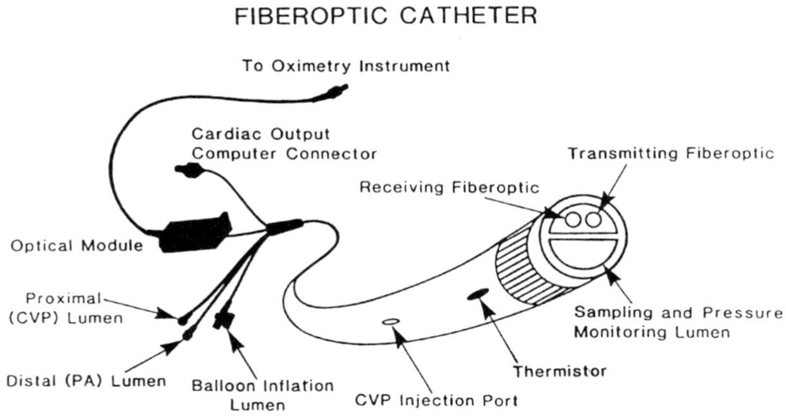

Figure 15.1 The fiber optic oximetry catheter. (From Sperinde J M and Seneli K M 1987 The Oximetrix™ Opticath™ oximetry system: theory and development *Continuous Measurement of Blood Oxygen Saturation in the High Risk Patient* vol 2, ed P J Fahey (Chicago: Abbott Laboratories) pp 59–75; with permission.)

three wavelengths or three fibers can accommodate a large change in red cell concentration (hematocrit) but two wavelength systems require the user to enter blood hemoglobin values into the instrument to obtain accurate saturation measurements. The success of *in vivo* oximetry is due to its demonstrated ability to provide continuous reliable information about the relative amount of oxygen in the blood entering the lungs and to its relative ease of calibration and use.

Fiber optic oximeters operate in the backscattering mode. That is, the source fiber and receiving fiber are adjacent so that only light that is scattered back to the receiving fiber is measured. Accurate and reproducible saturation measurement requires attention to the optical geometry of the measurement 'cell' or source and receiver fiber arrangement. Early fiber optic catheter oximeters used a 'random' arrangement of transmitting and receiving fibers: no two catheters were identical. These devices required *in vivo* calibration so that the particular geometry factor for each sensor could be established. The requirement of a blood sample to establish the initial calibration of an *in vivo* sensor detracts substantially from the value of the instrument because of the cost burden: time (and expense) that must be devoted to the instrument, not the patient. Development and manufacture of oximetry catheters require a good understanding of the optical properties of not only the instrument and fibers but also blood. The optical properties of whole blood, which depend on the cell concentration, cell size, cell orientation, etc, have been studied for many years; and the pertinent information is given in the article by Takatani and Ling [4].

In addition to establishing a reproducible fiber arrangement at the catheter tip, the light distribution within the fiber has to be controlled. Figure 15.3 shows

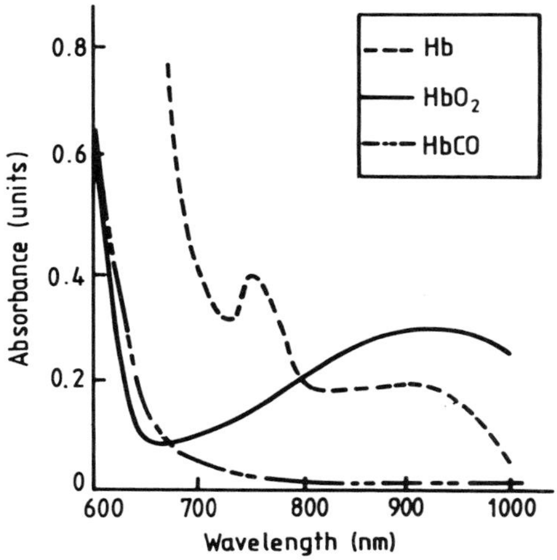

Figure 15.2 Absorption spectra of hemoglobin compounds at the visible and near-infrared wavelengths. (From [4]; copyright 1994 IEEE, with permission.)

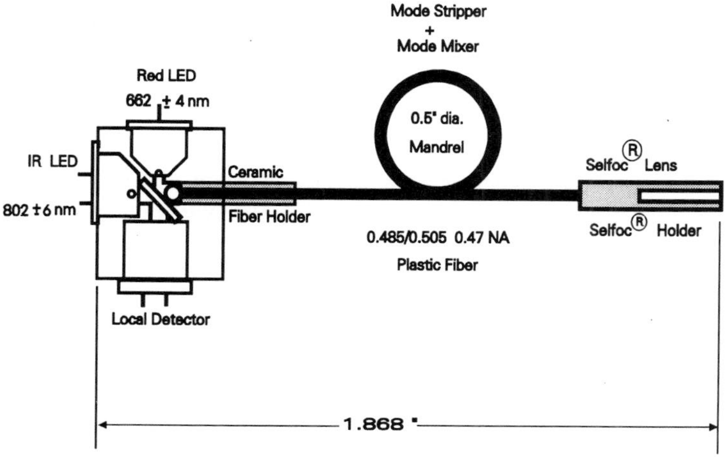

Figure 15.3 A schematic of the source portion of a fiber optic oximeter. The local detector monitors the source intensities; see the text for more details. (From the article by Amirkhanian and Lee [5], with permission.)

a diagram of the light source portion of a type of two-wavelength fiber optic oximeter optical module. On the left are light emitting diodes (LEDs) for the analytical wavelength of 662 nm and reference wavelength, 802 nm. There is a beam splitter and a local detector that is used to monitor the intensity of each LED. There is a small spherical lens that concentrates the light from each LED onto the end of a fiber held in a precision ceramic fiber holder. The Selfoc™ lens and holder are used to couple the light from the reusable optical module to the source fiber in the disposable catheter. Because the two LED sources are not identical, the light distribution entering the fiber from the LEDs is different for each wavelength. If that different distribution were to persist at the sensor end, the scattering function for each wavelength would be different causing the calibration to be imprecise. It is the role of the mode stripper and mixer to make the light distribution of the two wavelengths within the fiber more uniform and the system capable of precise calibration before the catheter is placed in a patient [5].

An oximeter catheter may have about 2 m of optical fiber. While high-silica fiber has very high optical transmission, it costs substantially more than plastic fiber. Lee and McCann recently summarized the properties of both plastic- and silica-based fibers [6]. They point out that polymethylmethacrylate (PMMA) fiber has the lowest cost of all fibers, higher flexibility, and better transmission in the visible range than polystyrene, the fiber commonly used in oximeters; they note PMMA absorbs water which causes shifts in the fiber absorbance during use. Figure 15.4 shows the transmission spectrum of a dry commercial PMMA fiber. The absorption bands of the intrinsic fiber material in the near-infrared limit the designer's choice of wavelengths that can be used for practical oximetry using plastic fibers. These absorption bands contribute to the path loss for the light and have to be considered during the decision about the particular light wavelengths to be used. For example, the loss in the material of figure 15.4 goes from about 5% m^{-1} to almost 50% m^{-1} over the wavelength range from 700 to 720 nm, approximately 2% nm^{-1}. If a light source had to be used in this range, it would be necessary to keep the wavelength constant within a fraction of a nanometer to keep the path loss change less than 1%. Note that monitoring only LED intensity, as in the module of figure 15.4, would not correct for wavelength shifts.

Fiber flexibility is important for clinical applications. The most likely place for fibers to be bent severely is where they pass through the skin. The wound at the insertion site is at risk of infection and it has to be accessible for inspection and care. This usually means the fibers (and whatever sheath covers them) must be able to move more freely outside the tissue than inside because the catheter may be moved during patient care; consequently that entry point is a place for high mechanical stress. An accidental sharp bend may cause a permanent change in plastic fiber transmission but may cause catastrophic failure of glass fibers.

There are at least three important considerations for fiber optics that are illustrated by the *in vivo* oximeter. First, optical fibers are not perfect 'light wires'. In contrast with electrons in insulated copper conductors, there is no rule that photons entering a fiber must exit at the end. Light can be scattered out of the fiber by defects or impurities and it can be absorbed by fiber substance itself. Conversely, external light impinging on the fibers can be scattered into the internal path and contribute to unexpected and potentially detector saturating

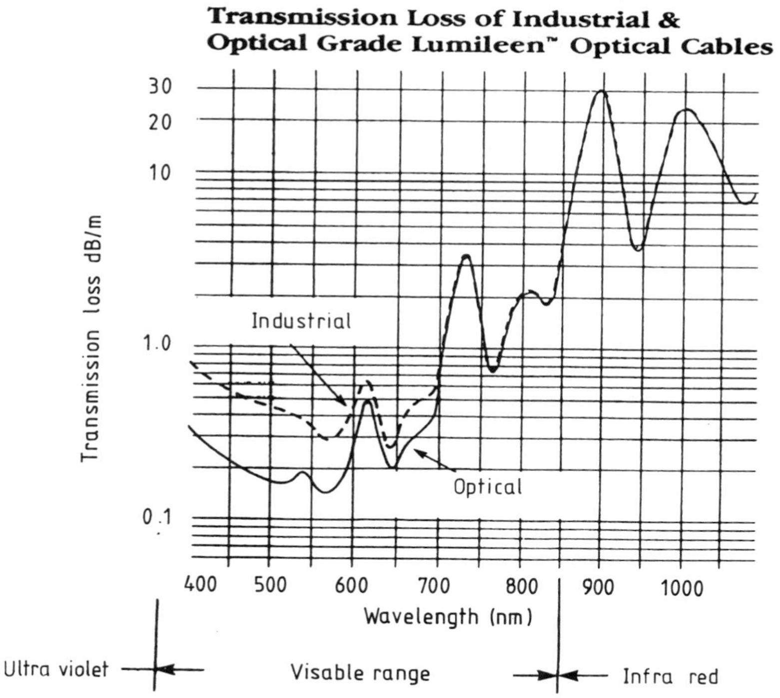

Figure 15.4 Transmission loss spectra of commercial plastic optical fiber. Lumileen™ is a trademark of Poly-Optical Products, Irvine, CA. (With permission.)

light levels. During an investigation of an early P_{CO_2} sensor, the present author observed that extraneous light from a small incandescent lamp could penetrate several centimeters of tissue and cause substantial signal shifts and increased detector shot noise [7]. Use of two or more wavelengths and signal modulation will compensate partially for fiber path losses and stray light as long as the latter does not shift the photodetector operating point out of its linear range. Second, controlling the light distribution within the fiber is important. Fibers are optical components and the geometric distribution of light within the fiber depends not only on how the light is put into the fiber but also how it is bent as it goes from source to detector. It is important to strive to have the path losses identical for each wavelength. If there is an optical connection between reusable and disposable components, the light for each wavelength should be distributed uniformly across the connection so losses at the connection are the same for all wavelengths. Third, the fiber has to be flexible enough to tolerate repeated bends with a radius of a fraction of a centimeter. These three considerations for the fiber portion of an *in vivo* optical sensor are applicable to IABG sensors as well.

15.3 INTRA-ARTERIAL BLOOD GAS SENSORS

The oxygen-sensitive optical properties of blood are exploited by the oximeter. However, IABG sensors use specific chemicals attached to the fibers to respond to their analytes. The development process for producing chemical sensors starts with a set of performance goals exemplified in table 15.1. As table 15.1 shows, the dynamic range over which good performance is needed covers a factor of 10–20 in analyte concentration. That is, the pH sensor should operate over a hydrogen ion activity from pH 6.8 to 7.8 (130–13 nmol l^{-1}), P_{O_2} from 4 to 80 kPa (30–600 mm Hg), and P_{CO_2} from 1.3 to 13 kPa (10–100 mm Hg). Sensitivity should be about 1% of the range and drift should be less than 10% d^{-1}. The stability of *in vivo* sensors is more important than accuracy because therapeutic activity usually results in bringing blood gas values toward the normal range. Trends towards or away from normal are used to indicate the need for possible therapy changes. The threshold of 10% d^{-1} is based on users' perception that 10% is a significant difference between a monitor value and a reference value.

15.4 SENSOR ATTRIBUTES AFFECTING PERFORMANCE

15.4.1 pH sensors

An optical pH sensor has a pH-sensitive dye in part of the optical path from light source to detector. The absorption spectra (for example) of the ionized and neutral forms of the dye are different. Figure 15.5 shows an idealized absorption spectrum of a dye, phenol red, that has been used as a pH sensor. A pH sensor has a means for transmitting light from a source to the pH-sensitive dye, a means for allowing the dye to equilibrate with the hydrogen ions of the blood, and a means to return the light that has interacted with the dye. Leiner and Wolfbeis [8] give detailed descriptions of the chemistry and responsivity of dye-based pH sensors. Figure 15.5 shows that the relative absorption of a given concentration of dye depends on wavelength and pH; increasing the amount of dye in the optical path by increasing the dye concentration or path length increases the absorbance, too. Phenol red absorbs relatively little light from 700 to 625 nm so light in this band may be used to establish optical path transmission losses. Near 560 nm, the absorbance of the dye is maximum at high pH and a peak with opposite sensitivity occurs near 430 nm. Near 475 nm is an isosbestic wavelength where the absorption is not sensitive to pH but depends on concentration and optical path length. At the 560 or 430 nm peaks, a small change in light source wavelength produces a negligible change in apparent absorbance but light absorption on the sides of the peaks is strongly wavelength dependent.

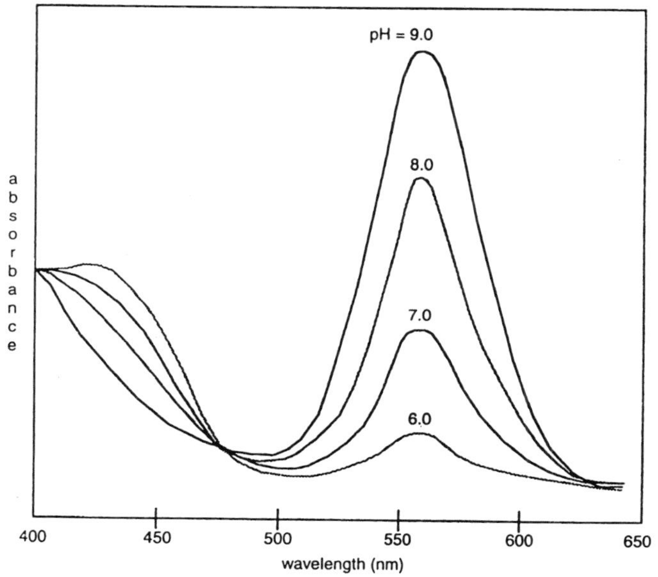

Figure 15.5 Absorption spectra of phenol red dye in solution at various pH values. (From the article by Soller [25]; copyright 1994 IEEE, with permission.)

The relation between pH and the concentration of the dye form which absorbs at 560 nm is given by

$$(A^-) = [10^{(\text{pH}-\text{p}K_a)}/(1 + 10^{(\text{pH}-\text{p}K_a)})](\text{dye})$$

where (A^-) is the concentration of the absorbing form, (dye) is the total dye concentration, and $\text{p}K_a$ is the pH at which the dye is half in the absorbing form. At pH values much greater than $\text{p}K_a$, the square-bracketed coefficient approaches unity; it is 0.5 when $\text{pH} = \text{p}K_a$; and it approaches zero at pH values much less than $\text{p}K_a$. The relation between pH and (A^-) is sigmoidal and the relation between light transmission and (A^-) is exponential:

$$I/I_0 = 10^{-((A-)\varepsilon l)}$$

where I is the transmitted light intensity, I_0 is the incident intensity, and εl is the product of the extinction coefficient (absorbance under standard conditions of concentration, path length, and wavelength) and optical path length. Dyes are relatively small molecules that are kept in the optical path by attaching them to a polymer that is then attached to the fiber. Typical dye concentrations needed to absorb from 10 to 90% of the light in the sensor may be 1–10 mM. At these concentrations, dye–dye interactions may shift the pH/absorbance properties. The concentration is high enough to influence the buffer capacity of the sensor, which affects how many hydronium ions are needed to equilibrate the sensor with its environment—a response time issue. The response time can be reduced by making the polymer more water permeable, but if that allows light absorbing molecules from the blood to enter the light path the sensor response curve will shift. Each lot of dyed polymer must be characterized to ensure it meets the sensor requirements. Total solution ion concentration also affects the absorption spectrum of most dyes, including phenol red, so the spectrum one obtains in solution is only a guide to the spectral properties of the same dye bound to a polymer exposed to blood. If the path length or dye concentration are considered parameters, then the pH–light signal relation becomes a multidimensional function or surface. Concatenated with this function are the effects of stray light and actual sensor geometry. It is the task of the system designer to identify enough parameters, at a known pH, to identify the location of any sensor on that response surface to make the sensor to meet specifications.

15.4.2 P_{CO_2} sensors

Sensors for P_{CO_2} are usually based on a pH sensor material with added bicarbonate ion (HCO_3^-) encased in a gas-permeable, ion-impermeable material [7]. Carbon dioxide reacts with water to form carbonic acid which dissociates and equilibrates with the sensor bicarbonate. The internal pH of the sensor depends on both the P_{CO_2} and the amount of bicarbonate. Establishing and maintaining an appropriate amount of bicarbonate in the sensor makes it more difficult to use than the pH sensor. If the sensor dries thoroughly, voids or crystals may form. These can scatter light out of the optical path and it may be very difficult to restore the material to a suitable optical form. In addition, blood has small amounts of organic acid anions which could eventually diffuse through the barrier separating the bicarbonate in the sensor, react with it, and thereby shift its concentration and the sensor operating point. The patents of Jackson and Hui and Nelson *et al* suggest ways to make satisfactory P_{CO_2} sensors [9, 10]. By design, this particular sensor is steam sterilized and stored in a buffer solution. The sensor is part of an IABG probe covered with a silicone compound. The disclosures mention that the pH values of the simple bicarbonate storage solutions can rise to levels that have deleterious effects on the biocompatibility properties of the silicone, particularly during high-temperature steam sterilization. Also, the P_{CO_2} sensor may drift after storage in such solutions. One of the disclosures describes a buffer solution made of a combination of non-volatile organic and inorganic salts and also carbonate salts [9]. This combination keeps the pH of the sterilized buffer from becoming too high and preventing alkaline-enhanced hydrolysis of the silicone. The other disclosure describes a poststerilization exposure to high P_{CO_2} to restore the internal composition of the P_{CO_2} sensor so that drift is minimized [10]. The actual problem that was addressed by these inventions is one of fairly simple chemistry and, like many problems in retrospect, the need for its solution could have been anticipated at the time that probe specifications were developed. However, going through the steps of determining the actual details of selecting the appropriate buffer, evaluating the safety of any residues that might be brought to the patient, confirming the biocompatibility of the probe surface after long storage in the buffer, and being sure that adequate probe calibration still occurs could have taken more than a staff year of effort. Solutions to a chemical problem like this example as well as the myriad other problems involved in IABG development become the knowledge base that leads to the success of a new technology.

15.4.3 P_{O_2} sensors

Optical P_{O_2} sensors are usually based on the principle of luminescence quenching [11]. Many excited state dyes can transfer their energy to oxygen rather than re-emitting it as fluorescence or phosphorescence; the luminescence is quenched. Quenching is inversely proportional to P_{O_2} so that the higher the P_{O_2}, the less light is released. The average time between light absorption by the dye and its re-emission also decreases with increased quenching. If one uses time-modulated excitation light, then the excitation–luminescence time relation changes with P_{O_2}. Both intensity and time have been exploited to measure P_{O_2}. The dye is usually in a polymer matrix through which oxygen must diffuse. The signal intensity then depends on polymer dye concentration and its oxygen permeability; permeability can change with temperature, dye–polymer lot variability, etc. If these are kept under control, the P_{O_2} sensor can be made robust and, if the decay time property is used, the sensitivity to the amount of dye in the path is very much reduced. The emission wavelength of phosphorescent dyes is usually well separated from the excitation wavelength so that it is easier to keep stray excitation light from the return path and the dye absorbs very little of its emitted light. These are some of the reasons P_{O_2} sensors have been relatively easy to develop. The following example illustrates optical oxygen measurement. Figure 15.6 shows a flashlamp system designed for oxygen measurement based on the reduction of phosphorescence intensity by oxygen quenching as described by Hauenstein *et al* [12]. The sensor dye at the end of the fiber is a combination of two components, one being terbium based and sensitive to oxygen and the second being europium based and relatively insensitive to oxygen. Both dyes are excited by the same ultraviolet wavelengths but each has narrow, well separated emission wavelengths. The signal from the europium dye can be used to monitor source intensity and path loss variations. Two photodetectors are used and are gated off until all the signal from the lamp and fiber luminescence has decayed and only dye signal is measured. The approach of Hauenstein *et al* has the advantage that calibration almost entirely depends on the sensor material so that sensors built with a given lot of material can be precalibrated at the factory. The disadvantages include complex instrumentation and expensive silica-based fibers.

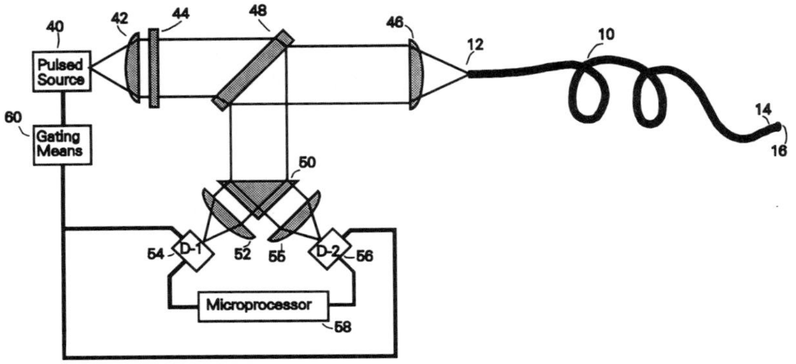

Figure 15.6 A schematic of a flashlamp-excited oxygen sensor system. Items are 10, optical fiber; 12, fiber input/output end; 14, fiber–sensor interface; 16, sensor; 54, 56, reference and signal detectors; 42, 46, 52, 55, lenses; 44, short-wavelength pass excitation filter; 48, beam splitter; 50, excitation wavelength rejecting and emission wavelength separation means; 40, 58, 60, electronic signal processing means. (From the patent of Hauenstein *et al* [12].)

15.4.4 General sensor attributes affecting performance

The dye is usually placed at the end of the fiber by some means such as dip coating. Calculations for luminescence-based sensors show that increasing sensor thickness by more than about one fiber diameter leads to diminishing incremental signal increase [13]; similar calculations can be performed for single-fiber absorbance sensors with retroreflectors. Modlin and Milanovitch give details on how to estimate the fraction of light that can be returned to the detector [13]. In the case of absorbance sensors, light has to pass through the sensor and some fraction of that light must be returned to the detector. Luminescent sensors may return 10^{-6} of the incident light and absorbance sensors may return 10^3 times as much; other losses in the system from the source to the detector may increase the overall path loss by another three orders of magnitude. Among the trade-offs between luminescence and absorbance sensors is the task of designing a system to detect small changes in a weak signal against a system to detect small changes in a modest signal. Dye layer thickness variations affect not only the amount of incident light that is absorbed but also the amount of light that is returned which, in turn, affects signal level and sensitivity. The greater dynamic range that can be accommodated by the optical detection system, the more tolerance can be allowed the sensor fabrication process. That process will include variables such as dissolved solid content, ambient temperature and humidity, solvent choice, and fiber size and surface preparation. Materials used for pH sensors are hydrophilic so that hydrogen ions can equilibrate with the dye. Changes in blood ionic composition can occur, particularly if a patient is acidemic. If the dye matrix swells or shrinks as the ionic environment changes, sensor geometry changes. The sensor may have to tolerate drying–wetting cycles which also can change its dimensions and put substantial strains on the bond between the sensor material and fiber.

Some dyes are photochemically unstable. When a molecule absorbs a photon, the molecule enters an excited state. It may lose that energy by converting it to heat, re-emitting it as luminescence, or causing a chemical reaction. The reaction could disrupt the bond between the dye and polymer and thus allow the dye to be lost from the sensor. The reaction could also convert the dye to a pH-insensitive form. These inactivating reactions may be facilitated by the presence of oxygen and be more prominent at elevated P_{O_2}. At least one oxygen sensor was based on a photochemical reaction that allowed at least 250 measurements before its performance became unacceptable [14].

Sensor temperature coefficient and time response are also variables to be understood. Temperature may shift the pK_a of the dye and change the cell thickness, and will certainly affect the actual value of the blood gas variables of the blood that is adjacent to the sensor. For these reasons, a complete blood gas sensor includes a local temperature sensor, particularly if the sensor is to be placed in a peripheral artery where local temperature may not be equal to central body temperature. The chemical sensor temperature coefficient must be well characterized so that it will accurately measure the local blood gas value. Bench analysers usually measure blood samples at 37 °C so the *in vivo* system must then adjust the measured value to that temperature. The temperature coefficient of the blood gas variables, in blood, may be several percent per degree, and, in the case of P_{O_2}, depend very strongly on the actual value. Thus, to make the *in vivo* sensor agree with bench analysers, local temperature sensing must have an accuracy of better than 1 °C. Size and accuracy requirements can be met by a miniature thermocouple. The system designer has to make sure that the temperature circuit can handle the microvolt signals with adequate accuracy and stability as well as meet patient electrical isolation requirements.

Response time is the time it takes the system to display a value that is some specified fraction of the final value after a sensor has been exposed to a change in environment. Analytes exchange with the sensor chemistry by diffusion so that response time will vary with sensor dimensions. Another contributor to the overall system response time is the time required by the instrument to obtain enough data for a satisfactorily small signal variation. There are two time periods when fast response is very desirable: during calibration and immediately after the sensor is placed in a patient. If calibration takes more than a few seconds, users will divert their attention during this period from other care giver tasks. The longer the calibration time, the longer it will take to identify the (hopefully rare) sensor that does not work 'out of box'. In the controlled calibration environment, it may be possible to extrapolate the final value and save time by terminating the calibration before equilibrium is reached. The reliability of that extrapolation, if necessary, is part of the error budget. Extrapolation may not be possible for the period between placement in the patient and the time when a believable value is displayed. If a user takes a blood sample to be measured with the local bench instrument during the period when the sensor is responding to the change from the final calibrant to the patient's blood gas values, then a difference may appear.

Sterilization and storage affect system performance. The three methods of steam, gaseous ethylene oxide, and radiation have the goal of reducing the probability of a viable organism to some small number typically fewer than 10^{-6} on a packaged device. Both pH and P_{CO_2} sensors need water to operate properly and, if more than 1 min is needed to activate the sensors for calibration and use, wet storage is needed. Although the manufacturing process may be designed to minimize the number of organisms on the product prior to sterilization, wet storage before sterilization can give those few organisms a chance to multiply beyond the capacity of the sterilization process. For that reason, it may be necessary to develop some sort of aseptic poststerilization filling process. This adds to the system cost. Ethylene oxide and radiation are the two processes that can be performed at temperatures that plastic fibers tolerate. Ethylene oxide sterilization is performed at controlled humidity. The gas reacts with liquid water to form a number of residues including sodium hydroxide. This material can shift the buffer state of a P_{CO_2} sensor. Extensive and expensive degassing may be required to reduce residues to acceptable levels before wetting the sensors. Radiation causes shifts in the absorption spectra of both plastic and glass fibers. Steam sterilization may be performed at 135 °C so all the exposed materials need to withstand not only the heat but also the chemical attack of very hot water vapor. The fibers in connectors should not move, glass fibers have to be protected from excessive bending stress, and so on. For various reasons, probes have to be designed to withstand at least two sterilization cycles. Sterilization effects on probe performance require considerable attention during the system design phase.

Although it would be desirable, it is unlikely that all the sensors in an optical blood gas/pH probe will work exactly as when made when the customer opens the package. Some sort of, hopefully brief, preuse calibration procedure is needed. There is a potential error associated with the preuse calibration process. Oximeter catheters have a single material in the sterile package to reflect light from the source fiber to the receiver fiber. This standard reflector establishes the relative signal level for each wavelength over the optical path. The internal chemistry and optical properties of IABG sensors may shift after sterilization and storage so the response at two or three known levels may be needed to achieve adequate accuracy. The usual method for generating a known gas partial pressure and pH is to equilibrate a buffered salt solution with a known mixture of O_2 and CO_2. The accuracy of the resulting calibrant depends on the accuracy of the gas mixtures, the accuracy of the solution mixture, and the local temperature and barometric pressure. Fortunately, the solution components can be prepared gravimetrically and, with adequate mixing and packaging, the result will be highly reliable. Local temperature and pressure measurements are straightforward.

15.5 ACCURACY COMPARED TO WHAT?

Customers for optical blood gas sensors expect them to give results comparable to values given by the customers' bench analysers. Laboratory bench blood gas/pH analysers are considered by users to provide true values. It is relatively easy to establish gas partial pressures in a salt solution with an uncertainty of less than 0.1 kPa and to estimate solution pH within 0.01 unit. However, bench blood gas instruments respond differently to salt solutions than to blood due to the blood proteins and cells that give blood very different transport properties from simple salt solutions. Bench instruments perform frequent automatic recalibrations so that their performance variations are kept within a narrow range. Blood is not a suitable calibrant because it is a living tissue, its properties change with time, and it can carry hazardous pathogens. The developers should be aware that agreement between an optical sensor and an instrument when tested with the same salt solution does not necessarily mean they will agree when challenged with a blood sample. The blood gas/pH values of a sample taken from a patient may be different from the values presented to the bench analyser sensors. The process of removing a sample from a patient may contaminate it with materials in the sampling syringe such as anticoagulants, air bubbles, and gases dissolved in the syringe polymers [15]. In addition to these transportation-related causes for differences between *in vivo* and bench analyser results, bench analyser responses vary from model to model for a given sample. The College of American Pathology (CAP) has set up a program by which laboratories at various locations can test their instruments' response to a common set of samples [16]. The results are pooled and reported so that each laboratory can learn how its instruments perform with regard to the mean of all laboratories. The *in vivo* system designer can hope that its performance will fall somewhere in the middle of the range of bench instrument performance [17].

15.6 TOOLS FOR SENSOR DEVELOPMENT

It would be nice to have the luxury of time to thoroughly investigate each of the variables that contribute to a sensor's performance. Efficient experimental design, particularly through the use of statistical design-of-experiment techniques, can be key to determining interactive effects [18]. Sensors are often built by artisans during the development phase. The skill and flexibility of the artisans allow process variables to be explored much more quickly than when changes must be made through a fully controlled production process. The number of sensors of a particular test configuration that can be built and tested in the development phase is limited, perhaps to a few dozen at a time. The small numbers mean that only major effects of process variables can be identified with confidence. There is a limit to the precision of hand-made sensors, so when a process variable is identified that could benefit from improved precision it is time to introduce process tools and mechanization for that process. IABG sensors are expected to function for more than 3 d so many tests will last more than that. To achieve statistically significant results, many tests must be run in parallel. A reliable instrument and software is used for each sensor so the developers need a small instrument factory to supply sensor test requirements. Packaged lifetime or shelf life should be at least 6 months and, if there are unacceptable performance changes during storage, it can take a long time to identify and rectify the problem. Efficient, statistically designed experiments, when properly interpreted, are the key to minimizing risk in making process development decisions.

15.7 EXAMPLES OF SENSOR FABRICATION TECHNIQUES

There are many ways to make optical sensors, as a search of the patent literature will show. Figure 15.7 illustrates two of these methods. The sensor of figure 15.7(*a*) shows a multilayer coating on the end of a fiber [19]. The dimensions of the coating play a substantial role in the sensor response; signal level and response time have already been mentioned. If the coating were to be placed by dipping the end of the fiber into the dye solution, the size of the drop retained by the fiber would be affected by surface tension, solvent evaporation rate and other variables. Second and succeeding layers have to be placed without changing the underlying layer. This geometry is suitable for luminescence sensors and, if one of the layers has reflective properties, absorbance sensors. The optical geometry may be difficult to control. Figure 15.7(*b*) shows another approach to making a sensor [20]. The fibers are bent double and a slot is cut in a fiber to make a cavity for the dye. This design offers the advantage of a well controlled sensor geometry for absorbance-based systems but has the potential disadvantages that the amount of light transmitted around the bend at the tip is small, two small fibers are needed for each sensor, and the sensors are on the side of the device, not necessarily in good contact with the patient's blood.

15.8 *IN VIVO* ISSUES

Now, suppose the IABG team has successfully shown that the system can be manufactured and it works both in bench studies and in animal models. The sensor set is deemed ready to be taken to the relatively unknown territory of the radial artery of a sick patient. The radial artery has been used for pressure measurement but only an intact blood column to the central arteries is needed for pressure measurement and there is not necessarily substantial blood flow. Blood sampling is often performed as a passive activity that allows a syringe to fill by the action of the pressure. IABGs require reasonable flow to bring sample to the sensors. Cold, peripheral vasoconstriction, hypotension, and mechanical

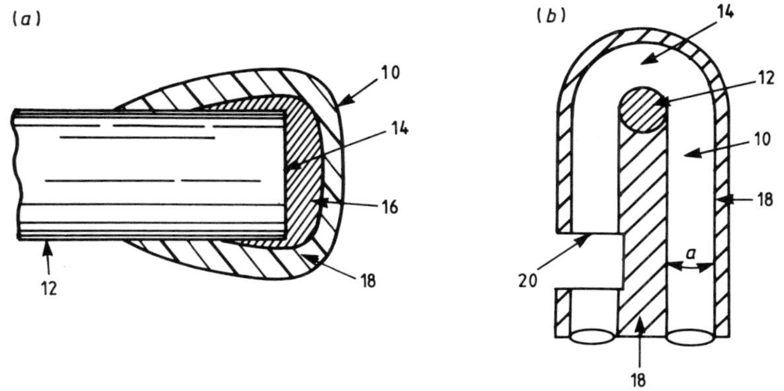

Figure 15.7 (*a*) A fiber optic sensor showing dye layers. Items are: 12, optical fiber; 14, optical surface; 16, sensing element; 18, overcoating. (From the patent of Yafuso *et al* [19].) (*b*) A fiber optic sensor with a bent fiber arrangement and gap for sensor reagent. Items are 10, optical fiber; 12, mandrel of diameter *a*; 14, the resulting bend; 18, tip support coating; 20, optical gap. (From the patent of Costello *et al* [20].)

trauma to the artery are some of the factors that can limit flow to sensors. In addition the location of the sensors on the probe is important. If the sensors are at the end, it is conceivable that the end could be buried partially in the vessel wall and be exposed to some mixture of blood and vascular tissue sample. If the sensors are on the side, that side may be against the vessel wall. According to the literature, several companies have taken different approaches to getting good samples to the sensors. Gravenstein *et al* reported results of a method to 'lift' the sensor part of the probe into the blood stream [21]. Maxwell described a method to measure blood gases by placing the sensors back inside the catheter tip and allowing pressure pulses to force blood past the sensors [22]. An additional use environment issue is the mechanical motion of the catheter hub caused by patient wrist motion. Wrist restraints can mitigate this problem but some patients can not tolerate the restraint and probe breakage in the catheter is a potential failure mode.

The lessons learned include the following: the literature is an uncertain guide to the environment IABGs face in the radial artery; there are no really good models that mimic the clinical environment; extensive clinical testing has to be part of the R&D plan; and the results of that testing will lead to several iterations of not only probe design but also instrument and software design. At least one company has been testing a set of fiber optic sensors placed on the wrist and connected to the radial artery catheter [23]. This near-patient or *ex vivo* approach allows much greater control over the sample and sensor environment. The *ex vivo* approach nearly achieves the goals of continuous monitoring but is still an intermediate technology. I believe a successful *in vivo* product will emerge from the past and ongoing efforts and patients will benefit from *in vivo* sensors.

15.9 SUMMARY

There is one successful fiber optic blood gas sensor, the oximetric catheter, and there have been many attempts to make complete *in vivo* blood gas/pH probes. Mahutte has stated that a conservative estimate of the investment exceeded $100 million [24]. This money was spread across many companies, as evidenced by the patent literature, so that there may have been multiple parallel developments and different solutions to similar problems. It takes an investment team with a cast iron stomach to swallow the cost of the initial investment to bring an IABG project to the clinical stage and then face the prospect of several more years of testing before any return on investment appears. The present concern about technology's contribution to the cost of health care adds additional uncertainty to the investment. This chapter has discussed some of the problems and solutions to the task of bringing fiber optic sensors to the market, and although the market has not embraced the present technology, the need for more timely, less labor intensive blood gas and pH measurement has not gone away.

REFERENCES

[1] Janata J, Josowiscz M and DeVaney M D 1994 Chemical sensors *Anal. Chem.* **66** 207R–228R

[2] Kleiber M 1961 *The Fire of Life* (New York: Wiley)

[3] Astrup P and Severinghaus J W 1986 *The History of Blood Gases, Acids and Bases* (Copenhagen: Munksgaard)

[4] Takatani S and Ling J 1994 Optical oximetry sensors for whole blood and tissues *IEEE Eng. Med. Biol.* **EMB-13** 347–57

[5] Amirkhanian V and Lee W I 1990 Mode mixing in fiber optic oximeter *SPIE Optical Fibers in Medicine V* vol 1201 (Bellingham, WA: SPIE) pp 330–7

[6] Lee W I and McCann B P 1993 Optical fibers for medical sensing *SPIE Fiber Optic Sensors in Medical Diagnostics* vol 1886 (Bellingham, WA: SPIE) pp 138–46

[7] Vurek G G, Fuestel P J and Severinghaus J W 1983 A fiber optic P_{CO_2} sensor *Ann. Biomed. Eng.* **11** 499–510

[8] Leiner M J P and Wolfbeis O S 1991 Fiber optic pH sensors *Fiber Optic Chemical Sensors and Biosensors* vol 1, ed O S Wolfbeis (Boca Raton, FL: Chemical Rubber Company) pp 359–84

[9] Jackson J T and Hui H K 1993 *US Patent* 5 212 092

[10] Nelson A, Hui H K, Bennett M, Hahn S and Bankert C S 1993 *US Patent* 5 204 265

[11] Wolfbeis O S 1991 Oxygen sensors *Fiber Optic Chemical Sensors and Biosensors* vol 2, ed O S Wolfbeis (Boca Raton, FL: Chemical Rubber Company) pp 19–53

[12] Hauenstein B L, Picerno R, Brittain H J and Nestor J A 1993 *US Patent* 5 190 729

[13] Modlin D N and Milanovitch F P 1991 Instrumentation for fiber optic chemical sensors *Fiber Optic Chemical Sensors and Biosensors* vol 1, ed O S Wolfbeis (Boca Raton, FL: Chemical Rubber Company) pp 237–302

[14] Wolthuis R, McCrae D and Hartl J C *et al* 1992 Development of a medical fiber optic oxygen sensor based on optical absorption change *Trans. IEEE Eng. Biol. Med.* **EBM-39** 185–93

[15] Eichhorn J H, Moran R F and Cormier A D 1985 *Blood Gas Pre-Analytical Considerations: Specimen Collection, Calibration, and Controls* **C27-P** (Villanova: National Committee for Clinical Laboratory Standards)
[16] Itano M 1983 CAP blood gas survey—1981 and 1982 *Am. J. Clin. Pathol.* **80** 554–62
[17] Khalil G, Yim J and Vurek G G 1994 In-vivo blood gases: problems and solutions *Biomedical Fiber Optic Instrumentation, SPIE* vol 2131, ed J A Harington, D M Harris, A Katzir and F P Milanovitch (Bellingham, WA: SPIE) pp 437–51
[18] Box G E P, Hunter W G and Hunter J S 1978 *Statistics for Experimenters* (New York: Wiley)
[19] Yafuso M *et al* 1991 *US Patent* 5 075 127
[20] Costello D J *et al* 1992 *US Patent* 5 124 130
[21] Gravenstein N *et al* 1992 *Anesth. Analg.* **74** S122
[22] Maxwell T P 1989 *US Patent* 4 830 013
[23] Gehrich J L, Maxwell T P and Hacker T G 1991 *US Patent* 4 989 606
[24] Mahutte C K 1994 Continuous intra-arterial blood gas monitoring *Intensive Care Med.* **20** 85–6
[25] Soller B R 1994 Design of intravascular fiber optic blood gas sensors *IEEE Eng. Med. Biol.* **EMB-13** 327–35

16

Electrochemical sensors: microfabrication techniques

Chung-Chiun Liu

Electrochemical sensors have been used as the basis or as an integral part of many chemical and biosensor developments. The introduction of microelectrode assembly added a new dimension to electrochemical sensors, and, consequently, to chemical and biosensor research. In recent years, the advancement of microelectronic fabrication technology has provided new impetus to the development of micro or miniature electrochemical sensors.

Electrochemical sensors can be classified according to their mode of operation, e.g. conductivity/capacitance sensors, potentiometric sensors, and voltammetric sensors. Amperometric sensors can be considered a specific type of voltammetric sensor. The general principles of electrochemical sensors have been extensively described in other chapters of this volume or elsewhere. This chapter will focus on the fabrication of electrochemical sensors of micro or miniature size.

An electrochemical sensor is generally an electrochemical cell with either a two-electrode or three-electrode configuration. Figure 16.1 shows the typical configuration of such a cell. For a two-electrode system, a reduction reaction occurs at the surface of the cathode and is referred to as the cathodic reaction. On the other hand, an oxidation reaction takes place at the anode and is referred to as the anodic reaction. The electrolyte and a container complete this two-electrode system. When this electrochemical cell is used as a sensor, either the reduction or oxidation of the sensing species will be the detection reaction. Consequently, the electrode at which this reaction takes place is called the working electrode, whereas the other electrode is termed the reference electrode. In a three-electrode system, there are working, counter, and reference electrodes. The polarizing voltage is applied between the working and reference electrodes, and the cell current flows between the working electrode and the counter electrode.

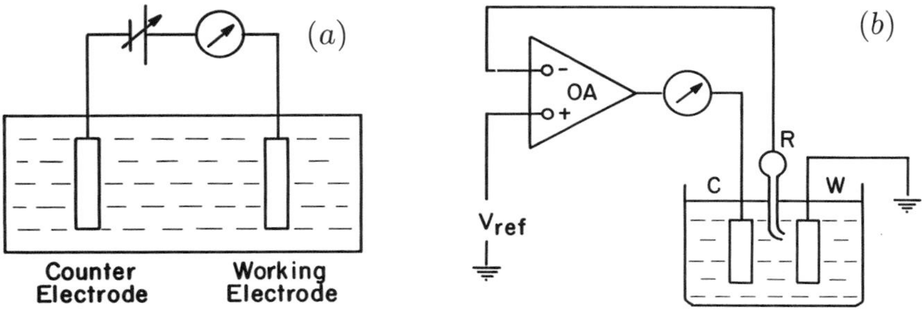

Figure 16.1 Typical electrochemical sensor configurations: (*a*) a two-electrode system; (*b*) a three-electrode system.

16.1 GENERAL DESIGN APPROACHES FOR MICROFABRICATED ELECTROCHEMICAL SENSORS

A reference electrode is used in both two- and three-electrode electrochemical sensors. The three-electrode configuration is used mainly for voltammetric or amperometric modes of sensing. The first design considerations for an electrochemical sensor should include the size, geometric shape, relative location, and material used for the electrode elements. A description of each of these design parameters is given below.

16.1.1 Size of electrode elements

Various microfabrication pattern reduction techniques permit extensive diversity in the size consideration of the electrode elements. Pattern reduction is accomplished using lithographic techniques. Photolithography can define a linewidth of less than 50–100 μm. Ion-beam lithography can achieve a linewidth below 0.1 μm, and x-ray lithography can produce a linewidth below 20 nm. Therefore, basic lithographic technology is sufficient for microfabrication of electrochemical sensor elements. In most cases, photolithography suffices in most electrochemical sensor applications. If thick-film metallization is used, however, a minimum linewidth size is imposed; normally a lower limit of 50 μm linewidth for a metallic film is considered appropriate. The gap or space between two metallic films should be 100 μm or more.

16.1.2 Geometric shape of electrode elements

The versatility of both the mask making and lithographic techniques can produce virtually any geometric shape desired for electrodes. Although most electrochemical sensing elements are circular or straight lines, other geometric patterns of sensing elements can be easily produced. For example, figure 16.2 shows the design of a microfabricated sensing element used in microtitration.

Figure 16.2 A microfabricated spiral shape sensing element used in microtitration application.

In traditional titration, reagent in small increments is added to the detecting species of unknown quantity or concentration. While titration is a highly reliable sensing method, there are many problems involved. They include precise reagent delivery, large sample size, and tedious, intense labor. Diffusional microtitration intends to overcome these shortcomings, using controlled diffusion of the reagent instead of mechanical delivery. The details of the principle and operational mode are described elsewhere [1, 2]. Microtitration is especially attractive for biomedical applications. In this approach, the construction of a microsensor structure is needed, and microfabrication techniques are well suited to produce such a structure with a functional design that is relatively simple to manufacture.

Currently, microfabricated sensing elements are mostly two dimensional. However, recent advancements in micromachining technology provide opportunities to produce three-dimensional structures. The incorporation of heater and temperature sensing elements into the sensor and the construction of semispherical electrodes are feasible. In any current measuring sensor, such as voltammetric and amperometric devices, a large current output is desirable. Consequently, a large electrode surface area or electrode elements connected in parallel may be a considered factor in the sensor design. If the geometric shape of the electrode is rectangular and parallel, the thickness or height of the electrode elements becomes critical, since the gap between the electrodes is the location where the electrochemical reaction takes place.

16.1.3 The relative location of the sensing elements

Certain design configurations of the electrode elements of an electrochemical cell are obvious and logical. For instance, as shown in figure 16.3, concentric ring and disk designs and an interdigitated structure are frequently used for two-electrode systems. For a three-electrode voltammetric or amperometric device, it

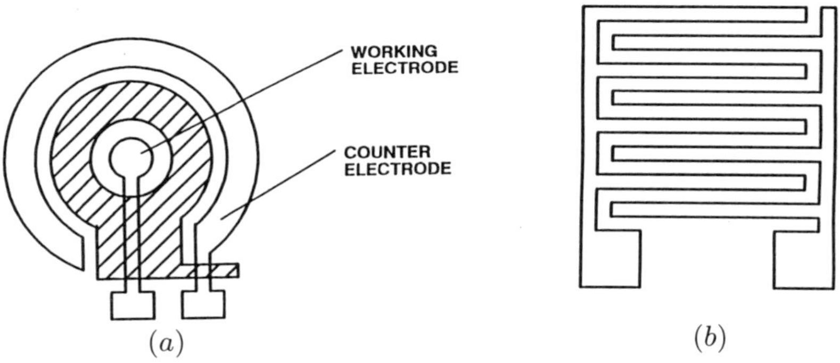

Figure 16.3 Commonly used microfabricated electrochemical sensor structures: (*a*) a ring and disk structure; (*b*) an interdigitated structure.

is desirable to have the reference electrode placed between the working electrode and counter electrode in order to minimize any potential effects.

In the case of an array of band electrodes or interdigitated electrode structures, the width of each single electrode element and the gap between the electrode elements must be considered carefully in the sensor design. Interactions between electrode elements and their effects on the transient response to a potential step perturbation will directly affect the overall sensor output. When an interdigitated electrode structure is used, chemical cross-talk among the reactants and products in both electrode elements (cathode and anode) may occur, which will then influence the sensor output [3]. Therefore, the relative location of the sensing elements is also an essential consideration.

16.1.4 Electrode material

The electrode elements of an electrochemical sensor are often metallic, semiconductive, or inert. For metallic electrode elements, noble metals such as gold, platinum, and silver are often used in conventional electrochemical sensors. Mercury or amalgam as electrode materials for microfabricated electrochemical sensors are seldom used due to the difficulty involved. Because mercury has a relatively high vapor pressure, it does not lend itself well in any fabrication process using a vacuum or low-pressure environment. The formation of amalgam requires the use of mercuric ion containing reagent. This step can be elaborate and complicated.

The use of a graphite electrode, particularly glassy carbon, is also relatively limited in microfabricated electrochemical sensors. However, thick-film silk-screened graphite electrodes have been used in chemical sensor development, and the use of carbon fiber in microsensor applications has been reported. The purity of the graphite ink for thick-film silk screening is very critical to the performance of the sensor.

The selection of a noble metallic material as the sensing element of a microfabricated electrochemical sensor is based on the catalytic and electrochemical properties of the metal. As mentioned, platinum, gold, and silver are often the metallic materials of choice. Palladium, iridium, and rhodium have also been used as electrochemical sensing elements.

The reference electrode in an electrochemical cell requires a high degree of reversibility. For traditional sensors, calomel, $Hg/HgCl_2$, and silver–silver chloride, Ag/AgCl, are the most commonly used. This is based on the high degree of reversibility of the oxidation–reduction reaction involving Hg–$HgCl_2$ and Ag–AgCl. For practical fabrication processing, Ag/AgCl is simpler compared to a calomel reference electrode. A silver film is first deposited by either thick- or thin-film metallization, serving as the base of the reference electrode. The silver film can then be chlorided electrochemically, forming the Ag/AgCl electrode.

16.2 METALLIZATION PROCESSES IN THE MICROFABRICATION OF ELECTROCHEMICAL SENSORS

Since the electrode elements of an electrochemical sensor are most often metallic, the metallization processes for an electrochemical sensor are of considerable importance. As mentioned, metallization of the electrode element can be accomplished by thick- or thin-film techniques. Each technique has its merits and limitations, and the choice between the two techniques depends on the requirements.

Thick-film technology is a silk screening process. Fine metal particles are mixed with an organic binder forming the metallic 'inks' or 'pastes' which can be used in the thick-film silk screening process. These inks or pastes can be applied to ceramic substrates, such as alumina, and fused at a high temperature forming a continuous, conductive structure. Glass-based or oxide fine particles are often added to the ink in order to enhance the adhesion of the metallic thick film on the substrate. This results in a relatively porous structure for the thick-film electrode element. Thick-film metallic electrodes are thermally and chemically stable and mechanically strong. Thus, thick-film electrodes are often used in high-temperature sensors.

A typical example includes the yttria-stabilized-zirconia-based high-temperature potentiometric oxygen sensor which is widely used in automotive applications. Platinum thick films are applied, forming both the cathode and anode of the sensor. The thick electrode has a porous structure which provides a larger electrode surface area compared to non-porous structures. For current measurement, a porous electrode is desirable since it leads to a larger current output. If the metallic film serves as the electrocatalyst, a porous structure is also desirable, for it provides more catalytic active sites. On the other hand, electrodes formed by the thick-film technique do not have an exact, identical geometric dimension, because the thick-film method cannot produce electrodes with identical porous microstructures. Consequently, deviation among electrodes exists. Such deviations may be acceptable depending on the sensor's required level of accuracy.

The thick-film metallization process requires modest capital equipment and the process is relatively simple. Typical capital equipment for thick-film metallization includes a silk screening printer and a furnace for annealing and removal of the organic binder. This equipment is relatively modest in cost. In addition to silk screening on a flat substrate, thick-film techniques can be applied by dipping, spraying, or roller coating the non-flat surface [4]. Thick-film techniques can also be applied in a continuous operation which can be cost effective.

Thin-film metallization techniques include thermal and electron-beam evaporation, ion-beam coating, plasma sputtering, and chemical vapor deposition. Thin-film deposition can be a few molecular layers thick, ångström sized, or as thick as 100 μm. Metallic electrode elements can be deposited by thin-film techniques. The thin-film electrode is geometrically well defined. Compared to thick-film electrodes, they are more uniform, reproducible, and can have a smaller structure. For thin-film sensing elements, a structure in the order of ångströms (Å) is feasible, whereas the thick-film sensing element has to be much larger. Thin-film electrodes are not as porous as thick-film ones, unless the thin-film element is extremely thin.

Physically, the adhesion of a thin metallic film on the substrate may not be strong, but the adhesion can be enhanced using an intermediate metallic film. For instance, the deposition of a platinum or gold thin film on a silicon substrate is often accomplished by first depositing a thin layer of titanium or chromium. The titanium or chromium film enhances the adhesion of the platinum or gold film onto the substrate. However, electrochemically, these titanium or chromium layers may participate in sensing element reactions. This may yield unexpected results, and careful examination of the electrode reaction(s) and experimental results is needed.

A thin-film electrode is relatively dense, as the metallic film does not have the electrocatalytic properties that a porous electrode has. Therefore, in many instances, the surface of the thin film is chemically or electrochemically modified to enhance its electrocatalytic activity. For instance, thin platinum film electrodes can be platinized electrochemically forming a porous platinum black layer. This platinum black layer is electrocatalytically more active than the thin platinum film. Thin-film processes are more capital and labor intensive and the process is more complicated than thick-film processes. Thin-film deposition is also a batch process which may produce sensors of limited numbers of silicon substrates. This is very desirable in prototype development, for it allows modification on prototypes with minimum cost.

In either thick-film or thin-film metallization for electrochemical sensor development, the following processing steps are generally used.

16.2.1 Preparation of the mask for the sensor design

General design approaches such as the size, geometric shape and relative location of the sensing element(s) are first determined based on the sensor's requirements. The sensor design can then be translated graphically using computer-aided design software. This design can then be used to produce the mask for either thick-film or thin-film application. The mask can be produced by optical, electron-beam, or x-ray lithography. For structure definition of electrochemical sensors, optical lithography is generally sufficient. In optical lithography, the sensor design is first scribed on a rubylith panel, either automatically or manually. This rubylith panel is then reduced by a high-precision camera. The photographic plate can then be further reduced using standard step-and-repeat camera processing. This results in an emulsion glass plate on which the sensor design is transferred through the optical processing steps described. This plate is the mask of the sensor which can be used for the metallization process.

16.2.2 The metallization process

In thick-film metallization, the formation of the sensor elements is achieved by silk screening. This involves preparation of the screen, printing with thick-film ink or paste, and drying the thick film. Preparation of the screen is accomplished using the mask and first transferring the sensor design to an emulsion film, then from the film onto the screen. The screen can be polymeric or metallic, and the mesh size of the screen defines the smallest linewidth the sensor design may have. Thick-film electrochemical sensors can be formed on various substrates, depending on the application. The most commonly used substrate is alumina, Al_2O_3. However, Pyrex glass, quartz, or yttria-stabilized zirconia (for high-temperature oxygen sensors) have also been used as substrates [5]. For most electrochemical sensors, gold, platinum, and silver inks are often used for thick-film printing. A silver–silver chloride electrode can be formed by first silk screening a silver thick film and then electrochemically chloriding it as described previously.

For thin-film metallization, a thin metallic film is first deposited onto the surface of the substrate. The deposition can be accomplished by thermal evaporation, electronic-beam- or plasma-assisted sputtering, or ion-beam coating techniques, all standard microelectronic processes. A silicon wafer is the most commonly used substrate for thin-film sensor fabrication. Other substrate materials such as glass, quartz, and alumina can also be used. The adhesion of the thin metallic film to the substrate can be enhanced by using a selected metallic film. For example, the formation of gold film on silicon can be enhanced by first depositing a thin layer of chromium onto the substrate. This procedure is also a common practice in microelectronic processing. However, as noted above, this thin chromium layer may unintentionally participate in the electrode reaction.

In both thick- and thin-film metallization processes, preparation of the surface of the substrate on which the sensor elements are formed is very critical. The cleanliness and degree of smoothness of the surface directly affect the topological structure and adhesive properties of the metallic film. Therefore, the substrate cleaning step is essential. Standard chemical and ultrasound cleaning procedures used in the microelectronic industry are employed in the metallization process of sensor fabrication.

16.2.3 The etching process

In thin-film metallization, the metallic sensor element is produced by etching. This step can be carried out using chemical and plasma etching. In chemical etching, one of the common approaches is the use of a photoresist coating. The photoresist material can be applied by spin coating and the thickness of the coating is governed by the spin rate and the viscosity of the material. The photoresist material can then be polymerized under UV light. The sensor design mask is then used to produce the pattern. Figure 16.4 shows the etching steps for formation of the sensor structure. In this etching process, the metal, in this example, is gold. In order to enhance the adhesion of gold onto silicon, a thin layer of chromium film of the order of 50 Å thick is normally first deposited. Gold film is then deposited by various thin-film metallization techniques, such as thermal evaporation, ion-beam coating, or electron-beam-assisted deposition.

Removal of the metal layer is accomplished by chemical etching. Most metallic films for chemical and electrochemical sensors, such as platinum, gold, and silver, are extensively used in microelectronic industries. Therefore, established procedures and reagents for chemical etching processing of metallic films can be readily adapted for sensor applications. Typically, photoresist etching will be used for the removal or defining of the metallic film structure. In the photoresist etching, as shown in figure 16.4, metallic film is first deposited then a photoresist layer which will be properly defined by polymerization with a mask and under UV light. The unpolymerized photoresist is then removed first chemically. Subsequently, the exposed metal film is removed as shown in figure 16.4.

In addition to the chemical etching method described above, other etching methods such as chemical anisotropic etching, and reactive plasma etching, may also be used to define the geometric configuration of the sensor element. These etching techniques are also established microelectronic processes and are extensively described elsewhere [6].

Another approach in the etching of thin-film processes is the 'lift-off' technique. In this approach, photoresist material is first deposited onto the substrate. A mask of inverse pattern is sized and the photoresist material is left on the substrate surface covering the area where metallization does not take place. Metal is then deposited over the substrate surface. The next step removes the photoresist chemically. Figure 16.5 shows the sequence of a typical lift-off

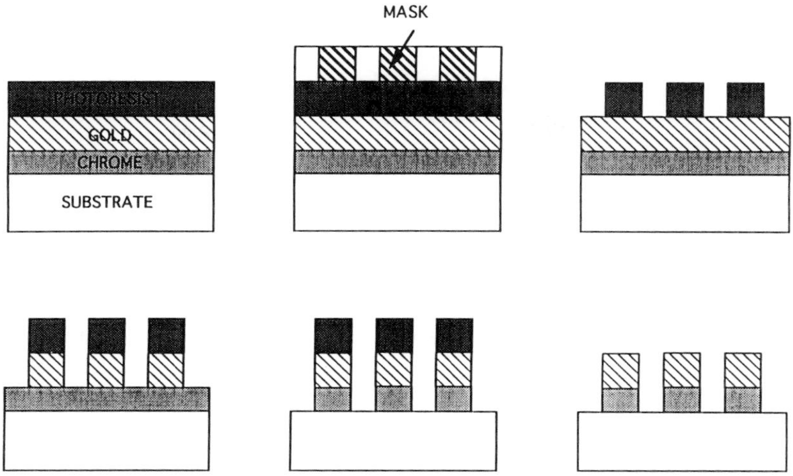

Figure 16.4 Typical chemical etching steps in microfabrication processing: step 1, deposition of chrome, gold metallic films, and photoresist film; step 2, application of mask under UV light to polymerize the photoresist film; step 3, chemical removal of unpolymerized photoresist; step 4, chemical etching of unprotected gold film; step 5, chemical etching of unprotected chromium film; step 6, chemical removal of photoresist film.

process. In this lift-off process, a patterned photoresist layer is first deposited, then followed by the deposition of the metal film. The removal of the photoresist layer is accomplished chemically using proper photoresist removal. The lift-off process is often used for metal films, such as platinum, which is more difficult to remove by chemical etching. The pattern of a lift-off may not be as well defined as chemical etching. Thus, the choice between chemical etching or lift-off process will depend on the properties of the metal film as well as the related structure and processes of the sensor.

16.3 PACKAGING

Fabrication of electrochemical sensors by microfabrication techniques, such as those described above, has been reported by many researchers in various publications and patent disclosures. However, practical electrochemical sensors fabricated by these technologies remain relatively limited in number. The major obstacle is proper packaging of the sensor.

Packaging of electrochemical sensors is an essential and critical issue in the overall fabrication of electrochemical sensors. The nature of any sensor requires the sensing element to be exposed to the sensing environment. This requirement differs from that of common microelectronic components which

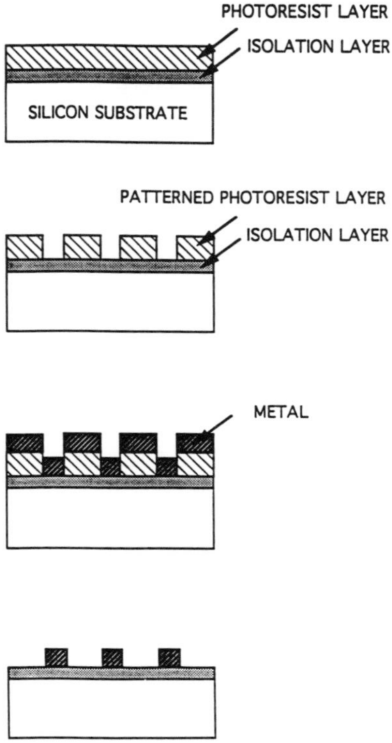

Figure 16.5 The sequence of typical lift-off processing in microfabrication.

are often completely encapsulated or sealed. Consequently, the material and the dimensions of the encapsulation are not as critical as those for a sensor.

When packaging a sensor, any area which should not be exposed must be covered properly. For thick-film metallic sensing elements, glass-based dielectric materials are often used to cover or insulate the exposed area defining the geometric dimensions of the sensing elements. These glass-based dielectric materials can be applied by thick-film techniques. Because silicon-based techniques are used in the formation of thin-film metallic sensing elements, silicon dioxide and silicon nitride are often the materials for covering or insulating any open space. The deposition of silicon dioxide and other materials is accomplished by conventional thin-film techniques. Silicon dioxide film can be deposited by thermal oxidation, which can be carried out in either a dry or a wet oxidation environment. Silicon nitride can be accomplished by chemical vapor deposition. In the development of chemical sensors, a multiple-layer structure is commonly used, Thus, the choice of a proper packaging layer, such as silicon dioxide, silicon nitride, or others, will consider the adhesive properties of these layers as well as the thermal match and mismatch of these layers in order to ensure the overall integrity of the sensor.

For both thick- and thin-film electrochemical sensors, the connection of the sensing element to the external electronic or measuring instrument requires good electrical contact. This contact is usually done with an external lead wire connecting the sensing element to a bonding pad where connection to the external electronic or measuring instrument is made. In sensor design, this bonding pad is included in the overall design on which the connection is required. The bonding of the lead wire onto the bonding pad is done by different methods, depending on whether the sensor element has a thick-film or thin-film metallic structure. For a thick-film sensor, the connection of the lead wire to the bonding pad is often completed by using relatively low-temperature indium soldering. Additional physical strength and electric conduction are ensured by applying an epoxy containing silver, gold, or other conductive metal over the connecting point of the lead wire and the bonding pad. A thicker lead wire is then used to make the connection with the thin aluminum, gold, or other wire from the bonding pad. The electrical insulation at the junction between the wires is accomplished by using epoxy, RTV, or other similar materials.

Electrochemical sensors are often used in an aqueous environment. Thus, the connection and insulation of the sensor are subjected to potential water penetration. This poses a critical challenge to the long-term operational life of the sensor, underscoring the importance of proper sensor packaging.

In most cases the sensing element of an electrochemical sensor will have to be exposed to the sensing medium, which can be hostile to the sensor. Thus, the package of an electrochemical sensor needs to protect the integrity of the overall sensor, covering connecting lead wires while allowing exposure of the sensing element. For thick-film electrochemical sensors, dielectric insulation materials are often silk screened, covering the areas that need protection. For thin-film sensors, the definition of the sensing element and the bonding pad is often made using silicon dioxide, silicon nitride, or other insulating materials. The connection between the sensing element and the external electronic or measuring equipment is made by thermal wire bonding with gold or aluminum thin wire.

Thin gold or aluminum wire, with a diameter of 0.025–0.075 mm, is used to make a connection between the bonding pad and the sensor. This wire bonding is performed using standard microelectronic techniques such as thermal compression or ultrasound bonding. The external lead wire is then bonded onto the bonding pad. Because of the thinness of the metallic film of the pad, the external lead wire connection is usually made by thermal compression. Similar to the connection of the thick-film sensor, the conductive epoxy is first applied to the connecting joint and then covered with insulation epoxy or silicone.

Microfabrication techniques can be used effectively in manufacturing thick- and thin-film electrochemical sensors. However, a well planned package procedure for the sensor is essential in rendering the sensor practical. The selection of the packaging material to protect the sensor's integrity in a testing environment is a key to its functioning properly.

16.4 PRACTICAL APPLICATIONS

Micromachined and microfabricated electrochemical sensors have been used either *per se*, or as part of a sensor system, in many practical applications. This includes various biosensors and chemical sensors reported in research literature. An example of a practical electrochemical sensor is the yttria-stabilized zirconium dioxide potentiometric oxygen sensor used for fuel–air control in the automotive industry. Thick-film metallization is used in the manufacture of this sensor. Even though the sensor is not microsize, this solid electrolyte oxygen sensor has proven to be reliable in a relatively hostile environment. It is reasonable to anticipate that a smaller sensor based on the same potentiometric or the voltammetric principle can be developed using advanced microfabrication and micromachining techniques.

A commercially successful microfabricated electrochemical sensor has remained elusive for a long time. The difficulty of advancing a laboratory prototype to a mass-manufactured product is recognized, but not yet resolved. Packaging problems must be overcome before sensors become commercially viable.

Interference is an important issue for a practical electrochemical or chemical sensor. Interference caused by the presence of any chemical species needs to be minimized. For electrochemical sensors, the proper selection of the voltage window, electrode material, catalyst, etc may somewhat alleviate the problems of interference. Use of a filter, selective membrane or other means in the packaging of the sensor will also enhance the selectivity of the sensor and reduce potential interference. The incorporation of a filter or membrane will add complexity to the package of the sensor. Nevertheless, this may be essential in the development of practical sensors.

The bonding of the lead wires to the sensor or the sensor bonding pads is often a problematic concern of practical sensors. The deterioration of the physical strength and integrity of the electrical connection of the wire bonding is often the major reason for sensor failure. Thus, the importance of packaging in order to avoid these problems cannot be overlooked.

16.5 EXAMPLES

The application of microfabrication and micromachining techniques for production of chemical and biosensor prototypes has been carried out by many investigators, including Schottky-diode-based sensors, metal oxide gas sensors, and electrochemical sensors. As a practical example, a Schottky-diode-based hydrogen gas sensor fabricated by microfabrication and micromachining techniques in our laboratory is described to illustrate the uniqueness of these fabrication procedures. Lundstrom *et al* [7] proposed using a palladium gate Schottky diode for hydrogen gas detection. However, palladium and silicon form Pd silicide, and the performance of this device suffers [8]. Modification of this sensor structure has been reported over the years in order to improve the performance of the sensor. Figure 16.6 shows a microfabrication Pd gate Schottky diode hydrogen gas sensor. An n-type (100) silicon wafer is used as the substrate and a heating and a temperature sensing element are integrated into the sensor. The incorporation of these two elements is important because it allows the sensor to be operated and controlled at elevated temperatures. Consequently, in most instances, the sensitivity and selectivity of the sensor are enhanced. The heating element can be a p^+-doped silicon resistance or a platinum thin-film resistance heater, whereas the temperature detector can be a p–n junction or a platinum thin-film thermometer. However, in order to minimize any thermal mass loss and energy consumption, the backside of the silicon wafer is selectively etched using the chemical anisotropic technique. The composition of the Pd gate is obtained by either thermal evaporation or electron-gun-assisted deposition. Silver, nickel, and chromium have been added to the Pd film to enhance its physical strength as well as the range of hydrogen detection. In our laboratory, silver is commonly used, and the thickness of the Pd–Ag film varies from 100 to 300 Å. The composition of the silver and the thickness influence the detectable range of hydrogen gas. Thus, for different applications, the composition of the silver and the thickness of the Pd–Ag film may alter. Figure 16.7 shows the processing of the microfabricated hydrogen sensor, which is considered relatively simple compared to the standard semiconductor microfabrication process. The advantages of using microfabrication and micromachining processing are evident. Devices with a high degree of reproducibility can be produced at relatively modest cost. This could even make disposable sensors a reality. This type of diode structure can also be applied to other gas or chemical sensing. Selecting the proper gate film will determine the selectivity of the chemical sensor.

Microfabrication and micromachining techniques have also been used in the manufacture of electrochemical sensors. This includes p_{O_2} and p_{CO_2} sensors. Zhou *et al* [9] describe an amperometric CO_2 sensor using microfabricated microelectrodes. In this development, silicon-based microfabrication techniques are used, including photolithographic reduction, chemical etching, and thin-film metallization. In Zhou's study, the working electrodes are in the shape of a microdisk, 10 μm in diameter, and are connected in parallel. In recent years, silicon-based microfabrication techniques have been applied to the development of microelectrochemical sensors for blood gases, i.e. p_{O_2}, p_{CO_2} and pH measurements.

In a biosensor application, Wolfson and his research group [10] employed microfabrication techniques to produce miniature electrodes for potential implantable, non-enzyme-based glucose sensors. Thin-film techniques, such as DC magnetron-sputtered processing, were used to deposit platinum film or platinum film enhanced with titanium film on quartz substrates. The geometric shapes of the quartz substrate include cylindrical rods and rectangular plates in a

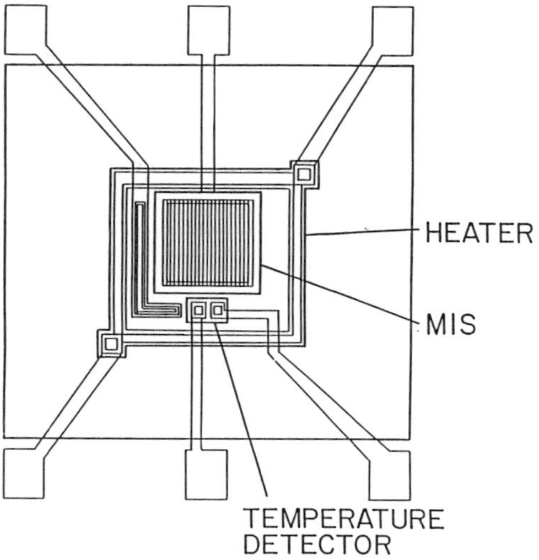

Figure 16.6 The schematic structure of a Pd–Ag gate Schottky diode hydrogen microsensor.

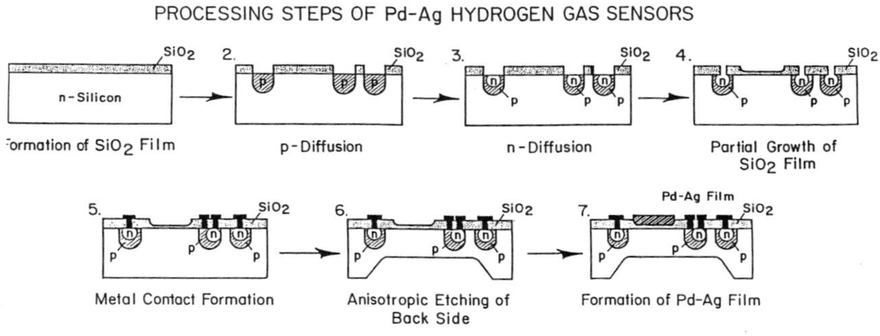

Figure 16.7 Microfabrication processing for the fabrication of a Schottky diode hydrogen microsensor.

surface area of approximately 5 mm × 5 mm. The rate of the thin-film formation is reported as 500 Å h^{-1} and the thickness of the film can be controlled by the total deposition time. The fabricated thin-film platinum sensor is then used in conjunction with an external Ag/AgCl reference electrode and a platinum disk counter electrode for glucose sensing. The sensor system operates in a three-electrode voltammetric mode. Good linearity between the peak current of this thin-film platinum sensor and the glucose concentration is observed over the concentration range of 50–300 mg dl^{-1}. In a more elegant approach, Wolfson and his research group placed the three electrodes, working, counter- and reference electrodes, on the same planar structure on a silicon substrate. The working electrode also employs a design in which individual sensing elements are relatively small and connected in parallel. This design minimizes the flow effect and retains a relatively large sensor signal output.

In summary, microfabrication and micromachining technologies provide a unique and practical means of manufacturing electrochemical and chemical sensors. There are technological challenges in producing practical sensors using these technologies. We believe, however, that the scientific and commercial rewards for successful development of microfabricated sensors are immense.

REFERENCES

[1] Gratzl M 1988 *Anal. Chem.* **60** 484–8

[2] Gratzl M 1988 *Anal. Chem.* **60** 2147–52

[3] Savinell R F, Chen C J and Liu C C 1990 *Solid State Microbatteries* ed J R Adkridge and M Balkanski (New York: Plenum) pp 329–42

[4] Rose A 1993 *Sensors* **10** 28–9

[5] Yamazoe N, Aizawa M, Yamauchi S and Egashira M 1993 *Proc. 4th Int. Meeting on Chemical Sensors (Tokyo, 1992)*; *Sensors Actuators* B **13–4**

[6] Wolf S and Tauber R N 1987 *Process Technology (Silicon Processing for the VLSI Era 1)* (Sunset Beach, CA: Lattice) pp 514–85

[7] Lundstrom I, Shivaraman M S and Svensson C M 1975 *J. Appl. Phys.* **46** 3876–81

[8] Shivaraman M S, Lundstrom I, Svensson C M and Hammarshin H 1976 *Electron. Lett.* **12** 483–4

[9] Zhou Z B, Wu Q H and Liu C C 1994 *Sensors Actuators* B **21** 101–8

[10] Yao S J, Guvench S, Guvench M G, Kuller A E, Chan L T and Wolfson S K 1991 *Bioelectrochem. Bioenerget.* **26** 211–22

17

Electrochemical sensors: enzyme electrodes and field effect transistors

Dorothea Pfeiffer, Florian Schubert,
Ulla Wollenberger and Frieder W Scheller

17.1 OVERVIEW OF DESIGN AND FUNCTION

Electrochemical sensors are well established tools in the determination of gases, nitrous compounds, ion activities, and oxidizable and reducible organic substances down to the submicromolar concentration range. The analysis of many other important substances by electrochemical sensors requires coupling with a chemical or biochemical reaction which generates an electroactive product (table 17.1). Traditionally, *enzymes* are used as analytical reagents to measure substrate molecules by catalyzing the turnover of these species to detectable products. In addition, compounds modifying the rate of the enzyme reaction, such as activators, prosthetic groups, inhibitors, and also enzymes themselves, are accessible to the measurement [1]. A problem of paramount importance to analytical chemistry is selectivity, particularly at trace concentrations where potential interferents might be present at higher amounts than the analyte. For example, the measurement of picomole per liter concentrations of adrenaline and noradrenaline has been realized in the presence of millimole per liter amounts of various metabolites using the coupled glucose dehydrogenase–laccase enzyme system and oxygen reduction current [2]. Owing to their excellent chemical specificity, enzymes allow the determination of minute amounts in complex media and thus avoid the need of highly sophisticated instrumentation, e.g. chromatography or mass spectroscopy. Furthermore, when enzymes are employed as labels in binding assays using antibodies, binding proteins, lectins, etc, the inherent chemical amplification properties of the enzyme's catalytic activity can be exploited to realize extremely sensitive assay methods. Finally, biologically related parameters, e.g. taste, odor, fatigue

Table 17.1 Types of enzyme electrodes.

	Measuring principle	
	Amperometric electrodes	Potentiometric electrodes Coulometric electrodes ISFET
Biocomponent	oxidases dehydrogenases	hydrolases transferases
Electrode material	noble metals (e.g. Pt, Au) carbon organic metals (e.g. $TCNQ^-$, NMP^+)	ion-sensitive membranes * ion exchange membranes, e.g., H^+ * polymer membranes containing ionophores e.g., valinomycin in PVC * solid state membranes, e.g., Al_2O_3, Ta_2O_5 * heterogeneous membranes, e.g., $AgCl^+$ polyfluorinated phosphazine metal oxides (e.g. Pd/SiO_2, IrO_2)
Indicated species	O_2, H_2O_2, NAD(P)H mediators prosthetic groups	H^+, HCO_3^-, NH_4^+, J^-, F^-

substances, mutagenicity, and nutritivity are quantifiable by using multienzyme systems, intact organelles, or cells.

17.2 DESCRIPTION OF DEVELOPMENT STEPS

According to their level of integration the enzyme electrodes described in the literature can be subdivided into three generations (figure 17.1). In the simplest approach (first generation) the biocatalyst is entrapped between or bound to membranes and this arrangement is fixed at the surface of the transducer. The direct adsorptive or covalent fixation at the electrode surface, sometimes together with reagents necessary for signal transfer, permits the elimination of semipermeable membranes (second generation). The direct binding of the biocatalyst to an electronic device transducing and amplifying the signal is the basis for a further miniaturization (third generation).

17.2.1 Immobilization of enzymes on electrodes

For the repeated use of enzymes, cells, and other biologically active agents in analytical devices numerous techniques for their fixation have been developed. In many cases the enzyme is stabilized by immobilization and may be easily

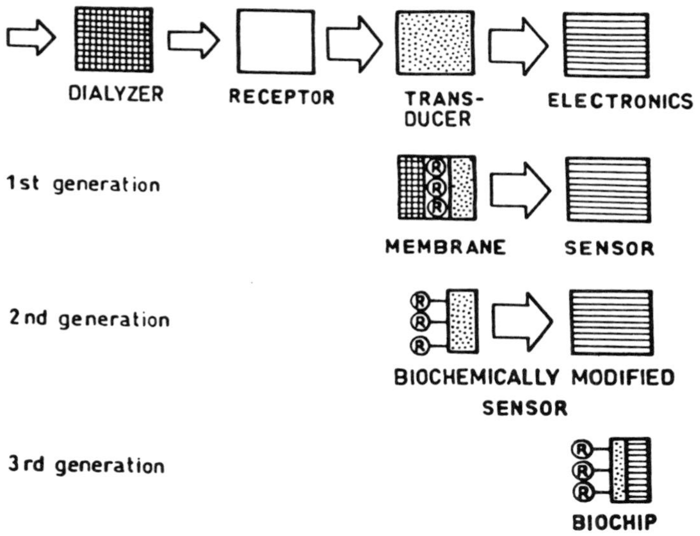

Figure 17.1 Biosensor generations: R, receptor component (reproduced with the permission of Elsevier Science Publishers BV).

separated from the sample. The stable and largely constant activity makes the enzyme an integral part of the analytical device. Methods to immobilize enzymes entail physical and chemical techniques and a combination thereof. The primary physical methods are direct adsorption to the electrode surface and entrapment in water-insoluble polymeric gels. Chemical immobilization is effected by covalent coupling to reactive groups at the transducer surface or by intermolecular cross-linking of the biomolecule using bifunctional reagents.

The structured deposition of enzymes on the surface of microelectrodes or ISFETs is a major goal on the way to mass production (see section 17.2.2). Layers of polymers, such as polypyrrole or polyaniline, are deposited on conducting areas by electropolymerization. Biomolecules are either entrapped in the polymer matrix during the layer formation or coupled to the layer via specific chemical reactions. A uniform layer containing a biocomponent can be structured by photolithography. Alternatively, enzymes are deposited only at the sensitive region by using an enzyme solution in a negative photoresist.

To assure the proper handling and optimal functional stability of enzyme sensors the following topics have to be considered.

17.2.1.1 Enzyme loading test. The minimum amount of enzyme required for maximum sensitivity is determined by varying the enzyme loading (figure 17.2). This test also reveals the enzyme limit of diffusion-controlled sensors. Owing to differences in K_M values and the layer thickness, the transition from the kinetic to diffusion control of different enzyme electrodes takes place at rather

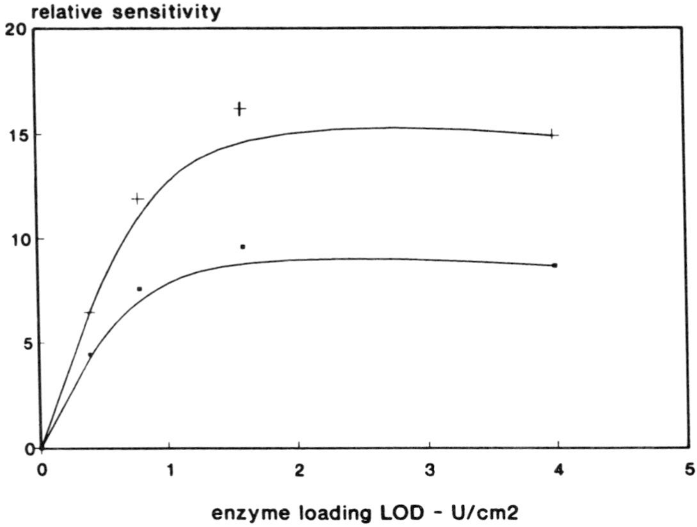

Figure 17.2 An enzyme loading test of a lactate oxidase membrane electrode: • 2.5 mmol l^{-1} lactate; + 5.0 mmol l^{-1} lactate (reproduced with the permission of Elsevier Science Publishers BV).

different enzyme activities. For example, the transient for lactate dehydrogenase is between 0.05 and 0.20 U cm^{-2} while lactate oxidase needs 1.00–1.80 U cm^{-2}. An average value using a gelatin matrix is 1 U cm^{-2} (1 U is the amount of enzyme consuming or forming 1 μmol substrate or 1 μmol product min^{-1} under standard conditions). The application of enzyme activity in excess of that needed to bring about the transition leads to functionally very stable enzyme sensors. For example, an excess of 40 U cm^{-2} of lactate oxidase results in a functional stability of more than 10 d and more than 2000 measurements [4].

17.2.1.2 pH dependence. In most cases the pH effect on immobilized enzymes differs from that in solution. Therefore, it is recommended to investigate the pH dependence. While kinetically controlled enzyme electrodes exhibit more or less sharp pH dependence, the pH profile of diffusion-controlled sensors with a high enzyme excess in the membrane is substantially less sharp [4]. This is illustrated in figure 17.3. Whereas glucose at a concentration of 0.14 mmol l^{-1} causes a dependence that is almost as flat as that for H_2O_2, with 10 mmol l^{-1} glucose a pronounced maximum is found. At this saturating concentration the signal depends on the enzyme activity and therefore on pH.

17.2.1.3 Apparent activity. The enzyme activity *acting* in the measuring process independently of the activity applied can be investigated by rate determination of product formation as well as substrate consumption by an

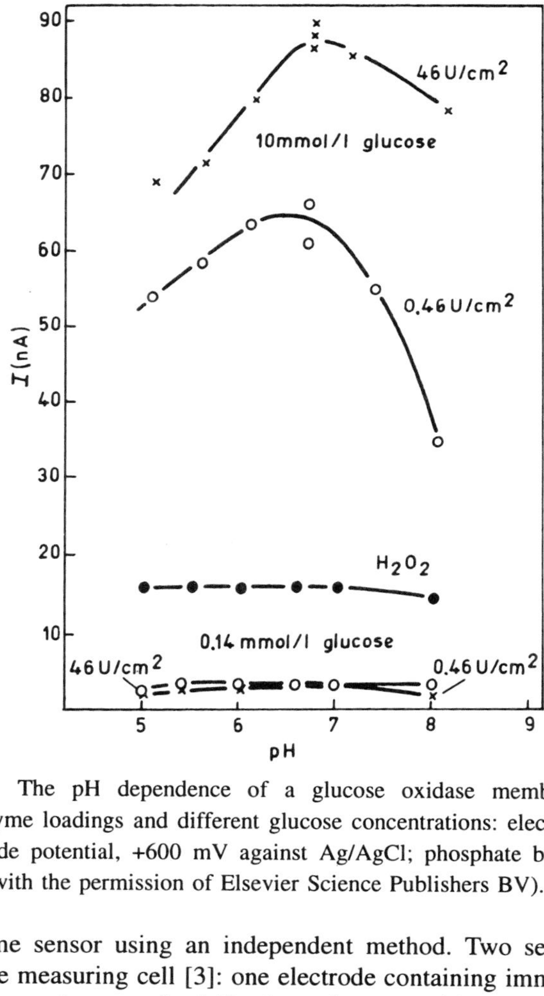

Figure 17.3 The pH dependence of a glucose oxidase membrane electrode at different enzyme loadings and different glucose concentrations: electrode surface, 0.22 mm^2; electrode potential, +600 mV against Ag/AgCl; phosphate buffer, 0.66 $mol\,l^{-1}$ (reproduced with the permission of Elsevier Science Publishers BV).

intact enzyme sensor using an independent method. Two sensors are placed into a double measuring cell [3]: one electrode containing immobilized glucose oxidase and one (enzyme-free) Pt electrode covered by a dialysis membrane. After calibration of the H_2O_2 indicating Pt electrode the monitoring of the H_2O_2 production with time after saturating glucose addition gives the acting concentration of glucose oxidase in the glucose electrode. Together with the result of an enzyme loading test the apparent enzyme activity reflects the relative importance of diffusion and enzyme kinetics in a given enzyme electrode.

17.2.2 Signal generation

Among biosensor researchers there is no unanimous view as to how the generations of biosensors should be defined. The classical approach introduced in section 17.2, however, appears to be a useful one in that it permits us to

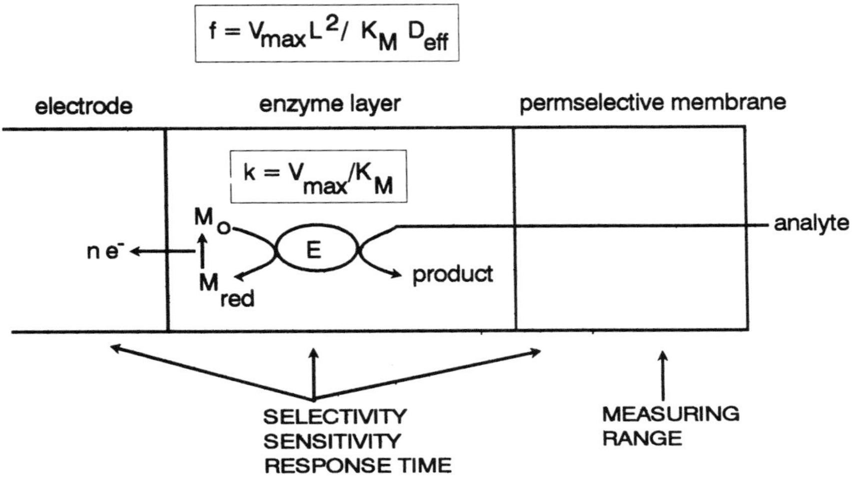

Figure 17.4 A scheme of an amperometric enzyme membrane electrode: f, effectiveness factor; V_{max}, maximum rate; L, thickness membrane; K_M, Michaelis constant; D_{eff}, diffusion coefficient analyte; k, rate constant; M_{red}, mediator oxidized/reduced; E, enzyme.

follow the development towards smaller size and higher integration. Of course one has to bear in mind that the limits between the generations so defined are not particularly strict. According to this concept, a first-generation electrochemical enzyme sensor (figure 17.4) is basically no more than a dense package of dialyzer, enzyme reactor, and detector with all signal processing separated. The idea behind this approach, however, created the new quality over conventional enzymatic analysis on which all biosensors are based. A typical example would be a glucose or lactate electrode comprising the appropriate oxidase entrapped in or bound to a membrane that is fixed at an oxygen or hydrogen peroxide detecting metal electrode and covered by a semipermeable inert membrane. The membrane directed towards the (usually ideally mixed or flowing) measuring solution fulfils a number of functions. Firstly it provides the sensor with a certain degree of selectivity. The pore size and, perhaps, charge permits the exclusion of deleterious or interfering molecules, such as proteins [5] or electroactive compounds [6] and may provide useful partitioning of other compounds (see, e.g., [7]). Furthermore, the thickness and pore size permit us to affect the measuring range of the sensor by controlling the actual amounts of reagents reaching the reaction layer [8]. Finally the response time can be strongly affected by the selection of the membrane material.

After permeation of analyte, cosubstrates, and effectors through the membrane to the underlying enzyme membrane the analyte is converted therein under formation or consumption of a detectable species. In the illustrated case this is the formation of reduced mediator (e.g. hydrogen peroxide from oxygen) that is oxidized at the electrode. The enzyme membrane is characterized by an enzyme loading factor reflecting the interplay of enzyme kinetics (V_{max} and K_M) and mass transport (effective diffusion coefficient, D_{eff}, and membrane thickness, L). The loading factor, f, is crucial for the response characteristics and the stability of the sensor (see section 17.2.1) whereas the choice of enzyme determines the chemical selectivity of the measurement.

The choice of the indicator electrode is largely determined by the species involved in the sensing reaction. Oxygen and H_2O_2, which are the cosubstrate and product of oxidases, as well as NAD(P)H—the cosubstrates of about 300 pyridine nucleotide-dependent dehydrogenases—can be determined amperometrically. Based on this principle, many enzyme sensors of the first generation have been developed and commercialized (see section 17.4 and table 17.2).

The oxygen electrode according to Clark [9] and its version modified for H_2O_2 indication [10] are the most widely used transducers in biosensors. The electrode potential is crucial for the selectivity of the sensor. Any electroactive substance being converted at lower potential contributes to the total current. Thus at an electrode potential of +600 mV for H_2O_2 measurement, ascorbic acid, uric acid, or paracetamol are oxidized as well.

Typical first-generation enzyme electrodes for glucose and lactate have been developed in the present authors' laboratory [11]. The appropriate oxidase is entrapped in a polyurethane layer and applied to the H_2O_2 probe and sandwiched between two dialysis membranes. Using regenerated cellulose membrane of MWCO 12 000 as the outer layer the sensors respond linearly to the analytes between 0.01 and 2.5 mmol l^{-1} lactate and 0.01 and 2.0 mmol l^{-1} glucose, respectively. They are thus well suited for application to diluted blood samples. Sample frequencies up to 120 h^{-1} with imprecision below 2% are possible. To ensure a linear concentration range that allows analysis of undiluted whole blood and serum, the diffusion behavior of the enzyme membranes was modified in such a way that the permeation of analyte was significantly reduced while that of oxygen was not influenced. Screening of different membranes revealed that the oxygen permeability of polycarbonate matrices impermeable to substrate was 20 times larger than that of polyester and 30 times larger than that of poly(vinylidene difluoride). The membrane optimization gave a polycarbonate membrane with a porosity below 0.025%. The incorporation of this diffusion barrier into the original enzyme membrane system significantly increased the response time and decreased the current output. At the same time, however, the membrane permitted sensors to be assembled that respond linearly to lactate up to 20 mmol l^{-1} and to glucose up to 24 mmol l^{-1}, i.e. the range relevant in whole-blood measurements. Furthermore, several substances which electrochemically interfere at +600 mV, e.g. ascorbic acid and paracetamol, could be excluded from the electrode up to concentrations of 1 mmol l^{-1}. Likewise, no interference of cysteine, creatinine, or uric acid was observed.

Another way to reduce interferences by cooxidizable sample constituents

Table 17.4 Analytical performance of enzyme sensors.

Analyte	Enzyme	Indicator	Measuring range (mM)	Detection limit (mM)	Response time	Functional stability
glucose	GOD	H_2O_2	0.1–48	0.015	6s	> 3000 samples (> 15 d)
glucose	GOD	ferrocinium	1.1–33	1.1	30s	disposables
lactate	LOD	H_2O_2	0.1–35	0.01	4s	> 3000 samples (> 10 d)
lactate	LOD+LDH	O_2	0.000 005–0.000 05	0.000 000 1	125s	5 d
lactose	GOD + β-galactosidase	H_2O_2/O_2	0.6–100	0.02	8s	> 3000 samples (> 15 d)
adrenaline	GDH+ laccase	O_2	0.000 000 5–0.000 010	0.000 000 05	600s	2 d
urea	urease	pH	0.1–10	0.1	3 min	n.d.[a]
penicillin	β-lactamase	pH	5–50	n.d.	2 min	n.d.

[a] n.d., not determined.

is by keeping the electrode potential as low as possible. Therefore a reaction partner is chosen to be electrochemically indicated that is converted at low potential. For this purpose, the natural electron acceptors of many oxidoreductases have been replaced by redox-active dyes or other reversible electron mediators. Among them are the ferricyanide/ferrocyanide couple [12], N-methylphenazinium sulfate [13], ferrocene [14], and benzoquinone [15]. With these mediators electrode potentials around +200 mV can be applied, which decrease electrochemical interferences and permit us to apply such enzymes coupled with electrodes also in oxygen-free solution. In analogy to the natural cosubstrate the mediator is often added to the sample solution.

The integration of redox enzymes at electrode surfaces together with such mediator compounds to act like electron transfer systems in biomembranes forms the basis of second-generation bioelectrodes. Fixation of biocatalyst and mediators as well as cosubstrates enables a reagentless measuring regime to be performed. The covalent binding or adsorption of mediator and enzyme to the electrode [16] or their integration into the electrode body itself, as in the case of carbon paste electrodes [17], have been shown to be successful concepts that lead to functioning glucose sensors. The direct immobilization of enzymes at electrode surfaces is facilitated by novel methods of electrodeposition of conducting or non-conducting polymer films that can be used to enfold or covalently bind the enzyme molecules [18]. The glucose electrode 'ExacTech' (Medisense, USA) represents the first commercialized enzyme electrode of the second generation. It is based on glucose oxidase adsorbed on a carbon electrode. The mediator ferrocene is incorporated in the carbon material [19]. Such sensors can be manufactured by mass production methods such as screen printing. Problems may arise from the oxygen influence at lower analyte concentrations (see section 17.4 and table 17.3).

Adsorption of redox polymers containing benzoquinone at carbon electrodes result in the catalysis of the electron transfer by wiring the enzyme molecules to the electrode. In a similar manner sensors for glucose, hydrogen peroxide, and NAD(P)H were developed by wiring enzymes to glassy carbon electrodes with an osmium complex containing a redox–conducting epoxy network [20]. It was shown that such a network is capable of connecting $FADH_2$/FAD centers of glucose oxidase and pyrroloquinoline quinone centers of glucose dehydrogenase to the electrode.

A symbiosis of mediated and direct electron transfer is obtained by covalent fixation of electron tunneling relays to the protein moiety of oxidoreductases [21]. The mediators are bound to amino acids near the prosthetic group. For fixation of the relays the protein has to be unfolded and renatured after the chemical modification procedure. The small distance between the bound mediator molecules (maximum 1 nm) provides a very fast tunneling process. Enzyme electrodes employing glucose oxidase or lactate oxidase modified in this way operate like mediator-modified electrodes, without reagent addition. Owing to their favorable structure, such sensors respond to the analyte in less than

Table 17.3 Enzyme-electrode-based portable devices.

Model	Company	Analyte	Measuring range (mM)	Functional stability
ExacTech	MediSense (USA)	glucose	1.1–33.3	disposables
Satellite G		glucose	2.0–33.3	disposables
Glucometer Elite	Bayer Diagnostics (Germany)	glucose		disposables
Glucocard	Kyoto Daiichi Kagaku Co. (Japan)	glucose	2.2–27.8	disposables
i-STAT PCA	i-STAT Corp. Princeton (USA)	glucose	2.9–23.6	disposables
		urea	1.0–43.0	disposables
BSE 5500	Orion Anal. Technol. Inc. (USA)/Dosivit (France)	glucose	1.6–16.0	disposables
		sucrose	1.6–16.0	disposables
		lactose	1.6–16.0	disposables
Biosen 6020 G	EKF Industrial Electronics (Germany)	glucose	0.5–20.0	24 d
Biosen 5020 L		lactate	0.5–20.0	10 d

1 s. However, with respect to stability they compare less well with immobilized enzyme-membrane-based sensors because the latter can be furnished with higher excess of the biocatalyst.

Many of the techniques mentioned can be easily applied to fabricate enzyme-modified planar electrode structures, such as chip electrodes and field effect transistors. For example, the gate area of pH-sensitive ISFETs has been covered with enzymes, such as urease and β-lactamase [22]. Furthermore, ATPase, glucose oxidase, and trypsin were used for ATP, glucose, and peptide ENFETs [23]. Other examples include maltose and lactate probes on the basis of polyurethane-entrapped enzymes in combination with pF-FETs [24]. Moreover, the use of lithographic techniques for deposition and patterning of the bioselective layer, e.g. based on photocurable gels, is a way to CMOS-compatible processing steps. Preparation of polymers by electropolymerization has become an attractive method for the spatially precise immobilization of enzyme molecules onto planar amperometric and conductimetric sensors. The process is governed by the electrode potential and the control of charges passed through the working electrode allows the determination of the film thickness and amount of enzyme immobilized.

Basic materials such as pyrrole and aniline have been used extensively [25]. By far most examples are concerned with immobilization of glucose oxidase in polypyrrole for the development of glucose sensors. In a typical experiment, the film is grown on a platinum electrode by oxidation of the monomer at a constant potential of +0.8 V (against Ag/AgCl) from an aqueous solution containing 0.2 mol l^{-1} pyrrole and 0.1 mol l^{-1} potassium chloride in the presence of 1–3 mg ml^{-1} glucose oxidase (*Aspergillus niger*) at neutral pH for about 5–60 min. The control of the electrode potential and current flow is facilitated with a potentiostat/galvanostat. The glucose response can be measured by indication of H_2O_2 generated in the enzyme reaction at +0.7 V (against Ag/AgCl). Due to the surface immobilization the sensor is responding within seconds to the addition of glucose and exhibits usually a working lifetime of more than 1 month.

Furthermore, the functionalization of conducting polypyrrole films has been used for formation of surfaces suitable to link enzymes or other proteins covalently [26].

For the patterned immobilization of enzyme membranes a uniform layer containing a biocomponent can be structured by the photolithographic method. Kimura and coworkers [27] developed a multi-ISFET enzyme sensor for glucose and urea using a lift-off method. In order to obtain thickness and characteristics controllable membrane positive photoresist was used, which consisted of phenol resin and a photosensitive reagent. To the ISFET wafer the positive photoresist is applied using spin coating, UV exposure, and development. Openings are created where serum albumin-based enzyme membranes can be deposited. In the lift-off step the photoresist is removed leaving precisely deposited (100 μm wide and 400 μm long) enzyme membranes on each ISFET gate. In this way urease and glucose oxidase were applied for realization of urea and glucose ENFETs. The precision of the membrane thickness (0.1μm) results from the spin coating step. However the thickness of the enzyme layer (1–5 μm) restricts the enzyme loading and is thus responsible for the limited lifetime of the enzyme sensor. To improve adhesion of the biocomponent containing layer to the silicon surface, surface silanization [28] is often used. Methacryl functionalities are covalently anchored to silicon dioxide surfaces [29]. The resulting polyhydroxymethacrylate (pHEMA) or polysiloxane layers show an improved adhesion to ISFETs or electrodes.

Alternatively, enzymes are deposited only at the sensitive region by using an enzyme solution in a negative photoresist [30]. Here as basic material water-soluble polymers such as polyvinylalcohol, polyvinylpyrrolidone, and pHEMA are used to which a photosensitizer is added. For example, after spin coating a 10% pHEMA–$(NH_4)_2Cr_2O_7$ solution the 0.8 μm thick hydrogel film is crosslinked by UV light through a photomask [31]. Typically, the exposure time is 60 s when a 5 mW cm^{-2} UV source with 254 nm wavelength is used. In the exposed regions an insoluble polymer is formed. The removal of the unexposed regions is carried out with water or methanol. Miniature size and the possibility to structure the sensors in such a way that multifunctionality results are but two of the advantages of this approach. The most attractive feature is the potential for direct integration with microelectronics, e.g. signal processing, which creates biosensors of the third generation. Thus, the configuration of a potentially implantable glucose sensor integrates, along with the sensor itself, the electronics needed for the conversion of the weak sensor signal to a plain low-impedance signal [31]. Altogether, this sensor system combines on one chip (i) a potentiostat operational amplifier, (ii) a current to voltage converter, (iii) a sample and hold circuit to obtain a DC output signal, (iv) a clock generator, (v) a temperature sensor, and (vi) the actual voltammetric sensor, which can be a two-electrode or a three-electrode system. This integrated potentiostat circuit with on-chip electrodes represents the basis for one of the first true third-generation biosensors. The chip dimensions of 0.75 mm × 4.5 mm render it suitable for catheter tip use.

17.2.3 Coupled enzyme reactions in electrochemical enzyme sensors

Since not all enzyme-catalyzed reactions involve changes in the level of compounds detectable at an electrode, such as H^+, oxygen, or hydrogen peroxide, only a limited number of substances can be determined with one-enzyme sensors. One way of overcoming this problem is to couple the catalytic activities of different enzymes either in sequence, in competing pathways, or in cycles. In general, enzymes to be used in multienzyme electrodes should fulfil the following requirements:

(i) their pH optima should be reasonably close to each other and to that of the indicator reaction,

(ii) they should not be inhibited by cofactors, effectors or intermediates required for sensing, and

(iii) their cofactors or effectors should not react with each other. In conjunction with appropriate measuring regimes by multienzyme systems not only does a much wider range of analyte species become accessible to measurement by the bioelectroanalytical approach but, in addition, the selectivity and sensitivity of the biosensor may be enhanced through appropriate choice of the coupling strategy. In enzyme sequences the primary product of the analyte conversion is further converted enzymatically with the formation of a measurable secondary product or reaction effect. On this basis, families of electrodes have been developed, which combine glucose, lactate or alcohol producing primary enzyme reactions with the respective oxidases. Such enzyme sequence electrodes are known for sucrose, lactose, maltose, glucose itself, gluconate, glucosinolate, bilirubin, ATP, glucose-6-phosphate, cholesterol esters, fatty acid esters, acetylcholine, creatine, hypoxanthine, glycerol, lactate, pyruvate, and inosine. It is clear that these sensors respond to all substrates of the sequence, which may be advantageous if not only the initial substrate is to be measured. In favorable cases, e.g. for a sensor using gelatin-entrapped lactate dehydrogenase and lactate oxidase for pyruvate and lactate assay, the sensitivity for the two substrates is virtually the same, so that the calibration is simplified to one step.

A sensor representative for the design and fabrication of multienzyme electrodes is a lactose electrode based on the sequential reaction of β-galactosidase and glucose oxidase coupled to a Clark-type electrode [32]. Using 46 $U\,cm^{-2}$ glucose oxidase the enzyme loading test for β-galactosidase gives

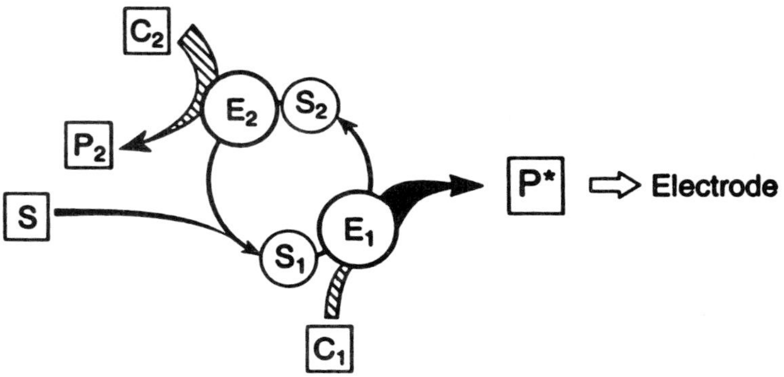

Figure 17.5 A schematic illustration of a substrate recycling biosensor: $S_{1/2}$ substrates; $C_{1/2}$ cosubstrates; $E_{1/2}$ enzymes; P product (reproduced with the permission of Elsevier Science Publishers BV).

the transient to diffusion control of the overall reaction at about 6 U cm^{-2}. To assure an enzyme reservoir and good functional stability excess of enzyme has to be used. The compromise between high enzyme reservoir and low membrane thickness (leading to high permeability and low response time) is 10 U cm^{-2} β-galactosidase. This results in a total membrane thickness of 100 μm and a response time of 8 s (kinetic measurement). A complete measuring cycle takes less than 1 min. The functional stability depends on the enzyme source applied: while sensors using β-galactosidase from *E. coli* and *Bifidobacterium adolescentes* are stable for a few days only, application of the enzyme from *Curvularia inaequalis* results in a stability of more than 15 d. An apparent activity of 1 U cm^{-2} has been determined by measuring the rate of hydrolytic glucose formation from lactose in the double measuring cell described above section (17.2.1). The pH optimum of the multienzyme system immobilized in gelatin is 6.0 with different buffer systems. The lactose sensor so designed is incorporated in the analyzers ESAT and Industrial Modul (Prüfgerätewerk Medingen, Dresden, Germany). These devices use H_2O_2 and O_2 detectors, respectively, as base sensors. Thus, the enzyme sequence sensor can be used for both foodstuff and urine analysis. The serial imprecision is below 1.5% ($n = 20$).

The application of competition schemes realized by coupled enzyme reactions also provides access to analytes not determinable with usual enzyme electrodes. Here the analytical information is gained either from the competitive action of two enzymes, of which one produces the signal, on the same substrate; or from the competition of the analyte with the substrate for the same (signal generating) enzyme, the latter approach resembling that of inhibitor determination.

The coupling of enzyme pairs in recycling schemes provides a means for signal amplification, i.e. the enhancement of the sensitivity of enzyme electrodes. This method works analogously to the cofactor recycling known from biochemical analysis using dissolved enzymes (figure 17.5). In a bienzyme electrode the analyte is converted in the reaction of enzyme 1 to a product which is the substrate of enzyme 2. The latter catalyzes the regeneration of the analyte which thus becomes available for enzyme 1 again, and so forth. One of the coreactants is detectable directly or via additional reactions. (It is evident that the terms 'analyte' and 'product' are interchangeable here; this may be useful, but may also cause problems in real samples.) If enzyme 1 is present in sufficiently high concentration to ensure diffusion control, a signal amplification is achieved by switching on enzyme 2 by addition of its cosubstrate. In such systems the analyte acts as catalyst, which is shuttled between both enzymes in the overall reaction of both cosubstrates. In this way significantly more cosubstrate will be converted than analyte is present in the enzyme membrane. Hence, the change in the parameter indicated at the electrode will greatly exceed that obtained with one-way analyte conversion. The ratio of the sensitivity in the linear measuring range of the amplified and the unamplified regime is termed the amplification factor. If an enzyme pair at an electrode catalyzes the following reactions:

$$\text{analyte} + \text{M}_0 \xrightarrow{k_1} \text{product} + \text{M}_r$$

$$\text{product} + \text{A} \xrightarrow{k_2} \text{analyte} + \text{B}$$

with M_0, M_r, A and B being cosubstrates, of which M_0 will provide the electrical signal, the behavior can be described by a set of differential equations [33]. A well studied example is a lactate sensor using coimmobilized lactate oxidase and lactate dehydrogenase and coupled to an oxygen electrode [34]. Assuming stationary initial conditions for M_0 (then oxygen) and equal diffusion coefficients in the enzyme membrane of analyte (lactate) and product (pyruvate) ($D_a = D_p = D_{eff}$) they can be explicitly solved. From the stationary solutions and under conditions of high enzyme loading a simple approximation for the amplification factor, G, can be derived [33, 36]:

$$G = k_1 k_2 L^2 / 2(k_1 + k_2) D_{eff}$$

where L is the membrane thickness.

In addition to lactate, enzyme electrodes based on analyte recycling have been developed for the following compounds: pyruvate, glucose, gluconolactone, NADH, NAD^+, ethanol, benzoquinone, hydroquinone, ADP, ATP, *p*-aminophenol, glutamate, leucine, and malate. Depending on the enzymes and membrane materials used, amplification factors range from three to 48 000. Detection limits down to 50 $\text{pmol}\,\text{l}^{-1}$ have been achieved. When dealing with extremely high amplification one has to bear in mind, however, that the sensor signal becomes highly susceptible to minute amounts of contaminants affecting the enzyme reactions or the diffusion of the reactants.

Another coupling principle used for amplifying the response of enzyme electrodes is based on biocatalytic accumulation of intermediates (figure 17.6). It is less effective in terms of sensitivity enhancement but bears the inherent

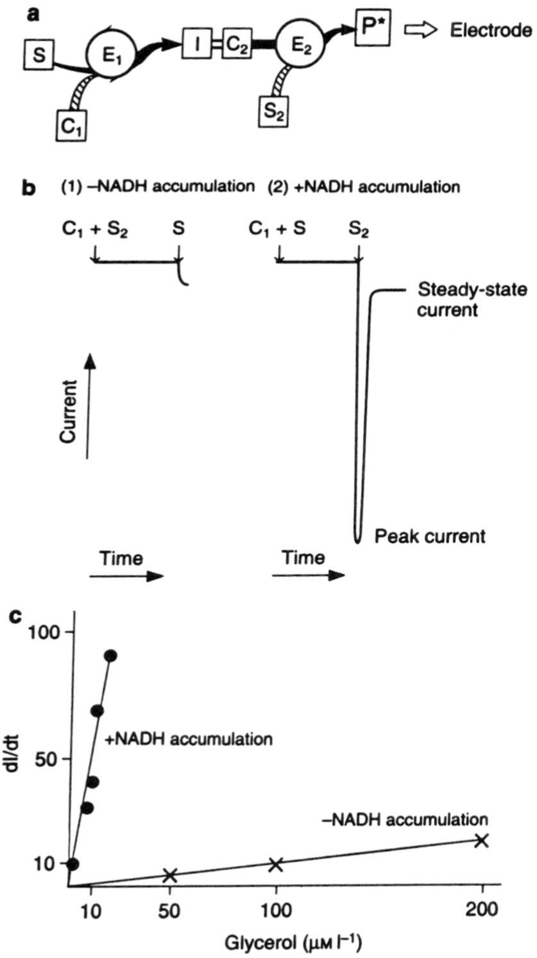

Figure 17.6 (*a*) The principle of an accumulation biosensor. (*b*) Current–time curves for a sensor operated without and with intermediate accumulation. The difference in the steady state and peak currents represents the amplification of the sensor response. (*c*) Glycerol measurement without and with NADH accumulation on an enzyme electrode containing immobilized glycerol dehydrogenase, lactate dehydrogenase, and lactate monooxygenase. NADH is stripped by pyruvate addition after 6 min accumulation (reproduced with the permission of Elsevier Science Publishers BV).

capability to increase the selectivity for a given analyte in a complex sample. The method relies on an enzyme sequence. In the first step of the measurement the conversion of the analyte by the first (accumulator) enzyme is permitted to proceed for a certain time (usually to equilibrium) during which an intermediate is preconcentrated in the enzyme layer. Then the second step, the actual measurement, is initiated by addition of an excess of cosubstrate (initiator) of the second (indicator) enzyme reaction. The conversion of the accumulated intermediate results in the analytical signal. This method yields amplification between twofold and 60-fold and has been used in the determination of glucose, glucose-6-phosphate, hypoxanthine, formate, glycerol, and $NADP^+$. It is applicable to all analytes being converted in sequential enzyme systems, provided the terminal reaction depends on a cosubstrate and can generate a signal at the transducer. Differences that are observed in the amplification factors may be attributed to the equilibrium constants of the generator enzymes, and the diffusion behavior of the intermediate and initiator substrate. The largest enhancement due to accumulation is obtained when the intermediate is a large molecule, and the initiator is a small molecule.

As mentioned above, an important point in the practical application of this approach is the possibility to eliminate any effects on the signal that are due to interfering molecular species. For this purpose, the sample is allowed to equilibrate with the sensor to establish a background signal before initiating the reaction that measures the analyte. This suppresses any contribution from interfering species to the signal and improves the selectivity of the analysis.

The selectivity of enzyme electrodes can also be improved by means of a further coupling principle which is capable of filtering chemical signals by eliminating disturbances of the enzyme or electrode reaction caused by constituents of the sample. Compounds that interfere with the signal transduction, e.g. ascorbic acid with anodic oxidation of hydrogen peroxide, can be transformed into inert products by reaction with an eliminator or anti-interference enzyme. Since the conversion of analyte and interferent proceed in parallel, both the eliminator and the indicator enzyme may be coimmobilized in one membrane. On the other hand, constituents of the sample that are at the same time intermediates of coupled enzyme reactions can be eliminated before they reach the indicator enzyme layer. For this purpose several different enzyme membranes have to be used. Of particular importance are enzymatic anti-interference systems for glucose, because this compound is contained in most real samples where disaccharides and polysaccharides have to be determined with enzyme sequence sensors based on saccharidases and glucose oxidase as indicator enzyme. Other systems have been devised to eliminate lactate, ascorbic acid, ammonia, and oxygen.

17.3 TRANSFER TO MANUFACTURING AND PRODUCTION

17.3.1 Measurable analytes and analytical characteristics

Electrochemical enzyme sensors for about 140 different analytes have been described. Among them are low-molecular-weight substances (metabolites, drugs, nutrients, gases, metal ions, coenzymes, enzyme activators, and vitamins) as well as macromolecules (enzymes, lectins, nucleic acids, and polymeric carbohydrates such as starch and cellulose), viruses, and microorganisms.

As with all methods and devices for quantitative analysis, the optimization of electrochemical biosensors is directed at the following features: (linear) measuring range, lower limit of detection, response time and sample frequency, precision, accuracy, stability, and selectivity for the target analyte.

The signal–concentration dependence for electrochemical biosensors is linear between one and three concentration decades. The lower limit of detection is at 0.2 $mmol\,l^{-1}$ with potentiometric and 1 $\mu mol\,l^{-1}$ with amperometric enzyme electrodes. The use of amplification reactions allows us to decrease the detection limit to the nanomolar to picomolar range. Whereas the response time of potentiometric enzyme electrodes averages 2–10 min, with amperometric ones an assay can be conducted within a few seconds up to 1 min. This permits up to several hundred determinations per hour to be performed. Increasing complexity of the biochemical reaction system, e.g. by coupled enzyme reactions, may bring about an increase in the overall measuring time.

Usually the imprecision found with enzyme electrodes ranges from 1 to 5%, which is of course dependent on the analytical set-up used (e.g. manual sample injection against flow injection). The lifetime of enzyme electrodes depends on several factors including the specific enzyme activity, the eventual formation of inactivating reaction products, and the operating conditions of the sensor. Generally, a high enzyme excess in the biocatalytic membrane is desirable in order to obtain a high stability. Lifetimes are in the range of several days to months or between some hundreds and ten thousand measurements per membrane.

Examples of well designed enzyme sensors are given in table 17.4.

17.4 PRACTICAL USE AND PERFORMANCE

The most relevant fields of practical application of enzyme electrodes are medical diagnostics, process control, food analysis, and environmental monitoring. In 1975 the first commercial enzyme-electrode-based analyzer was developed to meet the high demand for glucose determination in the blood of diabetic patients. Since then, analyzers for about 12 different analytes have been commercialized, most of them also for blood glucose, but with increasing importance of non-medical applications (table 17.2). All the sensors incorporated in the devices of table 17.2 represent enzyme sensors of the first generation where the enzymes are separated from the *measuring solution* by membranes. As compared with conventional enzymatic analysis, the main advantages of such analyzers are the extremely low enzyme demand (a few milliunits per sample), the simplicity to operate them, the high analysis speed, and the analytical quality. Thus, enzyme sensors of the first generation are characterized by a high analytical performance.

Table 17.2 Enzyme-electrode-based autoanalyzers.

Model	Company	Analyte	Measuring range (mM)	Sample throughput (h^{-1})	Functional stability
YSI 2300 G	Yellow Springs	glucose	0.0–27.8	45	7 d
YSI 2300 L	Instrum.	lactate	0.0–15.0	45	7 d
YSI 2700	(USA)	lactose	0.0–58.4	45	10 d
		ethanol	0.0–70.0	45	5 d
		sucrose	0.0–55.5	45	10 d
ESAT 6660	PGW Medingen,	glucose	0.6–45.0	120	15 d
ESAT 6661	Dresden	lactate	0.5–30.0	120	10 d
ECA 180	(Germany)	glucose	0.6–45.0	180	15 d
Industrial		lactose	0.5–100.0	60	15 d
modul		ascorbate	0.5–45.0	60	15 d
		lysine	1.0–100.0	60	15 d
EBIO 6666	Eppendorf–Netheler –Hinz Hamburg (Germany)	glucose	0.6–50.0	120	15 d
		lactate	0.5–30.0	120	10 d
Biosen 5030 L	EKF Industrial	lactate	0.5–30.0	80	10 d
Biosen 6030 G	Electronics Magdeburg (Germany)	glucose	0.6–50.0	80	15 d
Gluco 20	Fuji Electronic	glucose	0.0–27.0	90	>500 samples
UA-300 A	Corp. (Japan)	uric acid	5.0–60.0	3	
Auto-stat GA112	Daiichi (Japan)	glucose	1.0–40.0	120	
Stat Analyzer S80	Analytical Instrum. Corp. (Japan)	glucose	0.0–50.0	120	
Amperometric	Universal	glucose	0.003–3.0	30–60	>500 samples
biosensor	Sensors, Inc.	lactate	0.01–2.5	30–60	>500 samples
detector	New Orleans	urea	0.1–10.0	6	>500 samples
	(USA)	uric acid	0.006–0.6	20–60	>500 samples
		sucrose	0.006–2.0	30–60	>500 samples
		ethanol	0.01–10.0	30–60	>500 samples
		ascorbate	0.001–0.5	30–60	>500 samples
		glutamate	0.01–1.0	30–60	>500 samples
		lactose	0.01–3.0	20–60	>500 samples
		lysine	0.01–20.0	3–15	>500 samples
EXSAN	Acad. Sci. Lithuania,	glucose	2.0–30.0	45	
	Inst. Biochem.	lactate	0.1–15.0	45	
	Vilnius	urea	2.0–40.0	30	
	(Lithuania)	cholesterol	0.5–10.0	20	
Enzymat	Seres (France)	glucose	0.3–22.0	60	
		lysine	0.1–2.0	60	
		choline	1.0–29.0	60	
TOA-GLU 11	TOA Electronics Ltd	glucose	0.0–55.5	60	
Glucoprocesseur	Tacussel (France)	glucose	0.05–5.0	90	>2000 samples
Microzym-L	SGI, Toulouse	lactate	0.05–15.0	60	>200 samples
	(France)	glucose	0.25–27.0	60	>400 samples
OLGA	Biometra	glucose	0.6–50.0	120	15 d
	(Germany)	lactate	0.5–30.0	120	10 d
Nova-CRT	Nova Biomedical	glucose	1.1–30.0	50	7 d
Stat-Profile 5	(USA)	lactate	0.0–15.0	50	7 d
		urea	0.7–35.0	50	7 d
Sensomat	Gonotec Berlin (Germany)	ethanol	0.0–0.32	15	50 samples
Ionometer	Fresenius AG EG-HK (Germany)	glucose	2.0–50.0	60	> 2 month

Concerning the preanalytics that have to be used, two main concepts are available for the application of enzyme electrodes to different fields of analysis. Clinical chemists nowadays are interested in autoanalyzers characterized by high measuring frequency as well as in portable bedside-type analytical systems with extremely short lag time between sample withdrawal and analysis result. On the other hand the optimal control of food production as well as bioprocesses requires the analysis of the substances of interest in a continuous or quasicontinuous way. Therefore, enzyme-electrode-based analytical systems for the application of highly diluted as well as undiluted media have been developed and commercialized.

The majority of enzyme-electrode-based analyzers (tables 17.2 and 17.3) up to now operate on *highly diluted* samples. Development has begun with the measurement of discrete blood or serum samples using stirred measuring cells and internal dilution. Because of the good analytical performance of the most prominent representatives (YSI, Glukometer, Glucoprocesseur, Industrial Modul) this type of device is still usable for the determination of saccharides, lactate, alcohol, uric acid, amylases, amino acids, and ascorbic acid especially in small medical as well as biotechnological laboratories (see table 17.2). Using a sample volume of about 50 μl the serial imprecision is below 2% with a sample throughput of more than 40 h^{-1}.

An enhancement of sample frequency up to 130 h^{-1} has been achieved by the integration of segmented continuous flow systems. The resulting one-parameter autoanalyzers have been on the market since the early eighties for metabolites like glucose and lactate. They process prediluted samples and are especially well suited for centralized medical diagnostics. Representatives are ESAT (PGW Medingen, Germany), EBIO (Eppendorf–Netheler–Hinz Hamburg, Germany), Biosen 5030L and 5030G (EKF Industrial Electronics, Magdeburg, Germany) and Auto-Stat GA-112 (Daiichi, Japan).

An appropriate approach for comfortable on-line process monitoring is the combination of enzyme electrodes with flow injection analysis. For the determination of glucose and lactate this is incorporated in the OLGA system (Biometra Göttingen, Germany).

To assure rapid results immediately after sample withdrawal some companies have realized *undiluted* whole-blood analysis eliminating any preanalytical procedure, which is important especially with respect to fast metabolic substances.

The ExacTech (Medisense, USA) and Elite (Bayer, Germany) glucose analyzing systems are based on *disposable* strip electrodes representing enzyme electrodes of the second generation based on the immobilization of glucose oxidase together with the artificial mediators ferrocene and ferricyanide, respectively. Furthermore, i-Stat Corp. (USA) have commercialized a pocket device useful for analysis of glucose and urea, besides electrolytes. The base transducers are microfabricated thin-film metal sensors for H_2O_2 and ammonium ion-sensitive electrodes (see table 17.3).

Beside these disposable systems enzyme-membrane-based devices (first-generation enzyme sensors) are working with a biological component that can be repeatedly used for thousands of measurements. The functional stability depends on the quality of the enzyme membrane material used. The autoanalyzer Stat-Profile 5 (Nova Biomedical, USA) and the Ionometer (Fresenius, Germany) permit the analysis of metabolites, such as glucose and lactate, in addition to electrolytes and blood gases (see table 17.2).

On the other hand the *portable* enzyme-membrane-electrode-based systems Biosen 5020L and Biosen 6020G (EKF Industrial Electronics, Magdeburg, Germany), besides good analytical performance, are characterized by an extremely short lag time between sample withdrawal and measuring result. Therefore, and because of their mobility, these systems are particularly well suited for application to different sports facilities and intensive care away from centralized laboratories (see table 17.3).

Critical care as well as process control and food industry require enzyme electrodes for invasive and *in situ* analysis. However, direct application in bioreactors is associated with significant difficulties:

(i) the sensor has to be sterilized, which, to date, is virtually impossible,
(ii) calibration has to be performed by discrete measurements or *in situ*,
(iii) often the analyte concentration exceeds the linear range of the sensor,
(iv) various interfering compounds have to be dealt with,
(v) thermal and mechanical stress inactivate the sensor.

Therefore, so far no *in situ* system is commercially available.

In addition to the problems caused by *in situ* application on-line analysis in humans has a critical demand for hemocompatibility and must not cause any cytotoxic, carcinogenic, or genotoxic reactions, inflammation, irritation or sensitization.

Optimistic results concerning invasive glucose analysis have been obtained by groups from Neuchatel (Switzerland), Kansas (USA)/Paris (France), San Diego (USA), and Vienna (Austria) [35]. Nevertheless, no truly reliable implantable sensor has reached the market so far. In the meantime sampling by microdialysis seems to be helpful. Glucose monitoring systems based on microdialysis sampling have been brought to the market by Miles, USA (Biostator), Unitec ULM, Germany (Glucosensor), and Ampliscientifica Milano, Italy (Glucoday). All these systems have been successfully applied to diabetic monitoring over several hours resulting in the improvement of the assessment of labile diabetic patients.

REFERENCES

[1] Scheller F and Schubert F 1989 *Biosensoren* (Berlin: Akademie)
Hall E A H 1990 *Biosensors* (Cambridge: Open University Press)

Buck R P, Hatfield W E, Umana M and Bowden E F 1990 *Biosensor Technology—Fundamentals and Applications* (New York: Dekker)

Edelman P and Wang J 1992 *Biosensors and Chemical Sensors (ACS Symposium Series 487)* (Washington, DC: ACS)

[2] Ghindilis A L, Makower A, Bauer C G, Bier F F and Scheller F W 1994 Determination of p-aminophenol and catecholamines at picomolar concentrations based on recycling enzyme amplification *Anal. Chem. Acta.* **304** 25–31

[3] Scheller F W, Pfeiffer D, Hintsche R, Dransfeld I and Nentwig J 1989 Glucose measurement in diluted blood *Biomed. Biochim. Acta* **48** 891–6

[4] Pfeiffer D, Setz K, Klimes N, Makower A, Schulmeister T and Scheller F W 1991 Enzyme electrodes for medical applications *Biosensors: Fundamentals, Technologies and Applications* ed F Scheller and R Schmid (VCH) pp 11–8

[5] Sasso S V, Pierce R J, Walls R and Yacynych A 1990 Electropolymerized 1,2-diaminobenzene as a means to prevent interferences and fouling and to stabilize immobilized enzyme in electrochemical biosensors *Anal. Chem.* **62** 1111–17

[6] Tsuchida T and Yoda K 1981 Immobilization of D-glucose oxidase onto a hydrogen peroxide permselective membrane and application for an enzyme electrode *Enzyme Microb. Technol.* **3** 326–30

Palleschi G, Rahni M A N, Lubrano G, Ngwainbi J N and Guilbault G G 1986 A study of interferences in glucose measurements in blood by hydrogen peroxide based glucose probes *Anal. Biochem.* **159** 114–21

[7] Pfeiffer D, Wollenberger U, Makower A, Scheller F W, Risinger L and Johansson G 1990 Amperometric amino acid electrodes *Electroanalysis* **2** 517–23

[8] Mullen W H, Churchouse S J, Keedy F H and Vadgama P M 1986 Enzyme electrode for the measurement of lactate in undiluted blood *Clin. Chim. Acta* **157** 191–8

Pfeiffer D, Setz K, Schulmeister T, Scheller F W, Lück H-B and Pfeiffer D 1992 Development and characterization of an enzyme based lactate probe for undiluted media *Biosensors Bioelectron.* **7** 661–71

[9] Clark L C 1956 Monitor and control of blood tissue oxygen tensions *Trans. Am. Soc. Artif. Intern. Organs* **2** 41–8

[10] Clark L C and Lyons C 1962 Electrode systems for continuous monitoring in cardiovascular surgery *Ann. NY Acad. Sci.* **102** 29–45

[11] Pfeiffer D, Scheller F W, Setz K and Schubert F 1993 Amperometric enzyme electrodes for lactate and glucose determinations in highly diluted and undiluted media *Anal. Chim. Acta* **281** 489–502

[12] Williams D, Doig A and Korosi A 1970 Electrochemical–enzymatic analysis of blood glucose and lactate *Anal. Chem.* **46** 118–23

Racine P, Klemke H-J and Kochsiek K K 1975 Rapid lactate determination with an electrochemical enzymatic sensor: clinical usability and comparative measurements *Z. Klin. Chem. Klin. Biochem.* **13** 533–9

Durliat H, Comtat M and Mahenc J 1976 A device for the continuous assay of lactate *Clin. Chem.* **22** 1802–9

[13] Torstensson A and Gorton L 1981 Catalytic oxidation of NADH by surface modified graphite electrodes *J. Electroanal. Chem.* **130** 199–207

[14] Kulys J J and Cenas N K 1983 Oxidation of glucose oxidase Penicillium vitale by one- and two-electron acceptors *Biochim. Biophys. Acta* **744** 57–65

[15] Ikeda T, Hamada H and Senda M 1986 Electrocatalytic oxidation of glucose at a glucose oxidase–immobilized benzoquinone–mixed carbon paste electrode *Agric. Biol. Chem.* **50** 883–90

[16] Jönsson G and Gorton L 1985 An amperometric glucose sensor made by modification of a graphite electrode surface with immobilized glucose oxidase and adsorbed mediator *Biosensors* **1** 355–68

Palleschi G and Turner A P F 1990 Amperometric tetrathiafulvalene-mediated lactate electrode using lactate oxidase adsorbed on carbon foil *Anal. Chim. Acta* **234** 459–63

[17] Wang J, Wu L-H, Lu Z, Li R and Sanchez J 1990 Mixed ferrocene–glucose oxidase–carbon-paste electrode for amperometric determination of glucose *Anal. Chim. Acta* **228** 251–7

Amine A, Kauffmann J-M and Patriarche G J 1991 Long-term operational stability of a mixed glucose oxidase–redox mediator–carbon paste electrode *Anal. Lett.* **24/8** 1293–315

[18] Umana M and Waller J 1986 Protein-modified electrodes. The glucose oxidase/polypyrrole system *Anal. Chem.* **58** 2979–83

Malitesta C, Palmisano F, Torsi L and Zambonin P G 1990 Biosensor based on conducting polymers *Anal. Chem.* **62** 2735–40

Schuhmann W, Lammert R and Schmidt H-L 1990 Polypyrrole, a new possibility for covalent binding of oxidoreductases to electrode surfaces as a base for stable biosensors *Sensors Actuators* **1** 537–41

Hoa D T, Kumar S T N, Punekar N S, Srinivasa R S, Lal R and Contractor A Q 1992 Biosensor based on conducting polymers *Anal. Chem.* **64** 2645–6

Cooper J C and Schubert F 1994 A biosensor for L-amino acids using polytyramine for enzyme immobilization *Electroanalysis* **6** 957–61

[19] Cass A E G, Davis G, Francis G D, Hill H A O, Aston W J, Higgins I J, Plotkin E V, Scott L D L and Turner A P F 1984 Ferrocene-mediated enzyme electrode for amperometric determination of glucose *Anal. Chem.* **56** 667–71

Higgins I J, Hill H A O and Plotkin E V 1987 Sensor for components of a liquid mixture *US Patent* 4 711 245

[20] Ye L, Hämmerle M, Olsthoorn A J J, Schuhmann W, Schmidt H-L, Duine J A and Heller A 1993 High current density wired quinoprotein glucose dehydrogenase electrode *Anal. Chem.* **65** 238–41

Vreeke M, Maidan R and Heller A 1992 Hydrogen peroxide and beta-nicotinamide adenine dinucleotide sensing amperometric electrodes based on electrical connection of horseradish peroxidase redox centers to electrodes through a 3-dimensional electron relaying polymer network *Anal. Chem.* **64** 3084–90

[21] Degani Y and Heller A 1988 Direct electrical communication between chemically modified enzymes and metal electrodes. 2. Methods for bonding electron-transfer relays to glucose oxidase and D-amino acid oxidase *J. Am. Chem. Soc.* **110** 2357–8

Heller A 1990 Electrical wiring of redox enzymes *Accounts Chem. Res.* **23** 128–34

[22] Caras S and Janata J 1980 Field effect transistor sensitive to penicilin *Anal. Chem.* **52** 1935–7

van der Schoot B H and Bergveld P 1988 ISFET based enzyme sensors *Biosensors* **3** 161–85

[23] Anzai J, Tezuka S, Osa T, Nakajima H and Matsuo T 1987 Urea sensor based on an ion-sensitive field effect transistor IV. Determination of urea in human blood *Chem. Pharm. Bull.* **35** 693–8

[24] Hintsche R, Dransfeld I, Scheller F, Pham M T, Hoffmann W, Hueller J and Moritz W 1990 Integrated differential enzyme sensors using hydrogen and fluoride ion sensitive multigate FET *Biosensors Biolelectron.* **5** 327–34

[25] Foulds N C and Lowe C R 1986 Enzyme entrapment in electrically conducting polymers. Immobilization of glucose oxidase in poly-pyrrole and its application

in amperometric glucose sensors *Chem. J. Soc. Faraday Trans.* I **82** 1259–64

Bartlett P N and Whitaker R G 1987 Electrochemical immobilization of enzymes. Part II. Glucose oxidase immobilized in poly-N-methylpyrrole *J. Electroanal. Chem.* **224** 37–48

Fortier G, Brassard E and Belanger D 1988 Fast and easy preparation of an amperometric glucose biosensor *Biotechnol. Tech.* **2** 177–82

Bartlett P N and Cooper J M 1993 A review of the immobilization of enzymes in electropolymerized films *J. Electroanal. Chem.* **362** 1–12

Strike D J, deRooij N F and Koudelka-Hep M 1993 Electrodeposition of glucose oxidase for the fabrication of miniature sensors *Sensors Actuators* **13–14** 61–4

[26] Schuhmann W, Lammert R, Uhe B and Schmidt H-L 1990 Polypyrrole, a new possibility for covalent binding of oxidoreductases to electrode surfaces as a base for stable biosensors *Sensors Actuators* **1** 537–41

Pittner F, Mann-Buxbaum E, Hawa G, Schalkhammer T and Ogunyemi E O 1992 Construction of electrochemical biosensors: coupling techniques and surface interactions of proteins and nucleic acids on electrode surfaces *Proc. Conf. on Trends in Electrochemical Biosensors* (Singapore: World Scientific) pp 69–84

[27] Kimura J, Kuriyama T and Kawana Y 1986 An integrated SOS/FET multibiosensor and its application to medical use *Sensors Actuators* **9** 195–200

Nakamoto S, Ito N, Kuriyama T and Kimura J 1988 A lift-off method for patterning enzyme-immobilized membranes in multi-biosensors *Sensors Actuators* **13** 165–72

[28] Colowick S P and Kaplan N O 1976 *Methods in Enzymology* vol 44, ed K Mosbach (New York: Academic)

[29] Sudhölter E J R, van der Wal P M, Skowrowskaja-Ptasinska M, van der Berg A, Bergveld P and Reinhoudt D N 1990 Modification of ISFETs by covalent anchoring of poly(hydroxyethyl) methacrylate hydrogel. Introduction of a thermodynamically defined semiconductor–sensing membrane interface *Anal. Chim. Acta* **230** 59–65

[30] Hanazato Y, Nakako M, Maeda M and Shiono S 1987 Urea and glucose sensors based on ion sensitive FET with photolithographically patterned enzyme membranes *Anal. Chim. Acta* **19** 387–96

Nakako M, Hanazato Y, Maeda M and Shiono S 1986 Neutral lipid enzyme sensors based on ion-sensitive FET *Anal. Chim. Acta* **185** 179–85

[31] Lambrechts M and Sansen W 1992 *Biosensors: Microelectrochemical Devices* (Bristol: Institute of Physics)

[32] Pfeiffer D, Ralis E V, Makower A and Scheller F W 1990 Amperometric bi-enzyme based biosensor for the detection of lactose—characterization and application *J. Chem. Tech. Biotechnol.* **49** 255–65

[33] Schulmeister T 1990 Mathematical modelling of the dynamic behaviour of amperometric enzyme electrodes *Selective Electrochem. Rev.* **12** 203–60

[34] Wollenberger U, Schubert F, Scheller F, Danielsson B and Mosbach K 1987 Coupled reactions with immobilized enzymes in biosensors *Studia Biophys.* **119** 167–70

Mizutani F, Yamanaka T, Tanabe Y and Tsuda K 1985 An enzyme electrode for L-lactate with a chemically-amplified response *Anal. Chim. Acta* **177** 153–66

[35] Kulys J J, Sorochinskii V V and Vidziunaite R A 1986 Transient response of bienzyme electrodes *Biosensors* **2** 135–43

Schulmeister T 1990 Mathematical modelling of the dynamic behavior of amperometric enzyme electrodes *Selective Electrochem. Rev.* **12** 203–60

[36] Wilson G S, Zhiang Y, Reach G, Moatti-Sirat D, Poitout V, Thevenot D, Lemonnier F and Klein J-C 1992 Progress toward the development of an

implantable sensor for glucose *Clin. Chem.* **38/9** 1613–17
Armour J C, Lucisano J Y, McKean B D and Gough D A 1990 Application of chronic intravascular blood glucose sensor in dogs *Diabetes* **39** 1519–26
Urban G, Jobst G, Keplinger F, Aschauer E, Tilado O, Fasching R and Kohl F 1992 Miniaturized multi-enzyme biosensors integrated with pH sensors on flexible polymer carriers for *in vivo* applications *Biosensors Bioelectron.* **7** 612–19
Koudelka-Hep M, Strike D J and De Rooij N 1993 Miniature electrochemical glucose biosensors *Anal. Chim. Acta* **281/3** 461–6

18

Electrochemical sensors: capacitance

T L Fare, J C Silvia, J L Schwartz, M D Cabelli, C D T Dahlin, S M Dallas, C L Kichula, V Narayanswamy, P H Thompson and L J Van Houten

18.1 INTRODUCTION

18.1.1 Capacitance-based sensors

Capacitance-based chemical sensors are in the class of devices that transduce analytes into electrical currents. Such sensors are typically comprised of a dielectric, chemically-sensitive film coated onto a substrate electrode; these films pass low conduction current, making amperometric or conductimetric measurements less sensitive or attractive for signal transduction. To detect an analyte, changes in the chemically-sensitive film's capacitive properties (associated with its dielectric constant, charge uptake, or formation of interface dipole layers) are measured when an active species is present or generated.

Examples of capacitance-based sensors can be found in chemical, biological, and physical applications. One of the early successes in capacitance-based chemical sensors was the detection of hydrogen, ammonia, and alcohols using metal–oxide–semiconductor capacitors (MOSCAPs) with catalytic metals as the gates [1–3]. For these devices, it was proposed that the catalytic metal gate aids in the adsorption or dissociation of molecules at or near a metal–oxide interface. The adsorbed or dissociated species sets up a dipole layer with a coverage related to the gas or vapor pressure. The dipole field couples to the oxide–silicon interface, causing the voltage-dependent capacitance of the device to change as a function of the gas or vapor pressure [4].

Another successful application of capacitance-based sensing is in humidity and moisture monitoring. Humidity sensors have been developed for which the dielectric constant of the film or substrate changes upon the uptake of the highly polar water molecule [5–8]. Based on this principle, several different devices have been developed on a variety of insulating and dielectric substrates, including alumina [5], semiconducting materials [6], perovskites [7], porous silicon [8], and thermosetting polymers [9]. Humidity sensors using thermosetting polymer films are well understood, inexpensive, and commercially available.

Sensors based on adsorption of species onto or into lattice structures have been reported for molecules besides water. For example, devices based on the detection of carbon dioxide adsorption onto semiconductor materials have been developed [10]. In other cases, dielectric materials that have some degree of chemical specificity have been used for making chemically-sensitive layers. One such application is the use of the highly porous zeolite lattice to detect adsorbed hydrocarbons [11]. The specific dimensions and shape of the zeolite pores allows for size and chemical selectivity in the lattice. As in the case of the humidity devices, the adsorbed molecules' dipoles cause a local change in the electric fields that can be detected through a capacitive effect.

Antibodies, receptors, or enzymes can be used to confer a high degree of chemical specificity to a capacitance-based device [12–16]. The antibody- and receptor-based devices take advantage of the interaction of a molecule with the selective binding site of the protein to detect an adsorption event [12]. Enzyme-based capacitance devices have been used to generate charged by-products near the substrate space–charge region to change the electrode capacitance [13]. Simultaneous electrical and optical methods have been reported for monitoring direct adsorption of species onto antibody-coated surfaces [14, 15]. In this type of measurement, both the refractive index and the space–charge capacitance near the electrode surface are monitored during the formation of the analyte–antibody complex. This method is appropriate for relatively large proteins (as opposed to low-molecular-weight species) and has been applied to enzyme, antibody, and peptide detection [15]. Several different amplification schemes have been used in a capacitive affinity biosensor system developed for biological analytes [16].

Capacitive sensing has been used for monitoring cells and cell cultures. A method for following cell motility, growth, and spreading has been developed and adapted to study cell metabolism and viability [17]. Response of a cell culture to activating agents was reported in the study. Because the outer membranes of cells are composed primarily of lipids and glycoproteins, extracellular measurements performed on substrate electrodes are capacitive by their nature; models for cell membranes reflect the highly dielectric components of the proteins and lipid bilayer [18]. As an example, extracellular electrical signals from patterned neural networks in culture were recorded for chemical and electrical stimulation [19]; a comparison of the intracellular and extracellular recordings demonstrates the capacitive coupling of the cell to the planar substrate electrode. Non-planar electrodes for extracellular applications developed for stimulation and recording exhibit similar electrical behavior [20, 21].

For physical sensing applications, micromachined silicon has been used in applications for pressure sensing [22] and accelerometers [23]. Thin diaphragms or beams of silicon suspended over the base silicon are formed by chemical etching. Acceleration or applied pressure causes the suspended silicon to move closer to (or farther from) the base causing the capacitance to increase (or decrease).

18.1.2 SmartSense™ conducting polymer-based immunosensors

A capacitance-based method for specific detection of low levels of pesticides and environmental contaminants has been developed using conducting polymers [24, 25]. In this application, the conducting polymer serves as the chemically-sensitive film that transduces an immunoassay into an electrical signal. A major advantage in using conducting polymers for immunoassay-based biosensors (immunosensors) is that antibodies can be coated directly onto the active polymer surface with little degradation of antibody functionality.

A schematic of the electrode and the immunoassay steps is depicted in figure 18.1. The immunosensor makes use of standard techniques used in other competitive ELISAs (enzyme-linked immunosorbent assays): a glucose oxidase-labeled analyte (the conjugate) competes with the analyte of interest for a limited number of antibody sites on the conducting polymer surface. After discarding the unbound sample and standards, a proprietary catalyst buffer is added to each cell, followed sequentially by a glucose containing buffer. The conjugate generates hydrogen peroxide from the added glucose, triggering events that oxidize the film. As the polymer becomes oxidized, the current through the film increases. By supplying appropriate standards for the antibody-coated electrode system, the device response (defined as the change in current as a function of time) can be related to analyte concentration in the sample.

The response of this device depends on the physical and chemical properties of the conducting polymer. For example, the initial redox state of the polymer is dependent on polymer synthesis and processing. For the polythiophene formulation used in this device, the film is in its neutral (high-impedance) state; the film is exposed to water during antibody processing, which tends to keep the polymer in the neutral state [26]. Under these conditions, the dielectric nature of the polymer makes a substantial contribution to the admittance of the film and plays an important role in the device response. Conducting polymer films in the neutral state behave like charge sinks: that is, oxidizing species readily diffuse into the film under a driving chemical potential, similar to a battery charging up a capacitor. These oxidants contribute charge carriers both to the conduction process and to charge storage, which is reflected as a change in sensor capacitance. Sensor capacitance and conductance, both of which can be related to polymer film properties, are derived from the measured current flow.

To relate film response to the oxidant generated by the immunoassay requires: (i) relating solution oxidant concentration to oxidant concentration in the

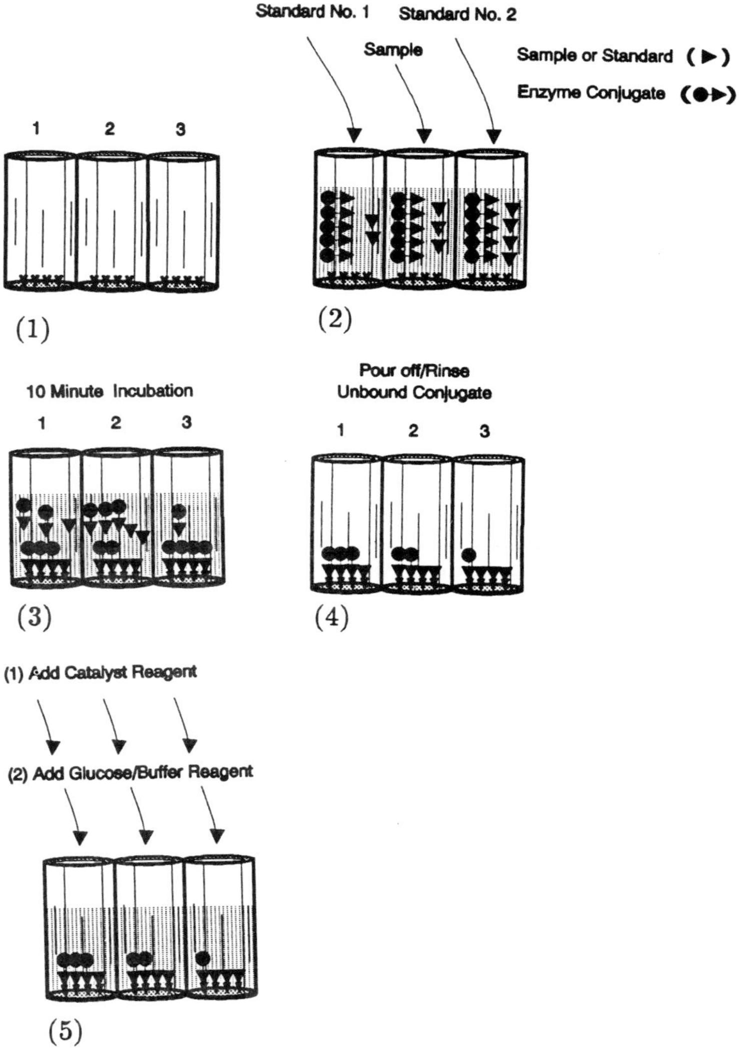

Figure 18.1 A depiction of the assay steps for the SmartSense^{TM} system: (1) start with an antibody-coated conducting polymer electrode in the base of wells 1, 2, and 3; (2) add a sample (unknown analyte concentration and known conjugate concentration) to well 2 and add two standards (fixed analyte and conjugate concentrations) to wells 1 and 3; (3) allow the conjugate and analyte to compete for the antibody binding sites; (4) after a 10 min incubation, discard the solutions; (5) add a catalyst containing reagent, then inject a glucose containing buffer to induce change in current during the response period.

polymer and (ii) relating oxidant concentration in the polymer to the electrical properties of the film. A model has been developed that can be used to relate polymer oxidant concentration to device response by way of an equivalent circuit for conducting polymer films [27, 28]. The basis of this model combines the analysis of a porous electrode with the capacitance charging. We use a modified version of this model for the film conductance and capacitance components to investigate film response and characterize the system illustrated in figure 18.1. The model is discussed further in sections 18.2 and 18.3 to relate sensor capacitance to the assay response for a given set of conditions.

18.1.3 Major advances for manufacturing conducting-polymer-based immunoassay devices

One of the problems in developing conducting-polymer-based films for immunosensor applications is reducing the variation in polymer characteristics over a large population of devices. A major advance in the realization of conducting polymer sensors has been the introduction of processible polymers such as polypyrrole, polyaniline, and polythiophene and their derivatives [29]. Compared with the early conducting polymers, the processible polymers are more chemically stable, have better physical properties, and are soluble in a variety of solvents. This ease of polymer handling has allowed the establishment of a scale-up process for coating and handling the electrodes in the SmartSense™ immunosensor system. Even with these recent improvements in polymer processing, there still may be a variation in polymer characteristics (e.g. degree of polymerization or redox state) from sample to sample. To address these issues, improved physical models, such as those derived for film admittance, can be used to account for variations in sensor response.

18.2 CONTRIBUTIONS TO CONDUCTANCE AND CAPACITANCE IN DEVICE RESPONSE

To obtain a response from a capacitance-based device, an AC voltage sinewave, $V = V_0 \sin(2\pi f t)$, of fixed frequency, f, and amplitude, V_0, is typically applied through an electrode to the chemically-sensitive film and the AC current is monitored. The current can be written as $I = I_0 \sin(2\pi f t + \phi)$, where I_0 is the current amplitude and ϕ is a phase shift induced by the sensor film. For a purely resistive sensor, there is no phase shift ($\phi = 0$) and for a purely capacitive sensor, the phase shift is 90°. In general, the device admittance is given by $Y = I/V$ and, for sinusoidal excitations, can be written using complex number notation as $Y = \mathrm{Re}[Y] + i\,\mathrm{Im}[Y]$, where $i = (-1)^{1/2}$ and $\phi = \tan^{-1}(\mathrm{Im}[Y]/\mathrm{Re}[Y])$ [30, see especially ch 9]. The impedance of the system, Z, is the inverse of admittance, Y ($Z = 1/Y$). For a parallel resistor, R, and capacitor, C, $\mathrm{Re}[Y] = 1/R$, $\mathrm{Im}[Y] = 2\pi f C$, and $\phi = \tan^{-1}(2\pi f RC)$. Capacitance-based sensors may have

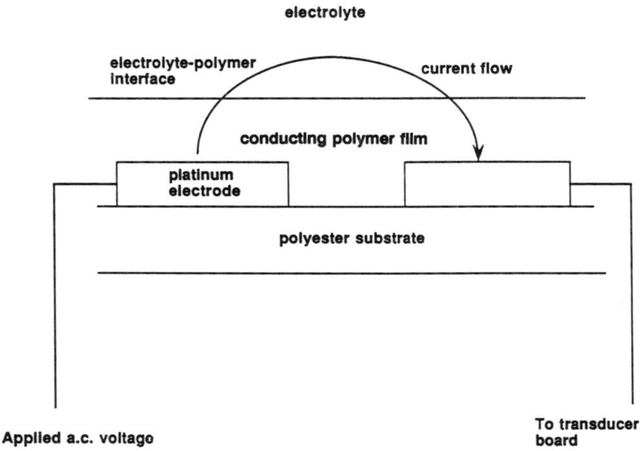

Figure 18.2 A cutaway view displaying the coated electrode system with electrolyte, conducting polymer, metal electrodes, and substrate. The current path across the interface between the electrolyte and the polymer determines the impedance.

a resistance component, but the device phase angle is generally close to 90° at the application frequency.

18.2.1 Interface admittance

Figure 18.2 depicts the electrode–polymer–electrolyte configuration that is used in the SmartSense™ system. Admittance in this type of electrode system is due to contributions from the interface and bulk film regions. For the electrodes under consideration, the polymer film thickness is of the order of 100 nm and the electrode separation is 0.05 cm; for these dimensions, the impedance through the film itself should be in excess of $10^6\ \Omega$. The measured impedance at 10 Hz, however, is of the order of $10^4\ \Omega$, implying that the predominant current path is across the polymer–electrolyte interface and through the electrolyte.

As shown in figure 18.2, the electrode–polymer system is symmetric, which allows a half-cell analysis. To simplify the analysis, three assumptions are made at the outset: (i) the electrolyte is highly conducting so that solution resistance is negligible; (ii) oxidation occurs primarily at the polymer–electrolyte interface; and (iii) the Pt–polymer interface admittance does not change substantially during an assay. Pt is the electrode material of choice because of its inertness in the system and its ability to form an ohmic contact with conducting polymers [28]. To ensure reproducibility of the Pt–polymer interface, the Pt electrodes are plasma cleaned prior to polymer deposition. We have observed that for highly oxidized polymer films, there is a residual impedance (5–10% of the initial impedance) that can be associated with the Pt–polymer interface.

For a homogeneous sample, film admittance is related to its physical dimensions and properties by

$$C = \frac{\epsilon A}{d}$$
$$R = \frac{d}{\sigma A} \tag{18.1}$$

where ϵ is the film dielectric constant, σ is the film conductivity, A is the electrode area, and d is the conducting polymer thickness. We have observed that film impedance increases weakly with film thickness and there is variation within a 'fixed' film thickness. Such variability may be the result of film thickness fluctuations over the macroscopic electrode area, however, it is more likely due to the irregularity of the porous polymer–electrolyte interface [27, 31].

Given the nature of the polymer and the conduction pathway, a simple homogeneous model cannot be applied to thin conducting polymer film–electrolyte systems [27, 28, 31]. For thin films (< 100 nm) with pore sizes estimated to range from 1 to 4 nm, the porous surface–electrolyte interface will dominate the electrical and physical properties of the sensor. Since the oxidation of the porous surface occurs first, the interface properties play a major role in determining device response. To make use of this information for the immunosensor response, the appropriate measurement frequency must be chosen to discriminate between bulk and interface phenomena. To determine the optimum frequency to probe the interface, the admittance spectra of the conducting polymer films in the frequency range of interest are required.

18.2.2 Admittance spectra of conducting polymer films

Admittance spectra have been used to characterize conducting polymer–electrolyte interfaces for polyaniline [32], polypyrrole [33], and polythiophene [27, 34]. Typical conducting polymer films exhibit low-frequency time constant dispersions that vary as a function of applied potential (affecting the redox state of the polymer). We have investigated the characteristics and response of protein-coated and untreated conducting polymer films from 0.5 Hz to 10 kHz with an EG&G PAR 5210 lock-in amplifier.

Data from a typical admittance spectrum are shown in figure 18.3(a) over a low-frequency range. Equivalent circuits have been proposed to account for low-frequency characteristics of polymer films [27, 35] and a version of these circuits modified for the SmartSense™ electrode is depicted in figure 18.3(b). A fit from this equivalent circuit is shown in figure 18.3(a) for comparison to the data. Interface and bulk admittance components are depicted and labeled in figure 18.3(b). These models reflect the contribution of a distributed space–charge region along the polymer pore walls, which yields a large interface capacitance and the time constant dispersions in admittance spectra.

Based on the physical processes in this system and the structure of the electrodes, the measurement frequency is chosen to minimize contributions

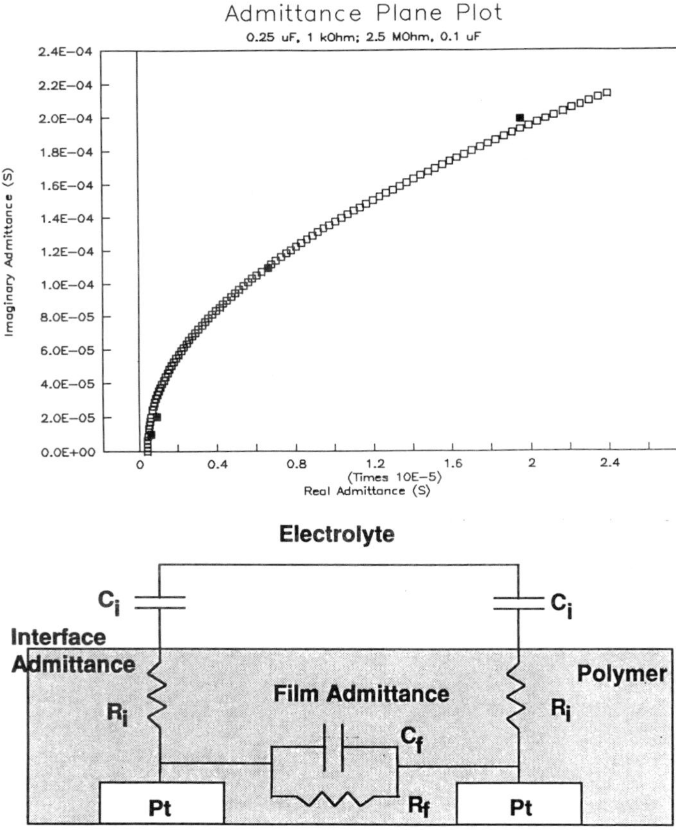

Figure 18.3 (*a*) The admittance spectrum of a typical film in the low-frequency range; the darkened squares are the measured data and the open squares are derived from (*b*) the circuit model for low-frequency characteristics containing the interface admittance, R_i, C_i, and the film admittance, R_f, C_f. For the curve fit, the interface capacitance, C_i, is 0.25 μF and the interface resistance, R_i, is 1 kΩ; the film capacitance, C_f, is 0.1 μF and the film resistance, R_f, is 2.5 MΩ.

of ionic diffusion across the electrolyte–polymer interface (lower frequencies, $f < 1\,\text{Hz}$) and geometric capacitance (higher frequencies, $f > 100\,\text{Hz}$) to device response. A sinusoidal waveform with a measurement frequency of 10 Hz was chosen to focus on the interface effects. To relate the assay response to the circuit model components, expressions derived for the real and imaginary terms of the admittance for the low-frequency model are given as

$$\text{Re}[Y] = 1/R_f + \frac{(2\pi f C_i R_i)^2}{1 + (2\pi f C_i R_i)^2} \frac{1}{R_i} \tag{18.2}$$

$$\mathrm{Im}[Y] = 2\pi f C_f + \frac{2\pi f C_i}{1 + (2\pi f R_i C_i)^2}. \tag{18.3}$$

It should be noted that there is a coupling of the real and imaginary admittance terms of the neutral film through the product $R_i C_i$. A consequence of setting the frequency to 10 Hz causes the the term $(2\pi f C_i R_i)^2$ to be negligible. This simplifies the analysis because the imaginary admittance can be related directly to the interface capacitance, C_i. In this model, the film capacitance, C_f, respresents a fixed value of the thin-film admittance and does not contribute directly to the response. In the next section, it is shown how the change in the interface capacitance, C_i, is related to the immunoassay response of the film.

18.3 MECHANISMS OF SENSOR RESPONSE: KINETICS, EQUILIBRIUM, AND MASS TRANSPORT

18.3.1 Reaction kinetics

For field applications, one of the primary requirements of an assay is that it is rapid. Hence, for the SmartSense™ system, the response in the initial stages of the reaction is used to determine analyte concentration. The sensor response is derived from a cascade of events; the reaction steps and their respective kinetics need to be identified and evaluated. For a competitive ELISA, such as the SmartSense™ system shown in figure 18.1, a labeled analyte (conjugate) is needed to determine the concentration of the analyte of interest. For this model system, the analyte of choice is atrazine, a compound for which antibodies and conjugate are readily available; the conjugate is atrazine labeled with glucose oxidase, GOD. In the first stage of the assay, a known amount of conjugate competes with atrazine for a fixed number of antibody sites, given schematically as

$$\text{Ab} + \text{atrazine} + \text{atrazine}^* \rightarrow \text{Ab–atrazine} + \text{Ab–atrazine}^* \tag{18.4}$$

where Ab is the immobilized antibody and atrazine* is GOD-labeled atrazine (figure 18.1(3)). After the competition stage, atrazine and the conjugate occupy the antibody sites in relative proportion to the concentration of these two species in solution; in other words, the more (less) atrazine in solution, the less (more) conjugate bound. Atrazine solution concentration is therefore inversely related to the amount of bound conjugate.

The amount of bound conjugate is determined by adding an excess of glucose to each cell in the cartridge. The conjugate produces hydrogen peroxide in direct relation to glucose consumption at fixed temperature and pH according to

$$\text{glucose} + O_2 \rightarrow \text{gluconic acid} + H_2O_2. \tag{18.5}$$

For GOD, apparent Michaelis–Menten constants (K'_m, the substrate concentration at which the reaction velocity is half the maximum velocity, $V_{max}/2$) ranging from 4 mM to 13 mM have been cited for different immobilization conditions [36]. In excess glucose (i.e. glucose concentrations sufficiently greater than K'_{max}), the rate of hydrogen peroxide generation can be approximated by

$$\frac{d[H_2O_2]}{dt} = -\frac{d[\text{glucose}]}{dt} \cong k[\text{atrazine}^*] \qquad (18.6)$$

where k is the forward rate constant for the conversion of the enzyme–substrate complex to free enzyme. From equation (18.6), the rate of hydrogen peroxide generation depends on the amount of conjugate bound for a given concentration of analyte in the sample.

Hydrogen peroxide alone will not generate a signal in the SmartSense™ system; rather, hydrogen peroxide is used to convert iodide to triiodide, which is the species that oxidizes the polymer. To form triiodide, a two-step process is proposed: in the first step, hydrogen peroxide combines with iodide in a development buffer to form iodine

$$H_2O_2 + 2I^- + 2H^+ \rightleftharpoons I_2 + 2H_2O \qquad (18.7)$$

and the iodine production rate for this reaction is given by [37]

$$\frac{d[I_2]}{dt} = k_1[H_2O_2][I^-](1 + k_2[H^+]). \qquad (18.8)$$

If hydrogen peroxide and iodide are in excess concentrations in the buffered solution, then the iodine rate production is constant. The second step in triiodide formation is the equilibrium conversion of iodine to triiodide given by

$$I_2 + I^- \rightleftharpoons I_3^-(s). \qquad (18.9)$$

In an excess of iodide, iodine is readily converted to triiodide, which is soluble in water.

Iodine production is the rate determining step in this multistep process and a variety of catalysts are available to speed the iodide to iodine formation. In fact, we have observed that triiodide formation in the presence of catalysts is rate limited solely by the hydrogen peroxide concentration. With the use of a catalyst, loss of hydrogen peroxide (e.g. by natural breakdown or peroxidase activity of contaminants) is less of a concern for reproducibility. Since iodine is almost completely converted to triiodide in the solution phase, the rate of triiodide production with catalyst present effectively follows the rate of hydrogen peroxide production or

$$\frac{d[I_3^-]}{dt} \cong \frac{d[H_2O_2]}{dt} \cong k'[\text{atrazine}^*] \qquad (18.10)$$

where k' is a combined rate constant. To a first approximation, therefore, the rate of triiodide formation is directly related to the amount of conjugate bound to the conducting polymer film and inversely related to the atrazine concentration in the sample.

18.3.2 Hydrogen peroxide–iodide film oxidation

To test the above-stated reactions, as well as qualify the polymer coating quality, an assay has been devised using triiodide generated from the hydrogen peroxide–potassium iodide reaction (equations (18.7)–(18.9)) to oxidize the film. Polymer-coated electrodes are set in a phosphate-buffered saline–potassium iodide (PBS–KI) solution. Current through the film is monitored to obtain a starting admittance and to determine whether the cell is stable (e.g. low baseline drift). At this point, the cells are dosed with a fixed H_2O_2 concentration and the generated triiodide oxidizes the film. The change in current as a function of time is recorded for a population of electrodes to determine average response and precision.

Equation (18.8) can be used to calculate the response for this assay. If we assume that in a buffered solution with an excess of iodide only the hydrogen peroxide concentration is significantly consumed, then we can obtain $I_2(t)$ as

$$[I_2(t)] = [H_2O_2]_i \left(1 - \exp(-k_1[I^-](1 + k_2[H^+])t)\right) \qquad (18.11)$$

where $[H_2O_2]_i$ is the injected hydrogen peroxide concentration. For the condition when $k_1[I^-](1 + k_2[H^+])t \ll 1$, the system has a linear response. Although this is not an identical simulation of the kinetics of the enzyme-mediated generation of hydrogen peroxide, it is a convenient way to test film response for a large sample size. Results for this test are reported in section 18.4.

18.3.3 Equilibrium considerations

To relate the triiodide concentration in solution to the oxidation of the polymer, an approach based on a Langmuir-like adsorption isotherm has been used [30, 31]. In this model, triiodide in solution is assumed to oxidize sites, S, in the film and contribute charge carriers to the polymer film. The electron exchange for the oxidation is given by [38]

$$I_3^-(s) + S \rightleftharpoons S^+I_3^- + e \qquad (18.12)$$

where $S^+I_3^-$ is the oxidized site formed in the polymer. The electron, e, is passed as a conduction current between the polymer and the platinum electrode.

An assumption in the Langmuir model is that a limited number of adsorption sites exist within the polymer. Since the number of sites acts as the saturation limit for polymer oxidation, the relationship for filling the available sites with triiodide at equilibrium can be obtained as follows:

$$K_{eq}[I_3^-]_0([S_0] - [S^+I_3^-]) = [S^+I_3^-] \tag{18.13}$$

where $[I_3^-]_0$ is the triiodide concentration at $x = 0$ (the polymer–solution interface), $[S_0]$ the initial concentration of available sites, and K_{eq} the system equilibrium constant. Solving this equation for the fraction of occupied sites relative to the available sites yields

$$\Gamma = \frac{[S^+I_3^-]}{[S_0]} = \frac{K_{eq}[I_3^-]_0}{1 + K_{eq}[I_3^-]_0} \tag{18.14}$$

where Γ is the fraction of oxidized sites at the porous polymer surface. Comparing the signals from a highly oxidized (saturated) hydrogen peroxide assay with those generated in the atrazine–conjugate system, the fraction of sites taken up in the immunoassay ranges from 0.01 to 0.1; this implies that $K_{eq}[I_3^-]_0 \ll 1$ for the duration of the assay.

18.3.4 Oxidant flux to the film

In the model system, device response ultimately depends on the rate of triiodide transfer to the polymer surface. In general, triiodide flux to the film can be calculated from [30]

$$\frac{i_{I_3^-}}{nFA} = -\Gamma_s \frac{\partial \Gamma}{\partial t} + D_0 \left[\frac{\partial [I_3^-(x,t)]_s}{\partial x} \right]_{x=0} \tag{18.15}$$

where Γ_s is the saturation number of oxidized sites per unit area of the polymer surface, D_0 the diffusion constant of triiodide in the solution, n the electrons exchanged by the film per triiodide ion reduced, F the charge on one mole of electrons, and A the area of the electrode. Because the conjugate is bound directly to the conducting polymer film surface, the gradient (diffusion) term in equation (18.15) can be neglected.

The rate of film oxidation is set by the rate determining step in the cascade reaction sequence. For this system, we have observed that the film oxidation rate essentially follows the hydrogen peroxide generation rate. We can use equations (18.14) and (18.15) for $K_{eq}[I_3^-]_0 \ll 1$ to relate the triiodide flux to the triiodide generation rate and, by equation (18.10), to the bound conjugate so that

$$\frac{i_{I_3^-}}{nFA} = -\Gamma_s \frac{\partial \Gamma}{\partial t} = -\Gamma_s K_{eq} \left[\frac{\partial [I_3^-]}{\partial t} \right]_{x=0} \cong -\alpha[\text{atrazine}^*] \tag{18.16}$$

where α is a system proportionality constant. This observation greatly simplifies analyzing the immunosensor response because the triiodide flux is then simply proportional to the amount of bound conjugate. To determine the triiodide flux and polymer oxidation rate exactly requires the simultaneous solution of the above non-linear system of equations. This includes accounting for effects such as the triiodide and hydrogen peroxide concentration gradients and the rate constant of each reaction.

18.3.5 Charge injection and film admittance

From the analysis of Feldberg and Tanguy, the capacitance of conducting polymer films is related to the extent of film oxidation [27, 31]. According to this treatment, triiodide oxidation of the polymer changes the charge concentration in the film; *hence, triiodide flux and film oxidation can be monitored directly through the AC capacitance current*. This model assumes an interface capacitance proportional to the oxidant concentration in the film given by

$$C_i = C_{max}\Gamma = C_{max}(\Gamma_0 + \delta\Gamma) \tag{18.17}$$

where C_{max} is the maximum capacitance of the oxidized polymer, Γ_0 the fractional charge concentration at $t = 0$, and $\delta\Gamma$ is the fractional change in polymer charge concentration due to oxidation. The initial charge, Γ_0, arises from the initial redox state of the polymer and the polymerization process itself, which may include the use of FeCl as a polymerization catalyst. We assume that the polymer lot is uniform in its chemical composition so that Γ_0 is the same for each electrode of a given production run.

With this assumption, the maximum capacitance can be written in terms of the initial interface capacitance and the initial charge concentration for $\delta\Gamma = 0$ at $t = 0$, so that

$$C_{max} = \frac{C_{i,0}}{\Gamma_0} \tag{18.18}$$

where $C_{i,0}$ is the interface capacitance at $t = 0$. The change in capacitance as a function of time can then be obtained from equations (18.16)–(18.18) as

$$\frac{\partial C_i}{\partial t} = \frac{C_{i,0}}{\Gamma_0}\frac{\partial \Gamma}{\partial t} = \frac{C_{i,0}\Gamma_s K_{eq}}{\Gamma_0}\left[\frac{\partial[\mathrm{I}_3^-]}{\partial t}\right]_{x=0}. \tag{18.19}$$

According to this analysis, the rate of change in sensor capacitance is proportional to the initial interface capacitance, $C_{i,0}$, and the triiodide generation rate at $x = 0$, which is related to the amount of bound conjugate by equation (18.16).

In the circuit model introduced in section 18.2, the capacitance, C_i, is included as part of the low-frequency imaginary admittance (equation (18.3)). Based on this model, a relationship between C_i and the amount of triiodide taken up by the film can be measured either in the immunoassay or the hydrogen peroxide assay. For an immunoassay, equations (18.10) and (18.19) can be used to correlate device response with amount of conjugate bound to the film and, hence, the atrazine concentration in the sample. For the hydrogen peroxide assay, we can relate changes in C_i directly to the hydrogen peroxide concentration and the initial capacitance by equations (18.8), (18.11) and (18.19). Data are presented in section 18.4 to investigate this model critically.

18.4 PRACTICAL EXAMPLE: FABRICATION AND TESTING OF SMARTSENSE™ IMMUNOSENSORS

18.4.1 Device fabrication

18.4.1.1 Metal electrode fabrication. An exploded view of the sensor assembly parts is shown in figures 18.4 and 18.5. The foundation for the sensor is a poly(3-alkylthiophene) coating on Pt metal electrodes. Electrodes are fabricated by sputter deposition of Pt in high vacuum onto polyester rolls. The metal is pattern defined using modified lithographic techniques, resulting in a parallel array of metal electrode stripes and spacings (figure 18.5(*a*)). Electrode stripe widths (0.68 cm) and spacings (0.05 cm) are patterned to be on center with standard electrical connectors. The metallized polyester roll is cleaned thoroughly in an aqueous-based solvent using a proprietary process. After solution cleaning, the roll is plasma cleaned to remove surface contaminants.

18.4.1.2 Polymer coating onto the metal electrodes. Conducting polymer is deposited onto the metallized polyester using a precision roll coating process. For this process, the neutral polymer is first dispersed in organic solvents and centrifuged. Concentration of the polymer in the organic solvent and the viscosity of the dispersion are determined prior to coating. A variety of solvent mixtures have been tested for this process and a mixture that optimizes solvent evaporation and polymer concentration was chosen. Once the roll is coated, it is heated to drive off excess solvent, then cut into sheets for further handling. The polymer films have an extremely hydrophobic surface, which is amenable to adsorbing protein hydrophobic sequences during the antibody immobilization step.

18.4.1.3 Coated electrode and cartridge assembly. Three sets of electrode pairs are cut into strips (2.21cm × 4.38 cm) from the coated sheets for assembly into individual cartridges (figure 18.5(*a*)). The cartridges that hold the electrodes are injection-molded parts consisting of two pieces (top and bottom). The design combines ease of manufacture and cartridge assembly and optimization of electrode surface area. Schematic drawings of the assembled pieces are shown in figures 18.4 and 18.5(*b*). For cartridge assembly, an electrode set is aligned in the bottom to registration posts and the top is snapped in place over the electrode set. As can be seen in figures 18.4 and 18.5, the top has three cell chambers; each chamber is defined by a 1.52 cm deep cylinder sealed to the electrode

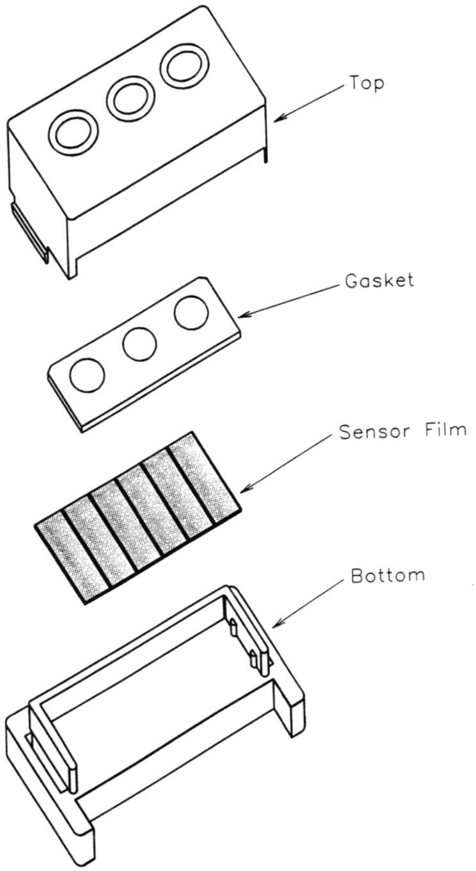

Figure 18.4 An exploded drawing of a cartridge for SmartSense™ assay with the three-well format.

substrate at the base of the cylinder. Each cylinder has a circular cross-section of radius 0.4 cm centered on one of the three electrode pairs.

18.4.1.4 Antibody and surface treatments. Once the cartridges are assembled, antibodies are coated directly onto the conducting polymer surface. In the case of an atrazine sensor, antibodies specific to atrazine are adsorbed onto the polymer from stock concentrations diluted in a buffered solution. System optimization has been achieved by investigating different buffers covering a range of pH values. Following a fixed period for antibody immobilization, the solution in the cells is decanted and a solution containing an albumin-based blocking agent to inhibit non-specific binding is added to each cell for approximately 1 h. After decanting the blocking agent solution, all cells are thoroughly rinsed and are ready for storage or immediate use in the immunoassay.

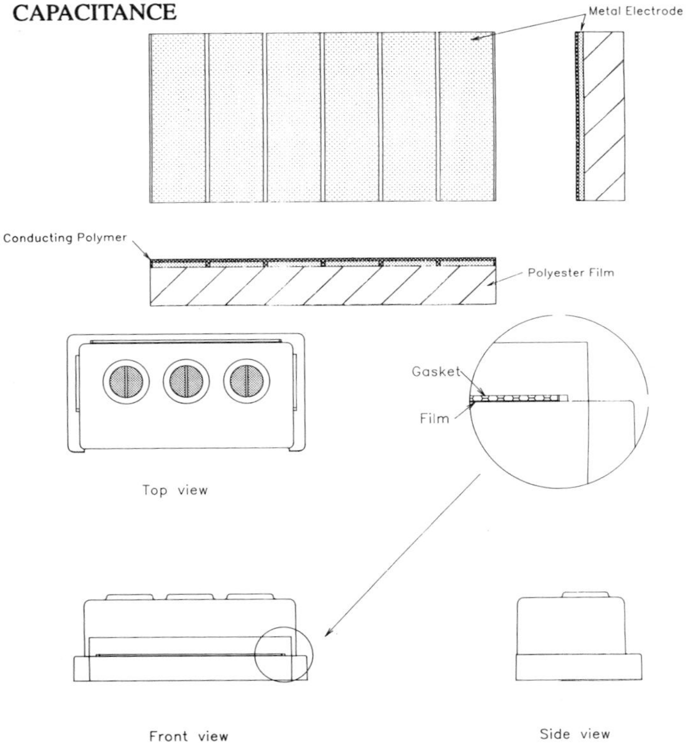

Figure 18.5 (*a*) A drawing of the film strips cut for use in the cartridge illustrates the detail of each electrode. (*b*) Perspective views of the assembled cartridge show the position of the wells in the cartridge.

Antibodies and blocking agents adsorbed to the film convert the hydrophobic surface of the polymer into a hydrophilic surface layer. It should be noted that the dimensions of antibodies and enzymes (5–10 nm) are typically larger than the pores in the conducting polymer surface (1–4 nm); the relative size difference minimizes the amount of antibody or conjugate entrapped inside the polymer itself. For a competitive assay, it is critical that the antibody surface concentration and non-specific binding of conjugate be well controlled to obtain reproducible standard curves; the polymer film surface should play as small a role as possible in binding excess antibody or conjugate.

18.4.2 Instrumentation

18.4.2.1 The transducer board. To monitor the signal from the cartridge, a portable electronic measurement unit has been built. The heart of the measurement electronics is the transducer board (shown in a simplified schematic in figure 18.6(*a*)), which reads the electrical signal from the cartridge. For a given cartridge in the semiquantitative format, the signal from the sample (center well) is compared to the signal from the two standards (side wells)

[39]. To excite the polymer film, an input voltage waveform causes current to flow through the cells in a cartridge. An AC signal was chosen to (i) facilitate the capacitance measurement; (ii) avoid electrode polarization; and (iii) lower susceptibility to noise.

The input waveform, common to all three cells, is buffered to reduce load distortion and filtered to remove noise. After this conditioning stage, the input waveform is applied to one of the electrodes in each cell; the second electrode is held at 0 V (virtual ground). Because of the voltage difference applied to the two electrodes, current flows through the polymer film, represented in figure 18.6(*a*) as resistors and capacitors. Amplifiers connected to the second electrode sense the current flow through the cell and generate an output voltage proportional to the current. These output voltages, V_1, V_2, and V_3, are the raw data fed into the control unit for further processing.

18.4.2.2 The control unit and software. While the assay is being run, the control unit (figure 18.6(*b*)) displays instructions on an LCD display for the user and directs the flow of data in the instrument. The control unit prompts the user to follow menu-driven instructions to set up the chemistry, run the assay, and collect and store data. A keypad allows the user to respond to the instructions at each step of the assay. As a qualification step, each cartridge undergoes a specification check during the test.

Once the cartridge is qualified, the user is given instructions for completing the assay. The unit keeps track of timing the competitive assay incubation (10 min) and glucose additions. Once the data collection from the transducer board begins, the central processing unit (CPU) instructs the analog to digital (A/D) converter to translate the output voltages, V_1, V_2, and V_3, to digital signals, which are then stored in memory. At the end of the assay, the responses from the three cells are passed to a comparison algorithm to determine whether the sample under test is above or below a user-selected level. A positive or negative result is displayed for the user to record; the test results are also automatically stored in the memory for recall. Results from a series of tests can be downloaded from the unit into standard graphics or spreadsheet packages.

18.4.2.3 Measurement hardware. An integral part of the system design is the measurement instrument itself. It can be used either as a stand-alone unit in the field for immediate results or connected to a personal computer for more intensive data handling. The unit houses all supporting electronics and hardware, such as rechargeable batteries, DC recharge plug, AC to DC adaptor, measurement circuitry with solid state memory, and an RS232 port to interface a computer or printer with the measurement electronics. To make electrical connection between the cells and the measurement electronics, an adapter plate that holds four cartridges is provided that inserts directly into the instrument. Software packages for both data manipulation and the user interface are available for the system and in memory. Data can be received, stored, and processed from

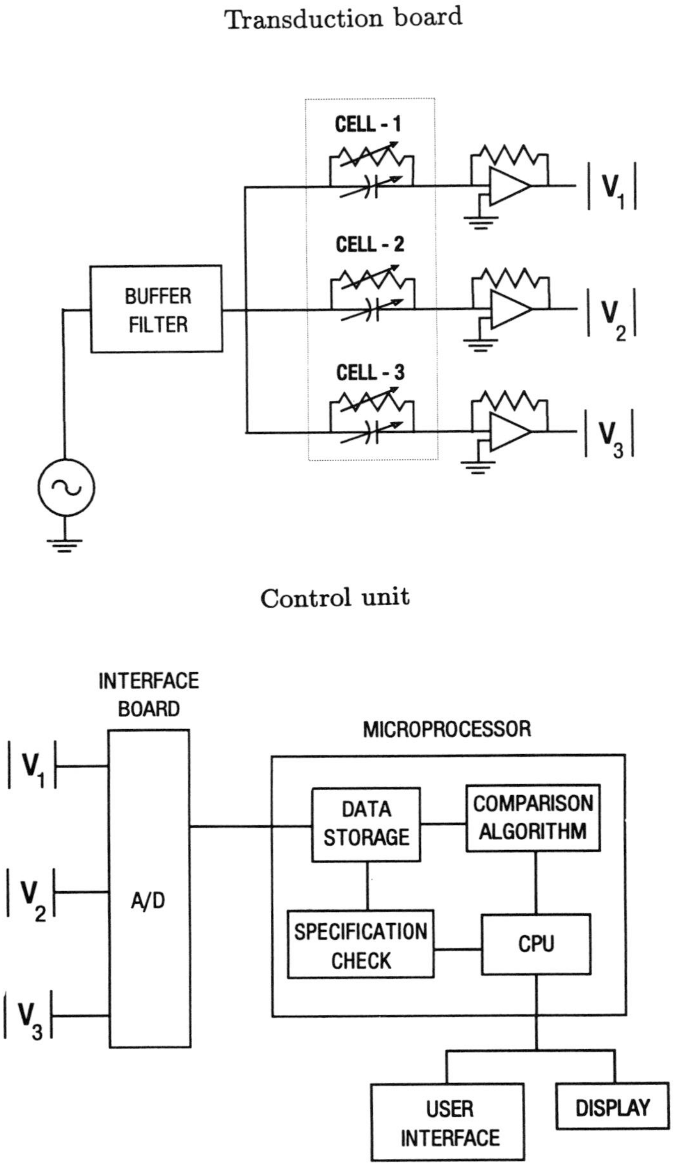

Figure 18.6 A simplified schematic of the electrical measurement system used to (*a*) transduce the signal and (*b*) convert the analog signal to digital format through the central processing unit (CPU).

more than one cartridge at a time. It is possible to obtain up to 16 results per hour with one instrument.

Table 18.1 The average response and the coefficient of variation for a hydrogen peroxide test from a polymer-coated electrode lot.

Number of samples	Response ($nF\,s^{-1}$)	Coefficient of variation
119	1.16	10.0%

18.4.3 Device testing: hydrogen peroxide response and immunoassays

18.4.3.1 Qualification testing and production quality control. Using the hydrogen peroxide–potassium iodide assay (section 18.3), we have tested polymer-coated electrode lots to obtain average response and coefficient of variation (CV, 100% times standard deviation divided by the mean for a large sampling). Results from a coated electrode lot are given in table 18.1; the CV serves to qualify process improvements as well as predict the film quality for the immunoassay. For process control, a low CV implies that all sensors have been prepared in an identical manner. We note that in addition to having achieved CVs of 5–10% across lots in our runs, the cell responses within a given cartridge are uniform.

For the samples under study, the film admittance, Y, has a phase angle at 10 Hz of approximately 86–88° in buffered, aqueous solutions. The large capacitive component is consistent with the neutral state (low conductance) of the film. An injection of 25 μl of 1mM hydrogen peroxide into 250 μl of buffer in a well causes the capacitance and conductance to increase; the phase angle reaches 55° in approximately 10 min. When a catalyst is used in the hydrogen peroxide tests, we observe two effects: (i) the current approaches an asymptotic value for a given hydrogen peroxide concentration and (ii) the current saturates to a limiting value for concentrations above 10 mM hydrogen peroxide. These observations qualitatively support the charge adsorption–saturation model discussed in section 18.3.

To evaluate these data quantitatively for an electrode lot, we have plotted the change in the current during its linear response region as a function of the initial capacitance in figure 18.7. During this interval, the current flow is dominated by the capacitance, so that change in current is proportional to the change in interface capacitance, C_i. Although there is scatter in the data in figure 18.7, there is a linear relationship between the change in capacitance and the initial capacitance, as predicted by equation (18.19).

18.4.3.2 Immunoassay standard curves. Immunosensors were prepared as described. Sample, standards, and conjugate were added sequentially to the respective cells to start the competitive assay. After a 10 min incubation, the solution was discarded and a proprietary buffer was added to each cell. A sufficient concentration of glucose was then added to each cell to drive the enzyme reaction to maximum velocity. As in the qualification tests, the hydrogen

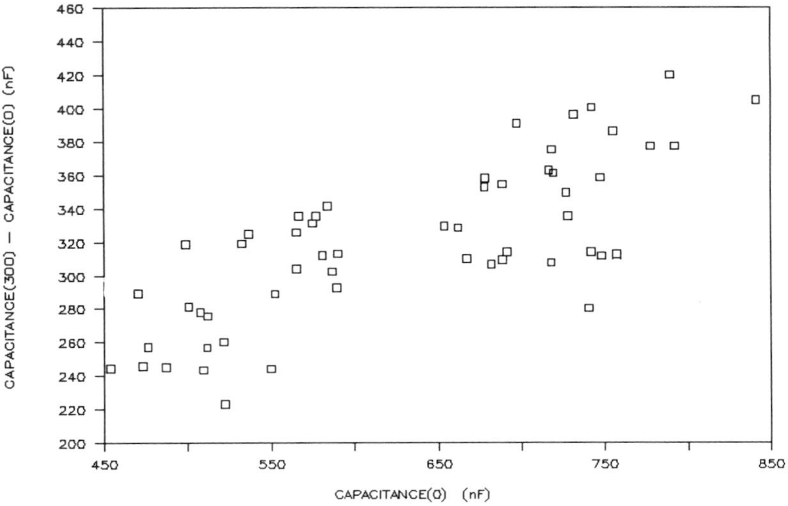

Figure 18.7 A plot of current change versus initial capacitance for the hydrogen peroxide assay during the linear region of response (t = 0–300 s).

peroxide generated during the reaction caused the film to oxidize and the time rate of change in the current through the film was recorded. The maximum response from the assay occurs when there is no analyte present; this is referred to as the B_0 response.

A standard curve was generated by incubating different concentrations of atrazine with a fixed concentration of conjugate for the competitive assay. The response is reported as a percentage of an averaged B_0 for a sample run. Figure 18.8 illustrates a standard curve for an atrazine competitive assay from an electrode lot with a low CV in the qualification test. The curve shows the inverse relation of immunosensor response to conjugate concentration. The data demonstrate the sensitivity of the assay to sub-ppb concentrations of atrazine. While there is scatter in the data at high and low concentration, the standard curve shows a good resolution of data points over the range of interest (around 1–10 ppb). Curves obtained from the atrazine assays are sigmoidal, i.e. a uniformly high response below a minimum detectable level and a uniformly low response above a maximum detectable level, setting the dynamic range of the assay.

Attempts were made to correlate the B_0 response with the initial capacitance for a given antibody coverage. There was, however, a lower correlation between the B_0 response and the initial capacitance than for the hydrogen peroxide assays. This was due, in part, to the lower signal obtained from the immunoassay compared to the hydrogen peroxide assay. Because of the variation in initial redox state and polymer porosity, a relatively large capacitance change compared with the initial capacitance is needed to observe this relationship.

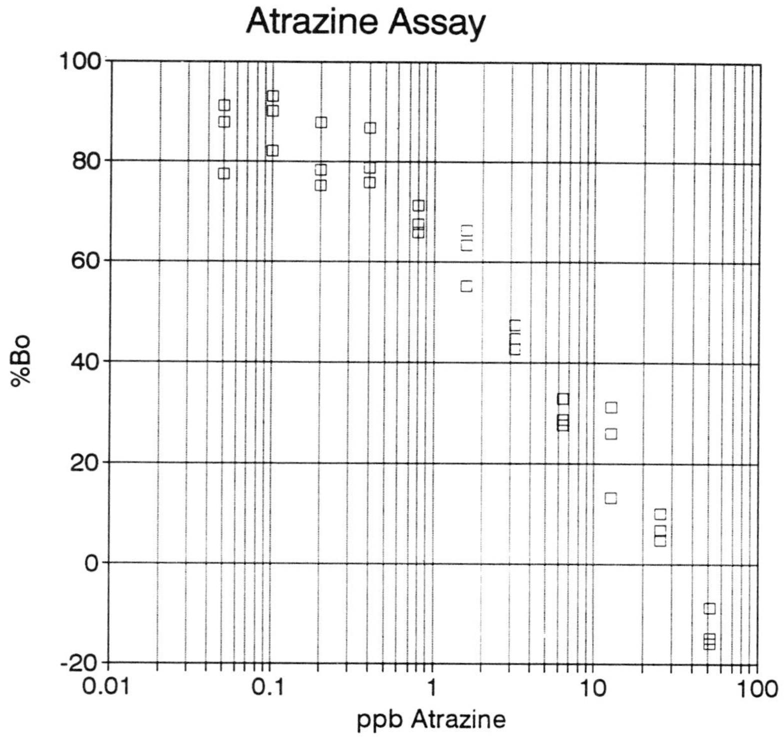

Figure 18.8 A standard curve for an atrazine assay showing sensitivity in the sub-ppb level for atrazine. Note that in the region of interest (1–10 ppb) there is little scatter of data points.

18.5 CONCLUSION

We have described a measurement that uses a conducting polymer film to transduce an immunoassay into an electrical signal. Following the model developed by Tanguy and Feldberg [27, 31], it was shown that the change in capacitance of the film can be related to the chemical oxidation of the polymer by triiodide generated in the assay. A hydrogen peroxide assay that mimics the immunoassay was used to qualify the chemistry and process control during the sensor development. Results from these assays showed that the film response can be related to the initial capacitance of the polymer.

The capacitance measurement technique applied to the antibody-coated sensor in our model system yielded a sensitivity to atrazine concentrations over the ppb to sub-ppb range. The immunoassay standard curve showed that the sensor is capable of the resolution needed for a semiquantitative assay [39]. Scatter in the data is sufficiently small to allow for a positive or negative result within a level of confidence. Several factors contribute to the scatter in the standard curve (e.g. cartridge quality, antibody coverage, non-specific binding) and work is under way to identify and reduce these effects.

Before a generalized measurement principle could be applied to these immunosensor electrodes, however, a reproducible fabrication technology had to be implemented. Manufacturing criteria strongly influenced the sensor design and testing format. For example, to use the three-cell format, a robust, continuous coating technology conducive to intracell comparison was developed. The choice of a continuous coating technology influenced the design of the electrode pattern, the cutting and handling of the coated film, and the film cartridge assembly process. By coordinating the design work with these processes, we now have the capability to produce electrode lots of 10 000 to 100 000 per coating run.

For the intended immunoassay application, a semiquantitative format is used. Sample response in the cartridge is compared to the responses from two standards to determine whether the sample is above (positive) or below (negative) a user-selected level. The total time for the atrazine assay was optimized at 15 min: a 10 min incubation and a 5 min response time. Recent results have demonstrated the efficacy of this format in discriminating positives and negatives around the decision point of 3 ppb atrazine [39].

The measurement system for the immunosensor was designed to accomodate the positive/negative decision in the commercial product. The electronics package includes the measurement circuitry and compatible software (menu-driven instructions and timers) to prompt the user through the assay and store data. Users can interface a personal computer directly to the instrument for downloading data. Upgrades will include multianalyte capability and supporting software as the antibody menu expands. In addition to antibodies developed uniquely for the immunosensor, antibodies developed for our RaPID Assay® kits [40–42] will be adapted for the electrode format. By combining immunoassays with the conducting polymer-based system, a technology has been developed that can be exploited for a variety of field portable screening applications.

REFERENCES

[1] Lundström I 1981 Hydrogen sensitive MOS-structures. Part 1.: Principles and applications *Sensors Actuators* **1** 403–26

[2] Ackelid U *et al* 1986 Ethanol sensitivity of palladium-gate metal–oxide–semiconductor structures *IEEE Electron Device Lett.* **EDL-7** 353–5

[3] Fare T L *et al* 1988 Quasi-static and high frequency C(V)-response of thin platinum metal–oxide–silicon structures to ammonia *Sensors Actuators* **14** 369–86

[4] Spetz A *et al* Hydgrogen and ammonia response of metal–silicon dioxide–silicon structures with thin platinum gates *J. Appl. Phys.* **64** 1274–83

[5] Chen Z *et al* 1990 Humidity sensors with reactively evaporated alumina films as porous dielectrics *Sensors Actuators* B **2** 161–71

[6] Shimizu Y *et al* 1989 The sensing mechanism in a semiconducting humidity sensor with platinum electrodes *J. Electrochem. Soc.* **136** 3868–71
[7] Lukaszewicz J P 1991 Diode-type humidity sensor using perovskite-type oxides operable at room temperature *Sensors Actuators* B **4** 227–32
[8] Anderson R C *et al* 1990 Investigations of porous silicon for vapor sensing *Sensors Actuators* A **23** 835–9
[9] Clayton W A 1993 Improved capacitive moisture sensors *Sensors* **10** 16–22
[10] Ishihara T *et al* 1991 Application of mixed oxide capacitor to the selective carbon dioxide sensor. I. Measurement of carbon dioxide sensing characteristics *J. Electrochem. Soc.* **138** 173–6
[11] Alberti K *et al* 1991 Zeolite coated interdigital capacitors as a new type of gas sensor *Catal. Today* **8** 509–13
[12] Taylor R F *et al* 1991 Antibody- and receptor-based biosensors for detection and process control *Anal. Chim. Acta* **249** 67–70
[13] Cullen D C *et al* 1990 Multianalyte miniature conductance biosensor *Anal. Chim. Acta* **231** 33–40
[14] Gebbert A *et al* 1992 Direct observation of antibody binding. Comparison of capacitance and refractive index measurement *DECHEMA Biotechnol. Conf. 1992* vol 5, part A; *Microbial Principles in Bioprocesses: Cell Culture Technology, Downstream Processing and Recovery* pp 459–62
[15] Stenberg M *et al* 1979 Silicon–silicon dioxide as an electrode for electrical and ellipsometrical measurements of adsorbed organic molecules *J. Colloid Interface Sci.* **72** 255–63
[16] Bresler H S *et al* 1992 Application of capacitive affinity biosensors: HIV antibody and glucose detection *Biosensor Design and Application* ed P R Mathewson and J W Finley (Washington, DC: American Chemical Society) pp 89–104
[17] Keese C R and Giaver I 1994 A biosensor that monitors cell morphology with electrical fields *IEEE Eng. Med. Biol. Mag.* **13** 402–8
[18] Bao J-Z *et al* 1993 Impedance spectroscopy of human erythrocytes: system calibration and nonlinear modelling *IEEE Trans. Biomed. Eng.* **BME-40** 364–78
[19] Jimbo Y *et al* 1993 Simultaneous measurement of intracellular calcium and electrical activity from patterned neural networks in culture *IEEE Trans. Biomed. Eng.* **BME-40** 804–10
[20] Najafi K 1994 Solid-state microsensors for cortical nerve recordings *IEEE Eng. Med. Biol. Mag.* **13** 375–87
[21] Kovacs G T A *et al* 1994 Silicon-substrate microelectrode arrays for parallel recording of neural activity in peripheral and cranial nerves *IEEE Trans. Biomed. Eng.* **BME-41** 567–77
[22] Bryzek J *et al* 1994 Micromachines on the march *IEEE Spectrum* **31** 20–31
[23] Cole J C and Braun D 1994 Applications for a capacitive accelerometer with digital output *Sensor* **11** 26–34
[24] Sandberg R G *et al* 1992 A conductive polymer-based immunosensor for the analysis of pesticide residues *Biosensor Design and Application* ed P R Mathewson and J W Finley (Washington, DC: American Chemical Society) pp 81–8
[25] Fare T L *et al* 1993 Portable biosensors for pesticides and toxic organics *Proc. 3rd Int. USEPA/A&WMA Symp. on Field Screening for Hazardous Wastes and Toxic Chemicals* VIP33 (Pittsburgh, PA: Air and Waste Management Association) pp 721–31
[26] Li Y and Qian R 1993 Stability of conducting polymers from the electrochemical point of view *Synth. Met.* **53** 149–54

[27] Tanguy J *et al* 1991 Poly(3-alkylthiophenes) and poly(4,4′-dialkyl-2,2′-bithiophenes): a comparative study by impedance spectroscopy and cyclic voltammetry *Synth. Met.* **45** 81–105
[28] Fletcher S 1993 Contribution to the theory of conducting-polymer electrode in electrolyte solutions *J. Chem. Soc. Faraday Trans.* **89** 311–20
[29] Kanatzidis M G 1990 Conductive Polymers *Chem. Eng. News* **68**(49) 36–54
[30] Bard A J and Faulkner L R 1980 *Electrochemical Methods: Fundamentals and Applications* (New York: Wiley)
[31] Feldberg S W 1984 Reinterpretation of polypyrrole electrochemistry. Consideration of capacitive currents in redox switching of conducting polymers *J. Am. Chem. Soc.* **1** 4671–4
[32] Glarum S H and Marshall J H 1987 The impedance of poly(aniline) electrode films *J. Electrochem. Soc.* **134** 142–7
[33] Burgmayer P and Murray R W 1984 Ion Gate Electrodes. Polypyrrole as a switchable ion conductor membrane *J. Phys. Chem.* **88** 2515–21
[34] Otero T F and de Larreta E 1988 Conductivity and capacity of polythiophene films: impedance study *J. Electroanal. Chem.* **244** 311–18
[35] Sunde S, Hagen G and Odegård R 1993 Impedance analysis of the electrochemical doping of poly(3-methyl-thiophene) from aqueous nitrate solutions *J. Electroanal. Chem.* **345** 59–82
[36] Zaborsky O R 1973 *Immobilized Enzymes* (Cleveland, OH: Chemical Rubber Company) (see especially pp 69, 87)
[37] Kolthoff I M and Belcher R 1957 *Volumetric Analysis* vol 3 (New York: Interscience) p 282
[38] Doblhofer K and Zhong C 1991 The mechanism of electrochemical charge-transfer reactions on conducting polymer films *Synth. Met.* **41–43** 2865–70
[39] Fare T L *et al* 1994 A commercial immunosensor system incorporating electroconductive polymers *Proc. Am. Chem. Soc., Div. Polym. Mater.: Science and Engineering* vol 71 (Washington, DC: American Chemical Society) pp 649–50
[40] Lawruk T S *et al* 1993 Quantification of cyanazine in water and soil by a magnetic particle-based ELISA *J. Agric. Food Chem.* **41** 747–52
[41] Lawruk T S *et al* 1992 Quantification of alachlor in water by a novel magnetic particle-based ELISA *Bull. Environ. Contam. Toxicol.* **48** 643–50
[42] Rubio F M *et al* 1991 Performance characteristics of a novel magnetic particle-based ELISA for the quantitative analysis of atrazine and related triazines in water samples *Food Agric. Immunol.* **3** 113–25

19

Piezoelectric and surface acoustic wave sensors

Ahmad A Suleiman and George G Guilbault

19.1 INTRODUCTION

The Curie brothers [1] are accredited with being the first to describe the phenomenon of piezoelectricity. However, piezoelectricity remained a scientific curiosity until the war in 1914, when the 'echo method' utilizing piezoelectric plates was developed to explore the ocean bottom. The development of other devices using crystals as control elements followed and eventually the art of crystals became one of the cornerstones of the network of communication systems.

The use of piezoelectric devices as potential chemical sensors was realized when Sauerbrey [2] described the relationship between the resonant frequency of an oscillating piezoelectric crystal and the mass deposited on the crystal surface

$$\Delta F = -2.3 \times 10^6 F^2 \Delta M/A$$

where ΔF is the frequency change due to the metal coating in hertz, F is the frequency of the quartz crystal in megahertz, ΔM is the weight of deposited film in grams, and A is the area of the quartz plate in square centimeters.

The most commonly known oscillator sensors are bulk acoustic wave (BAW) and surface acoustic wave (SAW) devices. The BAW devices operate according to the Sauerbrey principle: that very thin films on AT-cut crystals can be treated as equivalent mass changes of the crystal. The SAW devices can operate either on the Rayleigh wave propagation principle at solid thin-film boundaries [3] or as bulk wave devices [4].

The first application of a piezoelectric sensor was reported by King [5]. During the next two decades, research was directed to developing organic and inorganic coatings for the detection and determination of various toxic substances in the environment and the work place area [6, 7]. Since biologically active materials such as antibodies, enzymes, and antigens are highly specific,

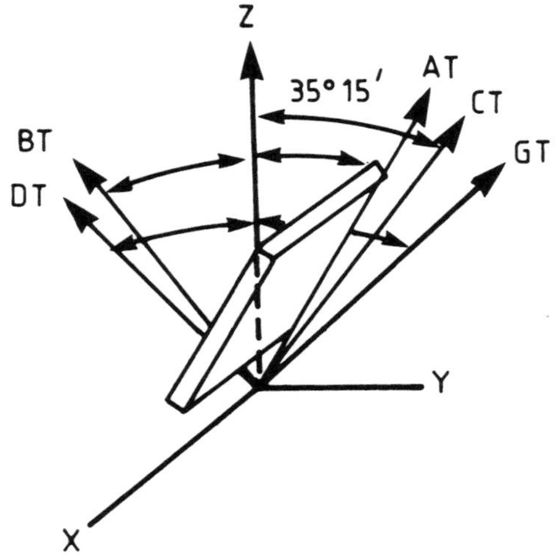

Figure 19.1 The angular orientation of some special cuts of quartz crystals.

their use as active coatings has been exploited leading to a new class of biosensors in liquid and gas phase analysis [8, 9]. The first reported BAW biosensors for the defection of gaseous substrates utilized formaldehyde dehydrogenase in addition to cofactors and a parathion antibody for the detection of formaldehyde and parathion, respectively [10, 11].

19.2 FUNDAMENTALS

19.2.1 Piezoelectric crystals

Alpha quartz (low quartz) is most often used for piezoelectric crystal detectors. It crystallizes at temperatures below 573 °C, and transforms to beta quartz (high quartz) between 573 and 870 °C. Generally, AT- and ST-cut quartz plates are used in BAW and SAW, respectively. The term 'cut' designates the direction of the normal to the major faces. A standard AT cut, for example, describes a plate with thickness in the *y* direction rotated 35° 15′ counterclockwise about the *x* axis (figure 19.1).

The most commonly used crystals in BAW devices are 5, 9, or 10 MHz quartz in the form of 10–16 mm disks that are approximately 0.15 mm thick. Metals are often evaporated directly onto the quartz plates to serve as electrodes. The metal electrodes are 3000–10 000 Å thick and 3–8 mm in diameter and can be made of gold, silver, aluminum, or nickel (figure 19.2). SAW devices, however, are capable of operating at much higher frequencies than the bulk devices and normally crystals of more than 100 MHz resonant frequency are used. Therefore,

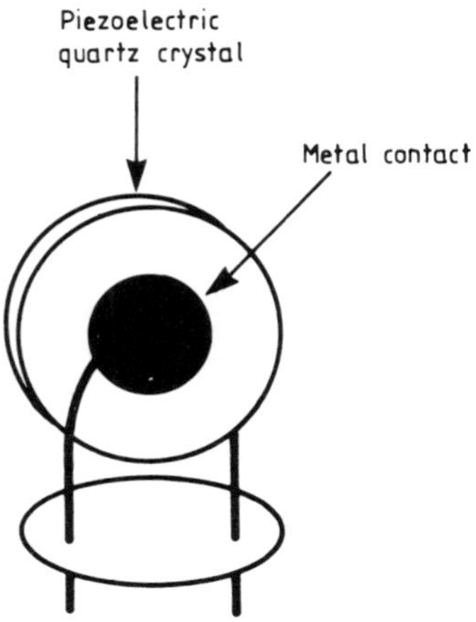

Figure 19.2 A typical piezoelectric crystal.

they potentially can offer greater sensitivity than BAW devices and are easily adaptable to microfabrication.

19.2.2 Measurement of the resonant frequency

When placed in an electronic oscillator circuit, the portion of the quartz wafer located between the electrodes vibrates at its fundamental frequency. The frequency output from the oscillator, which is identical to the resonant frequency of the crystal, can be measured by a frequency counter. Several oscillator circuits are commercially available and they are generally similar to the circuit described in figure 19.3. The choice of the oscillator circuit is essentially influenced by the potential applications. Recent advances in piezoelectric research have shown that quartz crystals can oscillate in solution and several pioneering studies have been reported addressing the theoretical aspects of the oscillating frequency of piezoelectric crystals in solution. Several recently published circuits were briefly evaluated by Barnes [12] and some suggestions were made for improvements. In addition, the author described two new circuits which allow total immersion of the crystal in different media, viscous liquids of viscosity up to 40 cP and buffer and electrolytes of several millimolar ionic strength.

There are generally two types of SAW device, the delay line and the resonator. The design of these devices has been addressed in several publications including papers by Wohltjen and coworkers [3, 13].

Much of the interest has been focused on the type of device that operates as a delay line device which can be configured to monitor either changes in amplitude or velocity. The device consists of two interdigital transducers, an emitter of the SAW, and and a detector of the wave. The resonant frequency

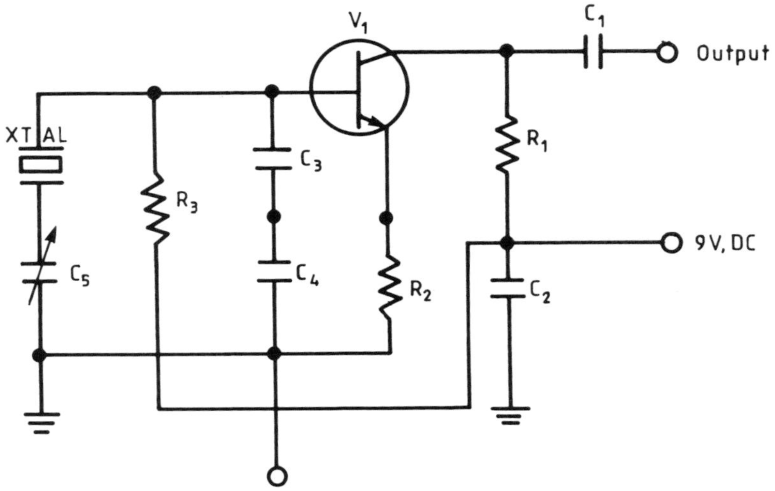

Figure 19.3 A schematic diagram of the oscillator circuit.

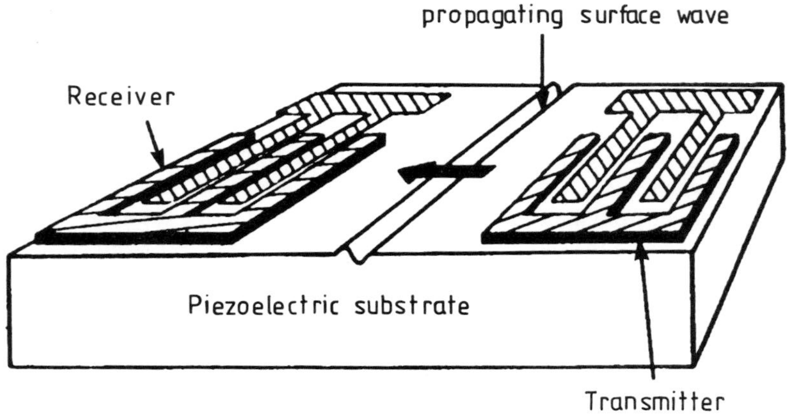

Figure 19.4 A model of a SAW device.

of the device is affected by changes in the velocity of the Rayleigh wave and subsequently can be monitored accurately with a frequency counter. A model of a SAW device is shown in figure 4.

19.2.3 Coating methods

Several techniques have been used for the application of the substrate to piezoelectric crystals. These include dropping, dipping, spraying, spin coating, and most recently using the Langmuir–Blodgett (LB) technique. The most critical factor in the coating process seems to be the ability to reproduce the operation, where the same amount of coating is applied in the same manner and covering the same area.

The rising interest in utilizing biocoatings especially for the detection of analytes in solution has also instigated exploitation of special immobilization methodologies that generally fall into the following categories.

(i) *Immobilization of biocatalyst on the crystal precoated with a suitable material.* This is the most popular method used to construct piezobiosensors. In this procedure, the piezoelectric crystal is precoated with a suitable material to create a thin layer capable of forming hydrophobic and/or covalent bonds with the biocatalyst. To date, several materials have been used to precoat the piezoelectric crystal (table 19.1).
(ii) *Immobilization via entrapment.* A solution containing acrylamide, N, N′-methylene-bis-acrylamide, potassium persulfate, and the biocatalyst is deposited onto the electrode of the piezoelectric crystal. The polymerization is initiated by addition of 3-dimethylaminopropionitrite. The method is limited to the assay of small substrates, which have greater ease of access than large substrates [14].
(iii) *Immobilization via cross-linking.* The biocatalyst is cross-linked with bovine serum albumin via glutaraldehyde. The resulting mixture is then deposited on the crystal's surface. The biocatalyst can also be cross-linked by itself via glutaraldehyde and then deposited onto the piezocrystal.
(iv) *Immobilization by irradiation.* DNA can be grafted to the piezoelectric crystal precoated with a solution of poly(butylmethacrylate) in ethylacetate and dimethoxyphenylacetophenone in acetone by irradiation with 365 nm UV light [15].

In biosensors, a prerequisite for the electrode surface is that it should be chemically stable during the measurement process, and a high density of the biological material must be immobilized onto it. However, overloading of the piezoelectric crystal with the biological material should be avoided, as this could restrict the free interaction between the immobilized material and the macromolecular analyte. This feature is particularly important in piezosensors where the interaction is between antibody and antigen. In addition leakage of the immobilized materials must not occur to any extent during the use of the biosensor, and this layer must also be stable during oscillation. Although the literature provides a plethora of information on the immobilization and the stability of an immobilized layer on the crystal surface, the choice of an appropriate immobilization procedure for a particular application still can not be predicted and must be determined through experimental trial and error.

19.2.4 Instrumental development

The simplest laboratory experimental set-up of a piezoelectric sensor consists of a piezoelectric crystal housed in a chamber and driven by a low-frequency

Table 19.1 Materials used to precoat the piezoelectric crystal.

Material	Procedure
Nyebar C	The crystal is dip coated with a low-surface-energy plastic (Nyebar C 30% solution in 1,3-di(trifluoromethyl)benzene). This coating layer is capable of forming hydrophobic bonds with enzymes, proteins, antibiotics, antigens, etc [28].
Polyethylenimine	The crystal is coated with a methanol solution containing 2–3% polyethylenimine. After treating the precoated crystal with glutaraldehyde, biocatalysts can be immobilized on the precoated crystal via the surface aldehyde groups [29].
APTES	The crystal is coated with γ-aminopropyltriethoxysilane (APTES) solution (2–5% in acetone, dry benzene). Biocatalysts are immobilized on the precoated crystal via glutaraldehyde activation [30].
Protein A	After treatment of the crystal with concentrated hydrochloride, protein A is deposited onto the surface of the crystal. The protein A-coated crystal is capable of reacting with antibodies and other biological materials [31].
Polyethylenimine	The crystal is first coated with a polyethylenimine and avidin 19 solution (2–3%). Avidin is then immobilized on the coated crystal via glutaraldehyde activation. The biotinylated biocatalyst is immobilized on the coated crystal via biotin–avidin interaction since one molecule of avidin binds four molecules of biotin [29].
PMBA	The crystal is coated with a poly(butylmethacrylate) (PBMA) solution (0.1% in ethylacetate) and dimethoxyphenylacetophenone (0.1% in acetone). DNA is grafted to the precoated crystal by irradiation with 365 nm UV light [15].
PHDMAB	The crystal is coated with a 0.06% aqueous solution of poly(2-hydroxy-3-dimethylamino)-1, 4-butane (PHDMAB). This coated crystal is capable of reacting with various antigens, including human gamma globulin (IgG), L-thyroxine, human serum albumin (HSA), and human alpha globulin [32].
Polyacrylamide	The crystal is coated with a thin layer of polyacrylamide gel. Antigens/antibodies are immobilized on the polyacrylamide via glutaraldehyde activation [33].

transistor oscillator. The frequency of the crystal is monitored by a frequency counter modified by a digital to analog converter. The change in the frequency of the vibrating crystal can be monitored directly from the frequency counter or can be translated to a permanent record. Subsequent improvements in the design include incorporating two crystals, a reference and a sampling crystal.

19.3 COMMERCIAL DEVICES

19.3.1 Gas/vapor and aerosol measurements

A coated piezoelectric crystal analyzer sensitive to water vapor as low as 0.1 ppm has been developed by King [5]. This hygrometer has been commercially available since 1964 (DuPont, Wilmington, DE, model 560). The instrument has a very high sensitivity, fast response, and long lifetime, with an accuracy of about 5%.

The qualitative and quantitative performance of a gas chromatography apparatus using a piezoelectric crystal detector was investigated for the determination of effluents vaporizable up to 200° C [16]. It was pointed out that piezoelectric detectors in chromatography can be interfaced to a larger, general purpose, time sharing computer for on-line data acquisition and processing with simple and inexpensive equipment [17]. A prototype piezoelectric crystal mass detector for GPC was compared with a refractometer detector. The performance of the piezoelectric crystal detector indicated that more developments are needed to develop a generally useful detector.

Two commercially available aerosol mass concentration analyzers have been described for the determination of aerosols and particulates [18]. The instruments are both portable and relatively rugged and operate on ordinary commercial power. A portable system utilizing an electrostatic precipitator was described for mass concentration measurements in the range of airborne dust particles smaller than 10 μm [19]. Chuan [20] also described a typical commercial instrument based on the piezoelectric quartz crystal microbalance for the rapid assessment of particular mass concentrations in the atmosphere. The adhesive coating used in the device is non-hygroscopic and non-reactive to the usual concentrations of pollutant gases in the atmosphere, such as CO, SO_2, NO_x, and hydrocarbons.

Detection of various gases in ambient air (sulfur dioxide, ammonia, hydrogen sulfide, hydrogen chloride, phosgene, etc), organophosphorus compounds (such as pesticides), and aromatic hydrocarbons has already been demonstrated with piezoelectric crystals [6, 7]. Portable instruments for monitoring gaseous components in air are commercially available in a unit that has rechargable batteries, a variable-speed pump, a digital display of Δf and concentration, alarm signals that can be preset for certain levels, and interfacing to both a computer and a recorder (Universal Sensors, Inc., PZ Model 105, New Orleans, LA).

An instrument the size of a hand calculator with a digital readout for measuring mercury vapor concentrations in air and/or personal exposure to mercury vapor has been developed and evaluated [21]. Measurements of concentrations below 15 ppb have a precision of 10%, with uncertainties of 5% above this level.

The performance of a portable instrument used as a dosimeter or personal monitor for ozone detection in parts per billion concentrations has been evaluated

[22]. The accuracy of the instrument is comparable to that of results obtained with the portable AID ozone analyzer and Drager detector tubes.

Turnham *et al* [23] designed and developed a working monitor for propylene glycol dinitrate (PGDN). The dual-crystal design, containing a trap for PGDN on the reference crystal, eliminates problems associated with frequency stability and selectivity, utilizing an autozero function. A detection limit of better than 0.05 ppm is possible with this device. It is highly selective for PGDN over all other normal atmospheric contaminants and is insensitive to fluctuations in humidity. The piezobalance aerosol mass monitor model 8510 (TSI, Inc., St Paul, MN) is reported to be capable of accurately detecting concentration changes as small as 0.001 mg m^{-3} on a real time basis.

QCM Research (Laguna Beach, CA) offers several crystal-based sensors including the mark 16m QCM surface film sensor. The sensor has a unique oscillator/mixer hybrid chip that allows operation at temperatures as low as 10 K. In addition, the company offers a computer-operated instrument that can handle and display up to four sensors.

A multigas monitor for anesthesia was reported by Engström of Sweden. The Engström EMMA is an inexpensive, on-line monitor for volatile anesthetic agents and can measure levels of halothane, enflurane, isoflurane, trichloroethylene, and methoxyflurane in any type of breathing circuit breath by breath. It is reported that no clinically significant interference was realized from nitrous oxide and water has only a small effect.

Microsensor Systems, Inc. (Fairfax, VA) offers SAW devices and coatings for organophosphorus compounds, aromatic hydrocarbons, water vapor, and others. A four-sensor array consisting of four 158 MHz dual-delay-line oscillators and a portable gas chromatograph with an SAW detector are also available.

19.4 EMERGING TECHNOLOGY

The future of piezoelectric sensors may be in liquid phase analysis and especially piezoelectric immunosensors where several companies have embarked on developing the adequate circuitry and the technology to achieve reproducible data in liquid media. Universal Sensors, Inc. (New Orleans, LA) and DuPont (Wilmington, DE) are developing BAW sensors utilizing enzyme and antibody coatings as described in table 19.1. BioResearch (Dublin) is developing an SAW device using a piezoelectric polyvinylidene fluoride membrane. The sensor is being developed principally as an immunosensor for application in antibody and antigen assays.

Another instrument has been advertised recently by ELCHEMA (Potsdam, NY). The electrochemical quartz crystal nanobalance system model EQCN-500 is capable of measuring simultaneously the voltamperometric characteristics and mass changes during an electrochemical process. The reported potential applications are numerous and range from simple adsorption studies to drug release.

Recently, amplification schemes analogous to sandwich enzyme-linked immunosorbent assay (ELISA) have been reported [24]. Two approaches using an amplified mass immunosorbent assay (AMISA) concept were described for the detection of adenosine 5′-phosphosulfate (APS) reductase and human chorionic gonadotropin (hCG). APS reductase detection was accomplished by binding APS reductase to an anti-APS reductase antibody immobilized on the crystal, followed by the addition of an anti-APS-alkaline phosphatase reductase conjugate and 5-bromo-4-chloro-3-indolylphosphate (BCIP). The enzymatically amplified deposition of the oxidized dimer of BCIP on the piezoelectric crystal results in a frequency change that corresponds to APS concentration and leads to significant enhancement of the detection limit. A reusable biosensor based on the detection of hCG with a horseradish peroxidase conjugate, in which a sandwich complex is immobilized on a separate nylon membrane positioned in close proximity to a polyvinyl–ferrocene (PV–FC) film coated on the crystal, has also been described. According to this approach, the enzyme peroxidase catalyzes the oxidation of I^- to I_2/I^{3-}, which then reacts with the PV–FC film with concomitant insertion of I^{3-}, resulting in a frequency shift related to the concentration of hCG. The original PV–FC film can be regenerated by electrochemically reducing the PV–FC + I_3^- film. This results in a rapid expulsion of I_3^- from the film rendering the sensor reusable.

In another study, a method for immunoassay of C-reactive protein (CRP) was developed [25]. The method was designated as latex piezoelectric immunoassay (LPEIA). In this method, the frequency shift was observed using antibody bearing latex without any film on the crystal. The proposed mechanism of the frequency change is that the crystal acts as a sensor for viscosity or density changes in the solution due to aggregation of latex particles.

The use of pattern recognition techniques in conjunction with sensor arrays constitutes a promising approach for multicomponent analysis and for improved selectivity [26]. Similarly, a smart sensor system that employs a temperature-controlled array of SAW sensors, automated sample preconcentration, and pattern recognition has been described [27]. The proposed technology seems to offer a satisfactory solution for the problems associated with non-selective adsorption by coating materials.

19.5 CONCLUSION

Generally, most of the piezoelectric crystal detectors described in the literature are laboratory prototypes where study conditions are limited to controlled temperature and humidity levels. One of the most serious limitations is non-selective adsorption of analytes, especially water vapor, which is probably the main reason for the current unacceptable performance of the respective detectors in real applications. The question of selectivity may be eventually resolved by utilizing arrays of sensors in addition to using statistical methods, an effort that has been already pursued by several investigators.

Further advances in sensor design and electronics will generally play a crucial role in expanding the versatility of the piezoelectric sensors, and will facilitate the overcoming of some of the limitations and drawbacks.

The future use of coated piezoelectric devices as immunochemical sensors, even directly in the liquid phase, is very promising and could be considered a very competitive alternative to other types of immunoassay. Only this technique and that of surface plasmon resonance provide labelless methods for the direct study of antigen–antibody reactions, and their analytical possibilities. The devices can be easily automated or combined with flow injection systems, extending their capability of continuous and repeated assays. This raises an exciting possibility of using crystal arrays to assay for different analytes in complex samples with an on-line display of the results.

Despite some shortcomings, many applications should be easily accessible with the piezoelectric immunosensors, especially since they are cost effective and experimentally very simple.

The change in oscillation frequency of immersed crystals in solutions must be better understood. So far, it is known only that the frequency shift depends principally on the density and viscosity of the solution. Further, the resonant frequency also decreases with increasing electrolyte concentration and specific conductivity, largely through their proportionality to solution density. Some researchers report a linear relationship between the frequency shift and the amount of material adsorbed on the surface of the crystal, but others insist that the frequency shift represents no association with the classic microgravimetric signal. Both the elasticity (shear modulus) and the viscosity of the medium adjacent to the crystal electrodes can influence the nature of the frequency response. A quantitative treatment of the influence of viscoelasticity is not yet satisfactory, but some strategies that allow an evaluation of its importance in a given system have been proposed. As the frequency response is not very specific, the interfacial chemistry must be adjusted to give specific and reproducible measurements.

The reproducible immobilization of chemical and biological materials on the crystal electrode must also be further developed. Problems remain in dissociating the bound analyte from the coated crystal because antibody–antigen interactions are very strong and irreversible. In some cases, antigens such as human IgG could by removed from the crystal coated with anti-human IgG by washing the crystal with high-ionic-strength solutions. However, the immobilized antibody is expected to gradually break away from the surface with repeated use, so that sensitivity of a given crystal decreases with age.

Whether or not the reusability of the crystal remains an important issue will depend on the production cost of the crystal and the coating. At the present time, disposable transducers are not yet feasible since each crystal still costs a few dollars and the cost of producing antibodies for some analytes is still prohibitive.

REFERENCES

[1] Curie J and Curie P 1880 Development, par pression, de l'electricite polarise dans les crystaux hemiedries et faces inclines *C. R. Acad. Sci., Paris* **91** 294

[2] Sauerbrey G Z 1959 Verwendung von schwingquarzen zur wagung dunner Schichten zur Mikrowagung *Z. Phys.* **155** 206

[3] Wohltjen H and Dessy R 1979 Surface acoustic wave probe for chemical analysis. I, introduction and instrument description *Anal. Chem.* **51** 1458

[4] Roederer J E and Bastiaans G J 1983 Microgravimetric immunoassay with piezoelectric crystals *Anal. Chem.* **55** 2333

[5] King W H 1964 Piezoelectric sorption detector *Anal. Chem.* **36** 1735

[6] Guilbault G G and Jordan J M 1988 Analytical uses of piezoelectric crystals: a review *CRC* **19** 1

[7] McCallum J J 1989 Piezoelectric devices for mass and chemical measurements: an update *Analyst* **114** 1173

[8] Ngeh-Ngwainbi J, Suleiman A A and Guilbault A A 1990 Piezoelectric crystal biosensors *Biosensors Bioelectron.* **5** 13

[9] Suleiman A A and Guilbault G G 1991 Piezoelectric (PZ) immunosensors and their applications *Anal. Lett.* **24** 1283

[10] Guilbault G G 1982 Determination of formaldehyde with an enzyme coated piezoelectric crystal detector *Anal. Chem.* **55** 1682

[11] Ngeh-Ngwainbi J, Foley P H, Kuan S S and Guilbault G G 1986 Parathion antibodies on piezoelectric crystals *J. Am. Chem. Soc.* **108** 5444

[12] Barnes C 1991 Development of quartz crystal for under-liquid sensing *Sensors Actuators* A **29** 59

[13] Wohltjen H 1984 Mechanism of operation and design considerations for surface acoustic wave device vapor sensors *Sensors Actuators* **5** 307

[14] Lasky S J and Buttry D A 1989 Sensors based on biomolecules immobilized on the piezoelectric quartz crystal microbalance *ACS Symp. Ser.* **403** 237

[15] Fawcett N C, Evans J A, Chien L C and Flowers N 1988 Nucleic acid hybridization detected by piezoelectric resonance *Anal. Lett.* **21** 1099

[16] Karasek F W, Guy P and Hill H H 1976 Chromatographic design and temperature-related characteristics of the piezoelectric detector *J. Chromatogr.* **124** 179

[17] Janghorbani M and Freund H 1973 Application of a piezoelectric crystal as a partition detector *Anal. Chem.* **45** 325

[18] Daley P S and Lundgren D A 1975 The performance of piezoelectric crystal sensors used to determine aerosol mass concentrations *Am. Ind. Hyg. Assoc. J.* **36** 518

[19] Sem G J and Tsurubayashi K 1975 A new mass sensor for respirable dust measurement *Am. Ind. Hyg. Assoc. J.* **36** 791

[20] Chuan R L 1975 Rapid assessment of particulate mass concentration in the atmosphere with a piezoelectric instrument *Adv. Instrum.* **30** 620

[21] Scheide E P and Taylor J K 1974 Piezoelectric sensor for mercury in air *Environ. Sci. Technol.* **8** 1097

[22] Fog H M and Rietz B 1985 Piezoelectric crystal detector for the monitoring of ozone in working environment *Anal. Chem.* **57** 634

[23] Turnham B D, Yee L K and Luoma G A 1985 Coated piezoelectric quartz crystal

monitor for determination of propylene clycoldinitrate vapor levels *Anal. Chem.* **57** 2120

[24] Ebersole R C and Ward M D 1988 Amplified mass immunosorbent assay with a quartz crystal microbalance *J. Am. Chem. Soc.* **110** 8623

[25] Kurosawa S, Tawara E, Kamo N, Ohta F and Hosokawa T 1990 Latex piezoelectric immunoassay: detection of agglutination of antibody-bearing latex using a piezxoelectric quartz crystal *Chem. Pharm. Bull. (Japan)* **38** 117

[26] Carey W P, Beebe K R, Kowalski B R, Illman D L and Hirschfeld T 1986 Selection of adsorbates for chemical sensor arrays by pattern recognition *Anal. Chem.* **58** 149

[27] Grate J W, Rose-Pehrsson S L, Venezky D L, Klusty M and Wohltjen H 1993 Smart sensor system for trace organophosphorus and organosulfur vapor detection employing a temperature-controlled array of surface acoustic wave sensors, automated sample preconcentration and pattern recognition *Anal. Chem.* **65** 1868

[28] Shons A, Dorman F and Najarian J 1972 An immunospecific microbalance *J. Biomed. Mater. Res.* **6** 565

[29] Prusak-Sochaczewski E, Luong J H T and Guilbault G G 1990 Development of a piezoelectric immunosensor for the detection of *Salmonella typhimurium* *Enzymol. Microb. Technol.* **12** 173

[30] Muramatsu H, Kajiwara K, Tamiya E and Karube I 1986 Piezoelectric immunosensor for the detection of *Candida albicans* microbes *Anal. Chim. Acta* **188** 257

[31] Davis K A and Leary T R 1989 Continuous liquid phase piezoelectric biosensor for kinetic immunoassays *Anal. Chem.* **61** 1227

[32] Oliveira R J and Silver S F 1980 Immunoassay for antigens *US Patent* 4 242 096

[33] Thompson M, Arthur C L and Dhaliwal G K 1986 Liquid-phase piezoelectric and acoustic transmission studies of interfacial immunochemistry *Anal. Chem.* **58** 1206

20

Thermistor-based biosensors

Bengt Danielsson and Bo Mattiasson

20.1 INTRODUCTION

In recent years biosensors have attracted a considerably increased interest due to an extended demand for sensors suitable for specific monitoring and control in biotechnology, especially of continuous processes. There is also a rapidly growing need for sensors in the biomedical field, for decentralized clinical chemistry, and for *in vivo* monitoring of, in particular, glucose in attempts to develop systems for diabetes control. In these fields heat-sensitive biosensors, such as the enzyme thermistor (ET), have definite advantages. Since biological processes generally are more or less exothermic, the combination of specific biocatalysts with a heat or temperature sensor leads to a biosensor with general applicability, which can be used for the determination of a particular compound present in a complex mixture, irrespective of the optical and electrochemical properties of the sample. In addition, the transducer, usually a thermistor, is virtually free from drift and fouling since it does not need to be mounted in direct contact with the sample or any other fluid.

Many different approaches to simplified calorimetric devices for use with immobilized enzymes were proposed in the 1970s [1]. Some of these were soon abandoned, while others, such as the so-called enzyme thermistor [2], have continued to develop. However, more widespread use or commercialization of thermistor-based sensors has to date not been realized in spite of their attractive features. This situation may change when the recent trend of miniaturization of calorimetric devices has become fully developed [3–6].

There are numerous ways to exploit the calorimetric detection principle in combination with biological materials, not only in metabolite assays, but also for analyses of proteins and other macromolecules as well as for whole cells. These include metabolite determination, bioprocess monitoring, measurements of enzymic activities in separation procedures, determinations in organic solvents, and miniaturized thermal biosensors.

20.2 INSTRUMENTATION

The thermistor-based calorimeter used in most of the studies described herein, with the exception of the studies on miniaturized systems, consists of a thermostated thick-walled aluminum jacket, 80 mm in diameter and 250 mm long, containing a cylindrical aluminum heat sink with heat exchangers and two column positions [2] (figure 20.1). The two columns can be used either with different enzymes for two different assays or with one column acting as a reference column (split flow) [7]. The columns are attached at the ends of the thermistor probes and are readily exchangeable. The thermistors are connected to a Wheatstone bridge with a maximum sensitivity of 100 mV m°C^{-1}. Thermistors are heat-sensitive resistive components with very high negative temperature coefficient, of the order of −4% °C^{-1}. Commonly used full-scale sensitivities are in the 10–50 m°C range permitting determinations in the 0.01–100 mM range. A large excess of enzyme (10–100 units or more per column, column volume up to 1 ml) bound to a mechanically stable, highly porous support, such as controlled pore glass (CPG) or Eupergit C, ensures an excellent operational stability. A continuous flow of 0.5–2 ml min^{-1} through the system is maintained with a peristaltic pump. Samples can be introduced with a chromatographic valve (0.1–1 ml or smaller). Such small sample volumes will not lead to a thermal steady state, but result in a temperature peak that is proportional to the substrate concentration. Thermograms are normally evaluated by peak height determination which is simple to perform either manually from the recorder diagram or automatically by an integrator or computer [8]. The peak area and the ascending slope of the peak are also linearly related to the substrate concentration [9]. An example of thermograms is given in the chapter on FIA (chapter 22, figure 22.4).

A number of instruments of the type described above has been built in the workshops at the Chemical Center at our university and sold on a non-profit basis to various laboratories over the world. This has resulted in many interesting contacts and feedback of useful ideas. Efforts to lift up the sale to a more commercial level have not yet been successful.

20.3 APPLICATIONS

20.3.1 Determination of metabolites

A large number of thermistor-based biosensor assays using immobilized enzyme reactors have been proposed for use in biotechnology, clinical chemistry, and food analysis (table 20.1). In the following examples the concentration ranges given have in general been obtained with 0.5 ml samples at a flow rate of 1 ml min^{-1}. The sensitivity can be adapted to higher concentrations by dilution and use of smaller sample volume. Oxidases generally offer higher sensitivity

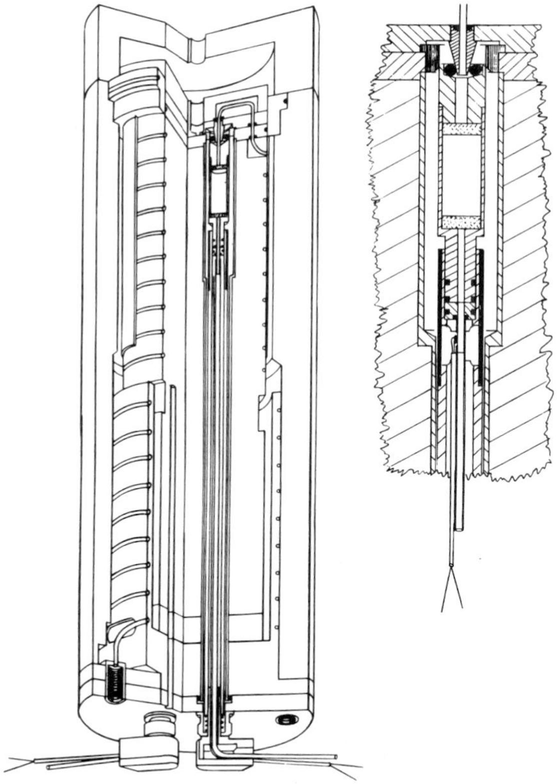

Figure 20.1 A cross-section of an enzyme thermistor calorimeter with an aluminum constant-temperature jacket and and aluminum heat sink. The enlargement shows the attachment of a column and the transducer arrangement (a thermistor fixed with heat conducting epoxy to a gold tube).

than dehydrogenases due to higher heat of reaction ($-\Delta H = 75$–100 kJ mol^{-1}) and have no extra cofactor requirement. Coimmobilization of oxidases with catalase has three additive effects: doubling the total reaction heat by adding the enthalpy change of the catalase reaction (-100 kJ mol^{-1}); removal of the hydrogen peroxide formed in the oxidase reaction avoiding protein damage by the hydrogen peroxide; and 50% improvement of the use of the oxygen available which extends the linear range to about 1 mM. The low solubility of oxygen in aqueous solutions is a serious drawback for use of oxidases. A solution to this problem, although not ideal, is to use an electron acceptor with higher solubility, such as benzoquinone [17]. Hydrolytic enzymes, such as disaccharidases, are usually associated with low enthalpy changes and have to be supplemented with secondary enzymes for practically useful assays. Cellobiose, for instance, can be determined with β-glucosidase in combination with glucose oxidase and

Table 20.1 Linear concentration ranges of substances measured with enthalpimetric sensors using immobilized enzymes.

Analyte	Enzyme(s) used	Linear range (mM)	Reference
Ascorbic acid	Ascorbate oxidase	0.01–0.6	[10]
ATP (or ADP)	Pyruvate kinase + hexokinase	10 nM–[a]	[11]
Cellobiose	β-glucosidase + glucose oxidase/catalase	0.05–5	[12]
Cephalosporins	Cephalosporinase (β-lactamase)	0.005–10	[13]
Creatinine	Creatinine iminohydrolase	0.01–10	[13]
Ethanol	Alcohol oxidase	0.005–1	[14]
Glucose	Hexokinase	0.5–25	[15]
Glucose	Glucose oxidase/catalase	0.001–0.8	[16]
		75[b]	[17]
L-lactate	Lactate-2-mono-oxygenase	0.005–2	[16]
L-lactate	Lactate oxidase/catalase	0.002–1	[16, 18]
L-lactate (or pyruvate)	Lactate oxidase/catalase + lactate dehydrogenase	10 nM–[a]	[18]
Oxalate	Oxalate oxidase	0.005–0.5	[19]
Penicillin	β-lactamase	0.005–200	[7, 8]
Sucrose	Invertase	0.05–100	[10]
Urea	Urease	0.005–200	[15, 20]

[a] With substrate recycling.
[b] With benzoquinone as electron acceptor.

catalase.

Another common way to increase the sensitivity of calorimetric measurements is to use buffers with high protonation enthalpy (such as Tris buffer), if a proteolytic reaction is connected with the enzymic reaction [21]. Substrate and coenzyme recycling is another way of increasing the sensitivity, in favorable cases up to several thousand-fold. As an example 5000-fold amplification was observed using coimmobilized lactate oxidase (oxidizing lactate to pyruvate), lactate dehydrogenase (reducing pyruvate to lactate), and catalase [18]. Lactate (or pyruvate) concentrations as low as 10 nM could be determined with this arrangement. Similar sensitivities for ATP (alternatively ADP) were obtained with the enzyme couple pyruvate kinase and hexokinase [11]. Highly sensitive detection of ATP/ADP, the same as with bioluminescence, can be accomplished by coupling the two cycles so that the pyruvate formed in the pyruvate kinase/hexokinase is recycled in the LDH/LOD cycle [13]. The practicality of this approach is unfortunately limited since it is directly influenced by the actual activity of all enzymes involved. This is in contrast with direct assays using immobilized enzyme reactors where the sensitivity is virtually unchanged as long as there is excess enzyme activity. The enzymes involved in the LDH/LOD system are, however, stable enough to make it practically useful, for instance as a detecting mechanism in enzyme immunoassays.

The most common enzyme support in our work has been propylamino-derivatized CPG (controlled pore glass from Corning) with a pore size in the range of 50–200 nm and a particle size up to 80 mesh (0.18 mm), loaded with a large excess of enzyme, often 100 units or more, immobilized with glutardialdehyde [16]. In more recent work we have been using a spherical CPG from Schuller GmbH (Steinach, Germany) with a particle size in the range of 125–140 μm and a pore size of 50 nm. In contrast to crushed CPG this beaded support material produces enzyme columns with remarkable resistance to clogging by particles in the sample. Even whole-blood samples can be used [22]. CPG offers high binding capacity, good mechanical and chemical as well as microbial stability, and relatively simple coupling procedures, but Eupergit C, (oxirane acrylic beads, from Röhm Pharma, Weiterstadt, Germany) and VA-Epoxy BIOSYNTH (Riedel-de Haën, Seelze, Germany) are good alternative support materials. The major limiting factor for column life is usually mechanical obstruction. If, however, the solutions used, as well as the samples, are filtered through at least a 1–5 μm filter, and if microbial growth in the solutions and the flow lines is prevented, good operational stability with unchanged performance for large series of samples (thousands) can be obtained and the column may be functional for several months.

(i) *Alcohols* can be measured with alcohol oxidase (EC 1.1.3.13) from *Candida boidinii* or *Pichia pastoris*. The latter enzyme has higher specific activity and a somewhat different substrate specificity. Coimmobilization with catalase increases the stability of the enzyme column to several months with an operating range of 0.005–1 mM (0.5 ml samples) using 0.1 M sodium phosphate, pH 7.0, as buffer. This assay is useful for the determination of ethanol in samples from beverages and blood and for monitoring of fermentations [14].

(ii) *Cellobiose.* As already mentioned, the heat produced by the hydrolysis of cellobiose with β-glucosidase is too low to give sufficient sensitivity. By measuring the glucose formed in a precolumn containing β-glucosidase with a glucose oxidase/catalase-loaded enzyme thermistor a typical operating range of about 0.05–5 mM can be obtained [12].

(iii) *Cholesterol and cholesterol esters.* Cholesterol has been determined in 0.16 M phosphate buffer, pH 6.5, containing 12% (v/v) ethanol and 8% (v/v) Triton X-100 using cholesterol oxidase (EC 1.1.3.6) from *Nocardia erythropolis*. Cholesterol esters can be measured by including a precolumn with cholesterol esterase (EC 3.1.1.13). The measuring ranges are adequate for clinical use [23].

(iv) *Glucose.* This is one of the most used assays based on calorimetry for various bioanalytical applications. It is usually carried out with glucose oxidase coimmobilized with catalase as mentioned above [16]. This procedure provides high sensitivity and specificity, it has no cofactor requirement, and the enzyme columns are very stable. Alternatively, the enzyme hexokinase can be used [15]. The enzyme, however, requires the cofactor ATP, but on the other hand a linear range of up to 25 mM can be obtained.

(v) *Hexokinase* can also be used in an indirect assay for *ATP*, if the sample solution contains an excess of glucose. Micromolar sensitivity can be obtained by this technique. In an analogous way *NADH* can be measured with the same sensitivity using a lactate dehydrogenase column and excess of pyruvate.

(vi) *L-lactate* can be determined down to micromolar concentrations with two different enzyme systems: the lactate-2-mono-oxygenase (EC 1.13.12.4) from *Mycobacterium smegmatis* and the lactate oxidase from *Pediococcus pseudomonas* (EC 1.1.3.2) together with catalase [16]. The latter enzyme is currently preferred because of lower price, but the monooxygenase could be interesting for removal and simultaneous determination of lactate in combination with the previously described recycling arrangement for lactate/pyruvate [18]. Since the end product of the monooxygenase reaction is acetate and not pyruvate, both metabolites could be determined in the same sample.

(vii) *Oxalate.* Oxalate was measured with oxalate oxidase (EC 1.2.3.4) from barley seedlings. A linear concentration range of 0.005–0.5 mM was observed in 0.1 M sodium citrate buffer, pH 3.5, containing 2 mM EDTA and 0.8 mM 8-hydroxyquinoline [18]. The assay was found to be suitable for the determination of oxalate in urine, beverages, and food samples. Urine samples had to be diluted 10-fold and passed through a C_{18} cartridge to remove interfering substances.

(viii) *Penicillins.* The procedures designed for the assay of β-lactams (for instance penicillin G and V) using β-lactamases, such as penicillinase type I from *Bacillus cereus* (EC 3.5.2.6) have been particularly successful [8]. The useful linear range is about 0.005–200 mM. Several industrial applications have been developed using both discrete samples and continuous monitoring on pilot plant and production scale fermentors. Alternatively, the more specific penicillin amidase (EC 3.5.1.11) can be used, especially in fermentation broths [7]. The sensitivity is, however, lower although sufficient for process monitoring. In both cases the enzyme columns are very stable and can be used for several months or for thousands of samples provided they are sterilely filtered.

(ix) *Sucrose.* In contrast to most other disaccharide splitting enzymes, invertase (EC 3.2.1.26) produces enough heat to allow direct determinations of sucrose in the range of 0.05–100 mM [10]. An important advantage of this procedure, compared to other biosensor assays, is that it is not disturbed by the presence of glucose. Invertase columns are extremely stable and useful in food and bioprocess analysis.

(x) *Triglycerides.* Practical routine methods for the determination of all the main blood lipid classes, cholesterol and cholesterol esters, phospholipids, and triglycerides using thermistor-based biosensors have been proposed [23]. Thus, triglycerides have been determined with lipoprotein lipase (EC 3.1.1.34) immobilized on CPG with a pore size of 2000 Å. The assay buffer was 0.1 M Tris buffer, pH 8.0, containing 0.5% Triton X-100. The linear response was 0.05–10 mM for tributyrin and 0.1–5 mM for triolein [24].

(xi) *Urea.* Urease gives a linear range of at least 0.01–200 mM and offers a clinically useful assay that is independent of the ammonium concentration in the sample [15]. Urease is very sensitive to inhibition by heavy metals, a fact that has been exploited in the design of a reversible procedure for heavy-metal determination. Addition of 1mM EDTA and 1 mM reduced glutathione to the buffer, on the other hand, protects the urease, leading to a very stable enzyme column. Acid urease from *Lactobacillus fermentum* has lower pH optimum and somewhat different properties than Jack bean urease and has been studied in reactors capable of removing urea from alcoholic beverages. This has attracted some interest, especially in Japan, since urea and ethanol upon standing or heating form ethylcarbamate, a carcinogenic compound. In this context acid urease has been shown to be suitable for urea determination [20].

Other metabolites that have been measured with thermistor-based biosensors include ascorbic acid, cephalosporins, creatinine, galactose, hydrogen peroxide, lactose, malate, phospholipids, uric acid, xanthine, and hypoxanthine.

The lifetime of an enzyme column depends to a large extent on the enzyme and the nature of the sample. With stable enzymes, such as invertase and glucose oxidase, and clean samples the column may last for several thousands of samples before a change in performance can be noticed. As mentioned above a large excess of enzyme is normally applied which results in constant response for a given concentration until maybe 90% of the original activity is lost. For a long column life crude samples, such as fermentation broth, milk, or blood, should be dialyzed or at least microfiltered (see subsections 20.3.2 and 20.3.8). It is important that the sample is free from microorganisms that may otherwise be trapped in the flow system and consume the analyte or give a non-specific contribution to the heat developed in the enzyme column. It was shown that β-lactamase columns used for determination of penicillin could last for well over 1000 samples if the broth samples were centrifuged and filtered (2 μm) before analysis [8]. If possible the enzyme should also be protected from inhibitors in the buffer or in the sample. Loss of urease activity is clearly prevented by addition of EDTA and reduced glutathione or DTT to protect from heavy metals and oxidation [16]. Coimmobilization of catalase with oxidases prolongs the column lifetime considerably by removing the hydrogen peroxide formed in the oxidase reaction [14]. Approximately doubled heat production and increased linearity are additional bonus effects.

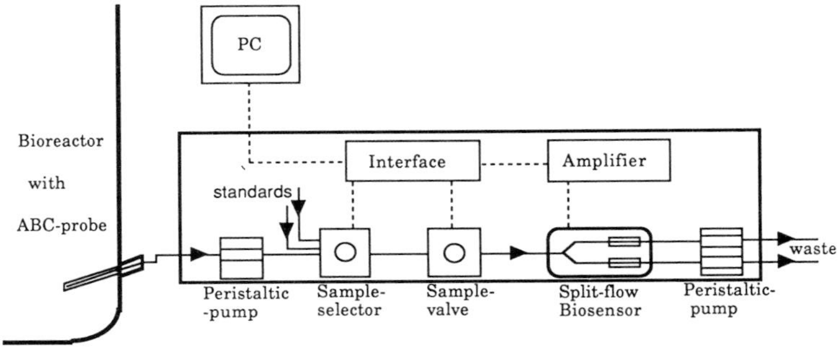

Figure 20.2 A schematic diagram of the set-up using a filtration probe for continuous withdrawal of sample from a fermentor. The filtered sample is led to the FIA system with the heat-sensitive split-flow biosensor. A personal computer was used for controlling the system and for calculation of concentrations. (Reproduced from [25] with permission.)

20.3.2 On-line monitoring of bioprocesses

Implementation of biosensors in the process environment assumes that the technical difficulties caused by high humidity, large temperature variations and ambient temperatures up to 40 °C, water and steam release, and continuous vibrations can be overcome. The complex composition of the fermentation media and high and variable concentrations of various components are additional problems.

For on-line monitoring of bioprocesses using thermistor-based sensors, the equipment is assembled inside a steel cabinet flushed with cool, filtered air to keep the temperature sufficiently constant. The enzyme thermistor is automated and equipped with a pneumatic sampling valve and a sample selector (figure 20.2). This instrumentation has been tested in penicillin fermentations at the fermentation pilot plant and the production plant at Novo-Nordisk A/S (Bagsvaerd and Kalundborg, Denmark). A sample stream of 0.5 ml min^{-1} was taken from a tangential flow filtration unit with a 0.1 μm microfilter placed in a sterile loop connected with the fermentor. In later studies sampling used a polypropylene hollow fiber filtration probe (Advanced Biotechnology Corp., Puchheim, Germany). A 0.1 ml sample was injected every 10–30 min over a time period of 1–2 weeks in the evaluation of the system. The flow through the ET unit (0.9 ml min^{-1}) was equally split between the enzyme column (β-lactamase (EC 3.5.2.6) bound to CPG) and an inactive reference column containing immobilized bovine serum albumin. The split-flow technique was considerably improved by a newly designed flow divider [25]. The column was protected against microbial growth by adding 1 mM sodium azide to the buffer solution.

Penicillin V could be measured during the entire fermentation run with the same enzyme column without serious problems in spite of rapid temperature variations between 20 and 40 °C, high humidity, and vibrations. The linear

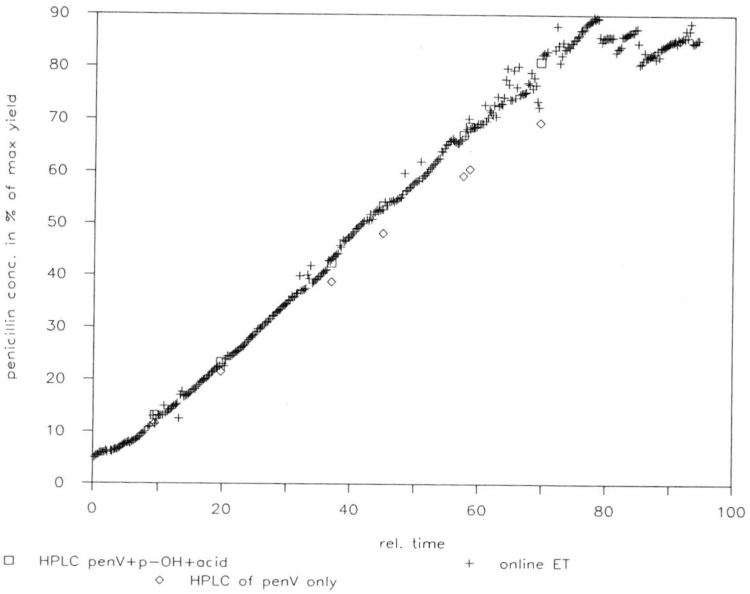

Figure 20.3 On-line monitoring of the penicillin V production in a 160 m^3 bioreactor over 1–2 weeks using a thermistor-based biosensor equipped with a column containing penicillin V acylase. +, on-line biosensor values; □, penicillin V, penicilloic acid, and *p*-hydroxypenicillin measured off-line by HPLC; ◇, off-line values of penicillin V measured by HPLC. (From [25] with permission.)

concentration range of penicillin V can be as large as 0.05–500 mM for 0.1 ml samples. The reference column efficiently compensates for non-specific heat effects. The results from one experiment are displayed in figure 20.3. The concentration determined by the ET was constantly 5–10% larger than the concentration of only penicillin V as determined by HPLC, since β-lactamase hydrolyzes 6-APA and *p*-hydroxypenicillin in addition to penicillin V. Isopenicillin N and penicilloic acid are, however, not detected [25]. A more specific assay for penicillin V can be obtained with penicillin amidase (EC 3.5.1.11). Presently, this assay is preferred, since it can also be used to follow the industrial conversion of penicillin to 6-APA. The amidase reaction is, however, less exothermic [25].

In addition, measurements have been performed on penicillin and *Saccharomyces* fermentations using alcohol oxidase for ethanol, glucose oxidase for glucose, and lactate oxidase for lactate [7]. In all these cases catalase was coimmobilized to increase sensitivity and linear range. All assays as well as the total system worked fully satisfactorily. Larger variations in the concentration registered could occasionally be seen, especially at the end of the fermentations when the viscosity of the broth was high due to very high cell mass, caused by improper function of the filtration unit. The general impression is, however, that the ABC filtration probe works better than the tangential flow unit at higher viscosities in smaller fermentors as well as in larger, production scale fermentors.

20.3.3 Characterization of immobilized biocatalysts

The usefulness of thermistor-based biosensors for the characterization of an immobilized enzyme (invertase) was demonstrated in a study in which the kinetic constants were directly determined without the need for postcolumn analysis [26]. An extension of this work allowed for the direct determination of the catalytic activity of immobilized cells as well [27]. *Trigonopsis variabilis* strains selected by mutagenesis for high cephalosporin transforming activity were used in a model system in which the yeast cells were immobilized by cross-linking with homobifunctional reagents or by physical entrapment in gels. The thermometric signal arising from the activity of one specific dominating enzymic step in the cells can be identified by comparative HPLC analysis of the reaction mixture. The cephalosporin transforming activity of D-amino acid oxidase isolated from selected yeast strains and immobilized by gel entrapment was identified in the same way. The thermometric signal was found to be proportional to the number of cells as well as to the amount of enzyme (DAAO) immobilized in the ET minicolumn.

Furthermore, the thermometric signal associated with cephalosporin transforming activity of both the immobilized cells and the isolated DAAO was found to be proportional to the height of the glutaryl 7-aminocephalosporanic acid peak in the HPLC chromatogram. The best correlation was obtained when the ET was operated at a flow rate of about 1 ml min^{-1}. The $K_{m(app)}$ values correlated in all cases very well with data computed for the (differential) ET column [27].

20.3.4 Recycling-amplified TELISA

In TELISA (thermometric enzyme-linked immunosorbent assay) [28] the ET column contains an immunosorbent. In a competitive assay mode, the sample is mixed with enzyme-labeled antigen and the concentration of bound antigen is determined by introduction of a substrate pulse, whereafter the column is regenerated by a pulse of glycine at low pH. The whole cycle takes less than 13 min. The sensitivity is adequate for the determination of hormones, antibodies, and other biomolecules produced by fermentation [29].

The use of alkaline phosphatase as an enzyme label allows enhancement of the sensitivity by using phosphoenolpyruvate as substrate and the utilization of a separate detection column in the ET unit for the determination of the product (pyruvate) by substrate recycling. This is accomplished by using the substrate recycling system described above [18] comprising the coimmobilized enzymes lactate dehydrogenase (reduces pyruvate to lactate under the consumption of

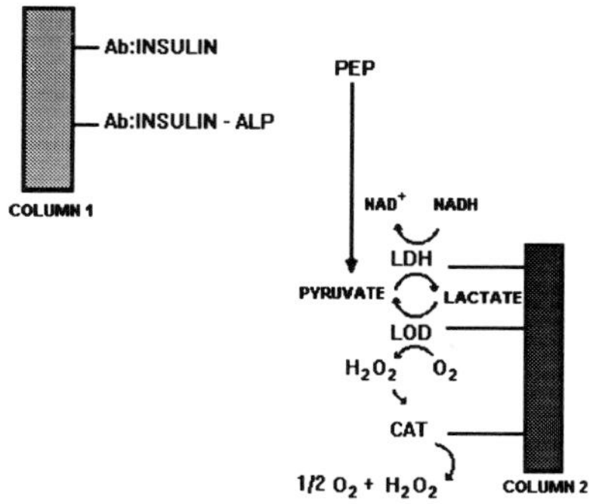

Figure 20.4 A principle scheme for a TELISA with enzymatic substrate recycling detection. Column 1 contains anti-insulin antibody (Ab) bound to Sepharose. Column 2 contains the three enzymes lactate dehydrogenase (LDH), lactate oxidase (LOD), and catalase (CAT) coimmobilized on CPG. PEP is phosphoenolpyruvate.

NADH), lactate oxidase (oxidizes lactate to pyruvate), and catalase (figure 20.4). In addition, genetically engineered enzyme conjugates have been used in immunoassays. Thus a human proinsulin–*E. coli* alkaline phosphatase conjugate was used for the determination of insulin or proinsulin. Concentrations lower than 1 μg ml^{-1} could be determined in less than 15 min [30].

20.3.5 Measurements in organic solvents

Much interest has been paid in recent years to the possibility of running enzymic reactions in organic solvents. In this context it would be valuable to be analyze reaction mixtures directly in organic media using biosensors. Thermistor-based biosensors, such as the ET, are of special interest since the temperature response depends on the heat capacity of the system and the specific heat is up to three times lower in some organic solvents than in water. In addition the solubility of some enzyme substrates (cholesterol and triglycerides for instance) is higher in organic solvents than in water. It has been possible to design potentially useful procedures for enzyme analysis in organic solvents, especially since the enzymes involved become stabilized by the immobilization. In cases where the enzymic activity is lost after some time, it is often possible to restore it fully by treatment with aqueous buffer. Since the enthalpy change may be different in organic solvents or in solvent–water mixtures and in pure buffer the temperature response is difficult to predict. A comparison was made of the temperature responses obtained for tributyrin in a buffer–detergent system and in cyclohexane with lipoprotein lipase immobilized on Celite. In the latter case the response was about 2.5 times higher (as would be expected from the specific heats) and linear up to higher concentrations. In other experiments, the increase in sensitivity was found to be much higher [31]. Thus, with peroxidase, an enzyme that seems to endure organic solvents particularly well, 45 times higher reaction heat was noted in toluene than in water. Addition of small amounts (<5%) of diethyl ether to the toluene increased the reaction heat further. The reason for this phenomenon has not been explored, but it was also found that, for instance, lower alcohols at lower concentrations (a few percent) in aqueous buffer increased the heat of the glucose oxidase and β-lactamase reactions significantly. Increased sensitivity was also found when the glucose reactor was operated in a flow system with photometric detection under the same conditions [31].

A practical demonstration of the usefulness of calorimetric sensors for work in different media was made in a study on immobilized α-chymotrypsin which was used for hydrolysis of peptide bonds in 0.05 M Tris-HCl, pH 7.8, containing 10% DMF and for syntheses of peptide bonds in 50% DMF + 50% 0.1 M Na borate, pH 10.0. With the α-chymotrypsin immobilized in the enzyme thermistor column both reactions could be followed, the hydrolysis giving an exothermic response while the synthetic route was endothermic [32].

20.3.6 Thermometric monitoring of soluble enzymes

On-line monitoring of chromatographic enzyme separations is usually restricted to registration of the UV absorption and determination of the pH or conductivity of the mobile phase. The normal procedure is to assay for the component of interest fractionwise collecting the fractions with the highest concentrations for further purification or concentration. This is time and labor consuming and may be damaging to labile components. Specific monitoring of proteins, for instance direct identification of a special enzyme, would greatly facilitate and speed up purification, since it would allow for an eluted enzyme fraction to be taken directly (on line) to a subsequent purification step. It has been shown that the ET has definite advantages as detector of enzymic activities. For chromatography monitoring the effluent or a suitable aliquot of the effluent from a chromatographic column is mixed with a stream of substrate to the enzyme of interest. The heat registered upon passing the mixture through an empty inert column in the ET unit is proportional to the enzymic activity. Furthermore, it has been demonstrated that the ET can be utilized in automated, rapid (10–15 min/sample) TELISA monitoring of biomolecules other than enzymes [33].

Another study demonstrated the control of an affinity adsorption procedure by the specific enzyme activity signal from an ET [34]. Lactate dehydrogenase (LDH) was recovered from a solution by affinity binding to an N^6-(6-aminohexyl)-AMP-Sepharose gel. The LDH activity signal from the ET was used in a PID controller (equipped with proportional, integrating, and derivating control functions) or a computer to regulate the addition of AMP-Sepharose suspension to the LDH solution. The rapid and precise control of the addition of adsorbent in our model experiments suggests that this technique should be attractive in pilot plant and industrial scale purifications of enzymes.

20.3.7 Environmental control applications

Since biological reactions in general are exothermic calorimetry has an interesting potential for studies of the biological effect of pollutants. Although these effects are clearly measureable [16], only a few routine applications of calorimetry have been described to date. One reason for this is the lack of suitable instrumentation. The inhibitory effect of pollutants such as pesticides on biological systems is usually irreversible which means that the column with immobilized enzyme or cells must be replaced after one positive sample. Therefore maybe only one to two samples per hour can be measured which is impractical. This drawback could be overcome by designing instruments with a magazine of columns or with several parallel columns.

Environmental studies using thermistor-based biosensors have mostly dealt with metal detection which basically can be performed in two ways: by measuring the inhibition of enzymic activity or by measuring the activation of apoenzymes by metal ions. In both cases a certain specificity can be obtained. An example of the first alternative is given by the very efficient inhibition of urease activity by heavy-metal ions (Hg^{2+}, Cu^{2+}, and Ag^{+}) [16]. By using enzyme columns with comparatively low activity, very sensitive determinations in the ppb range or lower can be made. The original activity of the urease column can be restored by washing with iodide and EDTA. The sampling frequency is three to four samples per hour.

Many enzymes require a certain metal in their active site to be active. It is often possible to remove this metal with strong chelating agents, which results in an inactive apoenzyme. Upon exposure to a sample containing the same (or related) metal ion, the activity is restored to an extent that is related to the concentration of the metal ion. This procedure can be repeated up to a couple of times per hour. Examples of this technique include determination of Co^{2+}, Cu^{2+}, and Zn^{2+} at nanomolar concentrations [35].

20.3.8 Miniaturized thermistor-based biosensors

A current research project in our laboratory deals with the development of miniaturized FIA systems with the final goal of constructing a complete system by micromachining. The calorimetric measuring principle has been found to work surprisingly well in these small devices. Encouraging results have been obtained with various silicon chip test structures [36].

As an intermediary step in the investigation of highly miniaturized constructions, plastic/aluminum devices in the size of 50 mm in length and 15–25 mm in diameter or smaller have been designed (figure 20.5). Due to high sensitivity, small dimensions, modest buffer consumption, and good operational stability these devices are very suitable for portable use, for instance for home monitoring of glucose in diabetes. To allow analysis directly on whole-blood samples three different approaches have been tested, all of them with good results.

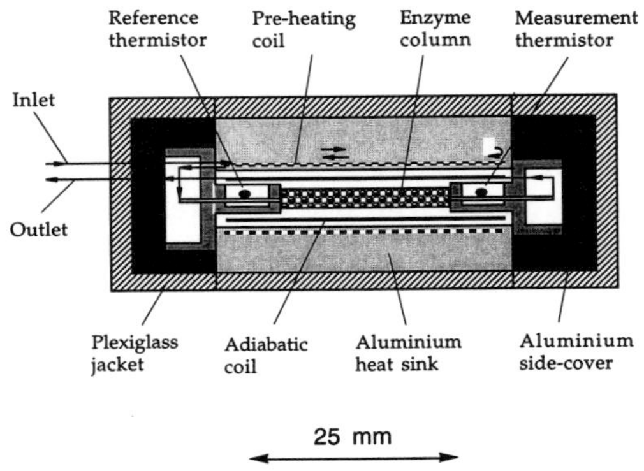

Figure 20.5 A schematic diagram of a miniaturized enzyme thermistor.

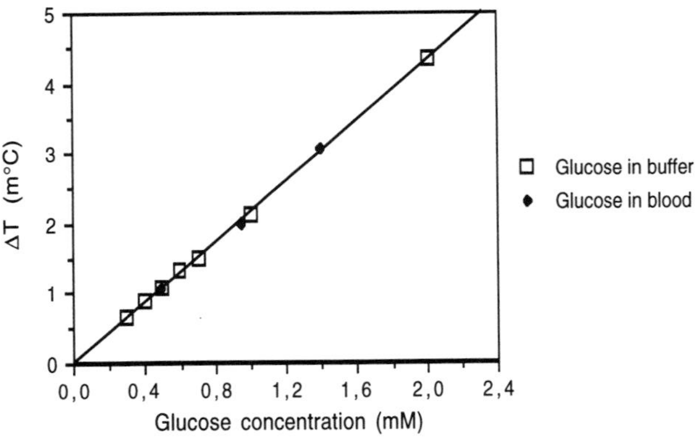

Figure 20.6 Calibration curve for glucose obtained with a miniaturized enzyme thermistor containing glucose oxidase/catalase on superporous agarose using 20 μl samples injected in a flow of 100 μl min^{-1}.

First, the blood cells can be removed by dialysis or filtration. Small coaxial dialysis units were constructed by attaching a 25 mm long 0.2 mm (i.d.) cuprophan hollow fibre inside a 0.5 mm PVC tubing. These units give about 5% yield of the glucose in the sample resulting in a linear range of 0–25 mM glucose with a 1.5 mm × 15 mm glucose oxidase/catalase column, which is adequate for diabetes monitoring. It is also possible to use a microdialysis probe (CMA/Microdialysis, Stockholm, Sweden) in which a thin needle (0.6 mm diameter) with a dialysis tubing (4–30 mm long) is inserted in a vein or under the skin and low-molecular-weight compounds are transported to the sample valve of the analytical device by a slow buffer stream (typically <5 μl min^{-1}). This gives a linear range for glucose of about 1–25 mM [37] and provides a reliable method for *ex vivo* monitoring of glucose [38].

Second, a superporous agarose material developed in our department [39] was

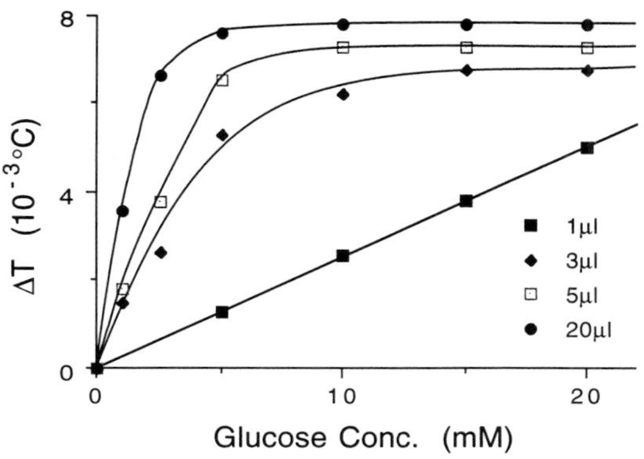

Figure 20.7 The effect of sample volume on the linear range of a thermometric glucose sensor with a 0.6 mm × 10 mm CPG column with glucose oxidase/catalase. The flow rate was 50 μl min^{-1}.

employed as enzyme carrier. This material allows a large number of whole-blood samples to be injected onto the enzyme column without any sign of clogging. Figure 20.6 shows a calibration curve obtained with 20 μl samples injected in a flow of 100 μl min^{-1}. For comparison, the response for 10-fold-diluted whole blood with glucose added to raise the concentration by 0, 0.45, and 0.9 mM is indicated in the diagram [40]. A suitable measuring range is chosen by proper selection of sample volume/flow rate.

Third, a very small column (0.6 mm × 10 mm) was filled with spherical 125–175 μm CPG particles loaded with glucose oxidase/catalase. The spaces between these particles are large enough to allow the cells to pass through without being trapped. A sample volume of 1 μl gave a suitable measuring range (1–25 mM) for blood glucose determination (figure 20.7). This type of column can be used with over 100 samples [22]. In conclusion all three approaches have been found to be practically useful for a home monitoring device. Similar assays for urea (0.2–50 mM) and lactate (0.2–14 mM) using 1 μl samples have been developed [41].

Further miniaturization of thermistor-based biosensor flow systems may be accomplished by micromachining in materials such as silicon and quartz [5, 36]. Integration of pumps and valves onto the thermal microchip is a feature that has not yet been fully realized. Recent developments include studies on multianalyte determination in a single sample stream by sequentially arranged enzyme systems as illustrated in figure 20.8. Glucose and urea or penicillin and urea were measured with a linear range of up to 20 mM for urea, 8 mM for glucose, and 40 mM for penicillin [42]. Designs permitting simultaneous assay of as many as four analytes are currently under development.

Another field under current investigation is hybrid sensors which combine two different measurement technologies into a hybrid that utilizes the most

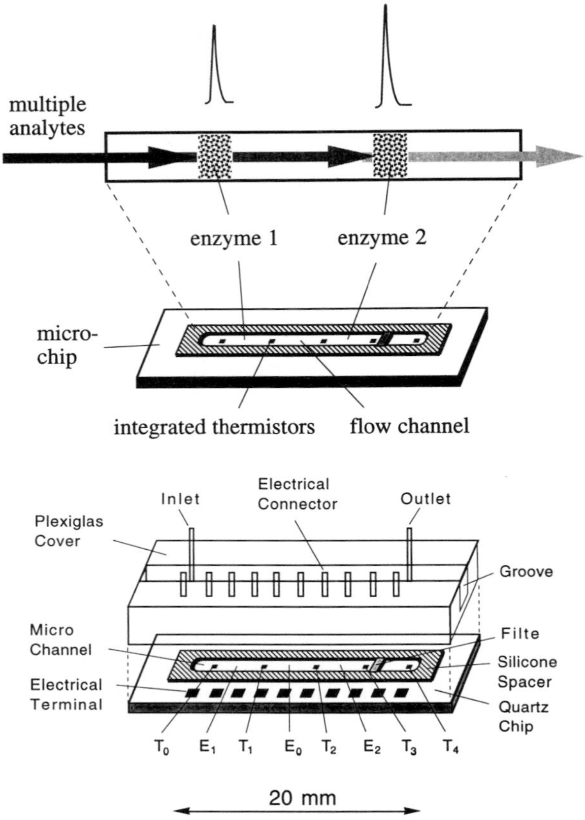

Figure 20.8 The working principle of a flow injection thermal microbiosensor for dual analyte determination. A pair of thermistors is placed upstream and downstream of each enzyme matrix. An enzyme-free region separates the two enzyme regions from each other.

advantageous properties of each technology. Thus a bioelectrocalorimetric device for glucose determination was developed that used calorimetric detection of the enzyme reaction (glucose oxidase) and electrochemical reoxidation of the system with ferrocene as mediator [43]. This sensor was less susceptible to interferences than a conventional electrochemical device and the linear range was independent of the oxygen concentration in contrast that of to a normal thermal biosensor.

ACKNOWLEDGMENTS

The continuous support form the Swedish National Board for Technical Developments and all who have contributed to this research field are gratefully acknowledged. The help in preparing the illustrations offered by Dr Bin Xie and Mr Henrik Danielsson is also gratefully acknowledged.

REFERENCES

[1] Mosbach K and Danielsson B 1981 Thermal bioanalyzers in flow streams. Enzyme thermistor devices *Anal. Chem.* **53** 83A

[2] Danielsson B 1990 Calorimetric biosensors *J. Biotechnol.* **15** 187–200

[3] Muramatsu H, Dicks J M and Karube I 1987 Integrated-circuit biocalorimetric sensor for glucose *Anal. Chim. Acta* **197** 347–52

[4] Muehlbauer M J, Guilbeau E J and Towe B C 1989 Model for a thermoelectric enzyme glucose sensor *Anal. Chem.* **61** 77–83

[5] Bataillard P, Steffgen E, Haemmerli S, Manz A and Widmer H M 1993 An integrated thermopile biosensor for the thermal monitoring of glucose *Biosensors Bioelectron.* **8** 89

[6] Xie B, Danielsson B and Winquist F 1993 Miniaturized thermal biosensors *Sensors Actuators* B **15–16** 443–7

[7] Rank M, Gram J and Danielsson B 1993 Industrial on-line monitoring of penicillin V, glucose and ethanol using a split-flow modified thermal biosensor *Anal. Chim. Acta* **281** 521–6

[8] Decristoforo G and Danielsson B 1984 Flow injection analysis with enzyme thermistor detector for automated detection of β-lactams *Anal. Chem.* **56** 263–8

[9] Danielsson B, Mattiasson B and Mosbach K 1981 Enzyme thermistor devices and their analytical applications *Appl. Biochem. Bioeng.* **3** 97–143

[10] Mattiasson B and Danielsson B 1982 Calorimetric analysis of sugars and sugar derivatives with aid of an enzyme thermistor *Carbohydr. Res.* **102** 273–82

[11] Kirstein D, Danielsson B, Scheller F and Mosbach K 1989 Highly sensitive enzyme thermistor determination of ADP and ATP by multiple recycling enzyme systems *Biosensors* **4** 231–9

[12] Danielsson B, Rieke E, Mattiasson B, Winquist F and Mosbach K 1981 Determination by the enzyme thermistor of cellobiose formed on the degradation of cellulose *Appl. Biochem. Biotechnol.* **6** 207–22

[13] Danielsson B, Mattiasson B and Mosbach K 1979 Enzyme thermistor analysis in clinical chemistry and biotechnology *Pure Appl. Chem.* **51** 1443–57

[14] Guilbault G G, Danielsson B, Mandenius C F and Mosbach K 1983 Enzyme electrode and thermistor probes for determination of alcohols with alcohol oxidase *Anal. Chem.* **55** 1582–5

[15] Bowers L D and Carr P W 1976 Immobilized-enzyme flow-enthalpimetric analyzer: application to glucose determination by direct phosphorylation catalyzed by catalase *Clin. Chem.* **22** 1427–33

[16] Danielsson B and Mosbach K 1988 Enzyme thermistors *Methods Enzymol.* **137** 181–97

[17] Kiba N, Tomiyasu T and Furusawa M 1984 Flow enthalpimetric determination of glucose based on oxidation of 1,4-benzoquinone with use of immobilized glucose oxidase column *Talanta* **31** 131–2

[18] Scheller F, Siegbahn N, Danielsson B and Mosbach K 1985 High-sensitivity enzyme thermistor assay of L-lactate by substrate recycling *Anal. Chem.* **57** 1740–3

[19] Winquist F, Danielsson B, Malpote J-Y, Persson L and Larsson M-B 1985

Determination of oxalate with immobilized oxalate oxidase in an enzyme thermistor *Anal. Lett.* **18** 573–88

[20] Satoh I, Akahane M and Matsumoto K 1991 Analytical applications of immobilized acid urease for urea in flow streams *Sensors Actuators* B **5** 241

[21] Danielsson B 1995 *Handbook of Analytical Sciences* (London: Academic) at press

[22] Xie B, Hedberg (Harborn) U, Mecklenburg M and Danielsson B 1993 Fast determination of whole blood glucose with a calorimetric micro-biosensor *Sensors Actuators* B **15–16** 141–4

[23] Satoh I 1988 Biomedical applications of the enzyme thermistor in lipid determination *Methods Enzymol.* **137** 217–25

[24] Satoh I, Danielsson B and Mosbach K 1981 Triglyceride determination with use of an enzyme thermistor *Anal. Chim. Acta* **131** 255–62

[25] Rank M, Danielsson B and Gram J 1992 Implementation of a thermal biosensor in a process environment: on-line monitoring of penicillin V in production-scale fermentations *Biosensors Bioelectron.* **7** 631–5

[26] Stefuca V, Gemeiner P, Kurillová L, Danielsson B and Báles V 1990 Application of the enzyme thermistor to the direct estimation of intrinsic kinetics using the saccharose-immobilized invertase system *Enzyme Microb. Technol.* **12** 830–5

[27] Gemeiner P, Stefuca V, Welwardová A, Michalková E, Welward L, Kurillová L and Danielsson B 1993 Direct determination of the cephalosporin transforming activity of immobilized cells with use of an enzyme thermistor. 1. Verification of the mathematical model *Enzyme Microb. Technol.* **15** 50–6

[28] Mattiasson B, Borrebaeck C, Sanfridsson B and Mosbach K 1977 Thermometric enzyme linked immunosorbent assay: TELISA. *Biochim. Biophys. Acta* **483** 221–7

[29] Birnbaum S, Bülow L, Hardy K, Danielsson B and Mosbach K 1986 Rapid automated analysis of human proinsulin produced by *Escherichia coli. Anal. Biochem.* **158** 12–9

[30] Mecklenburg M, Lindbladh C, Li H, Mosbach K and Danielsson B 1993 Enzymatic amplification of a flow-injected thermometric enzyme-linked immunoassay for human insulin *Anal. Biochem.* **212** 388–93

[31] Danielsson B, Flygare L and Velev T 1989 Biothermal analysis performed in organic solvents *Anal. Lett.* **22** 1417–28

[32] Stasinska B, Danielsson B and Mosbach K 1989 The use of biosensors in bioorganic synthesis: peptide synthesis by immobilized α-chymotrypsin assessed with an enzyme thermistor *Biotechnol. Tech.* **3** 281–8

[33] Danielsson B and Larsson P-O 1990 Specific monitoring of chromatographic procedures *Trends Anal. Chem.* **9** 223–7

[34] Flygare L, Larsson P-O and Danielsson B 1990 Control of an affinity purification using a thermal biosensor *Biotechnol. Bioeng.* **36** 723–6

[35] Satoh I 1989 Continuous biosensing of heavy metal ions with use of immobilized enzyme reactors as recognition elements *Mater. Res. Soc. Int. Meeting on Advanced Materials (Tokyo)* vol 14 (Pittsburgh, PA: Materials Research Society) p 45

[36] Xie B, Danielsson B, Norberg P, Winquist F and Lundström I 1992 Development of a thermal micro-biosensor fabricated on a silicon chip *Sensors Actuators* B **6** 127–30

[37] Harborn U, Xie B and Danielsson B 1995 unpublished

[38] Amine A, Xie B and Danielsson B 1995 A microdialysis probe coupled with a miniaturized thermal glucose sensor for *in vivo* monitoring *Anal. Lett.* **28** 2275–86

[39] Larsson P-O 1992 personal communication

[40] Harborn U, Xie B and Danielsson B 1994 Determination of glucose in diluted blood with a thermal flow injection analysis biosensor *Anal. Lett.* **27** 2639–45

[41] Xie B, Harborn U, Mecklenburg M and Danielsson B 1994 Urea and lactate determined in 1-μL whole blood with a miniaturized thermal biosensor *Clin. Chem.* **40** 2282–7

[42] Xie B, Mecklenburg M, Danielsson B, Öhman O, Norlin P and Winquist F 1995 Urea and lactate determined in 1-μL whole blood with a miniaturized thermal biosensor *Analyst* **120** 155–60

[43] Xie B, Khayyami M, Nwosu T, Larsson P-O and Danielsson B 1993 Ferrocene-mediated thermal biosensor *Analyst* **118** 845–8

21

On-line and flow injection analysis: physical and chemical sensors

Gil E Pacey

21.1 DEFINITIONS AND DESCRIPTIONS OF ON-LINE AND FLOW INJECTION

21.1.1 On-line, at-line, and off-line determinations

In process monitoring the point of sample acquisition is traditionally defined with the terms, on line, at line, and off line. All three sampling options can use automated sensor systems. Only on line requires some form of direct automated determinations and represents the 'ideal' situation for measurements.

From a practical perspective near-real-time measurements are made with on-line process monitors. These systems utilize sensors in the sample stream with calibration usually performed by moving the sensors into a separate calibration stream. This requires some redundancy in the system if continuous monitoring is needed. Direct measurements can be made in sample streams or in a slip stream, where all the sensing is performed.

Occasionally, analysis may require that the sample be removed from the process completely but measured at line. This approach is necessary when, for example, there is incompatibility between the process and the analysis system. This could be for safety or chemical reasons (e.g. solvent or pH incompatibility). While the measurement in such cases can be made at the process site, the actual monitoring system can be essentially the same as that used for the on-line measurements. Only the sample introduction needs to be modified.

The extreme case of process monitoring and analysis is off-line analysis where the sample is acquired and then taken to a measurement laboratory, although the actual measurement system can be the same as the on-line or at-line system. Usually calibration will be performed using traditional analytical procedures.

For the purpose of this chapter only on-line sensor systems will be discussed.

21.1.2 Continuous flow analysis

The scientific and engineering communities have developed improved ways to perform measurements. For example, new light sources for analysis include lasers capable of extremely well controlled power, spectral wavelength, coherence, and reproducibility. Wavelength dispersion is carried out using laser-blazed gratings and holographic gratings. Detectors range from single photodiodes to diode arrays or charge-coupled devices capable of collecting the entire light spectrum. The collection and evaluation of data, and the resulting reports, are handled by computers. However, strangely, the sample handling steps for most analyses are still at the flask and pipet stage. Modernization of sample handling requires a reduction in time and improvements in the types of sample handling step that can be automated, for example, matrix modification and dilution. Continuous flow analysis (CFA) is a group of techniques that has automated sample handling.

21.1.2.1 Components of CFA. All CFA techniques are similar in terms of components. All systems use some type of fluid pumping system, with peristaltic pumps the traditional choice. Piston-driven pumps have also been used; however, the pulsations created by these pumps must be dampened. If such pulsation is allowed to occur, the flow patterns inside the tubing will not be reproducible and will affect the precision of the sample analysis.

CFA requires some type of sample introduction. In segmented systems the sample is pumped. In non-segmented systems the sample is injected using low-pressure HPLC-type injectors.

The tubing material that connects CFA systems is either glass or Teflon tubing. Tubing is used for the mixing coils that make up the analytical manifold, which is the junction box that contains all the connections between the different streams. The output from the manifold is then passed through a flow-through detector.

For CFA systems containing sensors the only change will be in the design of the detection cell, since what is optimal for the sensor may not be best for the CFA system. In most cases the detection cell design is a compromise.

21.1.2.2 Rationale for automation. The arguments for automation by CFA systems are varied. These include reduction in personnel cost, high sample throughput, and situations which are too hazardous for human exposure. The single major problem that can be addressed by automation is the sample matrix. Real world samples are seldom in a clean, easy to handle matrix. Instead the sample is usually in a highly complex matrix that not only may contain potential chemical interferents, but also may exhibit properties such as turbidity, viscosity, or salt content that make the standard method of analysis useless. Sensors are on average even more sensitive to matrix problems.

In the case of sensors, the CFA system performs three functions. First, it presents to the sensor a reproducible sample plug that improves the precision

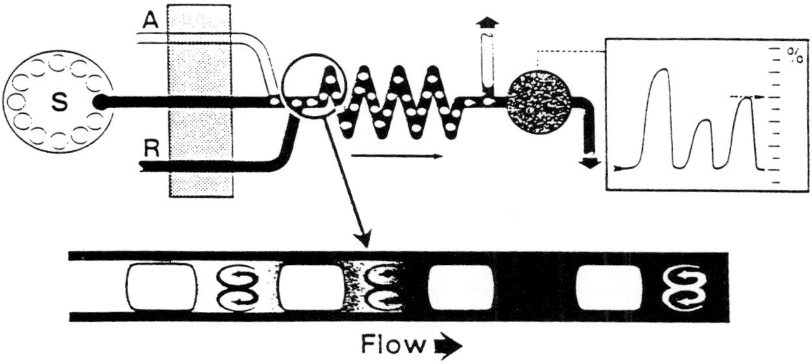

Figure 21.1 A typical continuous flow segmented manifold where sample, air, and reagent are pumped to provide segmentation between the air and sample–reagent segments. The bottom picture depicts the mixing process. (Reprinted by permission of John Wiley and Sons.)

of the measurement. Second, the sensor does not have to be in contact with the sample at all times. Indeed some type of regeneration or cleaning of the sensor can occur between sample introduction. Third, pretreatment of the sample is possible. Separation steps such as ion exchange, trapping, extraction, gas diffusion, or dialysis can precede the sensor in the analytical manifold, thereby improving the accuracy of the sensor.

21.1.2.3 Different CFA techniques. Confusion exists about the different CFA techniques. First, CFA is not chromatography where the objective is to separate and detect several constituents of a sample. CFA is usually used to create a detectable signal on only one analyte. The techniques are divided into two distinctly different approaches, segmented and unsegmented.

Similarities exist between segmented CFA systems such as the Technicon Autoanalyzer and flow injection analysis, FIA. The fundamental operational difference between the two types of system is the segmentation. For some time it was assumed that air segmentation was an absolute necessity (Skeggs 1957). In such systems segmenting allowed the identity of each individual sample plug to be preserved (limited sample dispersion) as shown in figure 21.1. Additionally, the segmented continuous flow analyzer is normally operated under steady-state chemical conditions.

The primary drawback to air-segmented systems is the bubble. These air bubbles are compressible, thereby creating pulsations in the flowing stream. In addition, for most detectors, the air bubble must be removed before the sample passes through the detection cell. In addition, the precision of bubble size has been difficult to control. This variation in bubble size adds to the irreproducibility of the system. If electrochemical or other static sensitive detectors are used, the segmented system, which builds up static charge, will present severe problems

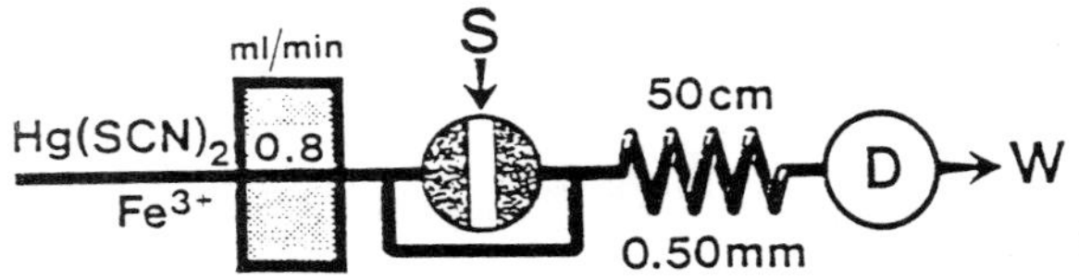

Figure 21.2 Typical FIA manifold for chloride determination where the sample is injected and the reagents are pumped as the carrier stream. (Reprinted by permission of John Wiley and Sons.)

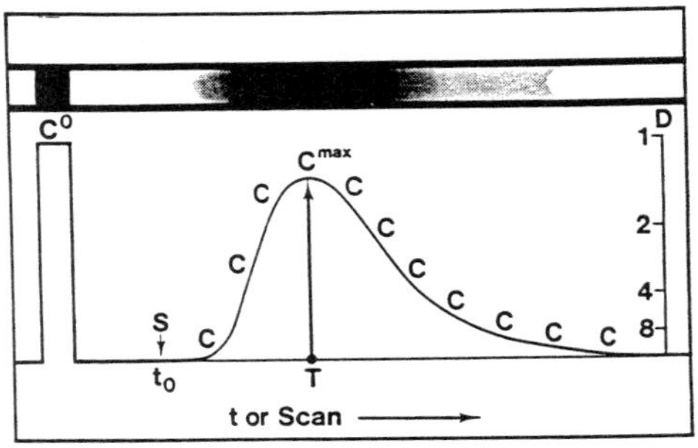

Figure 21.3 An originally homogenous sample zone (top left) disperses during its movement through a tubular reactor (top center), thus changing from an original square profile (bottom left) of original concentration C_0 to a continuous concentration gradient with maximum concentration C_{max} at the apex of the peak. (Reprinted by permission of John Wiley and Sons.)

in terms of baseline stability. More recent refinements of this technique make the removal of the air bubble unnecessary for some detectors. Additionally, computer predictions of the steady-state signal can be made by observing the initial portion of the rising section of the output thereby increasing the sample throughput.

We now consider non-segmented analyzers. The three basics of FIA, as defined by Ruzicka and Hansen (1988), are reproducible timing, sample injection, and controlled dispersion. As the FIA technique has been developed further these three basic principles have been modified. For example, samples are not always injected using a valve. Sample introduction is accomplished as long as a sample plug is reproducibly inserted into the flowing carrier stream. Without question reproducible timing is the critical principle, since it leads to reproducible physical conditions. The primary benefit of reproducible timing and controllable sample dispersion in FIA is that steady-state conditions are unnecessary. As a result, the relative standard deviations of signals from the same sample or standard are around 1%. A typical FIA manifold is shown in figure 21.2.

The FIA term controlled dispersion has created much confusion in the scientific community. It was observed that the injected sample zone always seems to spread out inside the manifold in a reproducible manner. By observing the peak shapes and heights it was clear that the sample zone reproducibly expanded or diluted in the flowing stream. This process is called dispersion (figure 21.3) and is defined as the amount of chemical signal that is reduced by injecting a sample plug into an FIA system. This is mathematically represented by:

$$D = C_0/C_{max}$$

where D is the dispersion coefficient at the peak maximum produced by the ratio between C_0, the concentration of a pure dye, and C_{max}, the concentration of the dye injected into the system as it passes through the detector.

The dispersion coefficient is useful in that it allows comparisons of different manifolds (Karlberg and Pacey 1989). Furthermore, it provides a means to verify and monitor the degree of sample dilution resulting from any changes made to the manifold during methods development. A dispersion coefficient of one to three is optimal when specific detectors such as ion selective electrodes, sensors, or atomic absorption are used. If some type of reaction chemistry is performed inside the manifold, then a dispersion coefficient between three to eight is optimal. Gradient techniques such as titrations require dispersion coefficients larger than 10. In dilution manifolds, dispersion coefficients higher than 10 000 have been observed.

In essence, what has been called 'controlled dispersion' is in fact the recognition that the sample is reproducibly diluted as it travels down the tubing. The reproducible timing allows for this reproducible physical mixing and dilution and the dispersion is controlled, but there is a second aspect to the controlled dispersion feature.

It is clear that the analyst has complete control over the amount of sample dispersion or dilution that occurs as the sample passes through the manifold. This control originates from the way in which the manifold is designed. What factors influence the sample dispersion in an FIA system? Minor factors are the inside diameter of the tubing, the flow rate inside any one line, and the reaction coil length. Major factors can be sample size and flow rates between the different fluid lines.

One negative factor associated with the combination of FIA with biosensors is the fouling of the detector and the manifold lines. Schugerl (1993) discusses the requirements for an FIA/biosensor system. Perhaps the biggest negative factor for FIA has been that the equipment manufacturers have been slow to respond to the customers' needs. In most process control situations, in-house expertise is used to design and build FIA systems. A word of warning is in order. Most designers overbuild their first FIA unit using knowledge and components from HPLC. After they realize that the system is overbuilt, they usually drastically underbuild the system. There are many people with expertise in the area and significant time can be saved by using their knowledge.

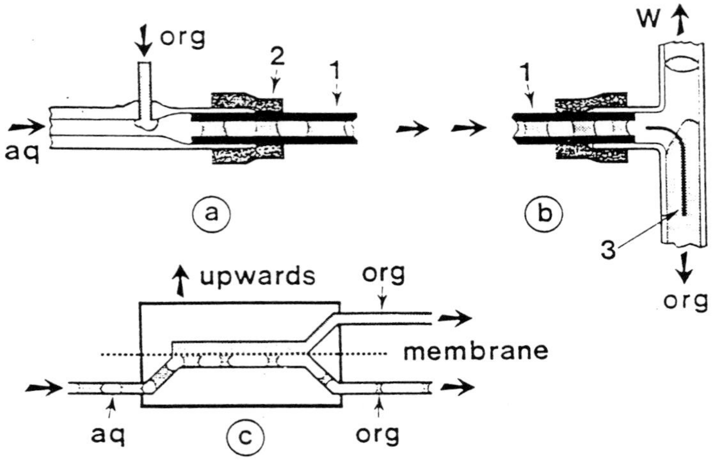

Figure 21.4 A separator for an extractor. (Reprinted by permission of John Wiley and Sons.)

The remaining part of this chapter will deal exclusively with FIA, since the simplicity of FIA makes it the best CFA technique for sensor operation.

21.2 SELECTIVITY ENHANCEMENTS, MATRIX MODIFICATION AND CONVERSION

The sample matrix for FIA often provides the greatest challenge to the analyst. Several techniques have been used to increase the accuracy of the measurement.

21.2.1 Extraction

Liquid–liquid extraction in continuous flow systems requires two immiscible phases which are brought together in a narrow tube in a controlled manner so that defined segments of each phase are formed. The choice of tubing material, tubing dimensions and mixing geometry is critical for the final result. In the extraction tube, often coiled, the contact area between the two phases is large, as is the ratio between the tube area and the tube volume.

The phase separator in a continuous flow system is designed to handle small volumes of either phase so that segments of the same phase are joined and completely separated from the segments of the other phase. The volume of the separator is often small to prevent dilution of the sample or to prevent deterioration of the original concentration gradient of the sample. A typical separator is shown in figure 21.4. The reader is referred to the article by Karlberg (1986).

21.2.2 Gas diffusion

The best transplantation technique for gaseous analytes, or those analytes which can be converted to a gaseous species, is gas diffusion.

The major advantage of the gas diffusion process is that it removes the analyte of interest from the sample matrix. The analyte is transferred to a new matrix, the acceptor stream, which contains no chemical or physical interferents. An additional advantage is that the acceptor stream can be configured so that the optimized conditions for the detection process are realized. For example, optimized reaction conditions for the formation of the analyte–reagent complex can be maintained in the acceptor stream.

Chemistry can be performed before the gas diffusion process. When all the necessary pretreatments of the donor stream have been accomplished, the donor stream is passed under the gas diffusion membrane where the gaseous analyte or high-vapor-pressure compounds can diffuse through the membrane into the acceptor stream. The acceptor stream may or may not contain reagents. However, it is important to realize that, since the donor stream and acceptor streams do not touch, the conditions of the acceptor stream can be optimized to maximize the observable signal. The most often used membrane is a 0.45 micrometer microporous Teflon from W L Gore. The is not the plumber's tape that is referred to in many published articles. The tape is a film and not a uniform microporous membrane. Silicon membranes from General Electric have also been used; however, they do have a breaking-in period which suggests that a memory effect could be observed for some samples. Additional information on gas diffusion can be found in the article by Canham *et al* (1988).

21.2.3 Dialysis

Dialysis is a selectivity enhancement technique where the observed selectivity is primarily dependent on the properties of the membrane used in the dialysis manifold (Martin and Meyerhoff 1986). The usual system will have a membrane which is capable of transporting ions. In theory, any ion of interest can be separated from the matrix. Unfortunately, most membranes which are currently available to analytical chemists are not very selective. The dialysis process can also be seen as a way to dilute a sample.

Essentially, the same design considerations as discussed for gas diffusion hold true for dialysis with the notable exception that the membranes are different. Whereas gas diffusion uses hydrophobic microporous membranes, dialysis uses hydrophilic membranes such as cellulose-based membranes. A variety of different-molecular-weight-cut-off membranes are also available.

21.2.4 Columns/packed bed reactors

Columns that are typically 3–4 cm long and 1 mm wide are used for a variety of FIA techniques; trapping, preconcentration, separations, and conversions. For example, by using a cation exchange resin potassium ions and other monovalent ions would be preferentially retained while multivalent cations remain primarily in the acceptor stream and pass on to waste.

What concerns should be raised placing a packed-bed reactor in an FIA manifold? The first concern is whether the dispersion properties of the FIA will change or not. It should be pointed out that the packed-bed reactors used in FIA are not the size of the packed-bed reactors used in HPLC since this leads to sample dispersion and an increase in dispersion means increased dilution of the sample zone. If the detection limits of the method are close to the desired levels, any additional dispersion is undesirable. However, all too often discussions about bed reactors center around mixing, which in turn leads to confusion between dispersion and mixing. In fact, the packed-bed reactors must create enough mixing to allow the reagents to come in contact with the analyte or the interferent with the reactive site on the column. Thus dispersion should be kept as low as possible. The amount of reaction which takes place depends on the kinetics of the reaction, the total residence time of the analyte in the reactor, and the concentration of reactive sites in the column. It is possible to create an appropriate residence time without significantly increasing dispersion.

Another concern with packed-bed reactors is reproducibility. There are two parts to reproducibility: can multiple packed-bed reactors be made reproducibly and what is the stability of the reactor? The first question concerning reproducibility can be answered as follows. The art of packing a column is at times difficult and each individual analyst will have different success. However, all packed-bed reactors can be calibrated in the FIA system. A variation in the performance level between reactors of less than 10% is acceptable. The second reproducibility question concerns the loss in reactivity due to reagent degradation or saturation of the reactive sites. In the ideal case the reactivity of the column should not change with respect to time, thereby producing a reproducible signal for the same concentration of analyte. This is a more difficult problem. Ideally, the analyst should find the reactor conditions that will minimize loss of activity. In practice, especially with enzymes, this condition will not be met. Frequent calibrations will be needed to insure the most accurate results.

Immobilized reagents are usually used when the reagent is expensive or when the presence of the first reagent in the manifold will adversely affect subsequent reactions of a method (Olson *et al* 1986). The classical example is the use of enzymes in an immobilized form. The expense of these reagents makes it imperative that a minimum amount is used.

The biggest difficulty in using immobilized enzymes is ensuring that the enzyme activity is stable. Commercially immobilized enzymes, usually on controlled pore glass, are becoming more available from suppliers of bioanalytical products and biochemical suppliers such as Pierce and Sigma. If the reagent reactivity decreases with time then the necessary and frequent recalibration of such a system becomes time consuming. Usually the reactivity can be stabilized by the proper choice of carrier stream. It is also possible to eliminate this problem by using easily replaceable reactors with reproducible reactivity.

21.3 SENSOR CELL DESIGN IN FIA

At first glance it would appear that the detector requirements commonly cited are the same for FIA as for HPLC. However, this is not true for all detectors. The differences include the flow characteristics, multiple detectors, and redundancy.

21.3.1 Flow characteristics

The flow characteristics of FIA create some problems that must be addressed such as the concentration gradient, stopped flow mode, read-out/software protocol, and peak height or peak area measurements made for quantitation.

21.3.1.1 Gradient. Any discussion about cell design begins with a discussion about a unique characteristic of FIA, the concentration gradient (figure 21.3). Both the dispersion in the system and the concentration gradient are shown in figure 21.3. On both the rising and falling portions of the peak there are points which have the same dispersion coefficient and are produced by the same concentration of signal producing species. Therefore, the peak represents a sequentially higher concentration of product until the peak height is reached, followed by a sequentially lower concentration of product until the return to the baseline. The reproducibility of individual points along the gradient is just as precise as that of the peak height.

The gradient is the most difficult feature to deal with in terms of cell design. The gradient is naturally formed as the sample plug is dispersed in the manifold. The front center is more concentrated than the material next to the walls (created by wall drag) and the concentration decreases from the front to the back of the dispersed sample plug. The peak shape indicates that the concentration suddenly rises then slowly lowers from front to back. When designing a cell the first question asked is do you need the gradient? If the answer is no then cell design is easy. If the answer is yes then maintaining the gradient is the first consideration.

The gradient is necessary if gradient dilution or additional detectors are to be used. In gradient dilution the numerous points (C in figure 21.3) along the concentration profile can be used for information. Each point corresponds to a specific dispersion coefficient for the dispersed sample. Therefore, it is possible to produce a series of calibration curves using different points on the FIA peak.

21.3.1.2 Stopped flow. Some special requirements for FIA detectors will be mentioned which relate to the stopping (and the restarting) of the flowing stream. If a stopped flow period is used, it is absolutely essential that no artifact transient signals appear since these can be interpreted as real peaks. Another important

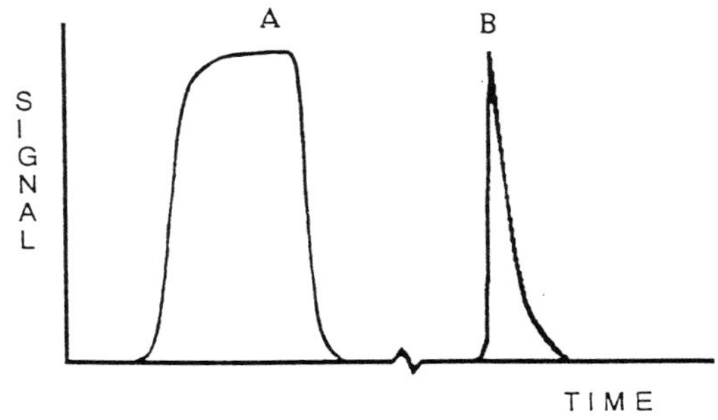

Figure 21.5 A comparison of signal output from an air-segmented continuous flow analyzer (A) and from a flow injection analyzer (B). (Reprinted by permission of Elsevier.)

aspect to be considered when stopping the sample stream is that the section of the stream that is situated inside the flow-through cell is heated by the light source of the detector. If this occurs for a stopped flow application, part of the sample zone will be leaving the detector during the stop period due to the expansion of the liquid. This effect can, to some extent, be prevented by feeding the outlet stream from the flow cell back into the pump. Thermostating of the flow cell can also be applied. If optical fibers are used the light source can be located at some distance from the cell which will minimize the heating problem. If the stopped flow period is used only to create a time delay, then the sample zone should be stopped before the cell.

21.3.1.3 Peak height versus peak area. An HPLC peak is a somewhat Gaussian peak that irreproducibly exits the column over a period of many seconds. Although the retention time, the center of the peak, remains reasonably constant, the overall peak is not that reproducible in most situations. This requires that HPLC detectors use peak area instead of peak heights to maintain acceptable reproducibility.

In FIA the peak is not a Gaussian peak, but is extremely reproducible (figure 21.5). Relative standard deviations of less than 1% are the norm rather than the exception. Since the tubing lengths are short and there is little retention of the sample peak inside the tubing, the normal FIA peak is only a few seconds long. Therefore, the peak height is the best way to measure an FIA peak. Only in a few special cases such as FIA titrations is peak height not used.

In the cases of some sensors used in FIA, longer residence times are needed for the sensor to respond (i.e. ion selective electrodes). There are two approaches to solve this problem, stopped flow and larger sample sizes. Stopped flow has already been discussed and would be used when the sample is precious, but larger sample sizes can create apparent longer retention times. Normally once a sample exceeds 350 microliters the peak behaves like a steady-state peak. That is, the dispersion coefficient will appear to be unity. If the sample is unlimited, it can be pumped continuously, allowing the sensor to respond at its own pace. If additional chemistry or separation techniques are part of the system, reverse FIA can be used, where the sample is continuously pumped and the reagents are injected or the separation technique run continuously.

21.3.1.4 The readout device/software. The speed of the normal FIA peak means that the detector must measure and output data at a rate of 25 ms or better. A slower rate results in missing the peak maximum. This is the first constraint on the sensor system. Since the normal FIA sample zone will reside near the sensor for only a few seconds, can the sensor respond fast enough to produce an analytical signal? Accumulation of data points like counting photons in fluorescence is possible, but significantly adds to the cost of the system. Many have chosen to get around this problem by using the previously mentioned continuous pumping of the sample.

Fast-responding sensors will pose no significant problems for the read-out system.

21.3.2 Multiple detectors

If multiple detectors are to be used in FIA, then the gradient must be left intact or complete and reproducible mixing must occur. Ideally all detectors must see a similar dispersed sample plug. The first detector will observe a reproducible concentration gradient. If that detector creates irreproducible turbulence, then subsequent detectors will not observe reproducible sample zones, which results in imprecision. This requirement complicates multiple-cell systems. The simplest approach is to split the stream into enough individual streams to feed all the detectors. This is more costly in terms of components, but avoids significant development time in sensor cell design.

To some the answer is to produce a sensor that can be applied to the surface of the tubing. However, since the analyte is not as concentrated next to the tubing walls as at the center of the sample zone, the sensitivity of this approach is decreased.

21.3.3 Redundancy

In most process situations, a significant level of redundancy must be built into the system. Two similar detectors may be used to allow for calibration and comparison of results. Even then a second method of checking the sensor's performance might be used. For example in a pH measurement system two electrodes in two different streams are used. Both electrodes are on line except when one is being calibrated. At all times not only the millivolts but also the impedance is observed, since a sudden change in impedance indicates a faulty electrode. Redundancy is an important feature in any process system.

21.4 MEASUREMENTS

Measurements are divided into two types, physical and chemical. Physical measurements such as temperature or viscosity usually do not require an FIA system. Sensors for these types of physical measurement are both simple and reliable. Chemical measurements involve additional sample manipulations before the analyte can be detected. Given that a sensor has been designed to measure a specific analyte, an automated system is used to pretreat and deliver a reproducible sample to the sensor.

Membrane based sensor detection is based on the transport of the analyte species towards the surface of a sensor. These sensors respond only to the species which are present in the absolute vicinity of the sensor surface. Thus, the species must be representative of the total solution which passes by the sensor. As a consequence, the design of the detector cell and of the hydrodynamic system becomes more critical.

The application of spectrophotometric sensors (UV–visible, IR, Raman, and luminescence) in FIA is possible assuming that the response time of the sensor–read-out device is fast enough to measure the FIA signal. For example, a membrane impregnated with a reagent was placed inside a cell similar in design to a gas diffusion unit and a fiber optic was placed behind the membrane. As the sample passed the membrane the analyte and reagent reacted to form the color that was monitored. The kinetics of this type of system must be fast and the reaction reversible. Although it is possible to use an irreversible reaction, the sensor's lifetime is limited: since the amount of reagent is finite eventually it will be consumed. Reversibility is achieved by using reagents that do not have high formation constants so that the plug of carrier stream behind the sample plug can wash/regenerate the membrane.

FIA systems do not affect any of the basic traits of batch spectrophotometric methods. Beer's law will still be obeyed with the sensitivity about the same. An exciting potential is the incorporation of diode array detectors into the FIA flow system. The high reproducibility of FIA will make the use of diode array detectors for multicomponent analysis, kinetics, and routine monitoring extremely easy to use.

Applications of FIA have been quite varied with some comparisons reported. One comparison of different biosensor systems, *in situ* versus flow-through FIA, was made by Bilitewski *et al* (1993). Others compared different sensors in the FIA system (Scheper *et al* 1993). Gebbert *et al* (1994) reported an on-line system for monitoring monoclonal antibodies. A multisensor FIA system operated with a software package FIACRE was reported by Busch *et al* (1993). A biosensor system to determine aspartame in food products was reported by Male *et al* (1993). Biofield effect transistors were reported by Kullick *et al* (1992). On-line measurements of medium components and enzymic products were reported by Spielman *et al* (1992). A micromachined immobilized column with electrochemical detection for glucose was reported by Murakami *et al*

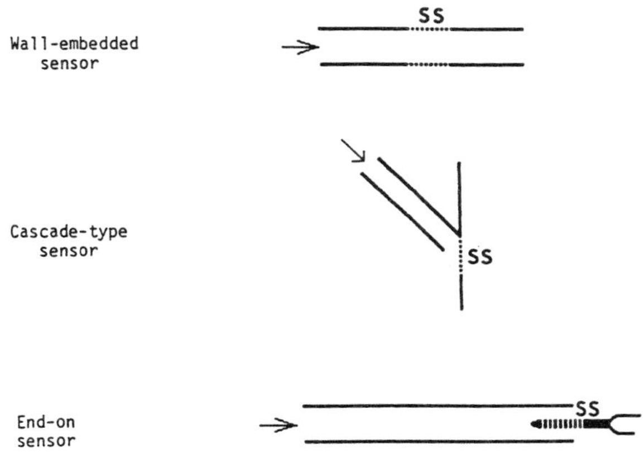

Figure 21.6 The three basic configurations for electrochemical detectors. SS, sensing surface. (Reprinted by permission of Elsevier.)

(1993). Thick-film electrodes have been used for food analysis bioprocess monitorings (Bilitewski *et al* 1992).

21.4.1 Electrochemical sensors

For electrochemical sensors cell configurations for FIA are almost as numerous as the number of papers describing electrochemical detection. Basically, electrochemical sensor design for FIA falls into one of the following three categories: (i) wall-embedded, (ii) cascade-type, and (iii) end-on sensors (see figure 21.6).

The wall-embedded sensor may comprise a cylinder inserted in place of tubing as part of the flow system or it may comprise one or several plates incorporated into the tubing wall. The cascade-type sensor is exposed to the carrier stream either in a tangential or frontal manner. If a tangential flow is used several sensors can be positioned after each other in series. The end-on category most frequently entails wire-shaped sensors positioned in the center of the flowing stream so they are 'piercing' the parabolic head of the sample plug. The positioning of reference and counter-electrodes is less critical. In most cases they are placed after the sensing electrode in the flowing stream, either in the stream itself or in a waste reservoir.

The so-called wall jet electrode is a hybrid of cascade and end-on electrochemical cell designs and has some potentially useful qualities for FIA application. As mentioned earlier, the possibilities for electrode and cell designs seem to be unlimited as long as the ohmic (IR) drop between the different electrodes matches the requirements of the electronics involved. A large IR drop should be avoided since it may lead to a non-linear response of the sensor. The magnitude of the IR drop is governed by the distance between the electrodes, by the hydrodynamic conditions, and by the electrolyte content

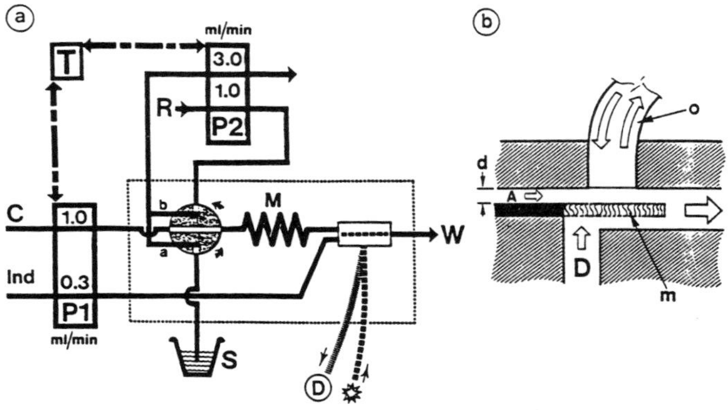

Figure 21.7 (a) A manifold for assay of ammonia and urea, comprising two peristaltic pumps (P1 and P2), a timer (T), reagent (R), carrier (C), and indicator (Ind) channels, a mixing coil (M, 70 μl), a sample channel (S), and a valve furnished with a split loop configuration comprising sample (a) and reagent (b) sections. The optosensor in (b) is connected to a light source and spectrophotometer (D) by means of optical fibers. W is the waste channel. The boxed area includes the components within the microconduit. (b) Details of the optosensor with gas diffusion membrane (m) separating a donor stream (D) and the indicator containing acceptor stream (A), the membrane being situated in that section of the channel that is monitored by the optical fiber (o). (Reprinted by permission of John Wiley and Sons.)

in the carrier stream. So the first, obvious measure would be to use a design where the distance between the electrodes is short. Then an inert electrolyte can be added to the carrier solution. If voltammetry is applied a counter-electrode can be implemented as an auxiliary, third electrode to carry electrical current thereby 'protecting' the reference electrode.

21.4.2 Optical sensors

One of the best interfaces between sensor cells and FIA streams has been reported by Ruzicka and Hansen (1985) and Petersson *et al* (1986). In both cases the concept of microconduits is powerfully demonstrated. A microconduit is a reduced-size FIA system. Early in the theoretical development of FIA, it was clear that a decrease in size was advantageous. The dispersion coefficient is lowered, the mixing is faster, and the manifold can be placed on a small 2 in by 4 in Plexiglass block. In fact all components of the FIA system with the exception of the pumps can be on the Plexiglass block.

In one example the determination of ammonia and urea is performed by combining a gas diffusion unit and a fiber-optic-based spectrophotometer (figure 21.7). This is an excellent example of how FIA improves the selectivity of the sensor system. The ammonia and the ammonia generated by the

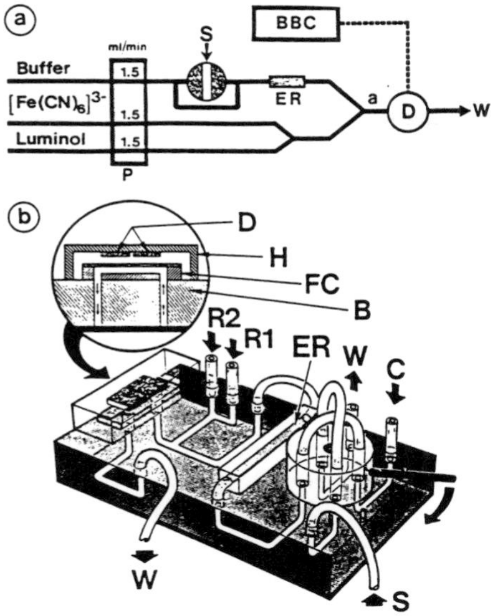

Figure 21.8 (a) A manifold for the determination of glucose, free cholesterol, creatinine, and lactic acid with detection by chemiluminescence. The sample (S) is injected into a carrier stream of buffer propelled forward by pump P and carried to a reactor (ER) containing immobilized enzyme (oxidase), in which the substrate of the sample is degraded, leading to the formation of hydrogen peroxide. Having been confluenced with luminol and hexacyanoferrate(III), the sample zone is finally guided through a short channel a (2 cm, corresponding to 16 μl) into the light detector D, the output from which is led to a computer (BBC); W is waste. (b) An integrated FIA microconduit accommodating the manifold components shown for the system in (a). C represents the carrier stream of buffer, R1 is luminol, and R2 is hexacyanoferrate(III). The flow cell (FC) comprises two photodiodes (D) contained in a house (H) mounted on the microconduit base plate (B). (Reprinted by permission of John Wiley and Sons.)

conversion of urea are passed through a gas diffusion membrane. The ammonia is transported to the acceptor stream where the colorimetric detection dye chemistry occurs. The fiber optics are actual a part of the gas diffusion cell's acceptor side. This system has a detection limit for both ammonia and urea of about 1 ppm. A more sensitive reagent system would improve the sensitivity.

A second example is the determination of glucose, free cholesterol, creatinine, and lactic acid (figure 21.8). This method uses on-line immobilized enzyme reactors followed by chemiluminescence detection of the hydrogen peroxide enzyme reaction product reaction with luminol. Detection limits of 0.1 mg dl^{-1} were observed.

An important point in each of these examples is that the cell was designed in such a way that the sensing surface is flat and part of the tubing wall. In both cases the cell is not cylindrical but flat. Whether the gradient still exists after the change in configuration from cylindrical to flat is not known; however, if the flow turbulence is kept at a minimum, the resulting analytical signal will be reproducible within less than ±2%. The microconduits do allow the sample zone to be longer, thereby creating a longer residence time with the sensor. The flat sensor surface not only decreases engineering time but increases sensing surface area.

21.5 CONCLUSION

The combination of chemical and biological sensors with flow injection has been demonstrated. Both more-traditional-type sensors such as pH electrodes and newer sensors such as fiber optics and surface acoustic wave detectors have been incorporated into FIA systems with success. An advantage that FIA brings to the sensor field is the possibility of turning a moderately selective sensor into a selective sensor by incorporating into the FIA system some type of selectivity enhancement technique such as gas diffusion, dialysis, and reactors. Finally the FIA systems permit renewable systems since sensor surfaces and reaction cells can be washed, surface regenerated, and reagents replenished on demand.

REFERENCES

Bilitewski U, Chemnitius G, Rueger P and Schmid R D 1992 Application of thick film electrodes to food analysis and bioprocess monitoring *Microbial Principles in Bioprocesses: Cell Culture Technology, Downstream Processing and Recovery (DECHEMA Biotechnology Conf. 5, Pt A)* pp 415–8

Bilitewski U, Drewes W, Neermann J, Schrader J, Surkow R, Schmid R D and Bradley J 1993 Comparison of different biosensor systems suitable for bioprocess monitoring *J. Biotechnol.* **31** 257–66

Busch M, Hobel W and Polster J 1993 Software FIACRE: bioprocess monitoring on the basis of flow injection analysis using simultaneously a urea optrode and a glucose luminescence sensor *J. Biotechnol.* **31** 327–43

Canham J, Gordon G and Pacey G 1988 Optimization of parameters for gas diffusion flow injection systems *Anal. Chim. Acta* **209** 157–63

Gebbert A, Alvarez-Icaza M, Peters H, Jaeger V, Bilitewski U and Schmid R 1994 Online monitoring of monoclonal antibody production with regenerable flow injection immuno systems *J. Biotechnol.* **32** 213–20

Karlberg B 1986 Flow injection analysis—or the art of controlling sample dispersion in a narrow tube *Anal. Chim. Acta* **180** 16–20

Karlberg B and Pacey G 1989 *Flow Injection Analysis: a Practical Guide* (Amersterdam: Elsevier)

Schugerl K 1993 Which requirements do flow injection analyzer/biosensor systems have to meet for controlling the bioprocess? *J. Biotechnol.* **31** 241–56

Kullick T, Quack R, Scheper T and Schugerl K 1992 Bio-field-effect transistors for monitoring and control of biotechnological production processes *Microbial Principles in Bioprocesses: Cell Culture Technology, Downstream Processing and Recovery (DECHEMA Biotechnology Conf. 5 Pt A)* pp 503–6

Male K, Luong J, Gibbs B and Konishi Y 1993 An improved FIA biosensor for the determination of aspartame *Appl. Biochem. Biotechnol.* **38** 189–201

Martin G and Meyerhoff M 1986 Membrane-dialyzer injection loop for enhancing the selectivity of anion-responsive liquid-membrane electrode in flow systems *Anal. Chim. Acta* **186** 71–80

Murakami Y, Takeuchi T, Yokoyama K, Tamiya E, Karube I and Suda M 1993 Integration of enzyme-immobilized column with electrochemical flow cell using micromachining techniques for a glucose detection system *Anal. Chem.* **65** 2731–5

Olson B, Stalbom B and Johansson G 1986 Determination of sucrose in the presence of glucose in a flow Injection system with immobilized multi-enzyme reactors *Anal. Chim. Acta* **179** 203–8

Petersson B, Hansen E and Ruzicka J 1986 Enzymic assay by flow injection analysis with detection by chemiluminescence: determination of glucose, creatinine, free cholesterol and lactic acid using an integrated FIA microconduit *Anal. Lett.* **19** 649–65

Ruzicka J and Hansen E 1975 Flow injection analysis. Part I. A new concept of fast continuous flow analysis *Anal. Chim. Acta* **78** 145–57

—— 1985 Optosensing of active surfaces—a new detection principle in flow injection analysis *Anal. Chim. Acta* **173** 3–21

—— 1988 *Flow Injection Analysis* (New York: Wiley)

Scheper T, Brandes W, Maschke H, Ploetz F and Mueller C 1993 Two FIA-based biosensor systems studies for bioprocess *J. Biotechnol.* **31** 345–56

Skeggs L 1957 Automated method for colorimetric analysis *Am. J. Clin. Pathol.* **28** 311–22

Spielman A, Garn M B, Aellen T, Haemmerlil S and Widmer H M 1992 Flow injection analysis methods for on-line bioprocess monitoring: the automated measurement of some basic medium components and an enzymic product *Microbial Principles in Bioprocesses: Cell Culture Technology, Downstream Processing and Recovery (DECHEMA Biotechnology Conf. 5, Pt A)* pp 367–70

22

Flow injection analysis in combination with biosensors

Bo Mattiasson and Bengt Danielsson

22.1 INTRODUCTION

There are numerous biosensors described in the literature. However, only a relatively small number of them have been tested under realistic conditions. To apply a biosensor in an analytical system involves integration of biosensor technology with sampling, sample handling, and data processing. The lack of this integration may to a large extent explain the so far limited success that biosensors have had on the market.

If a biosensor is to be applied to monitor a biotechnological process, it may either be applied in the reactor (in line), or a sample stream is taken out and the analysis is done off-reactor (at line). There are certain pros and cons for either of the two approaches [1].

It is obvious that biosensors *per se* give useful readings when applied to standard conditions and that the problems start to appear when real samples are introduced. Fouling of sampling as well as sample handling equipment and the biosensor itself may cause severe disturbances in the analyses. Besides such general complications, the presence of inhibiting substances may further distort the signal. Figure 22.1 shows a scheme of how an analytical process is actually carried out. What is obvious from the figure is that first sampling and then sample handling precede the bioanalytical step itself and the following data processing step. Thus, the biosensing step is just a small part of the total scheme for an analytical use of a biosensor.

It is from the figure quite obvious that the analytical process contains many more elements than just the biosensor. One has to realize that, in most cases, the biosensor is a part of an analytical system and that the whole system must be properly treated when designing an analysis.

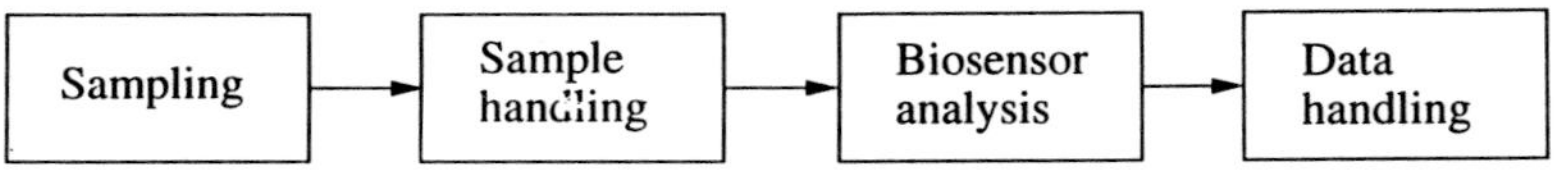

Figure 22.1 A schematic presentation of the different unit operations in a biosensor-mediated analysis.

22.2 FLOW INJECTION ANALYSIS

Off-reactor (at-line) analysis is very often carried out by applying flow injection analysis (FIA). The basic principle behind FIA is to operate with a continuous flow into which the sample is introduced as a liquid segment that then is transported at the same rate as the rest of the liquid (figure 22.2) [2]. FIA is an impulse–response technique, whereby the initial square wave input provided by sample injection is transformed into a response function by means of a modulator. As transport proceeds, some dispersion of the liquid segment takes place: a peak with a concentration profile is generated. Furthermore, most FIA applications involve a chemical reaction between the analyte and some suitable reagents prior to reading. Thus both dispersion and chemical reaction will influence the shape of the peak registered after injection of a sample pulse. Due to the high reproducibility in the flow rate and the volume of the sample pulse, the moving profile is also very reproducible. If a point (D in figure 22.2) is used for analysis, a substantially diluted sample is obtained. FIA offers superb possibilities for reproducible dilution, mixing and transport of liquid segments, and that forms the foundation for a technology that during the approximately 20 years that have passed since it was first presented has led to publication of approximately 60 000 scientific papers. Table 22.1 gives a list of some of the applications where FI bioanalysis has been used.

22.2.1 Equipment

The choice of pumps, valves, and sample loops strongly influences the performance of the FIA system. In most cases peristaltic pumps are used, and among these there are also large variations in flow properties. It is worth evaluating the pieces of equipment when setting up FIA systems, since the performance of the analysis is strongly dependent on the choice. Even when a reliable peristaltic pump is identified one has to keep control of the flow properties, since the tubings are aging and the flow may change because of this. From that point of view, piston pumps may offer an attractive, but often expensive alternative. The sample is introduced in the flow by use of a sample loop. By this arrangement variations in the volume of the sample added are kept at a minimum. The sample volume used is typically 100–200 μl and the flow rate often is approximately 1 ml min^{-1}.

FIA has been the target for miniaturization experiments. One can foresee that with smaller columns and lower dead volumes, a higher analytical throughput

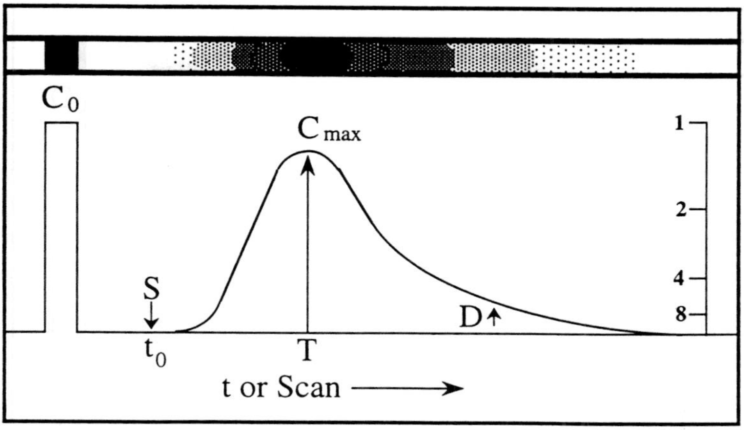

Figure 22.2 A schematic presentation of the performance of an originally homogeneous sample zone (at concentration C_0 to the left in the figure) as it is transported along the tubing system in the FIA equipment. The resulting peak represents a range of different concentrations with its peak value C_{max}.

Table 22.1 Areas where FIA has been implemented.

Area	Reference
Agricultural	
feed	[3]
pesticides	[4, 5]
Biochemical	
biotechnology	[6]
cell biology	[7]
Clinical	
blood/blood plasma	[8, 9]
urine	[10]
Environmental	
water	[11, 12]
Food	
beverages	[13]
juice	[13]
Industrial	
process control	[14, 15]

will be reached. A sacrifice one has to make is the limited amounts of active biomolecules that can be introduced in such miniaturized systems. The operational stability of a FIA system with immobilized enzymes is very much dependent on the fact that an excess of biomolecules is used, thereby making

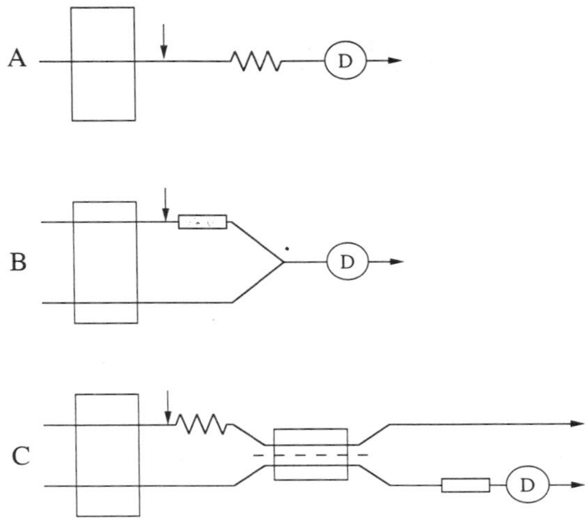

Figure 22.3 Schematic presentations of three different arrangements for FIA systems. (A) symbolizes the simple system where the sample is introduced and passed through a reaction coil through the detector. (B) contains two streams, one with a column reactor and one just for supplying media after passage of the reactor before entering the detector. (C) is a more complex arrangement with two liquid streams, one sample stream and one containing receiving buffer, a reaction coil, and a dialysis unit. After the dialysis step the sample passes through a column reactor before entering the detector zone.

Table 22.2 Enzyme content in biological solutions analysed by FIA.

Enzyme	Biological fluid	Assay procedure	Reference
Amyloglucosidase	enzyme solution	spectrophotometer	[16]
Alkaline protease	fermentation broth	spectrophotometer	[17]
Pullulanase	fermentation broth	spectrophotometer	[17]
Alkaline phosphatase	calf-bone extract	spectrophotometer	[18]
Cellobiose dehydrogenase	enzyme solution	spectrophotometer	[19]
α-amylase	fermentation broth	spectrophotometer	[20]
Hexokinase	yeast homogenate	calorimeter	[21]

the biocatalytic step diffusion controlled. Thus, even if the sensitivity is high in the reaction *per se*, the stability may be low where an insufficient amount of biomolecule is used.

Various unit operations can be done on line in a reproducible manner in an FIA system such as mixing, dialysis and reactions. Figure 22.3 illustrates a few schematic presentations of experimental set-ups for FIA.

FIA was first developed using conventional, wet chemistry and it was later realized that the technology was very suitable also to combine with the use of immobilized biochemical entities, initially mainly enzymes but recently also antibodies. However, there are applications, e.g. quantitation of soluble enzymes in a liquid sample, that still with preference are carried out with all the reagents in free solution. Table 22.2 lists some examples of analysis of activities of free enzymes.

22.2.2 FIA biological components

Two general concepts are used when applying immobilized enzymes in FI systems. One is the 'immobilized enzyme reactor' that according to a strict definition of the true biosensor is not a sensor. However, in most publications in the area it is regarded as being one. The other alternative is to use an enzyme probe applied as the detector in an FIA system.

When dealing with FIA in combination with reactors containing immobilized enzymes, usually very small reactors are used (100–250 μl packed bed volumes). The enzyme is immobilized to a solid support permitting a good flow rate and a high capacity for binding the enzyme. In most cases, controlled-pore glass (CPG) is used [11]. Other materials that have been reported are Eupergit [22, 23], VA Epoxy Biosynth [15], amino-Cellulofine [24], and Sepharose [25, 26]. It is important to operate with materials that have low non-specific adsorption and that have good flow properties. The most popular material is CPG since a surplus of convenient coupling alternatives are available and a rich literature is present on successful applications using CPG-bound enzymes. All the materials used are primarily selected from chromatographic media giving good flow properties. Furthermore, porous supports offer a large surface area and thus the option of immobilizing large numbers of biomolecules per unit volume of support material.

Typical loads of protein on the supports are 5–10 mg/ml support (sometimes even more has been used) depending on the purity of the enzyme preparation and on the turnover number for the enzyme. Sufficient enzyme is immobilized so that diffusion-controlled processes are at hand.

Typical column sizes are 50–200 μl bed volume and flow rates used are in the range 0.1–1.0 ml min^{-1}, often in the upper part of the range.

It is important to immobilize a large amount of enzyme on the support, since then the catalytic process becomes diffusion controlled, i.e., only a fraction of the enzyme population will be active in converting the substrate. On the other hand, when denaturation of the enzyme takes place, substrate will diffuse deeper into the support and meet enzyme molecules that earlier were resting because no substrate reached them. By operation with an excess of enzyme, one secures a constant observable catalytic output from the reactor and thus a constant read-out from the analysis. This property is important since then an efficient reuse of the analytical system can be achieved without having to control the catalytic capacity more than rarely. A high operational stability secures reproducibility in the catalytic part of the analytical process.

Table 22.3 Various detectors used in conjunction with FIA. Some applications and a reference are given.

Detector	Assay	Sensitivity/ operational range	Reference
Ammonia gas electrode	Hg^{2+}	0–0.7 nmol	[11]
Calorimeter	multiple recycling enzyme system	10^{-8} mol l^{-1}	[27]
Spectrophotometer	transferrin flow-ELISA	10^{-9} mol l^{-1}	[28]
Fluorescence	IgG	—	[29]
Fluorescence	monoclonal antibody	1–100 μg ml^{-1}	[30]
Chemiluminescence	thyroxine	$>10^{-11}$ mol l^{-1}	[31]
Chemiluminescence	monoclonal antibody	$>2.5 \times 10^{-12}$ mol l^{-1}	[32]
Electrochemical	HSA	10^{-6} mol l^{-1}	[33]
Surface plasmon	theophylline	10 pg	[34]
Ellipsometer	IgG	—	[35]
Optical waveguide	theophylline	$>10^{-10}$ mol l^{-1}	[36]

When dealing with samples that are well defined solutions, containing no particulate or inhibitory material, a simple process configuration may be applied. The throughput in most cases is equal to or better than 1 assay min^{-1}.

It is seen from figure 22.4 that the peaks are sharp and reproducible. The interpretation of the analytical result may be achieved by measuring either the peak height or the slope of the rising part of the peak, or by integrating the area under the peak. When evaluated manually peak height is the most convenient, and when computer-controlled evaluation is used peak area integration is preferred.

When running subsequent samples it is important to ensure that a previous sample is not interfering with the actual sample (so called carryover).

22.2.3 Detectors used

There is a large degree of freedom when choosing a detector to monitor the outcome. Most of the applications so far have been based on use of a small reactor with immobilized enzymes/immobilized antibodies in the flow system preceding the detection step. That the technology is versatile is illustrated by the broad spectrum of detectors that have been used (table 22.3).

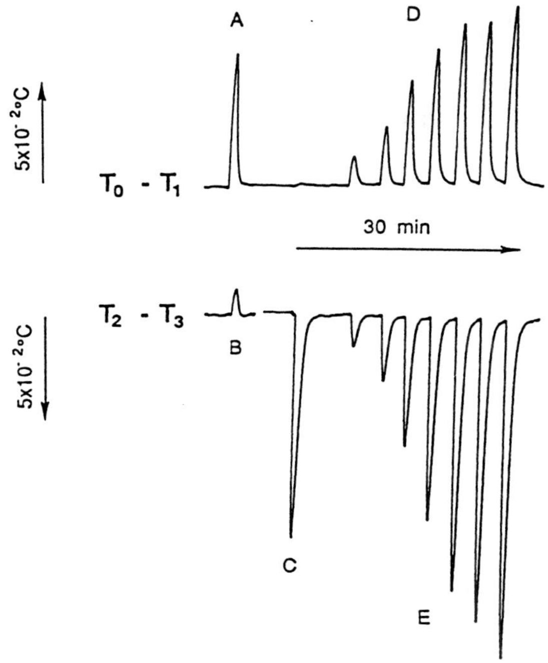

Figure 22.4 Typical peaks registered from repetitive assays using an FI system with thermal detectors. The responses were obtained after injection of urea, penicillin V, and a mixture of the two into a system with an integrated thermal biosensor for simultaneous determinations of multiple analytes. Both signals were exothermic. The upper trace was registered with a thermistor pair called T_0–T_1. The lower trace was registered using another thermistor pair, T_2–T_3. In this case, the direction of the penicillin peaks was reversed by alternating the recorder polarity. A, response to 30 mmol l^{-1} urea; B, influence of the urease reaction on on the thermistor pair T_2–T_3; C, response to 30 mmol l^{-1} penicillin V; D, simultaneous responses for urea (5–60 mmol l^{-1}) in the mixed samples; E, simultaneous responses for penicillin V (5–60 mmol l^{-1}) in the mixed samples (from [47], with permission).

22.2.4 Special arrangements to improve sensitivity

Using a computer-controlled FIA system, certain technical arrangements can be implemented in order to raise the sensitivity of the assays. Stopped flow procedures where the sample is kept in contact with the immobilized enzyme preparation for a longer time than the passage takes is one such arrangement being used [37]. Reversal of flow rate in order to improve mixing in the enzyme reactor has also been demonstrated to be successful [37].

On the chemical level, enzyme-catalyzed recycling of substrates/coenzymes to improve sensitivity of an assay has been successfully demonstrated [27, 38, 39]. The principle is illustrated in the reaction sequence shown in figure 22.5.

22.2.5 Sampling and sample handling

This is a crucial area when designing analyses for real samples [40]. Sampling must be done in such a way that a representative sample is analyzed. It must be removed from the reactor under conditions that do not involve risks of disturbance of the complete process. Moreover, care must be taken to stop any metabolic activity in the sample as soon as possible, since otherwise changes in the sample may take place while being transported in the FIA system. Dialysis probes have been applied when sampling from fermenters [41–43], whereas coaxial catheters are used both for fermenters and in medical applications [44–46]. The former approach involves the use of a dialysis membrane placed in the reactor. Sample molecules passing through the membrane are transported via the FIA system for subsequent analysis. The coaxial catheter (figure 22.6) involves additions of inhibitors to the sample at the sampling point and the mixture is then further processed for analysis [44]. In this case, a subsequent dialysis step is often used prior to analysis. The same coaxial catheter has also been used for

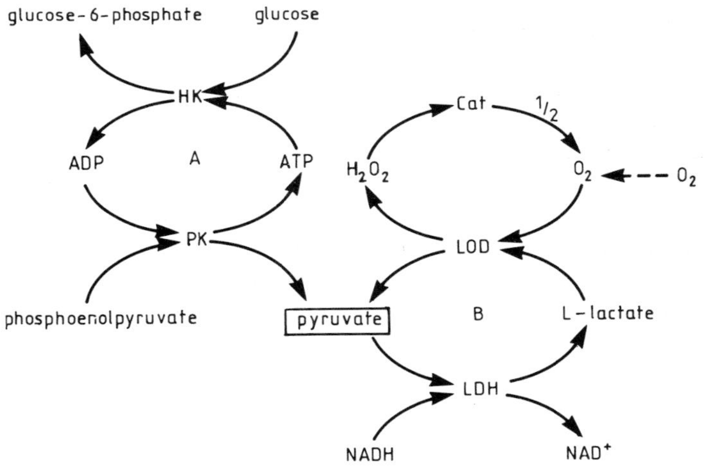

Figure 22.5 An example of reactions involved in an enzyme-catalyzed recycling processes for amplification of the sensitivity. In the left part (A) of the figure the enzyme-pair hexokinase and pyruvate kinase is used for recycling of the coenzyme ATP/ADP. In the right part (B), substrate recycling of pyruvate or lactate is accomplished using the enzyme-pair lactate oxidase/lactate dehydrogenase. A multiplication effect is obtained by combination of A and B resulting in a very high sensitivity [27]. The calorimetric sensitivity is further inreased by including catalase (cat).

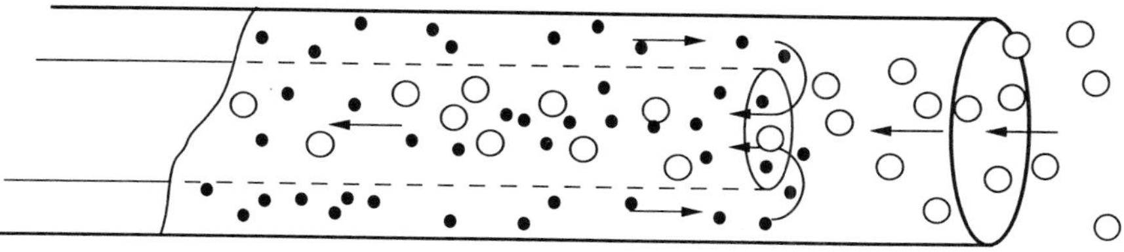

Figure 22.6 A coaxial catheter for sampling and mixing. ○ depicts the reaction mixture and • depicts the dilution solution containing an inhibitor. The flow rate in the inner lumen is typically set at twice the flow rate in the outer space.

Table 22.4 Examples of analyses carried out using immobilized preparations of a single enzyme in the FIA system.

Analyte	Enzyme	Operational range	Reference
Penicillin V	β-lactamase	5–40 mM	[47]
Fructose	fructose-5-dehydrogenase	0.02–2 mM	[24]
Urate	urate oxidase	up to 0.1 g l^{-1}	[48]
Glucose	glucose oxidase	up to 4 g l^{-1}	
Urea	urease	>0.2 μM	[10]
Sulfite	sulfite oxidase	10^{-5}–8×10^{-4} M	[49]
Glucose	glucose dehydrogenase	1 μM–10 mM	[50]
Hydrogen peroxide	peroxidase	>5 μM	[51]

on-line calibration of the analytical system, i.e. not only the sensor but also the integrated analytical system is calibrated [46]. This characteristic is especially important in situations when the analytical system can not be moved but still has to be calibrated. Thus, when running continuous registrations on patients by monitoring blood concentrations of metabolites such as glucose, lactate, or urea, one must be able to calibrate without disconnecting the system from the patient. It should be stressed that the calibration includes the whole system, i.e. also tubings etc, and thus will give a better view of how the analytical system works. The on-line calibration that was worked out for medical applications may very well turn out to be useful in fermentation monitoring.

Dialysis units have successfully been integrated into the system. By such an arrangement not only is it possible to remove particular matter in the stream, but one may also introduce a dilution step.

22.2.6 Applications involving simple enzymatic steps

A range of reactions (table 22.4) have been successfully carried out. The volume of sample, i.e., the pulse length, influences the sensitivity directly. From the table

Table 22.5 Analytical applications of immobilized enzyme sequences in conjunction with FIA.

Analyte	Enzymes	Sensitivity	Reference
Pullulan	pullulanase + amyloglucosidase + glucose dehydrogenase	10^{-5}–5×10^{-4} M	[13]
Galactose	galactose oxidase + peroxidase	10 μM–14 mM	[53]
Glucose	glucose oxidase + peroxidase	0.05–10 g l^{-1}	[54]
Sucrose	β-fructosidase + mutarotase + glucose dehydrogenase	>10 μmol l^{-1}	[13]
Maltose	α-glucosidase + glucose oxidase	0.2–5.0 g l^{-1}	[15]
Lactose	β-galactosidase + glucose oxidase	5.0–30.0 g l^{-1}	[15]
Sucrose	invertase + glucose oxidase	1.0–5.0 g l^{-1}	[15]
Glutamine	glutaminase + glutamate dehydrogenase	0.1–1.5 g l^{-1}	[15]
Starch	amyloglucosidase + glucose dehydrogenase	>0.1 mg ml^{-1}	[3]

is seen that such assays operate within a broad range of applications. If a product is formed in the first enzyme-catalyzed reaction that is difficult to quantify, then one or more subsequent enzymatic steps are introduced. Most enzyme reactions are possible to read in the concentration range down to 10 μM. Due to the excess of immobilized enzymes, one can predict a high operational stability and thus little need for calibration of the systems.

22.2.7 Analysis involving enzyme sequences

In many cases it is not possible or desirable to register the reaction catalyzed by an enzyme. It may be that the product is difficult to detect or that the sensitivity in the analysis one needs to apply is not high enough. Then the use of one or more additional enzymes is quite common. The strategy has been worked out for soluble enzymes, and, in the flow systems, the enzymes are either immobilized separately, or coimmobilized. The latter approach has certain advantages in the sense that a better kinetic performance can be observed in a coimmobilized enzyme sequence as compared to when the enzymes are immobilized separately [52]. However, since most assays are based on the use of an excess of immobilized enzymes, no dramatic differences are observed.

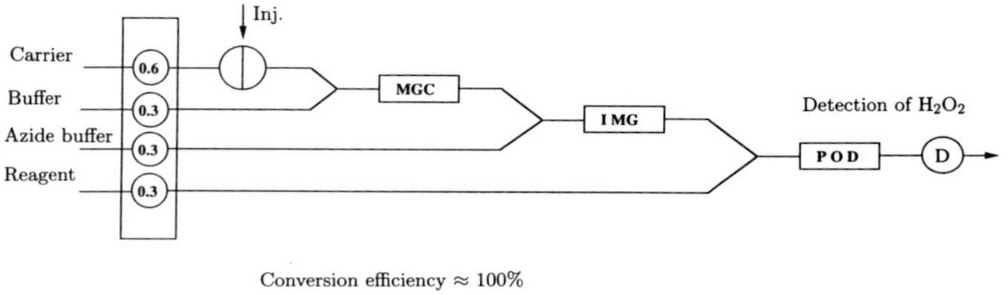

Figure 22.7 An experimental arrangement with three different enzyme columns used for analysis of sucrose in the presence of glucose (from [55], with permission). M = mutarotase; G = glucose oxidase; C = catalase; I = invertase; POD = peroxidase; D = detector.

In table 22.5 are listed several examples of analyses performed using enzyme sequences. Already discussed is the catalytic cycling of substrates/cofactors in order to improve the sensitivity of an assay. This can be regarded as a special case of utilizing enzyme sequences.

22.2.8 Sample pretreatment or differential analysis of a sample containing two or more components that can react in the enzyme assay

Analysis of sucrose in the presence of glucose is desired when evaluating the quality of sugar beet juice. Olsson *et al* [55] solved this problem by an intricate combination of three enzyme reactors (figure 22.7).

In the first reactor, mutarotase converts α-glucose into the β-form that is acceptable for the subsequently acting glucose oxidase. Glucose oxidase converts glucose into gluconolactone with a subsequent production of hydrogen peroxide. Catalase is the third enzyme in the reactor and it degrades the peroxide formed. Thus, after passing the first column, glucose is removed while sucrose is untouched. The second column contains invertase, glucose oxidase and catalase. Invertase hydrolyses sucrose thereby forming glucose and fructose. The outcome of the subsequent reactions is hydrogen peroxide that—after mixing with a suitable cosubstrate—in a subsequent reactor containing peroxidase is converted into a colored product that can be quantified in a photometric detector.

This example illustrates that, besides convenient analyses, one may also include sample pretreatment. This is a characteristic that seems to attract increasing attention as more complex samples are being analyzed [13].

22.2.9 Sequential injection assay (SIA)

SIA is a recent development in FIA [56]. SIA is based on the use of a pump (often a piston pump) connected via a spacing tubing with a multiport valve (figure 22.8). The strategy is to aspirate sample and reagents as small liquid segments in a desired order in the spacing tubing (figure 22.9). When the liquids

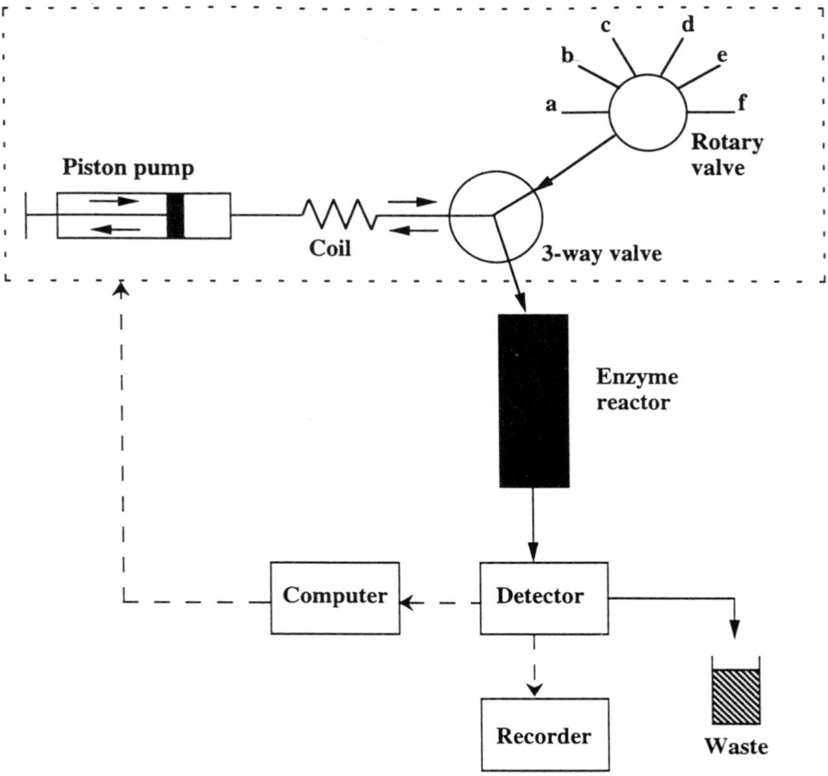

Figure 22.8 A schematic presentation of the sequential injection system including syringe pump, holding coil, three-way valve, six-way rotary valve, enzyme column, recorder, detector, and computer.

are arranged, the multiport valve is switched to connect the flow from the pump via an enzyme reactor to a detector. The pump is reversed and the train of segments of reagents is pushed through the enzyme reactor. Mixing and reaction take place and the outcome is registered in the detector.

The procedure and the equipment are simpler than those of FIA. The consumption of buffer is substantially reduced. When working with expensive reagents SIA seems to be preferred over FIA, since much less chemicals are used. There is a need for computer control of the pump in order to achieve a high reproducibility. Most of the work with this method was initially carried out on soluble reagents, but in analogy with FIA it also ought to be useful for immobilized preparations.

22.2.9.1 Determination of D-lactate using immobilized D-lactate dehydrogenase in an SIA system. D-lactate is of interest to quantify in different situations. As an indicator of freshness of meat, its concentration can be correlated with the

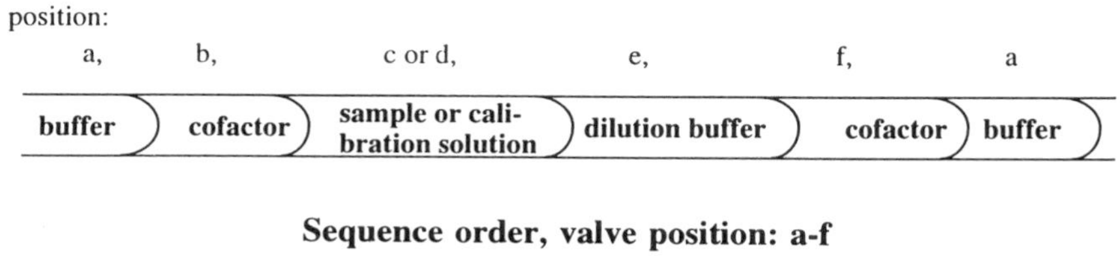

Figure 22.9 SIA operation principle with the segments of liquids arranged for assay of D-lactate using immobilized D-lactate dehydrogenase.

activity of lactic acid bacteria on meat. D-lactate dehydrogenase (EC 1.1.1.28) was immobilized on porous glass and the preparation was placed in a reactor (30 mm $\times$ 4 mm) and connected to the SIA system [37]. It soon turned out that the equilibrium for the enzymatic reaction is unfavorable and therefore a subsequent enzyme was also introduced, L-alanine aminotransferase (EC 2.6.1.2). The sample was loaded as a segment surrounded by small segments of cofactor solution. When passing the sample through the enzyme reactor unit, very little reaction took place. The flow was stopped in order to keep the sample pulse and the cofactor pulse in the reactor for conversion to take place; much improved response was registered. The time for incubation was optimized and it turned out that, with the conditions used, 90 s stop gave an optimal response. In order to investigate the effect of mixing, the flow was stopped, the flow reversed, stopped again, and reversed again. This procedure was repeated several times, but the improvements were negligible after the first cycle [37].

Using the same approach now in combination with a carbon paste electrode including D-lactate dehydrogenase, NAD^+, and a polymer-bound mediator, the concentration of D-lactate was followed during a fermentation [57, 58]. This system is reagentless and the response was faster as compared to that of the enzyme reactor described above. In both cases it was shown that the selectivity of the enzyme was very good; the only severely disturbing substance was pyruvate—the product of the reaction.

SIA opens new possibilities in cell biology research. The inventor of SIA as well as of FIA, Professor Ruzicka, has recently introduced SIA for studies on cell properties by measuring responses from individual cells to stimulants [7].

22.2.10 FI binding assays

Due to the high reproducibility in the experimental conditions FIA has proven very suitable for performing binding reactions. By applying a competitive ELISA in the FIA system, quick immunochemical binding assays can be set up. The flow systems give a very short contact time for the sample when passing the column of immobilized antibodies. The assay is performed far from equilibrium. This is possible to tolerate due to the high reproducibility of all experimental conditions [59, 60].

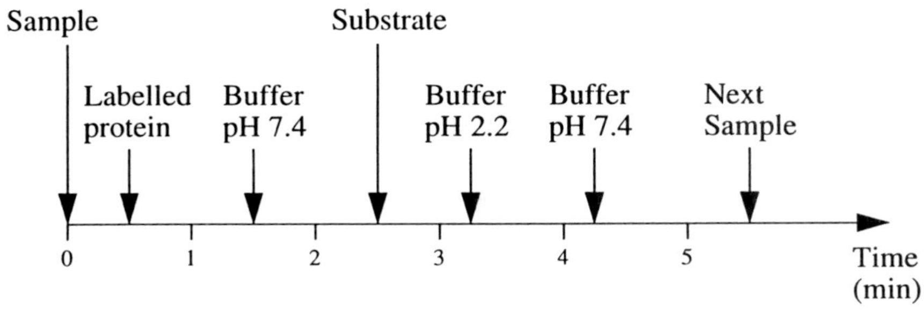

Figure 22.10 An assay scheme for a displacement binding assay. The sample is introduced and directly afterwards a pulse of labeled antigen is introduced. There is a buffer wash before addition of substrate for the labeling enzyme. Buffer pH 2.2 is used to dissociate the immunocomplex and afterwards the system is reconditioned with running buffer.

In figure 22.10 a schematic assay cycle is represented together with two calibration curves. The antigen containing sample may either be mixed with a fixed amount of enzyme-labeled antigen and then introduced into the flow system, or the enzyme-labeled antigen may be added directly after the sample pulse. According to the protocol of a competitive ELISA, the labeled and the native reagents are mixed first and then introduced. When carrying out the reaction in such a way, rather high variations were registered, and it became difficult to decrease the variation to an acceptable level. However, when applying the sequential mode of operation very low variation was observed. Obviously, it is possible first to expose the antibody column for the native antigens in the sample and directly afterwards to add the labeled antigen. So, even if no competitive binding takes place, a situation involving displacement chromatography may be at hand, and good calibration curves of the same shape as obtained from competitive assays can be registered.

The sensitivity of the assay will be dependent upon the antibodies used. A high-affinity antibody will give a higher sensitivity. The number of immobilized antibodies will influence the sensitivity as well as the ratio between labeled and native antigen in the competitive binding assay. Variations of the pulse length have also been shown to influence the sensitivity. Typical assays operated with a cycle time of around 6 min give a reading down to 10^{-9} mol l^{-1}. Lower concentrations can be read using a more dilute sample injected after a longer time [9]. The fact that the reading is available as early as approximately 2 min after injection of the sample makes flow-ELISA an attractive method for quick immunoassays for, e.g., process monitoring and in future also process control.

The calibration curve is expressed as the percentage of the response obtained when only labeled antigen is present and no native antigen is added. Since the competitive situation between labeled and native antigen stays constant, the calibration curve remains the same if correction is made for the loss of binding

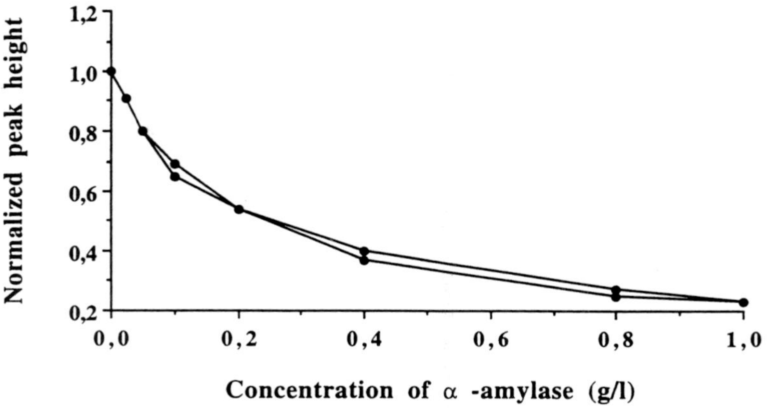

Figure 22.11 Two calibration curves for determination of α-amylase using flow-ELISA. The two curves are registered using the same antibody column; one curve is for the fresh column and the other is registered after some 280 assays. The readings are normalized with respect to the binding capacity of the antibody column. (From [26], with permission.)

capacity in the antibody column. In figure 22.11 two calibration curves obtained using the same antibody column are shown, one when the column is fresh and the other after 280 assays [26]. The same system was used for monitoring the production of α-amylase in a *Bacillus amyloliquifaciens* cultivation [14]. For that purpose a rather insensitive assay was set up, in order to avoid separate dilution steps.

Flow-ELISA has been used when the effluent of a chromatographic column has been followed. A specific immunological detection could be applied to identify the peak containing the target protein. The potential of using this kind of technology for convenient process monitoring in down-stream processing in biotechnology is very interesting [21, 61], but so far hardly any development has taken place.

A critical step is the regeneration of the antibody preparation since one needs to apply harsh conditions to achieve dissociation of the antigen–antibody complex. After binding and proper washing has taken place, incubation with substrate is carried out in order to quantify the amount of label that is bound. Then dissociation was performed, usually with 0.2 M glycine-HCl pH 2.2. This solution is known to be harmful to many antibody preparations and one can foresee that problems with stability may occur. When working with polyclonal antibodies, these are first affinity purified using a column with immobilized antigen and elution is achieved using glycine-HCl pH 2.2. By such a procedure we select antibodies that will dissociate from the antigen upon the treatment. In the assay the same conditions are used, but then the antibody is immobilized. It has been argued that it is not the low pH *per se* that causes the dissociation but rather the sharp pH gradient. Experience from a range of different preparations has shown that in most cases there is a slow, but fully observable denaturation. Often less than 1% of the binding capacity of the active antibodies is lost per cycle, but when dealing with more labile structures much higher denaturing rates are observed. This means that the lifetime of the column with the immobilized biochemical species becomes shorter, and then capacity measurements must be carried out more frequently.

One way to improve the stability of the antibody preparation would be to use a sandwich assay instead, since then a surplus of antibodies can be used in the immobilized preparation. This holds certain positive features, but also some drawbacks. It will be a slightly more laborious and time consuming assay, but the sandwich assay concept is attractive for the low concentration range.

In recent years a range of different groups have published on immunochemical assay in combination with FIA [30, 33, 62].

22.2.11 Commercialization

FIA as such is available from a few different suppliers. Reactors filled with immobilized enzymes are also commercially available. More sophisticated equipment based on the FIA concept is the BIACORE from Pharmacia Biosensor. Here a gold surface covered with a carboxydextran is used as reactor for binding reactions. The reading is done by means of evanescent wave technology [63]. The BIACORE illustrates one step in the direction of miniaturization of the analytical unit. As the understanding of the chemistry and the analytical protocol becomes available, much may be gained by miniaturization [64].

22.2.12 Conclusion

Flow injection analysis offers many attractive features to biosensor analysis. The reproducibility and the speed are two dominant characteristics when combining FIA with proper sampling and sample handling. It can be used for both enzyme-based assays and immunochemical binding assays.

FIA in all the potential configurations constitutes a very powerful tool when biosensors are set to work under realistic conditions.

ACKNOWLEDGMENTS

The continuous support from the National Swedish Board for Technical Development is gratefully acknowledged.

REFERENCES

[1] Mattiasson B 1986 Technological processes for biotechnological utilization of microorganisms *Biotechnology–Potentials and Limitations* ed S Silver (Berlin: Springer) pp 113–25

[2] Ruzicka J and Hansen E 1988 *Flow Injection Analysis* (New York: Wiley)

[3] Emnéus J, Appelqvist R, Marko-Varga G, Gorton L and Johansson G 1986 Determination of starch in a flow injection system using immobilized enzymes and a modified electrode *Anal. Chim. Acta* **180** 3–8

[4] Botrè F, Lorenti G, Mazzei F, Simonetti G, Porcelli F, Botrè C and Scibona G 1994 Cholinesterase based bioreactor for determination of pesticides *Sensors Actuators* B **18–19** 689–93

[5] Bier F F, Jockers R and Schmid R D 1994 Integrated optical immunosensor for s-triazine determination: regeneration, calibration and limitations *Analyst* **119** 437–41

[6] Mattiasson B 1994 Flow injection bioanalysis—a convenient tool in process monitoring and control *ECB 6, Proc. 6th Eur. Congress on Biotechnology (Firenze, 1993)* ed L Alberghina, L Frontali and P Sensi (Amsterdam: Elsevier) pp 869–72

[7] Ruzicka J 1994 Discovering flow injection: journey from sample to a live cell and from solution to suspension *Analyst* **119** 1925–34

[8] Kyröläinen M, Håkanson H and Mattiasson B 1993 Biosensor monitoring of blood lactate during open heart surgery *Anal. Chim. Acta* **279** 149–53

[9] Borrebaeck C, Börjesson J and Mattiasson B 1978 Thermometric enzyme linked immunoassay in continuous flow systems: optimization and evaluation using human serum albumin as a model system *Clin. Chem. Acta* **86** 267–78

[10] Winquist F, Spetz A, Lundström I and Danielsson B 1984 Determination of urea with an ammonia gas-sensitive semiconductor device in combination with urease *Anal. Chim. Acta* **163** 143–9

[11] Ögren L and Johansson G 1978 Determination of traces of mercury(II) by inhibition of an enzyme reactor electrode loaded with immobilized urease *Anal. Chim. Acta* **96** 1–11

[12] Svensson A, Hünning P Å and Mattiasson B 1979 Application of enzymatic processes for monitoring effluents. Measurements of primary amines using immobilized monoamine oxidase and the enzyme thermistor *J. Appl. Biochem.* **1** 318–24

[13] Ogbomo I, Kittsteiner-Eberle R, Engelbrecht U, Prinizing U, Danzer J and Schmidt H-L 1991 Flow-injection systems for the determination of oxidoreductase substrates: applications in food quality control and process monitoring *Anal. Chim. Acta* **249** 137–43

[14] Nilsson M, Vijayakumar A R, Holst O, Schornack C, Håkanson H and Mattiasson B 1994 On-line monitoring of product concentration by flow-ELISA in integrated fermentation and purification process *J. Ferment. Bioeng.* **78** 356–60

[15] Schügerl K, Brandes L, Dullau T, Holzhauer-Rieger K, Hotop S, Hübner U, Wu X and Zhou W 1991 Fermentation monitoring and control by on-line flow injection and liquid chromatography *Anal. Chim. Acta* **249** 87–100

[16] Holm K A 1986 Automated spectrophotometric determination of amyloglucosidase activity using p-nitrophenyl-α-D-glycopyranoside and a flow injection analyzer *Analyst* **111** 927–9

[17] Kroner K H and Kula M-R 1984 On-line measurement of extracellular enzymes

during fermentation by using membrane technique *Anal. Chim. Acta* **163** 3–15

[18] Takahashi K, Taniguchi S, Kuroishi T, Yusuda K and Sano T 1989 Automated stopped flow/continuous flow apparatus for serial measurement of enzyme reactions and its application as real-time analyzer for column chromatography *Anal. Chim. Acta* **220** 13–21

[19] Holm K A 1986 Spectrophotometric flow-injection determination of cellobiose dehydrogenase activity in fermentation samples with 2, 6-chlorophenolindophenol *Anal. Chim. Acta* **188** 285–8

[20] Hansen P W 1984 Determination of α-amylase by flow injection analysis *Anal. Chim. Acta* **158** 375–7

[21] Danielsson B and Larsson P-O 1990 Specific monitoring of chromatographic procedures *Trends Anal. Chem.* **9** 223–7

[22] Huck H, Schelter-Graf A and Schmidt H-L 1984 Measurement and calculation of the calibration graphs for flow injection analyses using enzyme reactors with immobilized dehydrogenases and an amperometric NADH detector *Biochim. Biophys. Acta* **632** 298–309

[23] Wehnert G, Sauerbrei A and Schügerl K 1985 Glucose oxidase immobilized on Eupergit C and CPG-10. A comparison *Biotechnol. Lett.* **7** 827–30

[24] Matsumoto K, Hamada O, Ukeda H and Osajiama Y 1986 Amperometric flow injection determination of fructose with an immobilized fructose-5-dehydrogenase reactor *Anal. Chem.* **58** 2732–4

[25] Mattiasson B, Borrebaeck C, Sanfridsson B and Mosbach K 1977 Thermometric enzyme linked immunosorbent assay: TELISA *Biochim. Biophys. Acta* **483** 221–7

[26] Nilsson M, Mattiasson G and Mattiasson B 1993 Automated immunochemical binding assay (FLOW-ELISA) based on repeated use of an antibody column placed in a flow-injection system *J. Biotechnol.* **31** 381–94

[27] Kirstein D, Danielsson B, Scheller F and Mosbach K 1989 Highly sensitive enzyme thermistor determination of ADP and ATP by multiple recycling enzyme systems *Biosensors* **4** 231–9

[28] Larsson K M, Olsson B and Mattiasson B 1987 Flow-injection enzyme immunoassay—a quick and convenient binding assay *Proc. 4th Eur. Congress on Biotechnology (Amsterdam, 1987)* vol 3, ed O M Neijssel, R R van der Meer and K A C M Luyben (Amsterdam: Elsevier) pp 196–9

[29] Kelly T A and Christian G D 1982 Homogeneous enzymatic fluorescence immunoassay of serum IgG by continuous flow-injection analysis *Talanta* **29** 1109–12

[30] Stöcklein W and Schmid R D 1990 Flow injection immunoanalysis for the on-line monitoring of monoclonal antibodies *Anal. Chim. Acta* **234** 83–8

[31] Arefyev A A, Vlasenko S B, Eremin S A, Osipov A P and Egorov A M 1990 Flow injection enzyme immunoassay of haptens with enhanced chemiluminescence detection *Anal. Chim. Acta* **237** 285–9

[32] Shellum C and Gübitz G 1989 Flow injection immunoassays with acridinium ester-based chemiluminescence detection *Anal. Chim. Acta* **227** 97–107

[33] Heineman W R and Halsall H B 1985 Strategies for electrochemical immunoassay *Anal. Chem.* **57** 1321A–31A

[34] Sjölander S and Urbaniczky C 1991 Integrated fluid handling system for biomolecular interaction analysis *Anal. Chem.* **63** 2338–45

[35] Jönsson U, Rönnberg I and Malmqvist H 1985 Flow injection ellipsometry—an in situ method for the study of biomolecular adsorption and interaction at solid surfaces *Colloid. Surf.* **13** 333–9

[36] Choquette S J, Locascio-Brown L and Durst R A 1992 Planar waveguide

immunosensor with fluorescent liposome amplification. *Anal. Chem.* **64** 55–60

[37] Shu H-C, Håkanson H and Mattiasson B 1993 D-lactic acid in pork as a freshness indicator monitored by immobilized D-lactate dehydrogenase using sequential injection analysis *Anal. Chim. Acta* **283** 727–37

[38] Schubert F, Kirstein D, Schröder K L and Scheller F 1985 Enzyme electrodes with substrate and coenzyme amplification. *Anal. Chim. Acta* **169** 391–6

[39] Scheller F, Siegbahn N, Danielsson B and Mosbach K 1985 High sensitivity enzyme thermistor determination of L-lactate by substrate recycling *Anal. Chem.* **57** 1740–3

[40] Mattiasson B and Håkanson H 1993 Sampling and sample handling - crucial steps in process monitoring and control *Trends Biotechnol.* **11** 136–42

[41] Graf H 1989 *PhD Thesis* Universität Hannover

[42] Ungerstedt U 1984 *Measurement of Neurotransmitter Release in vivo* ed C A Marsden (New York: Wiley) pp 81–105

[43] Mandenius C F, Danielsson B and Mattiasson B 1984 Evaluation of a dialysis probe for continuous sampling in fermenters and in complex media *Anal. Chim. Acta* **163** 135–41

[44] Holst O, Håkanson H, Miyabayashi A and Mattiasson B 1988 Monitoring of glucose in fermentation processes using a commercial glucose analyzer *Appl. Microbiol. Biotechnol.* **28** 335–9

[45] Thavarungkul P, Håkanson H, Holst O and Mattiasson B 1991 Continuous monitoring of urea in blood during dialysis *Biosensors Bioelectron.* **6** 101–7

[46] Kyröläinen M, Håkanson H and Mattiasson B 1995 On-line calibration in computerized biosensor system for continuous measurement of glucose and lactate *Biotechnol. Bioeng.* **45** 122–8

[47] Xie B, Mecklenburg M, Danielsson B, Öhman O, Norlin P and Winquist F 1995 Development of an integrated thermal biosensor for the simultaneous determination of multiple analytes *Analyst.* **120** 155–60

[48] Tabata M, Fukunaga C, Ohyaba M and Murachi T 1984 Highly sensitive flow injection analysis of glucose and uric acid in serum using an immobilized enzyme column and chemiluminescence *J. Appl. Biochem.* **6** 251–8

[49] Masoom M and Townsend A 1986 Flow-injection determination of sulphite and assay of sulphite oxidase *Anal. Chim. Acta* **179** 399–405

[50] Appelqvist R, Marko-Varga G, Gorton L, Torstensson A and Johansson G 1985 Enzymatic determination of glucose in a flow system by catalytic oxidation of the nicotinamide coenzyme at a modified electrode *Anal. Chim. Acta* **169** 237–47

[51] Olsson B and Ögren L 1983 Optimization of peroxidase immobilization and of the design of packed-bed enzyme reactors for flow injection analysis *Anal. Chim. Acta* **145** 87–99

[52] Mosbach K and Mattiasson B 1976 Multistep enzyme systems *Methods in Enzymology* vol 44, ed K Mosbach (New York: Academic) pp 453–79

[53] Olsson B, Lundbäck H and Johansson G 1985 Galactose determination in an automated flow-injection system containing enzyme reactors and an on-line dialyzer *Anal. Chim. Acta* **167** 123–36

[54] Yao T, Sato M, Kobayashi Y and Wasa T 1984 Flow injection analysis for glucose by the combined use of an immobilized glucose oxidase reactor and a peroxidase electrode *Anal. Chim. Acta* **165** 291–6

[55] Olsson B, Stålbom E and Johansson G 1986 Determination of sucrose in the presence of glucose in a flow-injection system with immobilized multi-enzyme

reactions *Anal. Chim. Acta* **179** 203–8

[56] Ruzicka J and Marshall G D 1990 Variable flow rates and a sinusoidal flow pump for flow injection analysis *Anal. Chem.* **62** 1861–6

[57] Shu H-C, Håkanson H and Mattiasson B 1995 On-line monitoring of D-lactic acid during a fermentation process using immobilized D-lactate dehydrogenase in a sequential injection analysis system *Anal. Chim. Acta* **300** 277–85

[58] Shu H-C, Persson B, Gorton L and Mattiasson B 1995 A reagentless amperometric electrode based on carbon paste, chemically modified with D-lactate dehydrogenase, NAD^+, and mediator containing polymer for D-lactic acid analysis. II. On-line monitoring of a fermentation process *Biotechnol. Bioeng.* **46** 280–4

[59] Mattiasson B 1984 Immunochemical assays for process control: potentials and limitations *Trends Anal. Chem.* **3** 245–50

[60] Mattiasson B, Nilsson M, Berdén P and Håkanson H 1990 Flow-ELISA: binding assays for process control *Trends Anal. Chem.* **9** 317–21

[61] Nilsson M, Håkanson H and Mattiasson B 1992 Process monitoring by flow-injection immunoassay. Evaluation of a sequential competitive binding assay *J. Chromatogr.* **597** 383–9

[62] de Alwis E and Wilson G S 1987 Rapid heterogeneous competitive electrochemical immunoassay for IgG in the picomole range *Anal. Chem* **59** 2786–9

[63] Karlsson R, Michaelsson A and Mattsson L 1991 Kinetic analysis of monoclonal antibody–antigen interactions with a new biosensor based analytical system *J. Immunol. Methods* **145** 229–40

[64] van der Linden W E 1987 Miniaturization in flow injection analysis. Practical limitations from a theoretical point of view *Trends Anal. Chem.* **6** 37–40

23

Chemical and biological sensors: markets and commercialization

Richard F Taylor

23.1 INTRODUCTION

The modern era for sensors began in the late 1970s and early 1980s: new materials, reproducible manufacturing methods, and microcomputing devices led to the development of reliable sensors which are being used in most industries today. These sensors are primarily *physical sensors*, sensors which measure time, temperature, electric and magnetic fields, acceleration, and other physical parameters.

The technologies developed for physical sensors provided the transduction and electronic basis for *chemical* and *biological* sensors. Chemical and biological sensors represent more complex extensions of physical sensor technology [1, 2]. While physical sensors are based on mature technology and have many commercial manifestations, chemical and biological sensor technology is still emerging and few commercial products are available. Market predictions and trends for chemical and biological sensors are thus dependent on calculated predictions of how the technology to build and mass produce these sensors will evolve.

For the purposes of establishing markets and describing the commercialization of chemical and biological sensor, it is first necessary to define these devices. For both types of sensor, the following general definition is used in this review:

> A chemical or biological sensor is a measurement device which utilizes chemical or biological reactions to detect and quantify a specific analyte or event. The device is self-contained and comprised of three basic components: (i) a chemically or biologically active surface which specifically interacts with the analyte to be measured; (ii) a transducer which detects the chemical/biochemical event occurring between the

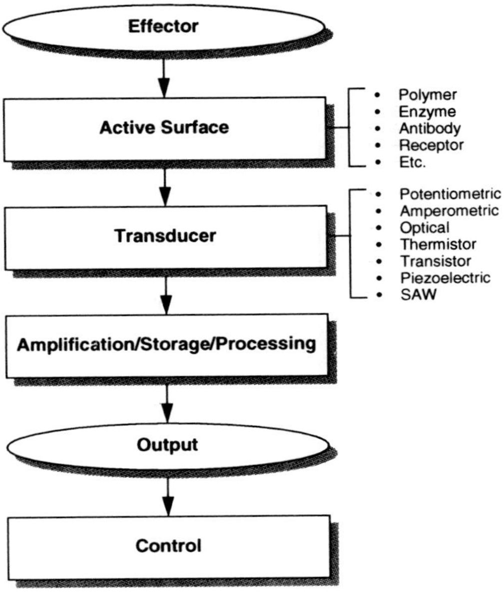

Figure 23.1 Basic components of a sensor.

surface and the analyte; and (iii) support electronics/software which amplify and report the output signal from the transducer.

The basic components of a chemical/biological sensor are illustrated in figure 23.1. The nature of the surface active layer determines the type of sensor. Chemical sensors utilize polymeric membranes *per se* or containing specific, low-molecular-weight doping agents. In most cases, the polymeric layer is the active component of the sensor which interacts with the target analyte. Biological sensors, or biosensors, can also contain polymeric layers and membranes, but all biosensors contain a biomolecule (such as an enzyme, antibody, or receptor) as the active component which interacts with the target analyte.

By this definition and as shown in figure 23.1, a true chemical sensor or biosensor thus incorporates both the detection and quantification systems into the same device.

This definition also clarifies what is *not* a chemical sensor or biosensor for purposes of market projections and commercialization. Passive assays such as, for example, colorimetric chemical reactions, immunoassays, and nucleic acid probes are not sensors. While such assays do result in a quantifiable entity (such as color production, fluorescence, etc), the assay itself does not provide the means to quantify the response. Rather, a separate quantification system is required.

23.2 DEVELOPMENT AND COMMERCIALIZATION

We estimate that there are approximately 100 companies, worldwide, actively marketing chemical sensors or biosensors or technology [3]. Commercial and academic laboratories actively working in chemical sensors or biosensors number many more than this and at least 500 laboratories or more are actively working in the area. The reasons for this activity have been illustrated throughout this text: chemical sensors and biosensors represent a new, improved, and often unique approach to real time measurements applicable to nearly every industry and market segment.

Table 23.1 lists a number of companies actively developing and/or marketing chemical or biosensors. Applications for these sensors are diverse, from medical applications to environmental monitoring and food quality determinations. This diversity illustrates the generic nature common to most chemical and biosensors. Once the basic transduction and electronic components of the sensor have been developed and manufactured, the application of the sensor to measurements for a wide range of analytes is only limited by the specificity and selectivity of its active chemical/biological surface. Thus while the initial development of chemical and biosensors requires high-risk investment, successful development of the basic sensor can result in a family of products applicable over broad markets.

While it may appear from table 23.1 that the sensor field is becoming crowded, this is not the case. In fact, the opportunities to enter the chemical sensor and biosensor market are wide open since few products have been commercialized. This slow pace of commercialization is due to a large extent on the pace of developments in sensor technology [4]. For example, chemical sensors are highly dependent on the discovery, characterization, and production of reproducible membranes and selective layers. Biosensors, which by definition utilize highly labile biomolecules, are dependent on reproducible immobilization and stabilization technologies. Both chemical sensors and biosensors are still faced with the challenges of linking the active detecting layer to the transducer, and in processing surface–analyte interactions into meaningful and useful outputs.

Chemical sensor and biosensor commercialization must also pass a difficult set of criteria and questions at every point of development. These questions are aimed at measuring current and future technical and market performance for the sensor and include the following.

(i) How specific is it? Can it match competing, non-sensor assays?
(ii) How sensitive is it? Can it meet the needs of the market?
(iii) How reliable and accurate is it? Can it be manufactured with high reproducibility?
(iv) How stable is it? What is its shelf-life? What restrictions must be placed on storage (e.g., refrigeration, desiccation, etc)?

(v) How cheap is it or will it be?
(vi) Who owns it? Is the technology protected?
(vii) What is (are) its target market(s)?

Not altogether satisfactory answers to these questions resulted in the slowing of biosensor investment and development in the late 1980s and early 1990s. The specificity and sensitivity of most chemical sensors and biosensors being developed are sufficient for their target markets, but background interferences and non-specific binding (especially in 'real' samples such as blood, other biological fluids, waste water, and dirt) still represent a major obstacle to increasing these parameters. Reliability and accuracy are largely dependent on standardized manufacturing processes and quality control. However, except for successes such as the MediSense ExactTech® blood glucose sensor and enzyme electrodes produced by companies such as YSI, few chemical sensors or biosensors have achieved great enough demand to justify large-scale manufacture. Thus reliability and accuracy questions can not yet be addressed for most chemical sensors and biosensors. Manufacturing itself remains a major hurdle for chemical sensors and biosensors. Most sensor fabrication involves a series of complex steps, especially for production of the active surface–transducer component. While such fabrication may be manageable in the laboratory or in small production runs, scale-up of these multistep, low-yield-per-step methods remains a significant problem to commercialization of a sensor. This problem also affects the final cost of the sensors, making most more costly than competing, non-sensor assays.

Market and business issues have also slowed chemical sensor and biosensor commercialization. As often occurs with technologies encompassing many disciplines, problems with patents and proprietary technology protection have appeared. For example, one common transducer used for both chemical sensors and biosensors is the integrated electrode capacitor (see chapter 8 for a description of this transducer). Although the design for this transducer has been in the public domain for over 25 years, chemical modification of the surface characteristics of the electrodes can lead to a new patent position. This then leads to complex claims and counterclaims about the use of the basic transducer technology.

Market focus remains another major problem in attracting investment funds to chemical sensor and biosensor development. Most biosensor technologies are developed by individual investigators in academia and industry. These technologies are then carried forward either as the basis for a start-up company, or as internal R&D projects in large companies. Unfortunately, the majority of such sensor R&D has not, historically, been focused on a specific application and market. Rather, broad claims are usually made concerning the sensor's generic application to all markets. This lack of focusing can quickly derail sensor development as companies are forced by their investors to reconsider and redirect their development path in midstream and to refocus their target

Table 23.1 Examples of chemical sensors and biosensors.

Company	Analyte(s)	Technology basis
Chemical sensors		
AAI-Abtech (Yardley, PA, USA)	Gases, organics	Semiconductive polymers on interdigitated (capacitance) electrode arrays
Argonne National Laboratory (Argonne, IL, USA)	Gases, volatile organics	Organic and metal film arrays on chemiresist transducers
AromaScan (Hollis, NH, USA)	Odors, aromas, volatile organics	Semiconducting polymers arrays on a chemiresist (capacitance) transducer
Cardiovascular Devices (Irving, CA, USA)	Blood pH, O_2, CO_2	Fluorescence quenching of dyes in semipermeable membranes
City Technology Centre (Portsmouth, UK)	O_2, CO, S, H_2, NO, NO_2, Cl_2, HCN, HCl	Electrochemical
Cygus Therapeutic Systems (Redwood City, CA, USA)	CO_2, O_2, NH_3, NO_x, SO_x	Polymer-(Nafion-, hydrogel-) coated amperometric and potentiometric sensors
DuPont Diagnostic Systems (Glasgow, DE, USA)	Blood electrolytes	PVC[a] membranes containing ionophores on ISEs
i-STAT Corp. (Princeton, NJ, USA)	Blood electrolytes, glucose, urea, nitrogen, hematocrit	Polymeric and enzyme membranes on ISE-based silicon arrays
Microsensor Systems Inc. (Bowling Green, KY, USA)	Organophosphates, volatile organics, ammonia	Polymeric (ethyl cellulose, fluorinated polyol, phthalocyanins) films on SAW[b] devices and chemiresist transducers
Naval Research Laboratory (Washington, DC, USA)	Organic vapors, gases; fire gases	Polymer-coated SAW devices
Neotronics Scientific (Flowery Beach, GA, USA)	Odors, aromas	Semiconducting polymer arrays on a chemiresist (capacitance) transducer
Oak Ridge National Laboratory (Oak Ridge, TN, USA)	Gases, volatile organics	Metal oxide film arrays on a chemiresist transducer
Sandia National Labs (Albuquerque, NM, USA)	Gases, mercury vapor	Thin metal films on optical fiber transducers
Seiko Instruments (Chiba, Japan)	Odorants (amyl acetate, citral)	Lipid membrane coated piezoelectric crystals
Transducer Research Inc. (Naperville, IL, USA)	CO, CO_2, O_3, H_2S, SO_2, NO_2, NO, Cl_2	Polymeric films on ISEs[c]

Table 23.1 *Continued.*

Company	Analyte(s)	Technology basis
Biosensors		
Universal Sensors (Metairie, LA, USA)	Organophosphorus compounds	Polymer-coated piezoelectric quartz crystals
Arthur D Little, Inc. (Cambridge, MA, USA)	Organophosphates, narcotics, cardiac glycosides, blood immunoglobulins	Receptor- and antibody-based membranes on interdigitated (capacitance) electrode transducers
Autoteam GmbH (Berlin, Germany)	Biological oxygen demand (BOD)	Microbial electrode
Bio-Technical Resources (Manitowoc, WI, USA)	Arsenic, lead, cadmium, PCBs (screening)	Fluorescence from luminescent bacteria
Cambridge Consultants Ltd (Cambridge, UK)	ATP (bacterial growth), antibody–analyte reactions	Bioluminescence from immobilized luciferase; evanescent wave changes due to antibody–antigen binding
Central Kagaku Corp. (Tokyo, Japan)	BOD, toxic organics	Redox, microbial-based electrode
Fujitsu, Ltd (Tokyo, Japan)	Glucose, urea, penicillin	Antibody- and enzyme-based FETs[d]
Idetek Inc. (Sunnyvale, CA, USA)	Antibiotics in milk, bacteria, pesticides	Optical diffraction changes due to antibody–analyte binding
Life Scan (Milpitas, CA, USA)	Blood glucose	Enzyme (GOD[e]/POD[f]) electrode
MediSense Inc. (Waltham, MA, USA)	Blood glucose (ExacTech®, Companion 2®)	Amplified enzyme electrode
Microbics Corp. (Carlsbad, CA, USA)	Toxins, mutagens (screening)	Fluorescence from luminescent bacteria
Mitsubishi Electric Corp. (Tokyo, Japan)	Glucose, triglycerides	pH-sensitive FET coated with an enzyme containing membranes
Molecular Devices Corp. (Menlo Park, CA, USA)	Total DNA (Threshold®); drugs, toxins (Cytosensor®)	Binding protein/antibody-based assay on a light addressable potentiometric transducer; pH changes in whole cells
Ohmicron Corp. (Newtown, PA)	Herbicides, PCBs, PAHs (SmartSense®)	Enzyme immunoassays on chemiresist (capacitance) transducers
Oriental Electric Co., Ltd (Tokyo, Japan)	ATP degradation (fish freshness indicator)	Enzyme electrode

Table 23.1 *Continued.*

Company	Analyte(s)	Technology basis
Pegasus Biotechnology (Agincourt, Ontario, Canada)	Fish and meat freshness indicators (MICROFRESH®)	Amperometric enzyme electrode
Pharmacia Biosensor AB (Uppsala, Sweden)	Multiple antigens (BIAcore®)	Antibody-coated surface plasmon resonance-based instrument
Prüfgerätewerk Medingen GmbH (Freital, Germany)	Blood glucose, lactate (ESAT®)	Enzyme electrode
ThermoMetric AB (Järfälla, Sweden)	Organic acids, urea, sugars, antibiotics, heavy metals, etc	Detection of changes in temperature due to immobilized enzyme–analyte reactions
TOA Electronics Ltd (Tokyo, Japan)	Glucose	Enzyme electrode
YSI (Yellow Springs, OH, USA)	Sugars, alcohols, starch	Enzyme electrodes

[a] Polyvinylchloride.
[b] Surface acoustical wave.
[c] Ion-selective electrode.
[d] Field-effect transistor.
[e] Glucose oxidase.
[f] Peroxidase.

products after significant investment funds have already been expended.

In spite of these limitations and problems, investment trends in sensors appear to have stabilized in the mid-1990s, and should increase steadily into the next century. The reason for this investment turnaround appears to be the maturing of chemical and biosensor companies and development efforts. Lead investigators in these efforts are more aware of the needs to develop sensors which are easily scaled up and mass produced. In addition, improved methods for the immobilization and stabilization of sensor active surfaces are being developed which promise both increased sensitivity/specificity for the sensors and simple and cost-effective mass production of highly stable sensors.

23.3 CURRENT AND FUTURE APPLICATIONS

Ultimately, chemical sensors and biosensors must compete in a business environment that is already heavily populated with other products and technologies, such as immunoassay and analytical monitoring. To be successful against such competition, chemical sensors and biosensors must be able to do the job faster and cheaper with the same or better sensitivity and performance as

Table 23.2 Current and potential applications for chemical sensors and biosensors.

Market	Applications	Competitive advantages
Medical/clinical	Diagnostics, point-of-care patient monitoring, drug monitoring, artificial organs and prostheses, new drug discovery	Real time monitoring, ease of use, portable, cost effective, diverse applications
Processing/industrial	Process monitoring and control, quality control, workplace monitoring, waste stream monitoring	In-line/at-line analysis, real time monitoring, designed to be 'smart' sensors
Environmental	Detection/monitoring of pollutants, toxic chemicals, waste water, and waste streams	Immediate/continuous monitoring, on-site analysis, portable, cost effective
Agricultural/veterinary	Plant/animal diagnostics, meat/poultry inspection, waste/sewage monitoring, soil and water testing	Portability, on-site real time monitoring, high specificity for pathogens (biosensors)
Defense/military	Detection of chemical, biological, and toxin warfare agents, treaty verification	Combined detection and alarm capabilities, ease of use, high specificity and selectivity, portable, real time output
Robotics/computers	Robotic controls, hybrid silicone/organic computing components	Taste, smell, and other sensory capabilities, analog/digital computing capabilities, low energy demands and heat generation, can be self-repairing (biosensors)

the product they are replacing. These criteria are already being met in biosensors such as the MediSense glucose sensors. In addition to direct competition with existing measurement products, chemical sensors and biosensors also present opportunities to create new markets, offering detection and measurement capabilities in new areas. These include on-site use in clinics and emergency medical facilities, in the field for environmental monitoring, and at food and chemical processing lines.

The basic characteristics of chemical sensors and biosensors—including high specificity and sensitivity, portability, real time output, cost effectiveness, and user friendliness—make them applicable to virtually every major product market. The immediate market targets for these sensors are in medical diagnostics, detection and alarm systems, environmental monitoring, and food processing. Future markets include application of chemical sensors and biosensors to organic computing/data handling devices, for intelligent control systems, as medical prosthesis devices, and to aid the discovery and treatment of diseases. These applications are summarized in table 23.2.

23.3.1 Industrial applications

To date, the primary use of chemical sensors in industry has been for the detection of gases, volatile organics, and odors/aromas [6, 7]. As shown in table 23.1, most commercialized or near-commercialized chemical sensors utilize multiple selective polymeric membranes in an array format to detect and characterize the target analytes. These sensors may be used to detect toxic gases in monitoring and as alarm systems (such as the detection of harmful gases in the workplace), or to characterize the aroma of a food as part of a quality control process. As chemical sensor technology advances in the next decade, such applications will become routine on chemical and food processing lines, and more sophisticated sensors will be incorporated into automated process monitoring and control systems.

Currently, biosensors are primarily used in the process and fermentation industries for monitoring the production of sugars, syrups, amino acids, antibiotics, specialty chemicals, and some genetically engineered proteins and drugs. These biosensors are almost exclusively enzyme electrode based, such as those marketed by YSI and Universal Sensors. This dominance by enzyme-electrode-based biosensors will change as bioaffinity sensors, such as antibody- and receptor-based biosensors (see chapter 8 of this text) are commercialized. Bioaffinity sensors will be broadly used for the monitoring and control of chemical, pharmaceutical, and food process streams for rapid measurement of ingredients and product, product quality control, and process control. In addition, these sensors are being developed for application to detection of toxic materials and pathogens in pharmaceuticals and foods, and for the monitoring and control of process waste streams. While chemical sensors are already being marketed for characterization of aromas and odorants in foods, bioaffinity sensors will actually incorporate functional receptor biomolecules which closely mimic human taste and smell, allowing ultimate, on-line quality control of food, cosmetic, and other products. Biosensors are already being marketed for determining the freshness of foods by detecting and quantifying food degradation products. For example, Pegasus Biotechnology markets a fish freshness, enzyme-electrode-based biosensor system which detects and quantifies ATP degradation products, which are directly related to freshness. The Oriental Freshness Meter of Oriental Electric Company also monitors ATP degradation using both an oxygen electrode and enzyme electrodes.

Currently, both chemical sensors and biosensors can be used at process lines in an at-line, FIA (flow injection analysis) mode. In FIA, a sample from the process stream is delivered to the sensor or measurement (see chapters 21 and 22 of this text). Currently, FIA is widely applied to process streams utilizing

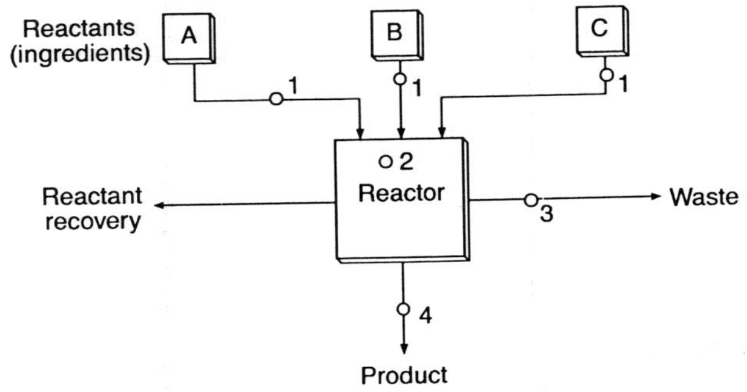

Figure 23.2 Automated process control using in-line/at-line sensors.

classical analytical methods including infrared and UV–visible spectroscopy and gas chromatography [8]. FIA systems utilizing chemical sensors and biosensors are also being developed [9, 10], but still are primarily laboratory prototype systems. Within the next 10 years, however, advances in chemical sensor and biosensor technology (primarily in active surface fabrication and stabilization) will allow real time, on-line monitoring of process streams. For example, figure 23.2 presents a conceptualized processing scheme that incorporates chemical sensors and biosensors for monitoring and control functions. In figure 23.2, sensor 2 monitors product concentration in the reactor/processing line and its output is used to control the addition of reactants by line feed sensors (sensors 1 in figure 23.2). Sensor 3 monitors the concentration of specific waste products and warns of concentration approaching regulatory limits for release or treatment. Sensor 4 detects and quantifies product purity and concentration in the end-process stream and feeds this information back to the central controller which may then stop the process or add more reactants/ingredients. This type of integrated, sensor-based system could result in near-complete automation of processing lines.

Systems such as the one illustrated in figure 23.2 will also incorporate artificial intelligence. The information from the sensors will be used with 'fuzzy logic' and neural networking to enable decisions by individual controllers based on the input from multiple sensors. Such systems will also incorporate sensor self-testing, self-calibration, and fault correction, resulting in reliable, automated systems applicable to any process or production line. These systems will significantly affect the productivity and profitability of food, chemical, and pharmaceutical production.

More futuristic uses of chemical sensors and biosensors for processing could include artificial vision. For example, specific molecular vision complexes, such as the rhodopsin–opsin complex, have been isolated from ocular tissue and could be reconstructed into artificial membrane systems. Integration of such 'vision complexes' with electronic transducers could result in artificial visioning sensors responding to specific wavelengths of light. Arrays of such sensors could result in sensors able to perceive and process color and brightness on processing lines.

23.3.2 Medical and clinical applications

Chemical sensors have been developed by companies such as DuPont and Cygus Therapeutic Systems for the measurement of blood electrolytes and gases, and ion selective membranes are common in many clinical analyzer systems. While the use of chemical sensors for such determinations will continue to increase, sensor applications in clinical diagnostics will favor development and application of biosensors due to the high specificity residing in the biological component of these sensors.

The ExactTech® blood glucose biosensor by MediSense is one of the first success stories for biosensors. Based on a ferrocene-amplified glucose oxidase reaction, the pen-shaped, digital read-out, home-use instrument and its disposable test strips was quickly accepted by a large number of consumers. Today the technology has evolved into a third generation of products for home and bedside use (Precision Q-I-D® and Precision G® systems) and MediSense has reportedly captured approximately 7% of the $2 billion blood glucose testing market [11].

Other biosensor-based diagnostic instruments, such as the enzyme-electrode-based analyzers of YSI for glucose and lactate, are utilized routinely in many clinical laboratories. i-Stat recently introduced a portable analyzer for bedside use which utilizes enzyme-electrode-based assays for glucose and urea, as well as chemical sensor tests for nitrogen, sodium, potassium, and chloride.

Unfortunately, such success stories are not the case for many other chemical sensors or biosensors. While biosensors have the potential for providing real time diagnostics for almost any analyte including drugs, toxins, bacteria, and viruses, they face the same cost and regulatory hurdles as any new diagnostic targeted for human use. Thus the optimistic viewpoint of the early 1980s that biosensors would revolutionize clinical point of care diagnostics has not been realized.

Biosensors will, however, impact clinical diagnostics by the next century. Immobilized antibody-, receptor-, and, recently, nucleic-acid-based biosensors will provide new diagnostic capabilities and allow broad screening for genetic diseases including cancer and heart disease. Immunobiosensors will be used to detect specific tumor markers and nucleic-acid-probe-based biosensors may be used to detect genes predictive of specific cancers, such as the familial breast cancer gene, BRCA-1, which is linked to breast cancer and is present in 10% of all women. Other biosensors may allow rapid testing for blood components, hormones, drugs of abuse, and infectious agents in sites which do not have access to clinical laboratories. This promise and the potential market for such sensors will continue to stimulate the development and commercialization of new sensors for clinical diagnostics.

In the future, chemical sensors and biosensors may provide new, revolutionary products for medical applications [12]. For example, since the mid-1980s, a number of pharmaceutical companies have sponsored research to develop implantable biosensors as therapeutic monitors and to control drug delivery, such as an implantable glucose biosensor which monitors blood glucose levels and controls the release of insulin as needed in diabetics. In another example, Cardiovascular Devices has developed implantable optical-fiber-based sensors containing fluorescence dyes for continuous measurement of blood pH and gases. Within the next 10 years, implantable measurement and control sensors may be available for a wide range of blood analytes.

Biosensors may also be used in medical prostheses and artificial organs. For example, a neural-receptor-based biosensor developed in our laboratories [13] is in effect an artificial nerve: it responds to specific, natural neurotransmitters to produce an electronic signal. If microsensors utilizing this technology could be integrated with body neurons, they could relay neural impulse signals across portions of damaged nerve tissue and restore functionality. The potential use of vision complexes for artificial sight has already been discussed.

23.3.3 Environmental applications

The industrial applications already discussed for chemical sensors and biosensors are also applicable to environmental applications for these sensors. As illustrated in table 23.1, many chemical sensors have been developed to detect harmful and polluting gases and organics, and these sensors are being applied as environmental monitors in many industries. Biosensors also have the potential for monitoring nearly any material of environmental concern including waste products, pollutants, toxic chemicals, pesticides, herbicides, and microbes. Such biosensors are already being used. For example, Ohmicron is marketing an immunosensor-based system for herbicides, PCBs and PAHs, and companies such as Autoteam GmbH, Bio-Technical Resources, Central Kagaku Corp., and Microbics Corp. have developed microbial-based systems for detection and characterization of heavy metals, PCBs, toxic organics, and mutagens (table 23.1).

The portability and real time output of chemical sensors and biosensors will be key to their success in environmental applications. Such sensors will provide rapid, easy to use, and cost-effective on-site testing. While these sensors will not displace confirmation of samples in a laboratory using classical analytical methods, sensors will decrease the number of samples requiring expensive laboratory analysis by eliminating samples which contain none of the analyte being screened for.

23.3.4 Other applications

Chemical sensors and biosensors will also find applications in agriculture, drug discovery, robotics, and organic computing devices. Agricultural applications will overlap with food and environmental applications. Chemical sensors and biosensors will be used for animal and plant disease diagnostics, detection of contaminants such as pesticides, drugs, and pathogens in milk, meat, and other foods, and determination of product quality, such as the ripeness (as measured by color and sugar level) and flavor of fruits and vegetables in the field.

Drug discovery is critical in the development and commercialization of new pharmaceuticals, providing the basis for new drugs and new therapies. Up to the late 1970s and early 1980s, the primary tool of drug discovery was mass screening using animal and culture methods. In the early 1980s, mass screening methods began to be replaced with molecular biology and molecular modeling technologies [14] and by the mid-1980s an industry had grown to serve the needs for computer-aided molecular design [15, 16]. During this period, biotechnology-based methods came to the forefront of drug design: proteins, receptors, and antibodies were used to predict the structures and activities of new drugs. The methods developed from this 'first generation' of biotechnologies for drug design have now led to a second generation which combines molecular biology with sensor technology. For example, companies such as Affymax (Palo Alto, CA, USA) have developed technology for the immobilization of DNA, RNA, and protein fragments onto solid state electronic surfaces [17]. The resulting 'chips' can be used for genome sequencing, detection/characterization of mutations, genetic screening, pathogen identification, and drug discovery. Such sensor arrays are already redirecting drug discovery and will significantly increase new drug entities within the next 10 years.

Biosensors may be key to advancing the capabilities of robotics. Advanced biosensors able to carry out functions such as neural stimulation, vision, hearing, tasting, and smelling may be integrated into robotic control devices for a wide variety of consumer products. For example, taste biosensors could standardize food flavor use, and optical and audio biosensors could be used to control production lines, home appliances, security systems, or automatic pilots in motor vehicles.

The merging of organic and silicon chip technology may lead to hybrid computing devices with unique characteristics. A key factor in such development will be the development and application of conductive and transport molecules and films, and the integration of such materials with sensor technology. For example, as shown in table 23.3, a variety of chemical and biochemical materials have been developed which have characteristics applicable to electron transfer, gating, valve, and switch functions in computing circuits. Polymers for molecular computing such as polysulfur nitride and *trans*-polyacetylene have been proposed since the late 1970s [18]. These chemical 'wires' were postulated as being able to propagate energy as solotons and branched structures were designed as soloton switches and valves. Later studies with enzymes illustrated that modification of active sites with appropriate electron transfer groups, such as ferrocenecarboxylate groups, resulted in biomolecules able to receive and transfer electrons [19]. Other studies, such as those listed in table 23.3, have now

Table 23.3 Examples of organic and biological materials for organic computing devices.

Material	Reference
Polysulfur nitride and *trans*-acetylene as molecular wires, switches, and valves	[18]
Ferrocene-modified enzymes as electron shuttles	[19]
Long-term storage of information in Ca^{2+}–calmodulin-dependent protein kinase molecules	[20]
Polyether-based molecular shuttle	[21]
Carbon and cyclic peptide nanotubes as nanometer scale wires and solenoids	[22, 23]
α-helix/β-sheet octadecapeptide-based switch	[24]
Photoinducible electron transfer using oligomeric DNA assemblies	[25]
Reversible pH- or redox-dependent molecular switches	[26]
Polyaniline wires in aluminosilicate crystals	[27]
DNA- and porphyrin-based, linked-ring molecular chains (catenanes) as information storage devices	[28, 29]
Porphyrin-based molecular photonic wire	[30]
Sexithiophene-based organic transistor	[31]
Poly(phenylene-ethynylene)s with cyclophane receptors as molecular wires	[32]

proved that polymers and biomolecules can form the basis of organic computing devices. When linked to a sensor system, these active layers could have a number of special characteristics: analog and digital computing potential, self-contained systems, potentially self-repairing, and potentially able to produce their own energy. These features would be applicable in a variety of products including bionic implants, memory-intensive systems, pattern recognition, 3D display screens, artificial intelligence and language processing, erasable memory disks, and hybrid organic–silicon computers. Such organic computing potential could be carried further using biosensors and the unique ability of biomolecules to generate and use energy efficiently with little heat loss, to transfer electrons, and to function as either single or linked components in a field of many biomolecules. The many variations in structure that a biomolecule can undergo during a reaction impart analog capabilities which may be applied to recognizing, storing, and transmitting complex patterns, similar to vision. These analog capabilities may also allow the creation of true artificial intelligence, resulting in computers able to utilize rational judgement to make decisions in situations with no clear-cut solutions. The information storage capacity of nucleic acids may also be applicable to memory-intensive systems. While these applications for chemical

Table 23.4 Classification of sensors by surface technology.

Enzyme electrode	Bioaffinity	Polymeric/chemical
MediSense	Molecular Devices	Microsensor Systems
Universal Sensors	Pharmacia	Cardiovascular Devices
YSI	Ohmicron	Neotronics
Pegasus Biotechnology	Arthur D Little, Inc.	Cygnus Therapeutic Systems
Oriental Electric	Idetek	AromaScan

and biosensor technology may be decades away, current developments in sensor technology are laying the groundwork for such applications.

23.4 CURRENT AND FUTURE MARKETS

In assessing and predicting markets for chemical sensors and biosensors, the definition cited above for chemical sensors and biosensors requires additional refinement since a functional (technological) definition does not adequately address specific applications and markets.

There have been a number of approaches to bridging technology to application. One has been to classify sensor applications and the participants in the sensor market by transducer and active surface type using the assumption that the electronic components of the sensors are non-technically limiting. Table 23.4, for example, illustrates the classification of a number of sensors on the basis of the detector chemical used in their active (measuring) layer: enzyme, bioaffinity (antibodies and receptors), or chemical (polymerics).

Although this approach toward sensor classification may be academically useful, it does not function well for market applications. For example, YSI's enzyme-electrode-based glucose and lactate analyzers are used primarily in industry process control and as laboratory instruments with individual instrument costs ranging from $5000 to $10 000. Similarly, the Threshold total DNA assay system developed by Molecular Devices is aimed at research and quality control laboratories and has a basic instrument cost in excess of $20 000. In contrast, portable and/or hand-held sensors such as those sold or being developed by companies such as Pegasus Biotechnology, Ohmicron, Microsensor Systems, and Arthur D Little, Inc., are targeted for use in remote field locations and are designed for automated, inexpensive, single-step use. Thus, classification of sensors on the basis of surface chemistry leads to a confusing mix of products and markets.

Sensors may also be classified by the transducer technology used. As shown in table 23.5, however, this further fragments sensor products and producers, creating more confusion.

A more logical method for sensor classification is to first define the characteristics of each sensor and its targeted market. Using this approach,

Table 23.5 Classification of sensors by transducer technology.

Ion selective electrodes	Capacitance	Optical
MediSense Universal Sensors YSI Pegasus	Ohmicron Molecular Devices Arthur D Little Microsensor Systems	Pharmacia Idetek Cambridge Consultants Ltd
Field effect transistors	**Thermistor**	**Piezoelectric**
Mitsubishi Electric Fujitsu	ThermoMetric	Universal Sensors Seiko Instruments
SAW devices		
Microsensor Systems Naval Research Laboratory		

Table 23.6 Classification of sensors by application/market.

Portable detectors	Laboratory instruments
Characteristics	
One or two step	Multiple steps
Low capital investment	High capital investment
Disposable test format	Disposable or reusable format
Qualitative to semiquantitative	Semiquantitative to quantitative
ppb to ppt sensitivity	ppb to ppt^{+} sensitivity
Portable	Laboratory/factory based
Primary market	
Remote/at-site testing	Clinical/laboratory testing
Examples of producers	
MediSense	YSI
Ohmicron	Universal Sensors
Pegasis Biotechnology	Pharmacia
Arthur D Little	Molecular Devices
i-STAT	ThermoMetric
Cambridge Consultants	AromaScan
	Neotronics

two major types of sensor product emerge as shown in table 23.6. The first, comprising portable sensors, is characterized by simplicity of use, portability, and low cost. The primary market for this type of sensor is the remote testing market such as point of care diagnostics, on-site environmental testing, and at-line/on-line process stream monitoring.

The second type of biosensor product is more accurately described as a

Table 23.7 Current and projected world markets for biosensors (millions of dollars).

Market	1994	1999[a]	2004[a]	AAGR[b] (%)
Medical/clinical	220	440	950	14–16
Processing	75	120	250	10–12
Government/defense	65	95	150	7–9
Environmental	25	40	75	10–12
Agriculture/veterinary	10	15	35	10–15
Other	5	10	20	9–12
Total:	400	720	1480	10–15

[a] Projected in 1994 dollars; includes both sales and contract research and development.
[b] Average annual growth rate.

bioanalytical instrument which utilizes chemical/biological sensor technology. The sensor technology (transducer and active surface) is packaged into an instrument which is designed for use in the laboratory and may cost from $2000 to as much as $100 000. Table 23.6 lists a number of companies according to market applications. This results in a more accurate description of sensor competitors and the markets they are pursuing.

Using this market approach, and the definitions already cited for chemical and biosensors, estimates can be made of current and future sales for these sensors. As shown in table 23.7, the total biosensor market for 1994 was approximately $400 million. Over 50% of all biosensor sales were in the medical area, primarily due to sales of glucose biosensors. For example, MediSense had sales of over $125 million in 1994 for its biosensor-based blood glucose monitors, capturing approximately 7% of the $2 billion blood glucose testing market [33]. The success of blood glucose biosensors will stimulate the commercialization of other biosensors for rapid blood diagnostics, resulting in sales of nearly $1 billion by the year 2004. These may include biosensors for LDL and HDL cholesterol; fitness indicators such as blood lactate, urea, and creatine kinase; implantable sensors for continuous blood monitoring; and point of care diagnostics including continuous neonatal monitoring. Other biosensor markets will also grow as biosensors are developed for and applied to food and chemical processing lines, environmental and workplace monitoring, and agriculture. In the processing industry, the systems integration of biosensors with neural networks will result in a new era of automated, decision making control systems.

Current and projected markets for chemical sensors are shown in table 23.8. The biomedical market consists primarily of chemical sensors for blood gases and electrolytes. This area is growing as portable, real time systems such as the i-STAT clinical analyzer are increasingly used at point of care and remote testing locations. The demand for faster, more reliable, and cheaper detectors for environmental and workplace monitoring for toxic gases and volatile organics

Table 23.8 Current and projected world markets for chemical sensors (millions of dollars).

Market	1994	1999[a]	2004[a]	AAGR[b] (%)
Biomedical	200	310	520	8–12
Environmental/workplace monitoring	250	420	760	10–14
Processing	50	70	100	6–8
Total:	500	800	1380	9–11

[a] Projected in 1994 dollars; includes both sales and contract research and development.
[a] Average annual growth rate.

is also growing and is stimulating the development and commercialization of chemical sensors. This area promises to be the most rapidly growing for chemical sensors. As for biosensors, chemical sensors will also play an increasing role in process and control systems, especially in the food, specialty chemical, and pharmaceutical industries.

These markets for chemical and biosensors will also drive further refinements and improvements in sensor design and performance. Within the next decade, chemical and biosensor fabrication and mass production will become as reliable and cost efficient as current production of physical sensors and will result in products which are faster responding, smaller, simpler to use, durable, and relatively inexpensive. With the addition of artificial intelligence and integration into more complex software systems, chemical and biosensors will also become smarter with capabilities applicable to almost any sensing and measurement need.

23.5 DEVELOPMENT AND COMMERCIALIZATION OF A CHEMICAL SENSOR OR BIOSENSOR

There is little doubt that chemical and biosensors will become major players in measurements and diagnostics by the turn of the century. This strong future will stimulate the constant investment of established companies into new sensors and sensor systems, and will encourage the start-up of new companies focusing on the development of new sensors.

The hurdles faced by both established and start-up companies in the development and commercialization of a new sensor are formidable and challenging. Initially, an assessment must be made of the potential success of the sensor. If it is to be successful, it should meet one or both of two basic criteria.

(i) If the new sensor is a competitor to an already established assay for an analyte, the new sensor must show improved productivity: lower cost per assay, greater efficiency and through-put, and, if possible, lower capital equipment investment and less dedicated technician time.

(ii) If the new sensor claims new measurement or diagnostic capabilities, it must address assays which have long-standing needs, provide an innovative technology, and/or offer the capability for measurement panels (multiple analyte assays).

New and/or innovative sensors and systems which do allow more rapid and accurate measurements are not, initially, cost limited. Cost becomes a factor as competitors offer similar assays or systems.

23.5.1 Commercialization of a new sensor

The development and commercialization of a new sensor requires an integrated team effort with interdisciplinary skills and experience. This may include technical expertise in biochemistry, electronics, engineering, polymerics, and ergometrics; quality control (QC), quality assurance (QA), and regulatory requirements; and manufacturing/process experience in scale-up and technology transfer, packaging, clinical evaluation and alternate site testing, and marketing/distribution. During commercialization, a number of key parameters crucial to the success of the sensor must be addressed by the team. These include

(i) technical performance to specifications,
(ii) QA/QC requirements and protocols,
(iii) projected scale-up and production problems and solutions,
(iv) sensor and sensor component costs and means to minimize such costs, and
(v) competitive design/use considerations.

When these requirements are taken into consideration together with the commercialization process, a six-phase program can be applied to both guide sensor commercialization and measure the success and stage of the development program. These six phases can be generalized as follows.

Phase 1. Initial R&D leading to a laboratory prototype. This includes

(i) definition of need(s) and the target product,
(ii) a technology audit and/or technology selection,
(iii) actual R&D resulting in the prototype, and
(iv) regular reviews and refocusing by the product team.

R&D is a function of available investment capital, allowable risk, and projected profitability. Careful planning and management in this program phase is critical for minimizing costs and time to commercialization for the sensor.

Phase 2. Technology review and design concept selection. This includes

(i) a technical audit of work to date,
(ii) definition of technology/product strengths and weaknesses with respect to competing products,

(iii) selection of the best design concept and setting of specification goals, and
(iv) formalization of the final development and commercialization program including projected cost and time to product release.

Phase 2 is critical for focusing the commercialization of a sensor and often utilizes resources outside the development company. It is at this phase that a sanity check should be carried out on the product concept and goals in order to decide whether to proceed with the program.

Phase 3. Final design and testing. This phase returns to the laboratory and includes

(i) optimization of product components and the total product package,
(ii) fabrication and testing of the final product prototype,
(iii) input of design and manufacturing considerations to minimize downstream manufacturing problems and costs,
(iv) formalization of the sensor format and the packaging design, and
(v) first draft documentation and engineering drawings.

Phase 3 represents the actual product shake-out: by the end of phase 3, the sensor product function and design is near finalization.

Phase 4. Manufacturing prototype assembly and testing. This phase focuses on the final production prototype of the sensor and includes

(i) limited production of the final sensor system and any associated support reagents and hardware,
(ii) sensor system evaluation and verification,
(iii) finalization of packaging design,
(iv) design revisions and documentation,
(v) regulatory application (e.g., 510(k), if required), and
(vi) an updating and revision of distribution and marketing plans.

At the end of phase 4, the sensor is ready for manufacturing.

Phase 5. Transfer to manufacturing. This phase includes

(i) initial production runs of the sensor system,
(ii) beta-site testing (if appropriate),
(iii) final documentation and 510(k) application submission (if necessary), and
(iv) finalization of distribution and marketing plans.

By the end of phase 5, the sensor system is ready for release.

Phase 6. Final production and product release. This phase includes

(i) reliability and QA testing,
(ii) finalization of manufacturing and QA/QC documents,
(iii) finalization of technical release documents/applications,
(iv) implementation of distribution and marketing plans, and
(v) the first, 6 month customer feedback review.

Table 23.9 Commercialization of a new sensor.

	Phase	Time (months)	Investment ($ millions)	Milestone
	1	12–36	Level of effort	Lab prototype
	2	4–6	0.3–0.5	Development/commercialization plan
	3	6–10	0.5–1	Final design
	4	6–10	0.5–1	Manufacturing prototype
	5	4–6	0.5–1	First production runs
	6	2–4	0.2–0.5	Market release
Total	2–6	22–36	2–4	

At the end of phase 6, the new assay or system is commercially available and being actively marketed.

The estimated time-lines and costs associated with the generic development and commercialization of a new sensor are shown in table 23.9. Such programs may take from 22 to 36 months and cost as much as $4 million. This does not include the costs and time required for the up-front R&D to develop the basic sensor technology.

23.5.2 Regulatory hurdles and design considerations for a new sensor

Sensors developed for purposes of determining the health status of humans are considered medical devices and thus must be approved for use by the FDA in the United States and appropriate regulatory agencies in other countries [33, 34]. This definition is far from ironclad. For example, it could be argued that tests for alcohol and nicotine in saliva fit the definition, yet the FDA position is that such tests are not medical devices. On the other hand, a compliance policy guide published by the FDA on 'RIA analysis of hair to detect the presence of drugs of abuse' states that the assay is a medical device.

The lead group at the FDA for review and approval of human diagnostics including sensors is the Division of Clinical Laboratory Devices (DCLD). Any new assay or sensor needs to go through a 510(k) (premarket notification) application and approval process. In some cases, the diagnostic must go through a PMA (premarket approval) process with the FDA. Significant changes in the manufacture, use, and indications of a diagnostic may require a new 510(k) application.

In order to be accepted by the FDA, an *in vitro* diagnostic sensor must measure a clinically meaningful parameter, and the clinical utility of the sensor must be demonstrated. Clinical utility can mean both the benefit of measuring a parameter to the patient's health and also the cost effectiveness of the assay performed by the sensor.

23.5.3 Risk *vs* benefit in new sensor development

Risk *vs* benefit considerations are central in the development and commercialization of a new sensor. These include not only risks connected with carrying out the test (e.g., the use of radioisotopes), but also those associated with the use of the results of the assay. The latter risks range from incorrect diagnosis to moral questions concerning diagnostic tests for diseases for which there is no cure. For example, there is considerable resistance in the USA to the use of diagnostic tests for genetic diseases such as cystic fibrosis and muscular dystrophy. The argument here is that a potential indication that parents are carriers of the disease or that a fetus will express the disease could lead to abortion. Conversely, if assays are available for such diseases but are not highly accurate, there are potential legal implications against the diagnostic company for failure of the test to accurately predict the disease. Such sociopolitical factors may, at least in the case of some diseases, outweigh the benefits of a specific clinical assay to patient management and improvement of life quality.

Risk–benefit must also address practical aspects of a new sensor assay, especially in the case of an assay for human diagnostics. For example, the rate of false positives and false negatives must be factored into decisions to market a new assay as well as the effect of the assay, and its failure rate, on the management of patients.

Since 1992, clinical assays must also address the requirements of the Clinical Laboratory Improvement Act (CLIA), which addresses the proficiency of laboratories handling and testing human samples. The current version of CLIA was passed in 1988 [35, 36] and implemented in 1992 as a response to allegations of misdiagnosis resulting from laboratory testing errors. The act is aimed at insuring that the quality of human testing is the same in all settings, including hospitals (laboratories, emergency rooms, ICUs, and at the bedside), doctors' offices, independent or reference laboratories, clinics, ambulatory-care facilities, nursing homes etc. Laboratories regulated by CLIA must register with HHS, pay a fee when applying for an operating certificate, and undergo inspections to ensure they meet specified quality standards. Only laboratories able to meet CLIA requirements are legally allowed to carry out human testing in the USA.

As a result of CLIA, new clinical assay applications to the FDA including sensors used to carry out specific assays must address overall assay QC and the training requirements of personnel using a sensor for a specific assay. QC includes both manufacturing and testing of the sensor prior to release, and the means for the user of a sensor to assess that the sensor is performing to specifications. The latter requirement is most easily addressed by building standard QC tests with reference materials into a sensor. Such tests should specify how the laboratory must perform QC tests and their frequency, calibration of the sensor, verification of performance specifications (accuracy, precision, and reproducibility), and assay range. Training requirements for using the sensor are addressed by defining the user market and preparing appropriate instructions for the assay to be carried out based on the assumed minimal training expected in the laboratories using the assay.

A sticking point in FDA enforcement of CLIA is the classification of all diagnostics used in clinical laboratories into one of three complexity categories: waived, moderately complex, and highly complex. Assignment to a category will dictate the degree of regulatory control on the diagnostic product: for example, highly complex diagnostic methods can only be carried out in laboratories certified to perform highly complex tests and not in laboratories certified for moderately complex tests. It is thus important in designing new sensors for clinical use to plan for simplicity of use. For example, performance criteria which should be addressed include

(i) high sensitivity,
(ii) high specificity,
(iii) rapid data output (seconds to a few minutes),
(iv) multiple analytes assayed in the same tests (panels),
(v) simplicity of use (homogenous),
(vi) reagentless format,
(vii) minimal or no sample preparation,
(viii) portable and/or hand-held,
(ix) low cost per assay,
(x) no or low risk to the user.

Design considerations must focus on the end-user and the place the sensor will be used: physician, nurse, technician, or layman; physician's office, clinical laboratory, critical care unit, or at home. This then defines the requirements for the instructions and any training material which must be included with the sensor.

ASLT (accelerated shelf life testing) must also be run on the sensor and its individual components to define realistic shelf life, storage conditions, and usage conditions. This information is also critical to the overall quality control for the sensor and the instructions for its use.

Finally, disposal methods must be considered and included in the sensor instructions. For example, RIAs require that all radioisotope regulations be followed for use and disposal. Usage must comply with not only federal regulations for disposal, but also for those of each state in which the sensor will be sold and used.

23.5.4 Clinical trials

All new clinical diagnostics must undergo clinical trials on targeted populations prior to acceptance by the FDA. The amount of testing required in the clinical trial depends on the novelty of the new assay. If the assay is to be the first on the market for the parameter it measures, then it will require much more extensive testing than an assay for an analyte which is already measured by other, commercially available assays.

In the case of a new sensor and if possible, the preferred approach to clinical trials is to select an already approved assay for the same parameter which is already on the market. In this case, all data on the approved assay (including the 510(k) or PMA application) can be used to guide clinical studies and claimed indications for the new sensor.

If the new sensor is the first of its kind, its clinical trials become more dependent on basic research and in-house data. In most cases, clinical trials for this type of assay will take longer and be costlier. The exception is a new assay or sensor which may provide a rapid means of diagnosis for a disease with high public health priorities, such as for the AIDS virus.

23.5.5 Non-regulated sensors

Sensors for non-medical human uses, veterinary applications, food and chemical analysis, and environmental analysis do not require approval by a government agency for commercialization and use. However, such sensors may target acceptance of their data by agencies such as the FDA, USDA, and EPA if they can attain performance characteristics of in-place analytical methodologies. This can then be used as part of their marketing strategy.

For example, standard accepted EPA methodology for pollutants is based on analytical methods such as GC, GC–MS, HPLC, and HPLC–MS. In-place EPA methods are rigorous and SOP directed, and laboratories carrying out EPA methods must subject themselves to EPA controls and inspections. In 1992, the EPA's Organics Methods Work Group (Office of Solid Wastes) announced that they had preliminarily approved and assigned draft methods for the immunoassay kits manufactured by EnSys Inc. (Research Triangle Park, NC) for environmental detection of PCP, PCBs, and TPH. This resulted in marketing literature by the company touting the approval as the '...only EPA approved on-site test methods for PCB, TPH and PCP'. Draft methods by the EPA were added in 1993 using the EnSys assays for PAH and TNT. Following close behind EnSys, Millipore announced in 1993 that their EnviroGard PCB test method was pending approval by the EPA's Organics Methods Work Group.

The key in the efforts of companies such as EnSys and Millipore to gain EPA approval lies in their willingness to submit their assays to extensive testing and QC analysis equivalent to a classical analytical methodology. The resulting approval by the EPA then opens the market for the tests to compete directly against established analytical methods. Once such assays meet regulatory requirements for detection limits, accuracy, and reproducibility, they can then compete against complex, laboratory-based methods based on their ease of use, rapid data output, portability (field use), and cost per test. This strategy is also applicable to manufacturers of sensors for environmental, food, chemical, and agricultural analysis and applications.

23.6 CONCLUSION

The current status and growth of chemical and biosensor products can be compared to the growth of immunoassays in the 1970s and 1980s. During the early 1970s, the emergence of immunoassay technology led to the explosive growth in the number of available assays for not only clinical diagnostics, but food, environmental, and agricultural analysis as well. As a result, hundreds of new companies were formed to develop and commercialize immunoassays. During the late 1970s and early 1980s, companies such as Abbott, Boehringer Mannheim, Baxter Diagnostics, Hybritech, and perhaps two dozen more emerged as dominant players in clinical immunoassays as they expanded product lines, simplified on-site assay kits, and developed captive automated immunoassay analyzer systems. Today, immunoassay-based clinical diagnostic products capture approximately 35% of the $15 billion *in vitro* clinical diagnostic market [37].

The chemical and biosensor market is currently at a stage very similar to immunoassay in the 1970s. Traditional analysis methods, as well as immunoassay, are being challenged by new sensors. The number of sensor products is currently limited, however, and the players are mainly small companies with limited product lines. As yet, there is no single company or group of companies who could be considered dominant in chemical or biosensors.

Technology advances and revitalization of investment indicates that sensor development and commercialization will change in the next 10 years. As sensor technology matures and consistent manufacturing processes are developed, chemical sensors and biosensors will become major players in diagnostic and measurement markets. Generic sensor technologies, now resource limited to one or two lead applications, will be expanded to product lines with multiple, cross-market applications. The success of such product lines will also lead to mergers and acquisitions with the emergence of a leading group of sensor companies. By the first decade of the next century, chemical sensors and biosensors will be established as reliable, valuable measurement products in all major industries.

REFERENCES

[1] Taylor R F 1991 Immobilized antibody- and receptor-based biosensors *Protein Immobilization: Fundamentals and Applications* ed R F Taylor (New York: Dekker) pp 263–303

[2] Janata J, Kosowicz M and DeVaney D M 1994 Chemical sensors *Anal. Chem.* **66** 207R–28R

[3] Taylor R F 1990 *Biosensors: Technology, Application and Markets* (Waltham, MA: Decision Resources)

[4] Alverez-Icaza M and Bilitewski U 1993 Mass production of biosensors *Anal. Chem.* **65** 525A–34A

[5] Taylor R F 1987 *Biosensors: Major Opportunities in the 1990s* (Waltham, MA: Spectrum/Biotechnology, Decision Resources)

[6] Gilby J 1994 Electrochemical sensors: a modern success story for an old idea *Sensor Rev.* **14** 30–2

[7] Kocache R 1994 Gas sensors *Sensor Rev.* **14** 8–12

[8] Taylor R F 1992 *Offline/Online Analysis and Control Methods for Food Processing* (Waltham, MA: Spectrum Food Industry, Decision Resources)

[9] Callis J B, Illman D L and Kowalski B R 1987 Process analytical chemistry *Anal. Chem.* **59** 624A–31A

[10] Lüdi H, Garn M B, Bataillard P and Widmer H M 1990 Flow injection analysis and biosensors: applications for biotechnology and environmental control *J. Biotechnol.* **14** 71–9

[11] Anon 1994 Best-selling test strips *Chain Drug Rev.* November **21** 20–1

[12] Hollingum J 1993 New scope for sensors in safety and health care *Sensor Rev.* **13** 32–3

[13] Taylor R F, Marenchic I G and Cook E J 1988 An acetylcholine receptor-based biosensor for the detection of cholinergic agents *Anal. Chim. Acta* **213** 131–8

[14] Venkataraghavan B and Feldmann R J (ed) 1985 Macromolecular structure and specificity: computer-assisted modeling and applications *Ann. NY Acad. Sci.* **439** 1–208

[15] Taylor R F 1988 *Computer-Aided Molecular Design* (Waltham, MA: Spectrum Advanced Materials, Decision Resources)

[16] Pramik M J 1989 Molecular modeling ushers in the age of rational biotech drug design *Genetic Eng. News* June 11

[17] Fodor S P A, Read J L, Pirrung M C, Stryer L, Lu A T and Solas D Light-directed, spatially addressable parallel chemical synthesis *Science* **252** 767–73

[18] Carter F L 1984 The molecular device computer: point of departure for large scale cellular automata *Physica* D **10** 175–94

[19] Heller A 1990 Electrical wiring of redox enzymes *Accounts Chem. Res.* **23** 128–34

[20] Lisman J E and Goldring M A 1988 Feasibility of long-term storage of graded information by the Ca^{+2}/calmodulin-dependent protein kinase molecules of the postsynaptic density *Proc. Natl Acad. Sci. USA* **85** 5320–4

[21] Anelli P L, Spencer N and Stoddart J F 1991 A molecular shuttle *J. Am. Chem. Soc.* **113** 5131–3

[22] Iijima S and Ichihashi T 1993 Single-shell carbon nanotubes of 1-nm diameter *Nature* **363** 603–5

[23] Ghadiri M R, Granja J R, Milligan R A, McRee D E and Khazanovich N 1993 Self-assembling organic nanotubes based on a cyclic peptide architecture *Nature* **366** 324–7

[24] Dado G P and Gellman S H 1993 Redox control of secondary structure in a designed peptide *J. Am. Chem. Soc.* **115** 12 609–10

[25] Murphy C J, Arkin M R, Jenkins Y, Ghatlia N D, Bossmann S H, Turro N J and Barton J K 1993 Long-range photoinduced electron transfer through a DNA helix *Science* **262** 1025–9

[26] Bissell R A, Cordova E, Kaifer A E and Stoddart J F 1994 A chemically and electrochemically switchable molecular shuttle *Nature* **369** 133–7

[27] Wu C G and Bein T 1994 Conducting polyaniline filaments in a mesoporous channel host *Science* **264** 1757–9

[28] Gynter M J, Hockless D C R, Johnston M R, Skelton B M and White A H 1994 Self-assembling porphyrin [2]-catenanes *J. Am. Chem. Soc.* **116** 4810–23

[29] Chen J and Seeman N C 1991 Synthesis from DNA of a molecule with the connectivity of a cube *Nature* **350** 631–3

[30] Wagner R W and Lindsey J S 1994 A molecular photonic wire *J. Am. Chem. Soc.* **116** 9759–60

[31] Garniew F, Hajlaoui R, Yassar A and Srivastava P 1994 All-polymer field-effect transistor realized by printing techniques *Science* **265** 1684–6

[32] Zhou Q and Swager T M 1995 Methodology for enhancing the sensitivity of fluorescent chemosensors *J. Am. Chem. Soc.* **117** 7017–8

[33] Kahan J S and Gibbs J N 1985 Food and Drug Administration regulation of medical device biotechnology, and food and food additive biotechnology *Appl. Biochem. Biotechnol.* **11** 507–16

[34] Miller H I 1988 FDA and biotechnology: update 1989 *Bio/Technology* **6** 1385–92

[35] Clinical Laboratory Improvement Act 1992 *Fed. Regist.* **57** 7002–288

[36] Zawisza J 1993 CLIA's impact: trouble brewing for IVD clearances? *Med. Dev. Diagnos. Ind.* **June** 55–9

[37] Taylor R F 1994 *Worldwide Immunoassay Markets, Technology and Applications, 1992–1998* (Waltham, MA: Decision Resources)

Index